Introduction to
Bioinformatics

Introduction to
Bioinformatics

THIRD EDITION

Arthur M. Lesk
The Pennsylvania State University

In nature's infinite book of secrecy
A little I can read.
–*Antony and Cleopatra*

OXFORD
UNIVERSITY PRESS

OXFORD

UNIVERSITY PRESS

Great Clarendon Street, Oxford OX2 6DP

Oxford University Press is a department of the University of Oxford.
It furthers the University's objective of excellence in research, scholarship,
and education by publishing worldwide in

Oxford New York

Auckland Cape Town Dar es Salaam Hong Kong Karachi
Kuala Lumpur Madrid Melbourne Mexico City Nairobi
New Delhi Shanghai Taipei Toronto

With offices in

Argentina Austria Brazil Chile Czech Republic France Greece
Guatemala Hungary Italy Japan Poland Portugal Singapore
South Korea Switzerland Thailand Turkey Ukraine Vietnam

Oxford is a registered trade mark of Oxford University Press
in the UK and in certain other countries

Published in the United States
by Oxford University Press Inc., New York

First edition published 2002
Second edition published 2005
Third edition published 2008

British Library Cataloguing in Publication Data
Data available

Library of Congress Cataloging in Publication Data
Data available

Typeset by Laserwords Private Limited, Chennai, India
Printed in Great Britain
on acid-free paper by
Ashford Colour Press Limited, Gosport, Hampshire

ISBN 978-0-19-920804-3

10 9 8 7 6 5 4 3 2

Dedicated to Eda,
with whom I have merged my genes.

Preface

Major changes in molecular biology since the second edition most prominently involve the great growth in new complete genome sequences that have become available. These are results of enhancements in methods of sequence determination. The new field of metagenomics—the survey of distributions of sequences in a region of the continents or ocean—has recently gained impressive momentum.

Major changes in information distribution involve the accelerating transition from paper to electronic libraries. A new chapter, treating this subject, appears in this edition. The implications for scientific research are only a part of the great social revolution that has flowed from the development of the Web.

In preparing the new edition I have tried to take into account new results, and to redistribute the extent of coverage of topics as now seems appropriate.

There are many different possible points of view from which to present molecular biology. Bioinformatics is one of them. I have also written about genomics, and about proteins, in companion volumes also published by Oxford University Press: *Introduction to Protein Architecture: The Structural Biology of Proteins,* 2001; *Introduction to Protein Science: Architecture, Function and Genomics,* 2004; and *Introduction to Genomics,* 2007. As a result, this book is focused more tightly on the applied science of bioinformatics. Numerous cross-references to the other volumes appear in this one. Readers are urged to put the books together for a more rounded appreciation of the pageant and mechanisms of life.

Preface to First Edition

On June 26, 2000, the sciences of biology and medicine changed forever. Prime Minister of the United Kingdom, Tony Blair, and President of the United States Bill Clinton, held a joint press conference, linked via satellite, to announce the completion of the draft of the Human Genome. *The New York Times* ran a banner headline: 'Genetic Code of Human Life is Cracked by Scientists'. The sequence of three billion bases was the culmination of over a decade of work, during which the goal was always clearly in sight and the only questions were how fast the technology could progress and how generously the funding would flow. The Box shows some of the landmarks along the way.

Next to the politicians stood the scientists. John Sulston, Director of The Sanger Centre in the UK, had been a key player since the beginning of high-throughput sequencing methods. He had grown with the project from the earliest 'one man and a dog' stages to the current international consortium. In the US, appearing with President Clinton were Francis Collins, director of the US National Human Genome Research Institute, representing the US publicly-funded efforts; and J. Craig Venter, President and Chief Scientific Officer of Celera Genomics Corporation, representing the commercial sector. It is difficult to introduce these two without thinking, 'In this corner...and in this corner...'. Although never actually coming to blows, there was certainly intense competition, in the later stages a race.

The race was more than an effort to finish first and receive scientific credit for priority. Indeed, it was a race after which the contestants would be tested not for whether they had taken drugs, but whether they and others could discover them. Clinical applications were a prime motive for support of the human genome project. Once the courts had held that gene sequences were patentable—with enormous potential payoffs for drugs based on them—the commercial sector rushed to submit patents on sets of sequences that they determined, and the academic groups rushed to place each bit of sequence that *they* determined into the public domain to *prevent* Celera—or anyone else—from applying for patents.

The academic groups lined up against Celera were a collaboration of laboratories primarily but not exclusively in the UK and USA. These included: The Sanger Centre in England, Washington University in St. Louis, Missouri, the Whitehead Institute at the Massachusetts Institute of Technology in Cambridge, Massachusetts, Baylor College of Medicine in Houston, Texas, the Joint Genome Institute at Lawrence Livermore National Laboratory in Livermore, California, and the RIKEN Genomic Sciences Center, now in Yokohama, Japan.

Both sides could dip into deep pockets. Celera had its original venture capitalists; its current parent company, PE Corporation; and, after going public, anyone who cared to take a flutter. The Sanger Centre was supported by the UK Medical Research Council and The Wellcome Trust. The US academic labs were supported by the US National Institutes of Health and Department of Energy.

On June 26, 2000 the contestants agreed to declare the race a tie, or at least a carefully out-of-focus photo finish.

Landmarks in the Human Genome Project

1953	Watson–Crick structure of DNA published.
1975	F. Sanger, and independently A. Maxam and W. Gilbert, develop methods for sequencing DNA.
1977	Bacteriophage ΦX-174 sequenced: first 'complete genome'.
1980	US Supreme Court holds that genetically-modified bacteria are patentable. This decision was the original basis for patenting of genes.
1981	Human mitochondrial DNA sequenced: 16 569 base pairs.
1984	Epstein-Barr virus genome sequenced: 172 281 base pairs.
1990	International Human Genome Project launched — target horizon 15 years.
1991	J. Craig Venter and colleagues identify active genes via Expressed Sequence Tags — sequences of initial portions of DNA complementary to messenger RNA.
1992	Complete low resolution linkage map of the human genome.
1992	Beginning of the *Caenorhabditis elegans* sequencing project.
1992	Wellcome Trust and United Kingdom Medical Research Council establish The Sanger Centre for large-scale genomic sequencing, directed by J. Sulston.
1992	J. Craig Venter forms The Institute for Genome Research (TIGR), associated with plans to exploit sequencing commercially through gene identification and drug discovery.
1995	First complete sequence of a bacterial genome, *Haemophilus influenzae*, by TIGR.
1996	High-resolution map of human genome — markers spaced by ∼600 000 base pairs.
1996	Completion of yeast genome, first eukaryotic genome sequence.
May 1998	Celera claims to be able to finish human genome by 2001. Wellcome responds by increasing funding to Sanger Centre.
1998	*Caenorhabditis elegans* genome sequence published.
September 1, 1999	*Drosophila melanogaster* genome sequence announced, by Celera Genomics; released Spring 2000.

1999	Human Genome Project states goal: working draft of human genome by 2001 (90% of genes sequenced to >95% accuracy).
December 1, 1999	Sequence of first complete human chromosome published.
June 26, 2000	Joint announcement of complete draft sequence of human genome.
2003	Fiftieth anniversary of discovery of the structure of DNA. Announcement of completion of human genome sequence.

The human genome is only one of the many complete genome sequences known. Taken together, genome sequences from organisms distributed widely among the branches of the tree of life give us a sense, only hinted at before, of the very great unity *in detail* of all life on Earth. They have changed our perceptions, much as the first pictures of the Earth from space engendered a unified view of our planet.

The sequencing of the human genome ranks with the Manhattan project that produced atomic weapons during the Second World War, and the space program that sent people to the Moon, as one of the great bursts of technological achievement of the last century. These projects share a grounding in fundamental science, and large-scale and expensive engineering development and support. For biology, neither the attitudes nor the budgets will ever be the same. Soon a 'one man and a dog project' will refer only to an afternoon's undergraduate practical experiment in sequencing and comparison of two mammalian genomes.

The human genome is fundamentally about information, and computers were essential both for the determination of the sequence and for the applications to biology and medicine that are already flowing from it. Computing contributed not only the raw capacity for processing and storage of data, but also the mathematically-sophisticated methods required to achieve the results. The marriage of biology and computer science has created a new field called bioinformatics.

Today bioinformatics is an applied science. We use computer programs to make inferences from the data archives of modern molecular biology, to make connections among them, and to derive useful and interesting predictions.

This book is aimed at students and practising scientists who need to know how to access the data archives of genomes and proteins, the tools that have been developed to work with these archives, and the kinds of questions that these data and tools can answer. In fact, there are a lot of sources of this information. Sites treating topics in bioinformatics are sprawled out all over the Web. The challenge is to select an essential core of this material and to describe it clearly and coherently, at an introductory level.

It is assumed that the reader already has some knowledge of modern molecular biology, and some facility at using a computer. The purpose of this book is to build on and develop this background. It is suitable as a textbook for advanced undergraduates or beginning postgraduate students. Many worked-out examples are integrated into the text, and references to useful web sites and recommended reading are provided.

Problems test and consolidate understanding, provide opportunities to practise skills, and explore additional subjects. Three types of problems appear at the ends of chapters. Exercises are short and straightforward applications of material in the text. Problems also require no information not contained in the text, but require lengthier answers or in some cases calculations. The third category, 'Weblems', requires access to the World Wide Web. Weblems are designed to give readers practice with the tools required for further study and research in the field.

What has made it possible to try to write such a book now is the extent to which the World Wide Web has made easily accessible both the archives themselves and the programs that deal with them. In the past, it was necessary to install programs and data on one's own system, and run calculations locally. Of course this meant that everything was dependent on the facilities available. Now it is possible to channel all the work through an interface to the Web. The web site linked with this book will ease the transition (see page xv). To ensure that readers will be able freely to pursue discussions in the book onto the Web, descriptions of and references to commercial software have been avoided, although many commercial packages are of very high quality.

A serious problem with the Web is its volatility. Sites come and go, leaving trails of dead links in their wake. There are so many sites that it is necessary to try to find a few gateways that are stable—not only continuing to exist but also kept up to date in both their contents and links. I have suggested some such sites, but many others are just as good. The problem is not to create a long list of useful sites—this has been done many times, and is relatively easy—but to create a short one—this is much harder!

My goal is that readers of this book will emerge with:

- An appreciation of the nature of the very large amount of detailed information about ourselves and other species that has become available.

- A sense of the range of applications of bioinformatics to molecular biology, clinical medicine, pharmacology, biotechnology, forensic science, anthropology and other disciplines.

- A useful knowledge of the techniques by which, through the World Wide Web, we gain access to the data and the methods for their analysis.

- An appreciation of the role of computers and computer science in the investigations and applications of the data.

- Confidence in the reader's basic skills in information retrieval, and calculations with the data, and in the ability to extend these skills by self-directed 'fieldwork' on the Web.

- A sense of optimism that the data and methods of bioinformatics will create profound advances in our understanding of life, and improvements in the health of humans and other living things.

Computer programming is introduced in this book based on the widely available language PERL. Examples of simple PERL programs appear in the context of biological problems. Many simple PERL tasks are assigned as exercises or problems at the ends of the chapters.

Where might the reader turn next? This book is designed as a companion volume—in current parlance, a 'prequel'—to *Introduction to Protein Architecture: The Structural Biology of Proteins* (Oxford University Press, 2001), and that title is of course recommended. Other books on sequence analysis range from those oriented towards biology to others in the field of computer science. The goal is that each reader will come to recognize his or her own interests, and be equipped to follow them up.

Preface to Second Edition

Bioinformatics has grown since the first edition of this book appeared.

The most striking change has been a refocus on integration; that is, of trying to see life processes as unified systems. As I said at the end of *Introduction to Protein Science: Architecture, Function and Genomics*, 'During the last century, molecular biologists have been taking living things apart. Our task now is to put them back together'. We have had large amounts of data. Now we are trying to see how they interrelate. At the heart of life processes are complicated patterns of interaction among the components, in space and in time. To understand these patterns the field has moved towards combining information into networks, and trying to understand their structures and dynamics.

Supporting this venture are the growing streams of data. The human genome, available in draft form when the first edition appeared, is now complete. It is joined by the complete genomes of 18 archaea, 155 bacteria, over 30 eukarya, and many other organelle and viral sequences. These genomes illuminate each other. One story that they tell is about unsuspected underlying unities of all living things, despite the obvious and profound differences in morphology and lifestyle.

Genomic sequences are supplemented by other data streams, notably the proteome. Patterns of gene expression, and networks of regulatory interactions, show how cells and organisms implement the information in the DNA. The potential for the life of an organism is contained in its genome, but it would be impossible to deduce a biography from it. Genomes are not formulas or scripts. It is in the proteins, and their interactions with themselves and with DNA, that we must seek the set of activities, contingent on and responsive to, the environment. Proteomics is giving us the information we need to see how the system works.

Research and applications require that the data be available in useful form. It is not enough to make the data public. The information must be subjected to quality control, annotation, and a logical structure must be imposed on it to make information retrieval possible. For this we are indebted to the institutions that archive, curate, organize and distribute the data. A recent trend has seen mergers of these groups into collaborative projects spanning the continents. In accord with the need to integrate the study of different types of data, we are moving in the direction of a single biological data repository. Individual scientists will be able to define 'virtual databanks', tailoring access to the information to suit particular needs and interests.

A gratifying consequence of academic bioinformatics is its contributions to applications in medicine, agriculture and technology. A better understanding of life processes empowers us to deal with them when they go wrong, and to extend them in novel and useful ways.

Plan of the book

Chapter 1 sets the stage and introduces all of the major players: DNA and protein sequences and structures, genomes and proteomes, databases and information retrieval, the World Wide Web and computer programming. Before developing individual topics in detail, it is important to see the framework of their interactions.

Chapter 2 presents the nature of individual genomes, including the human genome, and the relationships among them, from the biological point of view.

Chapter 3 describes the current state of the scientific literature as it makes the transition from paper to electronic form. This transition has many consequences, both intellectual and practical. It has had profound effects on research in bioinformatics.

Chapter 4 imparts basic skills in using the Web in bioinformatics. It describes archival databanks, and leads the reader through sample sessions involving information retrieval from some of the major databases in molecular biology.

Chapter 5 treats the analysis of relationships among sequences—alignments and phylogenetic trees. These methods underlie some of the major computational challenges of bioinformatics: detecting distant relatives; understanding relationships among genomes of different organisms; and tracing the course of evolution at the species and molecular levels.

Chapter 6 moves into three dimensions, treating protein structure and folding. Sequence and structure must be seen as full partners, with bioinformatics developing methods for moving back and forth between them as fluently as possible. Understanding protein structures in detail is essential for determining their mechanisms of action, and for clinical and pharmacological applications.

Chapter 7 treats proteomics and systems biology, including the new high-throughput sources of information about the expression and distribution of proteins in cells, and attempts to synthesize the information to reveal patterns of organization. The key idea of systems biology is integration: How do the pieces fit together? How do they interact? How do the individual molecules and processes together create a whole that so greatly transcends the parts in self-sufficiency?

Introduction to Bioinformatics on the Web

Bioinformatics is intimately bound up with the World Wide Web. Web resources appear throughout the book, to supplement treatments of specific topics. Some of these sites implement methods, such as sequence alignment, or homology modelling of protein structures. Others provide curated lists of other web sites specialized to particular subjects, such as expression databases. Links to all web resources referred to also appear on the book's accompanying Online Resource Centre, available at:

www.oxfordtextbooks.co.uk/orc/leskbioinf3e

This site contains the following online resources:

- Web link library of all web sites mentioned in the book, in context, giving the reader ready access to these resources.
- Guided tours of key websites, to help the reader get the most out of the vast array of information available online.
- Recommended reading from the book including active web links to primary literature articles. (Access to full text may require institutional subscription to the relevant journals.)
- Hints to end-of-chapter problems from the text, to support and encourage self-directed learning.
- All material from the book that the reader would find useful to have in computer-readable form, including data for exercises and problems, and all programs.
- Figures from the book available to download (for registered adopters only).

Acknowledgements

I am grateful to many colleagues for discussions and advice during the preparation of this book, and to the universities of Uppsala, Umeå, Rome 'Tor Vergata', Cambridge and Bologna for opportunities to try out this material.

I thank D.J. Abraham, S. Adhya, S. Aparicio, M.M. Babu, T. Baglin, D. Baker, S. Balaji, A. Bateman, A. Bench, J.M. Bollinger, M. Brand, A. Brazma, A. Buckle, C. Cantor, R.W. Carrell, C. Chothia, D. Crowther, T. Dafforn, I. Dodd, J.G. Ferry, R. Foley, A. Friday, M.B. Gerstein, T. Gibson, J. Irving, B. Jorden, J. Karn, K. Karplus, P. Klappa, A.S. Konagurthu, E.V. Koonin, M. Krichevsky, P. Lawrence, E.L. Lesk, M.E. Lesk, V.E. Lesk, V.I. Lesk, D. Liberles, A. Lister, L. Lo Conte, D.A. Lomas, A.D. MacKerell Jr, M. McFall-Ngai, J. McInerney, A. Madany Mamlouk, T. Madden, J. Magré, P. Miller, C. Mitchell, J. Moult, M. Münsterkötter, E. Nacheva, C. Notredame, C. Ouzounis, H. Parfrey, D. Parkinson, A. Pastore, M. Peitsch, D. Penny, J. Pettitt, C.A. Praul, F.W. Roberts, G.D. Rose, P.B. Rosenthal, B. Rost, M. Segal, E.J. Simon, O. Skovegaard, E.L. Sonnhammer, R. Srinivasan, R. Staden, J. Sulston, W. Swope, J.W. Thornton, I. Tickle, A. Tramontano, A.A. Travers, A.R. Venkitaraman, G. Vriend, P.L. Welcsh, J.C. Whisstock, S.H. White, M. Wildersten, A.S. Wilkins, R. Zhou and E.B. Ziff for advice and critical reading.

A.M.L.
August 2007

Contents

## CHAPTER 1

Introduction

Learning goals

1. To gain an overview of the subject—the topics, the questions, the point of view, and examples of specific problems and how to solve them. Many of the topics introduced in this chapter are developed elsewhere in the book.

2. To review and assemble the general principles of molecular biology necessary for dealing with data on sequences, structures and interactions.

3. To appreciate the very high capacity of the data streams that are producing data for molecular biology, notably but not limited to fast full-genome sequencing. The challenge of giving a manageable form to these data is the province of bioinformatics.

4. To understand the essential characteristics of a database: its coverage, its organization, and access routes to retrieve the information it contains.

5. To appreciate the importance of quality control and annotation in making data *useful*.

6. To understand the role of computer hardware and software in the infrastructure of bioinformatics. To evaluate your own talents, skills and interest, and to decide to what extent you want to create programs, and the extent to which you want only to become expert in using them.

7. To know the basic principles of protein structure, and the extent to which protein structures can be predicted from amino acid sequences.

8. To be familiar with the kinds of questions that the field of proteomics addresses, and the methods used to collect and analyse the data required to answer them. These methods include microarrays and mass spectrometry.

> Learning goals (*continued*)
>
> 9. To appreciate the clinical implications of discoveries in molecular biology, and the role of bioinformatics in forging the links between laboratory bench and clinical practice.

Biology has traditionally been an observational rather than a deductive science. Although recent developments have not altered this basic orientation, the nature of the data has changed radically. It is arguable that until recently most biological observations were fundamentally anecdotal—admittedly with varying degrees of precision, some very high indeed. However, in the last generation the data have become not only much more quantitative and precise, but, in the case of nucleotide and amino acid sequences, they have become *discrete*. It is possible to determine the genome sequence of an individual organism or clone not only completely, but in principle *exactly*. Experimental error can never be avoided entirely, but the quality of modern genomic sequencing methods is extremely high.

Not that this has converted biology into a deductive science. Life does obey principles of physics and chemistry, but for now life is too complex, and too dependent on historical contingency, for us to deduce its detailed properties from basic principles.

A second obvious property of the data of bioinformatics is their *very very large amount*. Currently the nucleotide sequence databanks contain 1.7×10^{12} bases, or 1.7 terabases (Tbp). If we use the approximate size of the human genome—3×10^9 letters—as a unit, this amounts to 567 Human Genome Equivalents (or 567 **huges**, an apt name). The database of macromolecular structures contains >50 000 entries, containing the full three-dimensional coordinates of proteins and nucleic acids, of average length ~400 residues. Not only are the individual databanks large, but their sizes are increasing at a very high rate. Figure 1.1 shows the growth of the International Nucleotide Sequence Database Collection and the world-wide Protein Data Bank (archiving macromolecular structures). It would be precarious to extrapolate.

This quality and quantity of data have encouraged scientists to aim at commensurately ambitious goals:

For a comprehensible standard of comparison, 1 *huge* is comparable with the number of characters appearing in 6 complete years of issues of *The New York Times.*

- To have it said that they 'saw life clearly and saw it whole'. That is, to understand integrated aspects of the biology of organisms, viewed as coherent complex organizations, at microscopic and macroscopic levels.

- To interrelate sequence, three-dimensional structure, expression pattern, interaction and function of individual proteins, nucleic acids and protein–nucleic acid complexes.

- To integrate the data on the different aspects of the life of a cell or organism into a 'systems' description of the structure and dynamics.

- To use data on contemporary organisms as a basis for travel backward and forward in time—back to deduce events in evolutionary history, forward to greater deliberate scientific modification of biological systems.

- To support applications to medicine, agriculture and technology.

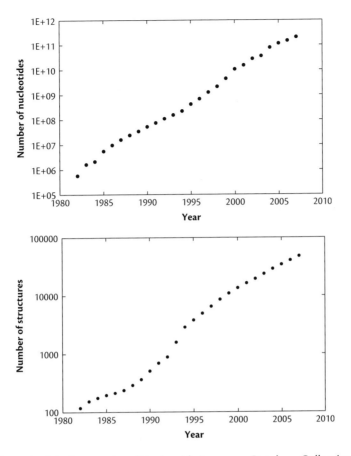

Fig. 1.1 (a) Growth of the International Nucleotide Sequence Database Collection. (b) Growth of the world-wide Protein Data Bank, archive of three-dimensional biological macromolecular structures, from the wwPDB, a collaboration between groups in the US, Europe and Japan. Note log scale on y-axes.

Life in space and time

It is difficult to define life, and it may be necessary to modify its definition—or to live, uncomfortably, with the old one—as computers grow in power and the silicon–life interface grows more intimate. For now, try this: a biological organism is a naturally occurring self-reproducing device that effects controlled manipulations of matter, energy and information.

From the most distant perspective, life on Earth is a complex self-perpetuating, evolving, system distributed in space and time. It is of the greatest significance that it is largely composed of *discrete* individual organisms, each with a finite lifetime and in many cases with unique features.

Spatially, starting far away and zooming in progressively, one can distinguish, within the biosphere, local **ecosystems,** stable until their environmental conditions change or they are invaded. Each species within an ecosystem is composed of **organisms** carrying out individual if not independent activities. Organisms are composed of **cells**. Every cell is an intimate local ecosystem, not isolated from its environment but interacting with it in specific and controlled ways. Eukaryotic cells contain a complex internal structure of their own; including nuclei and other subcellular organelles, and a cytoskeleton. And finally we come down to the level of molecules.

Life is extended not only in space but also in time. We see today a snapshot of one stage in a history of life that extends back in time for at least 3.5 billion years. The theory of natural selection has been extremely successful in rationalizing the process of life's development. However, historical accident plays too dominant a role in determining the course of events to allow much detailed prediction. DNA from extinct organisms affords only limited access to the historical record at the molecular level. Instead, we must try to read the past in contemporary genomes. US Supreme Court Justice Felix Frankfurter once wrote that '. . . the American constitution is not just a document, it is a historical stream'. This is also true of genomes, which contain records of their own development.

> We use
> 1 billion $= 10^9$.

Evolution is the change over time in the world of living things

The processes of evolution change distributions of genotypes and phenotypes in successive generations.

The **genotype** is an organism's genetic information, the sequence of its genome. All other observable features of an organism—macroscopic and biochemical—comprise its **phenotype**. The genotype is inherited from a parent or parents, subject to modification by mutation or by lateral transfer of genetic material. The phenotype depends on the genotype, which controls the development of the organism under the influence of its environment.

The asymmetry between genotype and phenotype is the engine of evolution:

- Changes in genotypes are inheritable. Effects on the phenotype, of the environment or lifestyle—for instance, better nutrition leading to larger body size, or debilitating effects of disease or injury—are not directly inheritable.

- During the development of any organism, genotype constrains phenotype. Phenotype does not influence genotype.

- Many genotypes can create the same phenotype. For example:
  - Many mutations in genes coding for proteins leave amino acid sequences unchanged, or make modifications with no apparent effect on function or expression.
  - *Alleles* are different forms (sequences) of the same gene. Any organism that contains two or more copies of a gene can repeat the same allele (homozygosity) or contain different alleles (heterozygosity). Homozygotes and heterozygotes have different genotypes, but, if a single gene has exclusive control over a trait, and one allele is dominant, homozygotes and heterozygotes may have the same phenotype.

> In mammals
> ~20% of loci are
> heterozygous.

At what levels does evolution operate? Most life consists of discrete *organisms*. A **population** is a group of similar organisms that interact. Populations of sexually reproducing organisms interbreed; individuals in all populations compete for resources. Evolution alters the composition and distribution of the gene pools and phenotypes in populations.

What is the mechanism of evolution? Within a population, individuals with a variety of genotypes arise, displaying a corresponding variety of phenotypes. Although selection has no *direct* leverage on genotype, individuals with different phenotypes show differential success at reproduction. As a result, the new generation may have an altered distribution of genotypes and phenotypes. **Natural selection**—enhanced reproduction by 'fitter' individuals—is the most important mechanism of evolution. Another mechanism of evolution is **genetic drift**, the random change in allelic frequencies, not in response to selection. Genetic drift is especially important in small, isolated populations.

Mechanisms that produce genetic variety create the *potential* for evolution:

- **Mutations,** such as point substitutions, insertions and deletions, and transpositions. Rates of generation of point mutations are estimated to be about 10^{-12}–10^{-10} per base pair per generation. (This is *not* the same as the rate of allelic replacement in a population. Mutations only propose candidates for evolutionary change.)

- **Recombination** can bring different loci together, or split them apart. Recombination within a gene can create a new allele. Recombination outside genes can affect the relationship between genes and regulatory elements.

- **Gene duplication,** followed by divergence.

- **Gene loss,** either by deletion, or by mutations that destroy expression or function.

- **Gene flow,** from mixing of populations, or gene transfer between species.

Evolution can increase or decrease the variety in gene pools. If a novel mutation confers selective advantage only in the homozygous state, the gene may spread throughout a population. Adoption of the allele by all members of a population can *decrease* the variety in the gene pool. If a gene arises that confers selective advantage in the heterozygous state only, the gene pool may move towards greater variety. Some mutations create recessive alleles that are deleterious only in the homozygous state. These are hard to remove from a population, especially if heterozygotes have some compensating advantage. An example is the gene for sickle-cell anaemia, which confers on heterozygotes enhanced resistance to malaria.

Microevolution refers to relatively small changes in a few genes, leading in most cases to relatively small changes in phenotypes. Microevolution affects the individuals *within* a population. Modern techniques allow us to follow microevolution at the molecular level, through measurements of genome sequences and protein expression patterns. **Macroevolution** refers to larger-scale changes in populations as a whole, including formation of new species. The fossil record provides a (partial) history of macroevolution, using geological methods to date events. Comparative anatomy and physiology, and embryology, provide additional clues.

Observations of micro- and macroevolution illuminate each other. Genome sequences help in the classification of species. The fossil record permits dating of past events that have had consequences on the molecular scale that we observe now. A major challenge to modern biology is to understand how large-scale events such as development of new species can occur as a result of microevolutionary events.

Dogmas: central and peripheral

The information archive within each organism—the blueprint of potential development and activity—is the genetic material, DNA, or, in some viruses, RNA. DNA and RNA molecules are long, *linear,* chain molecules containing a message in a four-letter alphabet (see Box). Even for microorganisms the message is long, typically 10^6 characters. Implicit in the structure of the DNA are mechanisms for self-replication and for translation of genes into proteins. The double helix, and its internal self-complementarity providing for accurate replication, are well known (see Plate I). Near-perfect replication is essential for stability of inheritance; but some imperfect replication, or mechanism for import of foreign genetic material, is also essential, else evolution could not take place in asexual organisms.

The strands in the double helix are antiparallel; directions along each strand are called 3′ and 5′ (for positions in the deoxyribose ring). In translation to protein, the DNA sequence is always read in the 5′ → 3′ direction.

For the history of the discovery of the double helix, see *Introduction to Genomics,* chapter 1.

The four naturally occurring nucleotides in DNA (RNA)

a   adenine   g   guanine   c   cytosine   t   thymine   (u   uracil)

The 20 naturally occurring amino acids in proteins

Non-polar amino acids

G	glycine	A	alanine	P	proline	V	valine
I	isoleucine	L	leucine	F	phenylalanine	M	methionine

Polar amino acids

S	serine	C	cysteine	T	threonine	N	asparagine
Q	glutamine	H	histidine	Y	tyrosine	W	tryptophan

Charged amino acids

D   aspartic acid   E   glutamic acid   K   lysine   R   arginine

Under typical physiological conditions, many histidines are charged.

→

Other classifications of amino acids can also be useful. For instance, histidine, phenylalanine, tyrosine and tryptophan are aromatic, and are observed to play special structural roles in membrane proteins.

In addition to the one-letter codes given above, amino acid names are frequently abbreviated to their first three letters, for instance Gly for glycine; except for isoleucine, asparagine, glutamine and tryptophan, which are abbreviated to Ile, Asn, Gln and Trp, respectively. The rare amino acid selenocysteine has the three-letter abbreviation Sec and the one-letter code U.

It is conventional to write nucleotides in lower case and amino acids in upper case. Thus atg = adenine-thymine-guanine and ATG = alanine-threonine-glycine.

The implementation of genetic information occurs, initially, through the synthesis of RNA and proteins. Proteins are the molecules responsible for much of the structure and activities of organisms. Our hair, muscle, digestive enzymes and antibodies are all proteins. Like nucleic acids, proteins are long, linear chain molecules. The genetic 'code' is in fact a cipher (see Box): successive triplets of letters from the DNA sequence specify successive amino acids; stretches of DNA sequences encipher amino acid sequences of proteins. Typically, proteins are 200–400 amino acids long, requiring 600–1200 letters of expressed DNA message to specify them. DNA sequences also direct the synthesis of RNA molecules, for instance the RNA components of the ribosome.

However, not all DNA is expressed as proteins or structural RNA. Most genes in higher organisms contain internal untranslated regions, or introns. Some regions of the DNA sequence are devoted to control mechanisms, and a substantial amount of the genomes of higher organisms appears to be 'junk'. Which may mean merely that we do not yet understand its function.

In DNA, the molecules comprising the alphabet are chemically similar, and the structure of DNA is, to a first approximation, uniform (although some DNA–protein interactions distort the DNA structure). Proteins, and structural RNAs, in contrast, show a great variety of three-dimensional conformations. These are necessary to support their very diverse structural and functional roles.

The amino acid sequence of a protein dictates its three-dimensional structure. For each natural amino acid sequence, there is a unique stable *native state* that under proper conditions is adopted spontaneously. If a purified protein is heated, or otherwise brought to conditions far from the normal physiological environment, it will 'unfold' to a disordered and biologically inactive structure. (This is why our bodies contain mechanisms to maintain nearly constant internal conditions.) When normal conditions are restored, protein molecules will generally readopt the native structure, indistinguishable from the original state. There are important exceptions, however. Irreversible denaturation leading to formation of insoluble aggregates is most familiar in hard-boiling an egg. Such aggregates are associated with many

Sydney Brenner distinguished 'garbage' from 'junk': garbage you throw away, junk you keep around.

. . . .
Alternative
genetic codes
appear in organ-
elles — chloro-
plasts and mito-
chondria — and
in some species.
. . . .

The standard genetic code

ttt	Phe	tct	Ser	tat	Tyr	tgt	Cys
ttc	Phe	tcc	Ser	tac	Tyr	tgc	Cys
tta	Leu	tca	Ser	taa	STOP	tga	STOP
ttg	Leu	tcg	Ser	tag	STOP	tgg	Trp
ctt	Leu	cct	Pro	cat	His	cgt	Arg
ctc	Leu	ccc	Pro	cac	His	cgc	Arg
cta	Leu	cca	Pro	caa	Gln	cga	Arg
ctg	Leu	ccg	Pro	cag	Gln	cgg	Arg
att	Ile	act	Thr	aat	Asn	agt	Ser
atc	Ile	acc	Thr	aac	Asn	agc	Ser
ata	Ile	aca	Thr	aaa	Lys	aga	Arg
atg	Met	acg	Thr	aag	Lys	agg	Arg
gtt	Val	gct	Ala	gat	Asp	ggt	Gly
gtc	Val	gcc	Ala	gac	Asp	ggc	Gly
gta	Val	gca	Ala	gaa	Glu	gga	Gly
gtg	Val	gcg	Ala	gag	Glu	ggg	Gly

diseases, including Alzheimer disease and bovine spongiform encephalopathies (such as 'mad-cow' disease).

From one dimension to three

The spontaneous folding of proteins to form their native states is the point at which Nature makes the giant leap from the one-dimensional world of gene and protein sequences to the three-dimensional world we inhabit. There is a paradox: the translation of DNA sequences to amino acid sequences is very simple to describe logically; it is specified by the genetic code. The folding of the polypeptide chain into a precise three-dimensional structure is very difficult to describe logically. However, translation requires the immensely complicated machinery of the ribosome, tRNAs and associated molecules; but protein folding occurs spontaneously.

The functions of proteins depend on their adopting the native three-dimensional structure. For example, the native structure of an enzyme may have a cavity on its surface that binds a small molecule and juxtaposes it to catalytic residues. Many regulatory mechanisms depend on the binding of proteins to other proteins or to DNA. We thus have the paradigm:

- DNA sequence determines protein sequence.

- Protein sequence determines protein structure.

- Protein structure determines protein function.

- Regulatory mechanisms, including but not limited to control of expression patterns, deliver the right amount of the right function to the right place at the right time.

Much of the organized activity of bioinformatics has been focused on the analysis of the data related to these processes.

So far, this paradigm does not include levels higher than the molecular levels of structure and organization, including, for example, such questions as how tissues become specialized during development or, more generally, how environmental effects exert control over genetic events. In some cases of simple feedback loops, it is understood at the molecular level how increasing the amount of a reactant causes an increase in the production of an enzyme that catalyses its transformation. More complex are the programmes of development that unfold during the lifetime of an organism. These fascinating problems about the information flow and control within an organism have now come within the scope of mainstream bioinformatics.

Systems biology, discussed in Chapter 7, focuses on integration and control of the activities of cells and organisms.

Observables and data archives

A databank includes (1) an archive of information, (2) a logical organization or 'structure' of that information—called a *schema*—and (3) tools to gain access to it. Databanks in molecular biology contain nucleic acid and protein sequences, macromolecular structures and functions, expression patterns and networks of metabolic pathways and control cascades. They include:

- Archival databanks of biological information

  - DNA and protein sequences, including annotation
  - Variations, such as compilations of haplotypes, or disease-associated mutations
  - Nucleic acid and protein structures, including annotation
  - Databanks focused on organisms, including genome databases
  - Databanks of protein expression patterns
  - Databanks of metabolic pathways

- Databases of interactions and of regulatory networks

◆ Derived databanks: these contain information collected from the archival databanks, inferred from analysis of their contents. For instance:

- sequence motifs (characteristic 'signature patterns' of families of proteins)
- classifications or relationships (connections between, and common features of, entries in archives). Examples include databanks of protein sequence families, or hierarchical classifications of protein folding patterns.

◆ Bibliographic databanks

◆ Databanks of web sites

- databanks of databanks containing biological information
- links between databanks

online resource centre

Web resources   Nucleic acid and protein sequences

The archive of nucleic acid sequences, the International Nucleotide Sequence Database Collection, is maintained by a triple partnership: *GenBank,* based at the US National Center for Biotechnology Information, in Bethesda, Maryland; *The EMBL Nucleotide Sequence Database,* or *EMBLBank* based at the European Bioinformatics Institute, in Hinxton, UK; and *The Center for Information Biology and DNA DataBank of Japan,* at the National Institute of Genetics in Mishima, Japan. The three sites exchange incoming submissions daily, to ensure common coverage. However, the format, annotation and embedded links differ among the corresponding entries released by the different databases.

The archive of amino acid sequences of proteins, now determinined almost exclusively from translation of gene sequences, is maintained by the *United Protein Database (UniProt),* a merger of the databases *SWISS-PROT, The Protein Identification Resource (PIR)* and *Translated EMBL (TrEMBL).*

Associated with the archives are tools for selection and retrieval of sequences. The *Sequence Retrieval System (SRS)* is a product of Lion Bioscience AG. The US National Center for Biotechnology Information offers *ENTREZ.* Both allow parallel searches in multiple data archives.

Many full-genome sequencing projects maintain databases focused on individual species. Notable are the Ensembl (Sanger Centre, Hinxton, UK) and University of California at Santa Cruz browsers, for the human and other genomes, and FlyBase.

Many derived databanks assemble families of proteins or subunits based on the similarities of their sequences. An 'umbrella' database, Interpro, integrates the contents, features and annotation of several individual databases of protein families, domains and functional sites; and contains links to others, including the Gene Ontology Consortium™ functional classification. Interpro intends to assimilate additional databases, including structural databases. (Resistance is futile.)

The mechanism of access to a databank is the set of tools for answering questions such as:

- 'Does the databank contain the information I require?' (Example: can I retrieve the amino acid sequence of human alcohol dehydrogenase?)

- 'How can I assemble selected information from the databank in a useful form?' (Example: compile a list of globin sequences; or even better, a table of aligned globin sequences.)

- Indices of databanks are useful in asking 'Where can I find some specific piece of information?' (Example: what databanks contain the amino acid sequence of porcupine trypsin?) Of course if I know and can specify exactly what I want, the problem is relatively straightforward.

How to achieve effective access is an issue of database design that ideally should remain hidden from users. It has become clear that effective access cannot be provided by bolting a query system onto an unstructured archive. Instead, the logical organization of the storage of the information must be designed with the access in mind—what kinds of questions users will want to ask—and the structure of the archive must mesh smoothly with the information-retrieval software.

A database without effective modes of access is merely a data graveyard.

A variety of database queries arise in bioinformatics. Compare the following typical examples:

1. Given a sequence, or fragment of a sequence, find sequences in the database that are similar to it. This is a central problem in bioinformatics. We share such string-matching problems with many fields of computer science. For instance, word processing and editing programs support string-search functions.

2. Given a protein structure, or fragment, find protein structures in the database that are similar to it. This is the generalization of the string-matching problem to three dimensions.

3. Given a sequence of a protein of unknown structure, find *structures* in the database that adopt similar three-dimensional structures. One might be tempted to cheat—to look in the sequence databanks for proteins with sequences similar to the probe sequence: for if two proteins have sufficiently similar sequences, they will have similar structures. However, the converse is not true, and one can hope to create more powerful search techniques that will find proteins of similar structure even though their sequences have diverged beyond the point where they can be recognized as similar by sequence comparison.

4. Given a protein structure, find *sequences* in the databank that correspond to similar structures. Again, one can cheat by using the structure to probe a structure databank, but this can give only limited success because there are so many more sequences known than structures. It is therefore desirable to have a method that can pick out the structure from the sequence.

Queries 1 and 2 are solved problems; such searches are carried out thousands of times a day. Queries 3 and 4 are active fields of research.

Tasks of greater subtlety arise when one wishes to study relationships between information contained in separate databanks. This requires links that facilitate simultaneous access to several databanks. Here is an example: 'For which proteins of known structure involved in diseases of purine biosynthesis in humans are there related proteins in yeast?' We are setting conditions on: known structure, specified function, detection of relatedness, correlation with disease, and specified species. Today, the quality of a database depends not only on the information it contains, but on the effectiveness of its links to other related sources of information. The growing importance of simultaneous access to databanks has led to research in databank interactivity: How can databanks 'talk to one another' without too great a sacrifice of the freedom of each one to structure its own data in ways appropriate to the individual features of the material it contains?

Information flow in bioinformatics

Data enter the bioinformatics establishment when a scientist deposits an experimental result in an archive. The archive curates and annotates the data, to create an entry of proper contents and format. The new entry appears in the public release of the archive. Note that the division of the archive into entries is determined by the provenance of the data; that is, an entry corresponds to one coherent set of experimental measurements, often corresponding to one published article. In some cases, fragments of a complete sequence appear in several articles. A databank can join the results to form an entry containing the complete biological entity.

Other information-retrieval projects, either associated with an archive or independent, may integrate newly released entries into their individual systems. They may select or reorganize the data structure, and provide novel tools for analysis.

Reorganization of the data may involve:

- Simply integrating the new entries into a general or specialized search engine.

- Extracting useful subsets of the data. Examples include (1) identification of genes in a connected DNA sequence, such as a bacterial genome or a eukaryotic, chromosome, and (2) the selection of a non-redundant set of protein sequences, both to shorten searches and to reduce statistical bias.

- Deriving new types of information from the original data. A simple example: release of a protein-coding gene by a DNA sequence archive will trigger the appearance of its amino acid sequence translation in databases of protein sequences.

- Recombining data in different ways. Many projects group sequences or structures of families of homologous proteins, or proteins that share function. Examples include the Protein Kinase Resource and the MEROPS protease database. (Archives tend to keep related entries separate to preserve clarity of provenance.)

- Reannotating the data, including provision of different constellations of links. The integration may be horizontal or vertical. That is, links may indicate relationships to other entries of the same type (for instance, correspondences within a genome among homologous genes or among genes associated with the same metabolic pathway). Alternatively, links may adduce a variety of information about a gene or protein (for instance, links between a gene and the clinical consequences of mutations.)

Many sites serve as gateways between the archives and the computational tools available for data analysis. Information retrieval permits selection and extraction of data to provide the ingredients of a research project. Many bioinformatics resources not only offer information retrieval, but facilitate the 'downstream' processing of the entries selected. A typical example would be to retrieve the sequences of a set of homologous genes, and then to align them. The goal is to provide smooth integration of all the data-processing steps required for a research project, by intimate links among the tools for data storage, retrieval and analysis.

Although there are good arguments for unique control over the archives, there is no need to limit the ways to access them—colloquially, the design of 'front ends'. Specialized user communities may extract subsets of the data, or recombine data from different sources, and provide specialized avenues of access. Such 'boutique' databases depend on the primary archives as the source of the information they contain, but redesign the organization and presentation. Indeed, different derived databases can slice and dice the same information in different ways.

On the other hand, there is a very strong trend towards merging and integration of data resources in bioinformatics. Only national or commercial rivalries impede fusion into a single world-wide database. Because of the danger that the result will prove unwieldly, it will be possible to tailor access to the needs of particular projects. The unification of the archives will be accompanied by a fragmentation of the routes of access.

A reasonable extrapolation suggests the concept of specialized 'virtual databases', (a concept first suggested in 1981), grounded in the archives but providing individual scope and function, tailored to the needs of individual research groups or even individual scientists.

Curation, annotation and quality control

The scientific and medical communities are dependent on the quality of databanks. Indices of quality, even if they do not permit correction of mistakes, may help us avoid arriving at wrong conclusions.

Databank entries comprise raw experimental results, and supplementary information, or annotations. Each of these has its own sources of error.

The most important determinant of the quality of the data themselves is the state of the art of the experiments. Older data were limited by older techniques; for instance, amino acid sequences of proteins were once determined by peptide sequencing, but are now translated from DNA sequences (except for partial sequencing by mass spectrometry; see Chapter 6). One consequence of the data explosion is that most data are new data, governed by current technology, which in most cases does quite a good job.

Annotations include information about the source of the data and the methods used to determine them. They identify the investigators responsible, and cite relevant publications. They provide links to related information in other databanks. In some sequence databanks, annotations include *feature tables:* lists of segments of the sequences that have biological significance—for instance, regions of a DNA sequence that code for proteins. These appear in computer-parsable formats, their contents restricted to a controlled vocabulary. Note that agreement among databases on a controlled vocabulary, and the definitions of the terms that appear, is essential for information-retrieval operations involving interactions between multiple databases, and distributed queries.

Formerly, a typical DNA sequence entry was produced by a single research group, investigating a gene and its products in a coherent way. Annotations were grounded in experimental data and written by specialists. In contrast, full-genome sequencing projects offer no experimental confirmation of the expression of most putative genes, nor characterization of their products. Curators at databanks base much of their annotation on the analysis of the sequences by computer programs.

Annotation is the weakest component of the genomics enterprise. Automation of annotation is possible only to a limited extent; getting it right remains labour-intensive; and allocated resources are inadequate. But the importance of proper annotation cannot be underestimated. P. Bork has commented that errors in gene assignments vitiate the high quality of the sequence data themselves.

Growth of genomic data will permit improvement in the quality of annotation, as statistical methods increase in accuracy. This will allow improved *reannotation* of entries. The improvement of annotations will be a good thing. It implies, however, the disturbing concomitant that annotation will be in flux. The problem is aggravated by the proliferation of web sites, with increasingly dense networks of links. Networks of web sites provide useful avenues for applications. But the Web is also a vector of contagion, propagating errors in raw data, in immature data subsequently corrected but the corrections not passed on, and in variant annotations.

Perhaps the only possible solution is a *distributed* and *dynamic* error correction and annotation process. Distributed, in that databank staff will have neither the time nor the expertise for the job; specialists will have to act as curators. Dynamic, in that progress in automation of annotation and error identification/correction will permit reannotation of databanks. We will have to give up the safe idea of a stable databank composed of entries that are correct when first distributed and stay fixed. Databanks will become a seething broth of information, growing in size, and maturing—we must hope—in quality.

The World Wide Web

All readers will have used the World Wide Web, for reference material, for news, for access to databases in molecular biology, for checking out personal information about individuals—friends or colleagues or celebrities—or just for browsing. The Web is a means of interpersonal and intercomputer contact over networks. It provides a complete global village, containing the equivalent of library, post office, shops and schools.

The Web can be thought of as a giant world-wide multimedia notice board. It contains text, images, cinema and sound. Virtually anything that can be stored on a computer can be made available and accessed via the Web. An interesting example is a site treating the poetry of Walt Whitman (www.whitmanarchive.org). The highest level page contains a table of contents. The site contains printed text of different poems. You can compare different editions. You can access critical analyses of the poems. You can see versions of some poems in manuscripts. There is even a link to an audio file, from which you can hear Whitman himself reading part of a poem.

Links embedded in a web site can be internal or external. Internal links take you to other portions of the text of a current document, or to associated images, cinema or sounds. External links may allow you to move *down* to more specialized documents, *up* to more general ones (perhaps providing background to technical material), *sideways* to parallel documents (other papers on the same subject) or *over* to directories that show what other relevant material is available.

The main thing to do, to get started using the Web effectively, is to find useful entry points. Once a session is launched, links will take you where you want to go. Among the most important sites are **search engines**, such as Google, that index the entire Web and permit retrieval by keywords. You can enter one or more terms, such as 'phosphorylase', 'allosteric change', 'crystal structure', and the search program will return a list of links to sites on the Web that contain these terms.

Once you have completed a successful session, when you next log in the intersession memory facilities of the browsers allow you to pick up cleanly where you left off. During any session, when you find yourself viewing a document to which you will want to return, you can save the link in a file of **bookmarks** or **favourites**. In a subsequent session you can return directly to any site on this list, not needing to follow the trail of links that led you there in the first place.

A personal **home page** is a short autobiographical sketch (with links, of course). You and your colleagues will have your own home pages which typically include name, institutional affiliation, addresses for paper and electronic mail, telephone and fax numbers, a list of publications and current research interests. It is not uncommon for home pages to include personal information, such as hobbies, pictures of the individual with his or her spouse and children, and even with the family dog!

Nor is the Web solely a one-way street. Many Web documents include forms in which you can enter information, and launch a program. Search engines are common examples. Many calculations in bioinformatics are now launched via such web servers. If the calculations are lengthy, the results may not be returned within the session, but sent by e-mail.

Electronic publication

We are in an era of a transition to paper-free publishing. More and more publications are appearing on the Web. A scientific journal may post only its table of contents, or a table of contents together with abstracts of articles, or complete articles. The recommended reading sections appearing in this book correspond to links in the associated web page. Many institutional publications—newsletters and technical reports—appear on the Web. Many other magazines and newspapers are showing up as well. You might want to try: http://www.nytimes.com Many printed publications now contain references to web links containing supplementary material that never appears on paper.

A major force in the conversion of paper to electronic libraries is Google's project to scan in the contents of a number of academic libraries. We shall develop this topic in Chapter 3.

Computers and computer science

Bioinformatics would not be possible without advances in computing hardware and software. Fast and high-capacity storage media are essential even to maintain the archives. Information retrieval and analysis require programs; some fairly straightforward and others extremely sophisticated. Distribution of the information requires the facilities of computer networks and the World Wide Web.

Computer science is a young and flourishing field with the goal of making most effective use of information technology hardware. Certain areas of computer science impinge most directly on bioinformatics. Consider their application to a specific biological problem: 'Retrieve from a database all sequences similar to the human PAX-6 sequence'. A good solution of this problem would appeal to computer science for:

- *Analysis of algorithms* *An algorithm is a complete and precise specification of a method for solving a problem.* For the retrieval of similar sequences, we need to measure the similarity of the probe sequence to every sequence in the database. It is possible to do much better than the naive approach of checking every pair of positions in every possible juxtaposition, a method that even without allowing gaps would require a time proportional to the product of the number of characters in the probe sequence times the number of characters in the database. A speciality in computer science known colloquially as 'stringology' focuses on developing efficient methods for this type of problem, and analysing their effective performance.

- *Data structures and information retrieval* How can we organize our data for efficient response to queries? For instance, are there ways to index or otherwise 'preprocess' the data to make our sequence-similarity searches more efficient? How can we provide interfaces that will assist the user in framing and executing queries?

- *Software engineering* Hardly ever anymore does anyone write programs in the native language of computers. Programmers work in higher-level languages, such as C, C++, PERL, JAVA or even FORTRAN. The choice of programming language depends on the nature of the algorithm and associated data structure, and the expected use of the program. Of course most complicated software used in bioinformatics is now written by specialists. Which brings up the question of how much programming expertise a bioinformatician needs.

Programming

Programming is to computer science what bricklaying is to architecture. Both are creative; one is an art and the other a craft.

Many students of bioinformatics ask whether it is essential to learn to write complicated computer programs. My advice (not agreed upon by everyone in the field) is: 'Don't. Unless you want to specialize in it'. To work in bioinformatics, you will need to develop expertise in using tools available on the Web. Learning how to create and maintain a web site is essential. And of course you will need facility in the use of the operating system of your computer, including general purpose applications programs such as word processors and presentation tools. Some skill in writing simple scripts in a language such as PERL provides an essential extension to the basic facilities of the operating system.

On the other hand, the size of the data archives, and the growing sophistication of the questions we wish to address, demand respect. Truly creative programming in the field is best left to specialists, with advanced training in computer science. Nor does *using* programs, via highly polished (not to say flashy) Web interfaces, provide any indication of the nature of the activity involved in writing and debugging programs. Bismarck once said: 'Those who love sausages or the law should not watch either being made'. Perhaps computer programs should be added to his list.

I recommend learning some basic skills with PERL, or with one of the related languages PYTHON or RUBY. PERL is a very powerful tool, and is available on most computer systems. PERL makes it very easy to carry out many very useful simple tasks, but can also be effective in projects demanding heavy computations.

How should you learn enough PERL to be useful in bioinformatics? Many institutions run courses. Learning from colleagues is fine, depending on the ratio of your aptitude to their patience. Books are available. A very useful approach is to find lessons on the Web—ask a search engine for 'PERL tutorial' and you will turn up many useful sites that will lead you by the hand through the basics. And of course use it as much as you can. This book will not teach you PERL, but it will provide opportunities to practise what you learn elsewhere. Should your programming ambitions go beyond simple tasks, check out the BioPERL Project, a source of freely available PERL programs and components in the field of bioinformatics (http://bio.perl.org).

Examples of *simple* PERL programs appear in this book. The strength of PERL at character-string handling makes it suitable for sequence analysis tasks in biology. Here is a very simple PERL program to translate a nucleotide sequence into an amino acid sequence according to the standard genetic code. The first line, #!/usr/bin/perl, is a signal to the UNIX (or LINUX) operating system that what follows is a PERL program. Within the program, all text commencing with a #, through to the end of the line on which it appears, is merely comment. The line __END__ signals that the program is finished and what follows is the input data (all material that the reader might find useful to have in computer-readable form, including all programs, appear in the web site associated with this book: http://www.oup.co.uk/xxx/yyy/zzz).

Case Study 1.1 Translation of a DNA sequence to an amino acid sequence using the standard genetic code

```perl
#!/usr/bin/perl
#translate.pl -- translate nucleic acid sequence to protein sequence
#                according to standard genetic code

#   set up table of standard genetic code

%standardgeneticcode = (
 "ttt"=> "Phe",   "tct"=> "Ser", "tat"=> "Tyr",   "tgt"=> "Cys",
 "ttc"=> "Phe",   "tcc"=> "Ser", "tac"=> "Tyr",   "tgc"=> "Cys",
 "tta"=> "Leu",   "tca"=> "Ser", "taa"=> "TER",   "tga"=> "TER",
 "ttg"=> "Leu",   "tcg"=> "Ser", "tag"=> "TER",   "tgg"=> "Trp",
 "ctt"=> "Leu",   "cct"=> "Pro", "cat"=> "His",   "cgt"=> "Arg",
 "ctc"=> "Leu",   "ccc"=> "Pro", "cac"=> "His",   "cgc"=> "Arg",
 "cta"=> "Leu",   "cca"=> "Pro", "caa"=> "Gln",   "cga"=> "Arg",
 "ctg"=> "Leu",   "ccg"=> "Pro", "cag"=> "Gln",   "cgg"=> "Arg",
 "att"=> "Ile",   "act"=> "Thr", "aat"=> "Asn",   "agt"=> "Ser",
 "atc"=> "Ile",   "acc"=> "Thr", "aac"=> "Asn",   "agc"=> "Ser",
 "ata"=> "Ile",   "aca"=> "Thr", "aaa"=> "Lys",   "aga"=> "Arg",
 "atg"=> "Met",   "acg"=> "Thr", "aag"=> "Lys",   "agg"=> "Arg",
 "gtt"=> "Val",   "gct"=> "Ala", "gat"=> "Asp",   "ggt"=> "Gly",
 "gtc"=> "Val",   "gcc"=> "Ala", "gac"=> "Asp",   "ggc"=> "Gly",
 "gta"=> "Val",   "gca"=> "Ala", "gaa"=> "Glu",   "gga"=> "Gly",
 "gtg"=> "Val",   "gcg"=> "Ala", "gag"=> "Glu",   "ggg"=> "Gly"
);


#   process input data

while ($line = <DATA>) {                                    # read in line of input
    print "$line";                                         # transcribe to output
    chop();                                                # remove end-of-line character
    @triplets = unpack("a3" x (length($line)/3), $line);  # pull out successive triplets
    foreach $codon (@triplets) {                           # loop over triplets
        print "$standardgeneticcode{$codon}";             # print out translation of each
    }                                                     # end loop on triplets
    print "\n\n";                                          # skip line on output
}                                                         # end loop on input lines

#   what follows is input data

__END__
atgcatccctttaat
tctgtctga
```

Running this program on the given input data produces the output:

```
atgcatccctttaat
MetHisProPheAsn

tctgtctga
SerValTER
```

Even this simple program displays several features of the PERL language. The file contains background data (the genetic code translation table), statements that tell the computer to do something, and the input data (appearing after the __END__ line). Comments summarize sections of the program and describe the effect of each statement.

The program is structured as blocks enclosed in curly brackets: {...}, which are useful in controlling the flow of execution. Within blocks, individual statements (each ending in ;) are executed in order of appearance. However, the outer block is a *loop:*

```
while ($line = <DATA>) {
      ...
}
```

Here <DATA> refers, successively, to the lines of input data (appearing after the __END__). The block is executed once for each line of input; that is, while there is any line of input remaining.

Three types of data structures appear in the program. The line of input data, referred to as $line, is a simple *character string*. It is split into an *array* or vector of triplets. An array stores several items in a linear order, and individual items of data can be retrieved from their positions in the array. Then, for ease of looking up the amino acid coded for by any triplet, the genetic code is stored as an *associative array*. An associative array, or hash table, is a generalization of a simple or sequential array. Whereas the elements of a simple array are indexed by consecutive integers, the elements of an associative array are indexed by *any* character strings, in this case the 64 triplets. We utilize the input triplets *in order of their appearance* in the nucleotide sequence, but we need to access the elements of the genetic code table *in an arbitrary order* as dictated by the succession of triplets in the input data. A simple array or vector of character strings is appropriate for processing successive triplets, and the associative array is appropriate for looking up the amino acids that correspond to them.

Case Study 1.2 Assembly of overlapping fragments

Here is another PERL program, that illustrates additional aspects of the language. It continues to emphasize the importance of descriptive comments as an *essential* part of good programming style. This program reassembles the sentence:

> All the world's a stage,
> And all the men and women merely players;
> They have their exits and their entrances,
> And one man in his time plays many parts.

after it has been chopped into random overlapping fragments (\n in the fragments represents end-of-line in the original):

> the men and women merely players;\n
> one man in his time
> All the world's
> their entrances,\nAnd one man
> stage,\nAnd all the men and women
> They have their exits and their entrances,\n
> world's a stage,\nAnd all
> their entrances,\nAnd one man

$\longrightarrow$

Case Study 1.2 *(continued)*

> in his time plays many parts.
> merely players;\nThey have

This kind of calculation is important in assembling DNA sequences from overlapping fragments.

```perl
#!/usr/bin/perl
#assemble.pl -- assemble overlapping fragments of strings

# input of fragments
while ($line = <DATA>) {              #   read in fragments, 1 per line
   chop($line);                       #   remove trailing carriage return
   push(@fragments,$line);           #   copy each fragment into array
}
# now array  @fragments  contains fragments

# we need two relationships between fragments:
# (1) which fragment shares no prefix with suffix of another fragment
#        * This tells us which fragment comes first
# (2) which fragment shares longest suffix with a prefix of another
#        * This tells us which fragment  follows  any fragment

# First set array of prefixes to the default value   "noprefixfound".
#      Later, change this default value when a prefix is found.
#      The one fragment that retains the default value must be come first.

# Then loop over pairs of fragments to determine maximal overlap.
#      This determines successor of each fragment
#      Note in passing that if a fragment has a successor then the
#          successor must have a prefix

foreach $i (@fragments) {             #   initially set  prefix of each fragment
    $prefix{$i} = "noprefixfound";    #      to  "noprefixfound"
    }                                 #   this will be overwritten when a prefix is found

# for each pair, find longest overlap of suffix of one with prefix of the other
#        This tells us which fragment FOLLOWS any fragment

foreach $i (@fragments) {             #   loop over fragments
    $longestsuffix = "";              #   initialize longest suffix to null

    foreach $j (@fragments) {         #   loop over fragment pairs
        unless ($i eq $j) {           #   don't check fragment against itself
            $combine = $i . "XXX" . $j;  #  concatenate fragments, with fence XXX
            $combine =~ /([\S ]{2,})XXX\1/;  #   check for repeated sequence
            if (length($1) > length($longestsuffix)) {   # keep longest overlap
                $longestsuffix = $1;  #   retain longest suffix
                $successor{$i} = $j;  #   record that $j follows $i
            }

        }
    }
    $prefix{$successor{$i}} = "found";   #  if $j follows $i then $j must have a prefix
}


foreach (@fragments) {                    # find fragment that has no prefix; that's the start
    if ($prefix{$_} eq "noprefixfound") {$outstring = $_;}
}
$test = $outstring;                       #   start with fragment without prefix
while ($successor{$test}) {               #   append fragments in order
```

Case Study 1.2 *(continued)*

```
    $test = $successor{$test};          #   choose next fragment
    $outstring = $outstring . "XXX" . $test;  # append to string
    $outstring =~ s/([\S ]+)XXX\1/\1/;  #   remove overlapping segment
}
$outstring =~ s/\\n/\n/g;               #   change signal \n to real carriage return
print "$outstring\n";                   #   print final result

__END__
the men and women merely players;\n
one man in his time
All the world's
their entrances,\nand one man
stage,\nAnd all the men and women
They have their exits and their entrances,\n
world's a stage,\nAnd all
their entrances,\nand one man
in his time plays many parts.
merely players;\nThey have
```

Biological classification and nomenclature

Back to the eighteenth century, when academic life at least was in some respects simpler.

Biological nomenclature is based on the idea that living things are divided into units called species—groups of similar organisms with a common gene pool. (Why living things should be 'quantized' into *discrete* species is a very complicated question.) Linnaeus, a Swedish naturalist, classified living things according to a hierarchy: Kingdom, Phylum, Class, Order, Family, Genus and Species (see below). Modern taxonomists have added additional levels. For identification it generally suffices to specify the **binomial**: Genus and Species; for instance *Homo sapiens* for human or *Drosophila melanogaster* for fruit fly. Each binomial uniquely specifies a species that may also be known by one or more common names; for instance, *Bos taurus* = cow. Of course, most species have no common names.

Classifications of human and fruit fly

	Human	Fruit fly
Kingdom	Animalia	Animalia
Phylum	Chordata	Arthropoda
Class	Mammalia	Insecta
Order	Primata	Diptera
Family	Hominidae	Drosophilidae
Genus	*Homo*	*Drosophila*
Species	*sapiens*	*melanogaster*

Originally the Linnaean system was only a classification based on observed similarities. With the discovery of evolution it emerged that the system largely reflects biological ancestry. But which similarities truly reflect common ancestry? Characteristics derived from a common ancestor are *homologous*; for instance an eagle's wing and a human's arm. Other apparently similar characteristics may have arisen independently by *convergent evolution*; for instance, an eagle's wing and a bee's wing: The most recent common ancestor of eagles and bees did not have wings. Conversely, truly homologous characters may have diverged to become very dissimilar in structure and function. The bones of our middle ears are homologous to bones in the jaws of primitive fishes; our eustachian tubes are homologues of gill slits. In most cases experts can distinguish true homologies from similarities resulting from convergent evolution.

Sequence analysis gives the most unambiguous evidence for the relationships among species. The system works well for higher organisms, for which sequence analysis and the classical tools of comparative anatomy, palaeontology and embryology usually give a consistent picture. Classification of microorganisms is more difficult, partly because it is less obvious how to select the features on which to classify them and partly because a large amount of lateral gene transfer threatens to overturn the picture entirely.

Ribosomal RNAs (rRNAs) turned out to have the essential feature of being present in all organisms, with the right degree of divergence. (Too much or too little divergence and relationships become invisible.)

On the basis of 15S rRNAs, C. Woese divided living things most fundamentally into three Domains (a level *above* Kingdom in the hierarchy): Bacteria, Archaea and Eukarya (see Fig. 1.2). Bacteria and Archaea are prokaryotes; their cells do not contain nuclei. Bacteria include the typical microorganisms responsible for many infectious diseases, and, of course, *Escherichia coli*, the mainstay of molecular biology. Archaea include, but are not limited to, extreme thermophiles and halophiles, sulphate reducers and methanogens. We ourselves are Eukarya—organisms containing cells with nuclei—as are yeast and all other multicellular organisms.

A census of the species with sequenced genomes reveals emphasis on bacteria, because of their clinical importance, and for the relative ease of sequencing genomes of prokaryotes. However, despite the obvious differences in lifestyle, and the absence of a nucleus, Archaea are in some ways more closely related on a molecular level to Eukarya than to Bacteria. It is also likely that the Archaea are the closest living organisms to the root of the tree of life.

Figure 1.2 shows the deepest levels of the tree. The Eukarya branch includes animals, plants and fungi. At the ends of the Eukarya branch are the metazoa (multicellular organisms) (Fig. 1.3). We and our closest relatives are deuterostomes (Fig. 1.4).

Use of sequences to determine phylogenetic relationships

Previous sections have introduced sequence databanks and biological relationships. Below are examples of the application of retrieval of sequences from databanks, and the use of sequence comparisons to analyse biological relationships.

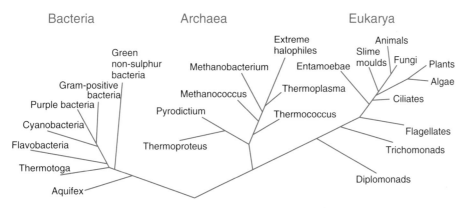
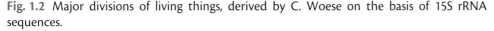

Fig. 1.2 Major divisions of living things, derived by C. Woese on the basis of 15S rRNA sequences.

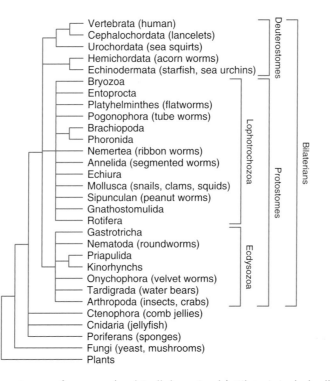

Fig. 1.3 Phylogenetic tree of metazoa (multicellular animals). Bilateria include all animals that share a left–right symmetry of body plan. Protostomes and deuterostomes are two major lineages that separated at an early stage of evolution, estimated at 670 million years ago. They show very different patterns of embryological development, including different early cleavage patterns, opposite orientations of the mature gut with respect to the earliest invagination of the blastula, and the origin of the skeleton from mesoderm (deuterostomes) or ectoderm (protostomes). Protostomes comprise two subgroups distinguished on the basis of 18S rRNA (from the small ribosomal subunit) and HOX gene sequences. Morphologically, Ecdysozoa have a moulting cuticle—a hard outer layer of organic material. Lophotrochozoa have soft bodies. (Based on Adouette, A., Balavoine, G., Lartillot, N., Lespinet, O., Prud'homme, B. and de Rosa, R. (2000). The new animal phylogeny: reliability and implications. *Proceedings of the National Academy of Sciences, USA*, **97**, 4453–4456.)

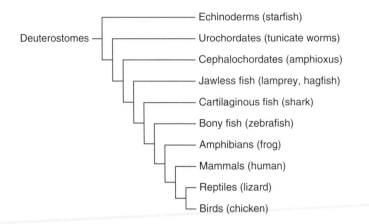

Deuterostomes — Echinoderms (starfish)
Urochordates (tunicate worms)
Cephalochordates (amphioxus)
Jawless fish (lamprey, hagfish)
Cartilaginous fish (shark)
Bony fish (zebrafish)
Amphibians (frog)
Mammals (human)
Reptiles (lizard)
Birds (chicken)

Fig. 1.4 Phylogenetic tree of vertebrates and our closest relatives. Chordates, including vertebrates, and echinoderms are all deuterostomes.

Case Study 1.3 Retrieve the amino acid sequence of horse pancreatic ribonuclease

Use the ExPASy server at the Swiss Institute for Bioinformatics: The URL is: http://www.expasy.org Type in the keywords

horse pancreatic ribonuclease

followed by the ENTER key. Select RNP_HORSE and then FASTA format (see Box: FASTA format). This will produce the following (the first line has been truncated):

```
>sp|P00674|RNP_HORSE RIBONUCLEASE PANCREATIC (EC 3.1.27.5) (RNASE 1) ...
KESPAMKFERQHMDSGSTSSSNPTYCNQMMKRRNMTQGWCKPVNTFVHEP
LADVQAICLQKNITCKNGQSNCYQSSSSMHITDCRLTSGSKYPNCAYQTS
QKERHIIVACEGNPYVPVHFDASVEVST
```

For example, we could retrieve several sequences and align them (see Box: Sequence alignment). Analysis of patterns of similarity among aligned sequences are useful properties in assessing closeness of relationships.

The ID code

RNP_HORSE

comprises abbreviations of the molecule and of the species.

FASTA format

A very common format for sequence data is derived from conventions of FASTA, a program for **FAST A**lignment by W. R. Pearson. Many programs use FASTA format for reading sequences, or for reporting results.

A sequence in FASTA format:

♦ Begins with a single-line description. A > must appear in the first column. The rest of the title line is arbitrary but should be informative.

⟶

→

FASTA format *(continued)*

♦ Subsequent lines contain the sequence, one character per residue.

♦ Use one-letter codes for nucleotides or amino acids specified by the International Union of Biochemistry and International Union of Pure and Applied Chemistry (IUB/IUPAC):

http://www.chem.qmw.ac.uk/iupac/misc/naabb.html and

http://www.chem.qmw.ac.uk/iupac/AminoAcid/

Use Sec and U as the three-letter and one-letter codes for selenocysteine:

http://www.chem.qmw.ac.uk/iubmb/newsletter/1999/item3.html

♦ Lines can have different lengths; that is, 'ragged right' margins.

♦ Most programs will accept lower case letters as amino acid codes.

An example of FASTA format: bovine glutathione peroxidase

```
>gi|121664|sp|P00435|GSHC_BOVIN GLUTATHIONE PEROXIDASE
MCAAQRSAAALAAAAPRTVYAFSARPLAGGEPFNLSSLRGKVLLIENVASLUGTTVRDYTQMNDLQRRLG
PRGLVVLGFPCNQFGHQENAKNEEILNCLKYVRPGGGFEPNFMLFEKCEVNGEKAHPLFAFLREVLPTPS
DDATALMTDPKFITWSPVCRNDVSWNFEKFLVGPDGVPVRRYSRRFLTIDIEPDIETLLSQGASA
```

The title line contains the following fields:

> is obligatory in column 1.

gi|121664 is the *geninfo number,* an identifier assigned by the US National Center for Biotechnology Information (NCBI) to every sequence in its ENTREZ databank. The NCBI collects sequences from a variety of sources, including primary archival data collections and patent applications. Its gi numbers provide a common and consistent 'umbrella' identifier, superimposed on different conventions of source databases. When a source database updates an entry, the NCBI creates a new entry with a new gi number if the changes affect the sequence, but updates and retains its entry if the changes affect only non-sequence information, such as a literature citation.

sp|P00435 indicates that the source database was SWISS-PROT, and that the accession number of the entry in SWISS-PROT was P00435.

GSHC_BOVIN GLUTATHIONE PEROXIDASE is the SWISS-PROT identifier of sequence and species (GSHC_BOVIN), followed by the name of the molecule.

Sequence alignment

Sequence alignment is the assignment of residue-residue correspondences. We may wish to find:

- a *Global match*: align all of one sequence with all of the other.

```
And.--so,.from.hour.to.hour,.we.ripe.and.ripe
||||   |||||||||||||||||||||||   ||||||
And.then,.from.hour.to.hour,.we.rot-.and.rot-
```

This illustrates mismatches, insertions and deletions.

- a *Local match*: find a region in one sequence that matches a region of the other.

```
My.care.is.loss.of.care,.by.old.care.done,
|||||||||    ||||||||||||||   |||||| ||
Your.care.is.gain.of.care,.by.new.care.won
```

For local matching, overhangs at the ends are not treated as gaps. In addition to mismatches, seen in this example, insertions and deletions within the matched region are also possible.

- a *Motif match*: find matches of a short sequence in one or more regions internal to a long one.

A perfect match:

```
    match
    |||||
The match is made; she seals it with a curtsy.
```

One can allow mismatching characters:

```
       match
       ||||
for the watch to babble and to talk is most tolerable
```

```
or:   match                        match
      |||                          || |
And witch the world with noble horsemanship.
```

or insertions and/or deletions:

```
        mat--ch   mat-ch
        ||   |    ||  |
Fear not, Macbeth; no man that's born of woman
Shall e'er have power upon thee.
```

- a *Multiple alignment*: a mutual alignment of many sequences.

```
no.sooner.---met.---------but.they.-look'd
no.sooner.look'd.---------but.they.-lo-v'd
no.sooner.lo-v'd.---------but.they.-sigh'd
no.sooner.sigh'd.---------but.they.--asked.one.another.the.reason
no.sooner.knew.the.reason.but.they.-------------sought.the.remedy
no.sooner.               .but.they.
```

The last line shows characters conserved in all sequences in the alignment.

See Chapter 5 for an extended discussion of alignment.

Case Study 1.4 **Determine, from the sequences of pancreatic ribonuclease from horse (*Equus caballus*), minke whale (*Balaenoptera acutorostrata*) and red kangaroo (*Macropus rufus*), which two of these species are most closely related**

Knowing that horse and whale are placental mammals and kangaroo is a marsupial, we expect horse and whale to be the closest pair. Retrieving the three sequences as in the previous example, and pasting the following:

```
>RNP_HORSE
KESPAMKFERQHMDSGSTSSSNPTYCNQMMKRRNMTQGWCKPVNTFVHEP
LADVQAICLQKNITCKNGQSNCYQSSSSMHITDCRLTSGSKYPNCAYQTS
QKERHIIVACEGNPYVPVHFDASVEVST
>RNP_BALAC
RESPAMKFQRQHMDSGNSPGNNPNYCNQMMMRRKMTQGRCKPVNTFVHES
LEDVKAVCSQKNVLCKNGRTNCYESNSTMHITDCRQTGSSKYPNCAYKTS
QKEKHIIVACEGNPYVPVHFDNSV
>RNP_MACRU
ETPAEKFQRQHMDTEHSTASSSNYCNLMMKARDMTSGRCKPLNTFIHEPK
SVVDAVCHQENVTCKNGRTNCYKSNSRLSITNCRQTGASKYPNCQYETSN
LNKQIIVACEGQYVPVHFDAYV
```

into the multiple sequence alignment program CLUSTAL-W http://www.ebi.ac.uk/clustalw/ (or alternatively, T-Coffee: http://www.ch.embnet.org/software/TCoffee.html) produces the following:

```
CLUSTAL W (1.8) multiple sequence alignment

RNP_HORSE     KESPAMKFERQHMDSGSTSSSNPTYCNQMMKRRNMTQGWCKPVNTFVHEPLADVQAICLQ 60
RNP_BALAC     RESPAMKFQRQHMDSGNSPGNNPNYCNQMMMRRKMTQGRCKPVNTFVHESLEDVKAVCSQ 60
RNP_MACRU     -ETPAEKFQRQHMDTEHSTASSSNYCNLMMKARDMTSGRCKPLNTFIHEPKSVVDAVCHQ 59
              *:** **:*****:  :......*** **  *.**.* ***:***:**.  *.*:* *

RNP_HORSE     KNITCKNGQSNCYQSSSSMHITDCRLTSGSKYPNCAYQTSQKERHIIVACEGNPYVPVHF 120
RNP_BALAC     KNVLCKNGRTNCYESNSTMHITDCRQTGSSKYPNCAYKTSQKEKHIIVACEGNPYVPVHF 120
RNP_MACRU     ENVTCKNGRTNCYKSNSRLSITNCRQTGASKYPNCQYETSNLNKQIIVACEG-QYVPVHF 118
              :*: ****::***:*.* : **:** *..****** *:**: :::******* ******

RNP_HORSE     DASVEVST 128
RNP_BALAC     DNSV---- 124
RNP_MACRU     DAYV---- 122
              *  *
```

In this table, an * under the sequences indicates a position that is conserved (the same in all sequences), and : and . indicate positions at which all sequences contain residues of very similar physicochemical character (:), or somewhat similar physicochemical character (.).

Large patches of the sequences are identical. There are numerous substitutions but only one internal deletion. By comparing the sequences *in pairs,* the number of identical residues shared among pairs in this alignment (not the same as counting *'s) is:

Number of identical residues in aligned ribonuclease A sequences (out of a total of 122–128 residues)			
Horse	and	Minke whale	95
Minke whale	and	Red kangaroo	82
Horse	and	Red kangaroo	75

→

→

Case Study 1.4 *(continued)*

Horse and whale share the most identical residues. The result appears significant, and therefore confirms our expectations. *Warning: or is the logic really the other way round?*

Let's try a harder one:

Case Study 1.5 Phylogeny of Elephantidae

The two living genera of elephant are represented by the African elephant (*Loxodonta africana*) and the Indian (*Elephas maximus*). It has been possible to sequence the mitochondrial cytochrome *b* from a specimen of the Siberian woolly mammoth (*Mammuthus primigenius*) preserved in the Arctic permafrost. To which modern elephant is this mammoth more closely related?

Retrieving the sequences and running CLUSTAL–W:

```
Indian Elephant    MTHTRKFHPLFKIINKSFIDLPTPSNISTWWNFGSLLGACLITQILTGLFLAMHYTPDTM 60
Siberian Mammoth   MTHIRKSHPLLKILNKSFIDLPTPSNISTWWNFGSLLGACLITQILTGLFLAMHYTPDTM 60
African Elephant   MTHIRKSHPLLKIINKSFIDLPTPSNISTWWNFGSLLGACLITQILTGLFLAMHYTPDTM 60

                   *** ** ***:**.*********************************************

Indian Elephant    TAFSSMSHICRDVNYGWIIRQLHSNGASIFFLCLYTHIGRNIYYGSYLYSETWNTGIMLL 120
Siberian Mammoth   TAFSSMSHICRDVNYGWIIRQLHSNGASIFFLCLYTHIGRNIYYGSYLYSETWNTGIMLL 120
African Elephant   TAFSSMSHICRDVNYGWIIRQLHSNGASIFFLCLYTHIGRNIYYGSYLYSETWNTGIMLL 120
                   ***********************************************************

Indian Elephant    LITMATAFMGYVLPWGQMSFWGATVITNLFSAIPYIGTNLVEWIWGGFSVDKATLNRFFA 180
Siberian Mammoth   LITMATAFMGYVLPWGQMSFWGATVITNLFSAIPYIGTDLVEWIWGGFSVDKATLNRFFA 180
African Elephant   LITMATAFMGYVLPWGQMSFWGATVITNLFSAIPYIGTNLVEWIWGGFSVDKATLNRFFA 180
                   **************************************:*********************

Indian Elephant    FHFILPFTMVALAGVHLTFLHETGSNNPLGLTSDSDKIPFHPYYTIKDFLGLLILILLLL 240
Siberian Mammoth   LHFILPFTMIALAGVHLTFLHETGSNNPLGLTSDSDKIPFHPYYTIKDFLGLLILILFLL 240
African Elephant   LHFILPFTMIALAGVHLTFLHETGSNNPLGLTSDSDKIPFHPYYTIKDFLGLLILILLLL 240
                   :********:**********************************************:**

Indian Elephant    LLALLSPDMLGDPDNYMPADPLNTPLHIKPEWYFLFAYAILRSVPNKLGGVLALFLSILI 300
Siberian Mammoth   LLALLSPDMLGDPDNYMPADPLNTPLHIKPEWYFLFAYAILRSVPNKLGGVLALLLSILI 300
African Elephant   LLALLSPDMLGDPDNYMPADPLNTPLHIKPEWYFLFAYAILRSVPNKLGGVLALLLSILI 300
                   *******************************************************:*****

Indian Elephant    LGLMPLLHTSKHRSMMLRPLSQVLFWTLTMDLLTLTWIGSQPVEHPYIIIGQMASILYFS 360
Siberian Mammoth   LGIMPLLHTSKHRSMMLRPLSQVLFWTLATDLLMLTWIGSQPVEYPYIIIGQMASILYFS 360
African Elephant   LGLMPLLHTSKHRSMMLRPLSQVLFWTLTMDLLTLTWIGSQPVEYPYIIIGQMASILYFS 360
                   **:.*********************:. *** **********:***************

Indian Elephant    IILAFLPIAGMIENYLIK 378
Siberian Mammoth   IILAFLPIAGMIENYLIK 378
African Elephant   IILAFLPIAGVIENYLIK 378
                   **********:*******
```

The mammoth and African elephant sequences have four mismatches (at which positions?) and the mammoth and Indian elephant sequences have 10 mismatches.

→

Case Study 1.5 *(continued)*

It appears that the mammoth is more closely related to African elephants. However, this result is less satisfying than the previous one. There are fewer differences. Are they significant? (It is harder to decide whether the differences are significant because we have no preconceived idea of what the answer should be.)

This example raises a number of questions:

1. We 'know' that African and Indian elephants and mammoths must be close relatives—just look at them. But could we tell *from these sequences alone* that they are from closely related species?

2. Given that the differences are few, do they represent true evolutionary divergence or merely random noise or drift? We need sensitive statistical criteria for judging the significance of the similarities and differences.

As background to such questions, let us reemphasize the distinction between **similarity** and **homology**. *Similarity* is the observation or measurement of resemblance and difference, independent of the source of the resemblance. *Homology* means, specifically, that the sequences and the organisms in which they occur are descended from a common ancestor, with the implication that the similarities are shared ancestral characteristics. Similarity of sequences (or of macroscopic biological characters) is observable in data collectable *now,* and involves no historical hypotheses. In contrast, assertions of homology are statements of historical events that are almost always unobservable. Homology must be an *inference* from observations of similarity. Only in a few cases is homology directly observable; for instance in pedigrees of families showing unusual phenotypes such as the Hapsburg lip, or in laboratory populations, or in clinical studies that follow the course of viral infections at the sequence level in individual patients. The new field of metagenomics will provide other examples (see Chapter 2, and *Introduction to Genomics*).

The assertion that the cytochromes *b* from African and Indian elephants and mammoths are homologous *means* that there was a common ancestor, presumably containing a unique cytochrome *b*, that by alternative mutations gave rise to the proteins of mammoths and modern elephants. Does the very high degree of similarity of the sequences justify the conclusion that they are homologous; or are there other explanations?

◆ It might be that a functional cytochrome *b requires* so many conserved residues that cytochromes *b* from all animals must be as similar to one another as the elephant and mammoth proteins are, whether or not they are homologues. We can test this by looking at cytochrome *b* sequences from other species. The result is that cytochromes *b* from other animals differ substantially from those of elephants and mammoths.

◆ A second possibility is that there are special physiological requirements for a cytochrome *b* to function well in an animal with the size and form of an elephant, that the three cytochrome *b* sequences started out from independent ancestors and that common selective pressures forced them to become similar. (Remember that we are asking what can be deduced from cytochrome *b* sequences alone.)

◆ The mammoth may be more closely related to the Indian elephant, but since the time of the last common ancestor the cytochrome *b* sequence of the Indian elephant has evolved faster than that of the African elephant or the mammoth, accumulating more mutations.

◆ Still a fourth hypothesis is that all common ancestors of elephants and mammoths had very dissimilar cytochromes *b*, but that living elephants and mammoths gained a common gene by transfer from an unrelated organism via a virus.

Suppose, however, that we are satisfied that the similarity of the elephant and mammoth sequences is high enough to imply homology, what then about the ribonuclease sequences in the previous example? Are the *larger* differences among the pancreatic ribonucleases of horse, whale and kangaroo evidence that they are *not* homologues?

How can we answer these questions? Specialists have undertaken careful calibrations of sequence similarities and divergences, among many proteins from many species for which the taxonomic relationships have been worked out by classical methods. In the example of pancreatic ribonucleases, the reasoning from similarity to homology is justified. In the second edition of this book, I wrote: 'The question of whether mammoths are closer to African or Indian elephants was decided only recently, in favour of African elephants'. Since then, expert opinion—including that of some of the same experts—has shifted to the conclusion that *Indian* elephants are the closest extant relatives of mammoths.

Why has this question proved so difficult? It reflects the limited power of our tools, applied to the available data, to resolve events that happened very close to each other, very long ago.

At the family level in our lineage, humans, chimpanzees, gorillas and orangutans comprise the hominidae.

The three major groups of elephants are: African elephants, Asian elephants, and mammoths. These taxa comprise a family, the Elephantidae, containing three main genera: *Loxodonta*, including the African species *L. africana*; *Elephas*, including the Asian species *E. maximus*; and *Mammuthus*, including the Siberian species *M. primigenius*. These genera diverged about 6 million years ago (Mya) in Africa, at approximately the same time as the divergence of human and chimpanzee ancestors. Today, 'mammoth' connotes an extinct arctic animal. However, our ancestors hunted mammoths in southern Europe, as depicted in cave-wall paintings.

Mammoth fossils helped shape ideas about species extinction

Cuvier himself first distinguished the African, Asian and mammoth lineages, in a 1796 paper. Cuvier accepted the idea that species could become extinct, a prerequisite to development of ideas of evolution. Many contemporaries believed

→

→

Mammoth fossils helped shape ideas about species extinction *(continued)*

instead in the immutability of species. US President Thomas Jefferson was one. He instructed Meriwether Lewis and William Clark, explorers of the Louisiana Territory purchased from France in 1803, to keep on the lookout for living mammoths.

The challenging phylogenetic problem is to determine the branching order of Asian and African elephants and mammoths. Which group split off first? It took only ~500 000 years to establish the three lineages. The shortness of this time, relative to the available data, makes great demands on our analytic tools.

Sequence data from mammoths are, unsurprisingly, sparse. Preservation of DNA in bone and hair, even in frozen specimens, is imperfect. (In contrast, the *L. africana* genome is complete, although at low coverage). We know, for non-extinct species with large amounts of available sequence data, that different genes from the same set of species may follow different phylogenetic trees. Therefore the limited data from extinct species prevent us from seeing the whole picture.

Other factors that make the identification of the true branching pattern difficult include:

♦ No relatively close relative can serve for comparison (as an outgroup). The species used, dugong or hyrax, diverged from elephants ~65 million years ago. Closer relatives would be preferable.

♦ Small population sizes may increase the importance of fluctuations.

♦ The assumption of constant rates of evolution in the different lineages may be unjustified.

The conclusion is that although current data and analysis suggest that mammoths are more closely related to Asian elephants, the significance of the result is very near—or, some authors feel, beyond—the limit of confidence.

Despite the difficulty of the elephant–mammoth problem, analysis of sequence similarities in genomes and proteins is now sufficiently well established that it is considered the most reliable method for establishing phylogenetic relationships, even though sometimes the results may not be significant, and in other cases they even give incorrect answers. Except for attempts to treat extinct species, there are copious data available, effective tools for retrieving what is necessary to bring to bear on a specific question, and powerful analytic tools. None of this replaces the need for thoughtful scientific judgement.

Use of SINES and LINES to derive phylogenetic relationships

Major problems with inferring phylogenies from comparisons of gene and protein sequences are (1) the wide range of variation of similarity, which may dip below statistical significance, and (2) the effects of different rates of evolution along different branches of the evolutionary tree. In many cases, even if sequence similarities confidently establish relationships, it may be very difficult or impossible to decide the *order* in which sets of taxa have split. (The Elephantidae are an example.) The phylogeneticist's dream—features that have an 'all-or-none' character, and the appearance of which is irreversible so that the order of branching events can be decided—is in some cases afforded by certain non-coding sequences in genomes.

SINES and LINES (Short and Long Interspersed Nuclear Elements) are repetitive non-coding sequences that form large fractions of eukaryotic genomes—at least 30% of human chromosomal DNA, and >50% of some higher plant genomes. Typically, SINES are ~70–500 bp long, and up to 10^6 copies may appear. LINES may be up to 7000 bp long, and up to 10^5 copies may appear. SINES enter the genome by reverse transcription of RNA. Most SINES contain a 5′ region homologous to tRNA, a central region unrelated to tRNA and a 3′ AT-rich region.

Features of SINES that make them useful for phylogenetic studies include:

- A SINE is either present or absent. The presence of a SINE at any particular position is a property that entails no complicated and variable measure of similarity.

- SINES are inserted at random in the non-coding portion of a genome. Therefore, the appearance of similar SINES at the same locus in two species implies that the species share a common ancestor in which the insertion event occurred. No analogue of convergent evolution muddies this picture, because there is no selection for the *site* of insertion.

- SINE insertion appears to be irreversible: no mechanism for *loss* of SINES is known, other than rare large-scale deletions that include the SINE site. Therefore, if two species share a SINE at a common locus, *absence* of this SINE in a third species implies that the first two species must be more closely related to each other than either is to the third.

- Not only do SINES show relationships, they imply which species branched off first. The last common ancestor of species containing a common SINE must have come *after* the last common ancestor linking these species and another that lacks this SINE.

N. Okada and colleagues applied SINE sequences to questions of phylogeny.

Whales, like Australians, are mammals that have adopted an aquatic lifestyle. But what—in the case of the whales—are their closest land-based relatives? Classical palaeontology linked the order *Cetacea*—comprising whales, dolphins and porpoises—with the order *Artiodactyla*—even-toed ungulates (including, for instance, cows and sheep). Cetaceans were thought to have diverged before the common ancestor of the three extant Artiodactyl suborders: *Suiformes* (pigs), *Tylopoda* (including camels and llamas) and *Ruminantia* (including deer, cows, goats, sheep, antelopes, giraffes, etc.). To place cetaceans properly among these groups, several studies were carried out with DNA sequences. Comparisons of mitochondrial DNA, and genes for pancreatic ribonuclease, γ-fibrinogen and other proteins, suggested that the closest relatives of the whales are hippopotamuses, and that cetaceans and hippopotamuses form a separate group within the artiodactyls, most closely related to the *Ruminantia* (see Weblem 1.7).

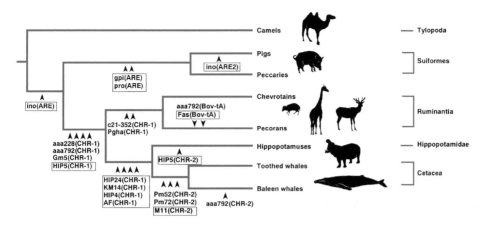

Fig. 1.5 Phylogenetic relationships among cetaceans and other artiodactyl subgroups, derived from analysis of SINE sequences. Small arrowheads mark insertion events. Each arrowhead indicates the presence of a particular SINE or LINE at a specific locus in all species to the *right* of the arrowhead. Lower-case letters identify loci, upper-case letters identify sequence patterns. For instance, the ARE2 pattern appears only in pigs, at the ino locus. The ARE pattern appears twice in the pig genome, at loci gpi and pro, and in the peccary genome at the same loci. The ARE insertion occurred in a species that was ancestral to pigs and peccaries but to no other species in the diagram. This implies that pigs and peccaries are more closely related to each other than to any of the other animals studied. (From Nikaido, M., Rooney, A.P. and Okada, N. (1999). Phylogenetic relationships among cetartiodactyls based on insertions of short and long interpersed elements: hippopotamuses are the closest extant relatives of whales. *Proceedings of the National Academy of Sciences, USA,* **96**, 10261–10266 (Copyright 1999, National Academy of Sciences, USA. Reproduced by permission).)

Analysis of SINES confirms this relationship. Several SINES are common to *Ruminantia*, hippopotamuses and cetaceans. Four SINES appear in hippopotamuses and cetaceans only. These observations imply the phylogenetic tree shown in Fig. 1.5, in which the SINE insertion events are marked.

Recently discovered fossils of land-based ancestors of whales confirm the link between whales and artiodactyls. This is a good example of the complementarity between molecular and palaeontological methods: DNA sequence analysis can specify relationships among living species quite precisely, but fossils reveal relationships among their *extinct* ancestors.

Searching for similar sequences in databases: PSI-BLAST

A common theme of the examples we have treated is the search of a database for items similar to a probe. For instance, if you are studying a novel gene, or if you identify within the human genome a gene responsible for some disease, you will wish to determine whether related genes appear in other species. The ideal method is both *sensitive*—that is, it picks up even very distant relationships—and *selective*—that is, all the relationships that it reports are true.

Sensitivity and selectivity

Database search methods involve a trade-off between *sensitivity* and *selectivity.* Does the method find all or most of the examples that are actually present, or does it miss a large fraction? Conversely, how many of the 'hits' that it reports are incorrect? Suppose a database contains 1000 globin sequences. Suppose a search of this database for globins reported 900 results, 700 of which were really globin sequences and 200 of which were not. This result would be said to have 300 false negatives (misses) and 200 false positives. There is a trade-off between sensitivity and selectivity: lowering a tolerance threshold will increase the numbers of *both* false negatives and false positives. Often one is willing to work with low thresholds to make sure of not missing anything that might be important; but this requires detailed examination of the results to eliminate the resulting false positives.

A powerful tool for searching sequence databases with a probe sequence is PSI-BLAST, from the US National Center for Biotechnology Information (NCBI). PSI-BLAST stands for 'Position Sensitive Iterated-Basic Local Alignment Search Tool'. A previous program, BLAST, worked by identifying local regions of similarity without gaps and then piecing them together. The PSI in PSI-BLAST refers to enhancements that identify patterns within the sequences at preliminary stages of the database search, and then progressively refine them. Recognition of conserved patterns can sharpen both the selectivity and sensitivity of the search. PSI-BLAST involves a repetitive (or iterative) process, as the emergent pattern becomes better defined in successive stages of the search (see Chapter 5).

Case Study 1.6   Homologues of the human PAX-6 gene

PAX-6 genes control eye development in a widely divergent set of species (see Box, p. 35). The human PAX-6 gene encodes the protein appearing in SWISS-PROT entry P26367. To run PSI-BLAST, go to the following URL:

http://www.ncbi.nlm.nih.gov/BLAST

Enter the sequence, and use the default options for selections of the database to search, and the similarity matrix to use.

   The program returns a list of entries similar to the probe, sorted in decreasing order of statistical significance. (Extracts from the response are shown in the Box, *Results of PSI-BLAST search for human PAX-6 protein.* Only a few lines are shown, merely to illustrate the format.) A typical line appears as follows:

```
pir||I45557 eyeless, long form - fruit fly (Drosophila melano...  250 2e-64
```

$\longrightarrow$

→

Case Study 1.6 *(continued)*

The first item on the line is the database and corresponding entry number (separated by | |) in this case the PIR (Protein Identification Resource) entry I45557. It is the *Drosophila* homologue *eyeless*. The number 250 is a score for the match detected, and the significance of this match is measured by $E = 2 \times 10^{-64}$. E is related to the probability that the observed degree of similarity could have arisen by chance: E is the number of sequences that would be expected to match as well or better than the one being considered, if the same database were probed with random sequences. $E = 2 \times 10^{-64}$ means that it is *extremely* unlikely that even *one* random sequence would match as well as the *Drosophila* homologue. Values of E below ~0.05 would be considered significant; at least they might be worth considering. For borderline cases, you would ask: are the mismatches conservative? Is there any pattern or are the matches and mismatches distributed randomly throughout the sequences? There is an elusive concept, the *texture* of an alignment, that you will become sensitive to. The court of last resort is whether the structures are similar, but often this information is not available.

Note that if there are many sequences in the databank that are very similar to the probe sequence, they will head the list. In this example, there are many very similar *PAX* genes in other mammals. You may have to scan far down the list to find a distant relative that you consider interesting.

Even in the case of *Drosophila* eyeless, a very close relative of the probe sequence, the program reports only a local match to a portion of the sequences. The full alignment is shown in the Box. *Complete pairwise sequence alignment of human PAX-6 protein and* Drosophila melanogaster *eyeless.*

Et in terra PAX hominibus, muscisque . . .

The eyes of the human, fly and octopus are very different in structure. Conventional wisdom, noting the immense selective advantage conferred by the ability to see, held that eyes arose independently in different phyla. It therefore came as a great surprise that a gene controlling human eye development has a homologue governing eye development in *Drosophila.*

The *PAX-6* gene was first cloned in the mouse and human. It is a master regulatory gene, controlling a complex cascade of events in eye development. Mutations in the human gene cause the clinical condition *aniridia,* a developmental defect in which the iris of the eye is absent or deformed. The *PAX-6* homologue in *Drosophila*—called the *eyeless* gene—has a similar function of control over eye development. Flies mutated in this gene develop without eyes; conversely, expression of this gene in a fly's wing, leg or antenna produces ectopic (= out of place) eyes. (The *Drosophila eyeless* mutant was first described in 1915. Little did anyone then suspect a relationship to a mammalian gene.)

→

→ ───

Et in terra PAX hominibus, muscisque . . . (continued)

Not only are the insect and mammalian genes similar in sequence, they are so closely related that their function crosses species boundaries. Expression of the mouse *PAX-6* gene in the fly causes ectopic eye development just as expression of the fly's own *eyeless* gene does.

PAX-6 has homologues in other phyla, including flatworms, ascidians, sea urchins and nematodes. The observation that rhodopsins—a family of proteins containing retinal as a common chromophore—function as light-sensitive pigments in different phyla is supporting evidence for a common origin of different photoreceptor systems. The genuine structural differences in the macroscopic anatomy of different eyes reflect the divergence and independent evolution of higher-order structure.

Results of PSI-BLAST search for human PAX-6 protein

Five iterations of PSI-BLAST were run, using human PAX-6 as the query sequence, searching the non-redundant (nr) database. The NCBI nr database is a set of unique sequences selected from the full databases to eliminate multiple hits to very similar sequences. The output contains a list of sequences identified. It also contains pairwise alignments of well-matching regions from the query and the retrieved sequences. Three selected alignments are shown: one from near the top of the list, PAX-6 from *Danio rerio*, $E = 3 \times 10^{-171}$; *Drosophila* eyeless, $E = 8 \times 10^{-59}$; and a *Drosophila* circadian clock protein with $E = 0.42$, for which the matching is both shorter and less perfect.

```
Query= sp|P26367|PAX6_HUMAN Paired box protein Pax-6(Oculorhombin)
(Aniridia, type II protein) - Homo sapiens (Human).
         (422 letters)

Database: All non-redundant GenBank CDS
translations+PDB+SwissProt+PIR+PRF
           2,738,511 sequences; 768,166,133 total letters

Results of PSI-Blast iteration 5
                                                        Score      E
Sequences producing significant alignments:            (Bits)  Value

gb|AAA59962.1|  oculorhombin >gb|AAA59963.1| oculorhombin      600    6e-170
ref|NP_000271.1|  paired box gene 6 isoform a [Homo sapiens] >...  600    7e-170
gb|ABI98848.1|  paired box 6 transcript variant 3 [Columba livia]  599    9e-170
ref|NP_001035735.1|  paired box gene 6 (aniridia, keratitis) [...  599    1e-169
gb|EAW68233.1|  paired box gene 6 (aniridia, keratitis), isofo...  599    1e-169
gb|ABA90484.1|  paired box protein PAX6 isoform a [Oryctolagus cu   598    2e-169
ref|NP_037133.1|  paired box gene 6 [Rattus norvegicus] >sp|P6...  598    2e-169
dbj|BAA24025.1|  PAX6 SL [Cynops pyrrhogaster]                   596    8e-169
ref|NP_038655.1|  paired box gene 6 [Mus musculus] >emb|CAA453...  596    1e-168
dbj|BAC25729.1|  unnamed protein product [Mus musculus]          595    2e-168
emb|CAC80516.1|  paired box protein [Mus musculus]              594    2e-168
ref|NP_001595.2|  paired box gene 6 isoform b [Homo sapiens] >...  594    3e-168
```

→

Results of PSI-BLAST search for human PAX-6 protein *(continued)*

```
gb|AAH41712.1|  MGC52531 protein [Xenopus laevis]                  594    3e-168
ref|NP_990397.1|  paired box gene 6 [Gallus gallus] >dbj|BAA23...  594    3e-168
gb|ABO70134.1|  PAX6 [Canis familiaris]                            593    5e-168
gb|EAW68236.1|  paired box gene 6 (aniridia, keratitis), isofo...  593    6e-168
emb|CAF29075.1|  putative pax6 isoform 5a [Rattus norvegicus]      593    8e-168
ref|NP_001075686.1|  paired box protein PAX6 isoform b [Orycto...  593    9e-168
emb|CAE45868.1|  hypothetical protein [Homo sapiens]               593    9e-168
prf||1902328A  PAX6 gene                                           592    1e-167
gb|AAS48919.1|  paired box 6 isoform 5a [Rattus norvegicus] >g...  592    1e-167
gb|AAB36681.1|  paired-type homeodomain Pax-6 protein [Xenopus la  592    1e-167
gb|AAF73271.1|AF154555_1  paired domain transcription factor v...  590    5e-167
gb|AAB05932.1|  Xpax6 [Xenopus laevis]                             589    7e-167
sp|P47238|PAX6_COTJA  Paired box protein Pax-6 (Pax-QNR) >pir|...  589    1e-166
dbj|BAA24024.1|  PAX6 LL [Cynops pyrrhogaster]                     589    1e-166
sp|P55864|PAX6_XENLA  Paired box protein Pax-6 >gb|AAB36683.1| Pa  588    2e-166
emb|CAA68838.1|  PAX-6 protein [Astyanax mexicanus]                588    3e-166
```

...

```
emb|CAE66896.1|  Hypothetical protein CBG12277 [Caenorhabditis br  44.7   0.010
gb|AAP79287.2|  hox 7 [Saccoglossus kowalevskii]                   44.7   0.010
gb|AAS07621.1|  homeobox protein Lox18 [Perionyx excavatus]        44.7   0.010
gb|AAL04488.1|AF365974_1  transcription factor SOHo [Oryzias lati  44.7   0.010
ref|NP_186796.1|  ATHB-1 (Homeobox-leucine zipper protein HAT5...  44.7   0.010
ref|XP_001076009.1|  PREDICTED: similar to gooseberry-neuro CG...  44.7   0.010
ref|XP_001060443.1|  PREDICTED: similar to double homeobox 4c [Ra  44.7   0.010
gb|EAT37245.1|  lim homeobox protein [Aedes aegypti]               44.7   0.010
gb|AAW70293.1|  invected [Heliconius pachinus]                     44.7   0.010
ref|NP_174164.1|  HB-1 (homeobox-1); transcription factor [Arabid  44.7   0.010
ref|NP_001029316.1|  NK-3 transcription factor, locus 1 [Rattu...  44.7   0.010
dbj|BAE44266.1|  hoxB3a [Oryzias latipes] >dbj|BAE53473.1| hox...  44.7   0.010
dbj|BAE06563.1|  transcription factor protein [Ciona intestinalis  44.7   0.010
gb|EAT43388.1|  homeobox protein [Aedes aegypti]                   44.7   0.010
gb|AAS21413.1|  HOX11 [Oikopleura dioica]                          44.7   0.010
```

... additional 'hits' deleted ...

Three selected alignments follow:

```
ref|NP_571379.1| paired box gene 6a [Danio rerio]
 emb|CAA44867.1| pax-6 [Danio rerio]
 emb|CAM16650.1| paired box gene 6a [Danio rerio]
Length=451

 Score =  662 bits (1707),  Expect = 0.0, Method: Composition-based stats.
 Identities = 404/436 (92%), Positives = 409/436 (93%), Gaps = 18/436 (4%)

Query  1    MQNSHSGVNQLGGVFVNGRPLPDSTRQKIVELAHSGARPCDISRILQ------------   47
            MQNSHSGVNQLGGVFVNGRPLPDSTRQKIVELAHSGARPCDISRILQ
Sbjct  20   MQNSHSGVNQLGGVFVNGRPLPDSTRQKIVELAHSGARPCDISRILQTHADAKVQVLDNE  79

Query  48   -VSNGCVSKILGRYYETGSIRPRAIGGSKPRVATPEVVSKIAQYKRECPSIFAWEIRDRL  106
             VSNGCVSKILGRYYETGSIRPRAIGGSKPRVATPEVV KIAQYKRECPSIFAWEIRDRL
Sbjct  80   NVSNGCVSKILGRYYETGSIRPRAIGGSKPRVATPEVVGKIAQYKRECPSIFAWEIRDRL  139

Query  107  LSEGVCTNDNIPSVSSINRVLRNLASEKQQMGADGMYDKLRMLNGQTGSWGTRPGWYPGT  166
            LSEGVCTNDNIPSVSSINRVLRNLASEKQQMGADGMY+KLRMLNGQTG+WGTRPGWYPGT
Sbjct  140  LSEGVCTNDNIPSVSSINRVLRNLASEKQQMGADGMYEKLRMLNGQTGTWGTRPGWYPGT  199

Query  167  SVPGQPTQDGCQQQEGGGENTNSISSNGEDSDEAQMRLQLKRKLQRNRTSFTQEQIEALE  226
            SVPGQP QDGCQQ +GGGENTNSISSNGEDSDE QMRLQLKRKLQRNRTSFTQEQIEALE
Sbjct  200  SVPGQPNQDGCQQSDGGGENTNSISSNGEDSDETQMRLQLKRKLQRNRTSFTQEQIEALE  259
```

→

Results of PSI-BLAST search for human PAX-6 protein *(continued)*

```
Query  227  KEFERTHYPDVFARERLAAKIDLPEARIQVWFSNRRAKWRREEKLRNQRRQASNTPSHIP  286
            KEFERTHYPDVFARERLAAKIDLPEARIQVWFSNRRAKWRREEKLRNQRRQASN+ SHIP
Sbjct  260  KEFERTHYPDVFARERLAAKIDLPEARIQVWFSNRRAKWRREEKLRNQRRQASNSSSHIP  319

Query  287  ISSSFSTSVYQPIPQPTTPVSSFTSGSMLGRTDTALTNTYSALPPMPSFTMANNLPMQPP  346
            ISSSFSTSVYQPIPQPTTPV SFTSGSMLGR+DTALTNTYSALPPMPSFTMANNLPMQP
Sbjct  320  ISSSFSTSVYQPIPQPTTPV-SFTSGSMLGRSDTALTNTYSALPPMPSFTMANNLPMQP-  377

Query  347  VPSQTSSYSCMLPTSPSVNGRSYDTYTPPHMQTHMNSQPMGTSGTTSTGLISPGVSVPVQ  406
               SQTSSYSCMLPTSPSVNGRSYDTYTPPHMQ HMNSQ M  SGTTSTGLISPGVSVPVQ
Sbjct  378  --SQTSSYSCMLPTSPSVNGRSYDTYTPPHMQAHMNSQSMAASGTTSTGLISPGVSVPVQ  435

Query  407  VPGSEPDMSQYWPRLQ  422
            VPGSEPDMSQYWPRLQ
Sbjct  436  VPGSEPDMSQYWPRLQ  451
```

```
>pir||I45557  eyeless, long form - fruit fly (Drosophila melanogaster)
 emb|CAA56038.1| UniGene info transcription factor [Drosophila melanogaster]
Length=838
```

```
 Score =  224 bits (572),  Expect = 8e-59, Method: Composition-based stats.
 Identities = 133/212 (62%), Positives = 143/212 (67%), Gaps = 2/212 (0%)
```

```
Query    2  QNSHSGVNQLGGVFVNGRPLPDSTRQKIVELAHSGARPCDISRILQVSNGCVSKILGRYY   61
               HSGVNQLGGVFV GRPLPDSTRQKIVELAHSGARPCDISRILQVSNGCVSKILGRYY
Sbjct   35  HKGHSGVNQLGGVFVGGRPLPDSTRQKIVELAHSGARPCDISRILQVSNGCVSKILGRYY   94

Query   62  ETGSIRPRAIGGSKPRVATPEVVSKIAQYKRECPSIFAWEIRDRLLSEGVCTNDNIPSVS  121
            ETGSIRPRAIGGSKPRVAT EVVSKI+QYKRECPSIFAWEIRDRLL E VCTNDNIPSVS
Sbjct   95  ETGSIRPRAIGGSKPRVATAEVVSKISQYKRECPSIFAWEIRDRLLQENVCTNDNIPSVS  154

Query  122  SINRVLRNLASEKQQMGADGMYDKLRMLNGQTGSWGTRPGWYPGTSVPGQPTQDGCQQQE  181
            SINRVLRNLA++K+Q           N +      G       G
Sbjct  155  SINRVLRNLAAQKEQQSTGSGSSSTSAGNSISAKVSVSIGGNVSNVASGSRGTLSSSTDL  214

Query  182  GGGENT--NSISSNGEDSDEAQMRLQLKRKLQ  211
                 +S S  +S E  +  + KL+
Sbjct  215  MQTATPLNSSESGGATNSGEGSEQEAIYEKLR  246
```

```
>gb|AAB94890.1|  circadian clock protein [Drosophila melanogaster]
Length=1398
```

```
 Score = 33.5 bits (75),  Expect = 0.42, Method: Composition-based stats.
 Identities = 22/145 (15%), Positives = 37/145 (25%), Gaps = 31/145 (21%)
```

```
Query  113  TNDNIPSVSSINRVLRN-------LASEKQQMGADGMYDK----LRMLNGQTGSWGTRPG  161
            N  P+ S+   L N        +   A  K        +          G +
Sbjct  411  NNTTNPTSSAPQGCLGNEPFKPPPPLPVRASTSAHAQMQKFNESSYASHVSAVKLGQKSP  470

Query  162  WYPGTSV-------------PGQPTQDGCQQQEGGGENTNSISSNGEDSDEAQMRLQLKR  208
              +            Q   C     EN SIS++  D D  Q + Q ++
Sbjct  471  HAGQLQLTKGKCCPQKRECPSSQSELSDCGYGT-QVENQESISTSSNDDDGPQGKPQHQK  529

Query  209  K------LQRNRTSFTQEQIEALEK  227
                 + RT +    + L +
Sbjct  530  PPCNTKPRNKPRTIMSPMDKKELRR  554
```

Complete pairwise sequence alignment of human PAX-6 protein and *Drosophila melanogaster* eyeless

```
PAX6_human   -----------------------------MQNSHSGVNQLGGVFVNGRPLPDSTRQ  27
eyeless      MFTLQPTPTAIGTVVPPWSAGTLIERLPSLEDMAHKGHSGVNQLGGVFVGGRPLPDSTRQ  60
                                        ::.************.**********

PAX6_human   KIVELAHSGARPCDISRILQVSNGCVSKILGRYYETGSIRPRAIGGSKPRVATPEVVSKI  87
eyeless      KIVELAHSGARPCDISRILQVSNGCVSKILGRYYETGSIRPRAIGGSKPRVATAEVVSKI 120
             ********************************************************.******

PAX6_human   AQYKRECPSIFAWEIRDRLLSEGVCTNDNIPSVSSINRVLRNLASEKQQ----------- 136
eyeless      SQYKRECPSIFAWEIRDRLLQENVCTNDNIPSVSSINRVLRNLAAQKEQQSTGSGSSSTS 180
             :*****************.*.******************::*:*

PAX6_human   ------------MG----------------------------------------ADG 141
eyeless      AGNSISAKVSVSIGGNVSNVASGSRGTLSSSTDLMQTATPLNSSESGGATNSGEGSEQEA 240
                         :*                                           :.

PAX6_human   MYDKLRMLNGQTGS-------------------WGTRP------------------- 160
eyeless      IYEKLRLLNTQHAAGPGPLEPARAAPLVGQSPNHLGTRSSHPQLVHGNHQALQQHQQQSW 300
             :*:***:** * .:                   ***.

PAX6_human   -------GWYPG-------TSVP----------------------------GQP---- 172
eyeless      PPRHYSGSWYPTSLSEIPISSAPNIASVTAYASGPSLAHSLSPPNDIKSLASIGHQRNCP 360
                    .***       :*.*                              *:

PAX6_human   ----------TQDGCQQQEGG---GENTNSISSNGEDSDEAQMRLQLKRKLQRNRTSFTQ 219
eyeless      VATEDIHLKKELDGHQSDETGSGEGENSNGGASNIGNTEDDQARLILKRKLQRNRTSFTN 420
                       ** *.:* *    ***:*. :**   ::::  * ** *************:

PAX6_human   EQIEALEKEFERTHYPDVFARERLAAKIDLPEARIQVWFSNRRAKWRREEKLRNQRRQAS 279
eyeless      DQIDSLEKEFERTHYPDVFARERLAGKIGLPEARIQVWFSNRRAKWRREEKLRNQRRTPN 480
             :**::***************.**.********************************  ..

PAX6_human   NTPSHIPISSSFSTSVYQPIPQPTTPVSSFTSGSMLG---------------------- 316
eyeless      STGASATSSSTSATASLTDSPNSLSACSSLLSGSAGGPSVSTINGLSSPSTLSTNVNAPT 540
             .* : . **: :*:       *:. :. **: *** *

PAX6_human   ------------------------------------------------------------ 
eyeless      LGAGIDSSESPTPIPHIRPSCTSDNDNGRQSEDCRRVCSPCPLGVGGHQNTHHIQSNGHA 600

PAX6_human   -------------------------RTDTALTNTYSALPPMPSFTMANNLPMQPPVP 348
eyeless      QGHALVPAISPRLNFNSGSFGAMYSNMHHTALSMSDSYGAVTPIPSFNHSAVGPLAPPSP 660
                                      :*  ::::*.*:.*:***. :   *: ** *

PAX6_human   S-------QTSSYSCMLPTSP--------------------------------SVNGRS 368
eyeless      IPQQGDLTPSSLYPCHMTLRPPPMAPAHHHIVPGDGGRPAGVGLGSGQSANLGASCSGSG 720
                     :* *.* :. *                                * .* .

PAX6_human   YDTYTP--------------------------PHMQTHMNSQP----------MGTS 389
eyeless      YEVLSAYALPPPPMASSSAADSSFSAASSASANVTPHHTIAQESCPSPCSSASHFGVAHS 780
             *:. :.                          **   :* *         :. *

PAX6_human   GTTSTGLISPGVS---------------VPVQVPGS----EPDMSQYWPRLQ----- 422
eyeless      SGFSSDPISPAVSSYAHMSYNYASSANTMTPSSASGTSAHVAPGKQQFFASCFYSPWV 838
             . *:. ***.**               .* ...*:    *. .*::.
```

The few PSI-BLAST hits to the probe sequence PAX-6 excerpted in this section include contributions from the following species:

Aedes aegypti	*Astyanax mexicanus*	*Canis familiaris*
Columba livia	*Cynops pyrrhogaster*	*Gallus gallus*
Heliconius pachinus	*Homo sapiens*	*Mus musculus*
Oikopleura dioica	*Oryzias latipes*	*Perionyx excavatus*
Rattus norvegicus	*Saccoglossus kowalevskii*	*Xenopus laevis*

A longer list of hits would of course include sequences from more species (see Weblem 1.14).

How would we extract these species names from the results? The following is a typical example of the pattern-identification facilities of PERL.

Case Study 1.7 What species contain homologues to human PAX-6 detectable by PSI-BLAST?

PSI–BLAST reports the species in which the identified sequences occur (see Box, *Results of PSI-BLAST search for human PAX-6 protein*). These appear, embedded in the text of the output, in square brackets; for instance:

```
emb|CAA56038.1| (X79493) transcription factor [Drosophila melanogaster]
```

(In the section reporting *E*-values, the species names may be truncated.)

The following PERL program extracts species names from the PSI-BLAST output:

```
#!/usr/bin/perl
#extract species from psiblast output

# Method:
#   For each line of input, check for a pattern of form [Drosophila melanogaster]
#   Use each pattern found as the index in an associative array
#   The value corresponding to this index is irrelevant
#   By using an associative array, subsequent instances of the same
#      species will overwrite the first instance, keeping only a unique set
#   After processing of input complete, sort results and print.


while (<>) {                             # read line of input
    if (/\[([A-Z][a-z]+ [a-z]+)\]/) {    # select lines containing strings of form
                                         #    [Drosophila melanogaster]
       $species{$1} = 1;                 # make or overwrite entry in
    }                                    #          associative array
}

foreach (sort(keys(%species))){          # in alphabetical order,
    print "$_\n";                        #    print species names
}
```

The program makes use of PERL's rich pattern recognition resources to search for character strings of the *form* `[Drosophila melanogaster]`. We want to specify the following pattern:

• a square bracket,

• followed by a word beginning with an upper case letter,

→

- followed by a variable number of lower case letters,

- then a space between words,

- then a word all in lowercase letters,

- then a closing square bracket.

This kind of pattern is called a ***regular expression*** and appears in the PERL program in the following form: `[([A-Z][a-z]+ [a-z]+)]`

Building blocks of the pattern specify ranges of characters:

`[A-Z]` = any letter in the range A, B, C,...Z

`[a-z]` = any letter in the range a, b, c,...z

We can specify repetitions:

`[A-Z]` = *one* upper case letter

`[a-z]+` = *one or more* lower case letters

and combine the results:

`[A-Z][a-z]+ [a-z]+` = an upper case letter followed by one or more lower case letters (the genus name), followed by a blank, followed by one or more lower case letters (the species name).

Enclosing these in parentheses: `([A-Z][a-z]+ [a-z]+)` tells PERL to save the material that matched the pattern for future reference. In PERL this matched material is designated by the variable $1. Thus if the input line contained `[Drosophila melanogaster]`, the statement:

`$species{$1} = 1;`

would effectively be:

`$species{"Drosophila melanogaster"} = 1;`

Finally, we want to include the brackets surrounding the genus and species name, but brackets signify character ranges. Therefore we must precede the brackets by backslashes: `[...]`, to give the final pattern: `\[([A-Z][a-z]+ [a-z]+)\]`

The use of the associative array to retain only a unique set of species is another instructive aspect of the program. Recall that an associative array is a generalization of an ordinary array or vector, in which the elements are not indexed by integers but by arbitrary strings. A second reference to an associative array with a previously encountered index string could change the value in the array but not the list of index strings. In this case we do not care about the value but just use the index strings to compile a unique list of species detected. Multiple references to the same species will merely overwrite the first reference, *not* make a repetitive list. The set of indices (or 'keys') in the associated array %species collects the names of the species found.

Newer versions of PSI-BLAST report the taxonomic distribution of the hits. However, the program in this example would be useful if one wanted to retrieve the alignments, or do other kinds of analyses of the results.

Introduction to protein structure

With protein structures we leave behind the one-dimensional world of nucleotide and amino acid sequences and enter the spatial world of molecular structures. Some of the facilities for archiving and retrieving molecular biological information survive this change pretty well intact, some must be substantially altered, and others do not make it at all.

Biochemically, proteins play a variety of roles in life processes: there are structural proteins (for example, viral coat proteins, the horny outer layer of human and animal skin, and proteins of the cytoskeleton); proteins that catalyse chemical reactions (the enzymes); transport and storage proteins (haemoglobin); regulatory proteins, including hormones and receptor/signal-transduction proteins; proteins that control gene transcription; and proteins involved in recognition, including cell adhesion molecules, and antibodies and other proteins of the immune system.

Proteins are large molecules. In many cases only a small part of the structure—an *active site*—is directly functional, the rest existing only to create and fix the spatial relationship among the active site residues. Proteins evolve by structural changes produced by mutations in the amino acid sequence and genetic rearrangements that bring together different combinations of structural subunits.

Approximately 50 000 protein structures are now known. Most were determined by X-ray crystallography or nuclear magnetic resonance (NMR). From these we have derived our understanding both of the functions of individual proteins—for example, the chemical explanation of catalytic activity of enzymes—and of the general principles of protein structure and folding.

Chemically, protein molecules are long polymers typically containing several thousand atoms, composed of a uniform repetitive *backbone* (or *mainchain*) with a particular *sidechain* attached to each residue (see Fig. 1.6). The amino acid sequence of a protein records the succession of sidechains.

The polypeptide chain folds into a curve in space; the course of the chain defining a *folding pattern*. Proteins show a great variety of folding patterns. Underlying these are a number of common structural features. These include the recurrence of explicit structural paradigms—for example, α-helices and β-sheets (Fig. 1.7)—and common principles or features such as the dense packing of the atoms in protein interiors. Folding may be thought of as a kind of intramolecular condensation or crystallization.

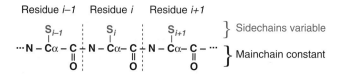

Fig. 1.6 The polypeptide chains of proteins have a mainchain of constant structure and sidechains that vary in sequence. Here S_{i-1}, S_i and S_{i+1} represent sidechains. The sidechains may be chosen, independently, from the set of 20 standard amino acids. It is the sequence of the sidechains that gives each protein its individual structural and functional characteristics.

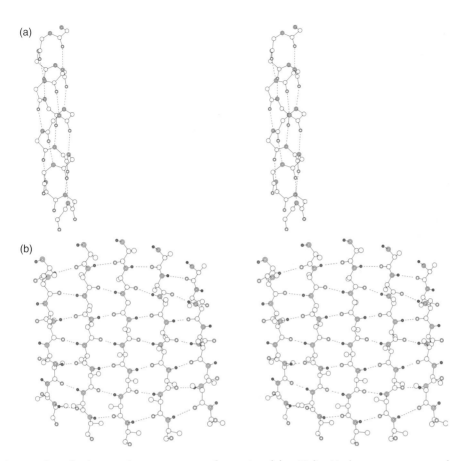

Fig. 1.7 Standard secondary structures of proteins. (a) α-Helix. Hydrogen atoms not shown. (b) β-Sheet. Hydrogen atoms shown. (b) illustrates a parallel β-sheet, in which all strands point in the same direction. Antiparallel β-sheets, in which all pairs of adjacent strands point in opposite directions, are also common. In fact, β-sheets can be formed by any combination of parallel and antiparallel strands.

The hierarchical nature of protein architecture

The Danish protein chemist K.U. Linderstrøm-Lang described the following levels of protein structure: The amino acid sequence—the set of primary chemical bonds—is called the *primary structure*. The assignment of helices and sheets—the hydrogen-bonding pattern of the mainchain—is called the *secondary structure*. The assembly and interactions of the helices and sheets is called the *tertiary structure*. For proteins composed of more than one subunit, J.D. Bernal called the assembly of the monomers the *quaternary structure*. In some cases, evolution can merge proteins—changing quaternary to tertiary structure. For example, five separate enzymes in the bacterium *E. coli*, that catalyse successive steps in the pathway of biosynthesis of aromatic amino acids, underwent a *gene fusion*. These separate genes in *E. coli* correspond to five regions of a single protein in the fungus *Aspergillus nidulans*. Sometimes homologous monomers form oligomers in different ways; for instance, globins form tetramers in mammalian haemoglobins, and dimers—using a different interface—in the ark clam *Scapharca inaequivalvis*.

It has proved useful to add additional levels to the hierarchy:

- *Supersecondary structures*: Proteins show recurrent patterns of interaction between helices and strands of sheet close together in the sequence. These *supersecondary structures* include the α-helix hairpin, the β-hairpin, and the β-α-β unit (Fig. 1. 8).

- *Domains*: Many proteins contain compact units within the folding pattern of a single chain, that look as if they should have independent stability. These are called domains. (Do not confuse domains as substructures of proteins with domains as general classes of living things: Archaea, Bacteria and Eukarya.) The RNA-binding protein L1 (Fig. 1.9) has features typical of multidomain proteins: the binding site appears in a cleft between the two domains, and the relative geometry of the two domains is flexible, allowing ligand-induced conformational changes. In

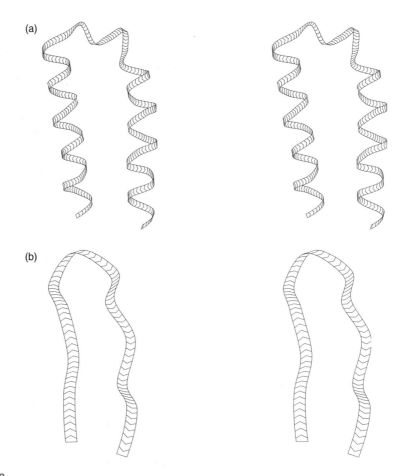

(a)

(b)

Fig. 1.8

(c)

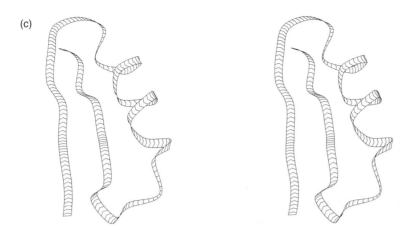

Fig. 1.8 Common supersecondary structures. (a) α-Helix hairpin. (b) β-Hairpin. (c) $\beta - \alpha - \beta$ unit. The chevrons indicate the direction of the chain.

the hierarchy, domains fall between supersecondary structures and the tertiary structure of a complete monomer.

◆ *Modular proteins*: Modular proteins are multidomain proteins which often contain many copies of closely related domains. Domains recur in many proteins in different structural contexts; that is, different modular proteins can 'mix and match' sets of domains. For example, fibronectin, a large extracellular protein involved in cell adhesion and migration, contains 29 domains including multiple tandem repeats of three types of domains called F1, F2 and F3. It is a linear array of the form: $(F1)_6(F2)_2(F1)_3(F3)_{15}(F1)_3$. Fibronectin domains also appear in other modular proteins. (See http://www.bork.embl-heidelberg.de/Modules/ for pictures and nomenclature.)

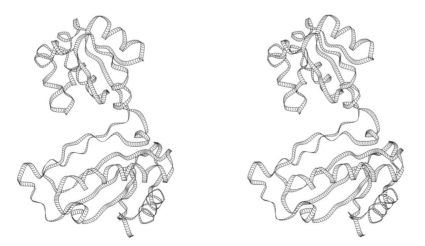

Fig. 1.9 Ribosomal protein L1 from *Methanococcus jannaschii* [1CJS]. ([1CJS] is the Protein Data Bank identification code for the entry.)

Classification of protein structures

The most general classification of families of protein structures is based on the secondary and tertiary structures of proteins.

Class	Characteristic
α-helical	Secondary structure exclusively or almost exclusively α-helical
β-sheet	Secondary structure exclusively or almost exclusively β-sheet
α + β	α-helices and β-sheets separated in different parts of the molecule; absence of $\beta - \alpha - \beta$ supersecondary structure
α/β	Helices and sheets assembled from $\beta - \alpha - \beta$ units
α/β–linear	Line through centres of strands of sheet roughly linear
α/β–barrels	Line through centres of strands of sheet roughly circular
Proteins with little or no secondary structure	

Within these broad categories, protein structures show a variety of folding patterns. Among proteins with similar folding patterns, there are families that share enough features of structure, sequence and function to suggest an evolutionary relationship. However, unrelated proteins often show similar structural themes.

Classification of protein structures occupies a key position in bioinformatics, not least as a bridge between sequence and function. We shall return to this theme, to describe results and relevant web sites. Meanwhile, an album of small structures provides opportunities for practising visual analysis and recognition of the important spatial patterns (Fig. 1.10). Trace the chains visually, picking out helices and sheets. (The chevrons indicate the direction of the chain.) Can you see supersecondary structures? Into which general classes do these structures fall? (See Exercises 1.13 and 1.14, and Problem 1.2.) Many other examples appear in *Introduction to Protein Architecture: The Structural Biology of Proteins* and *Introduction to Protein Science*.

(a)

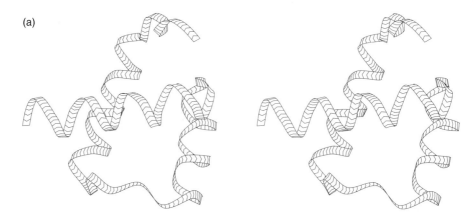

Fig. 1.10

(b)

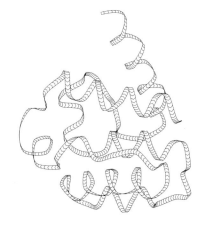
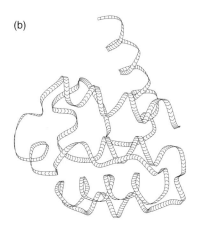

(c)

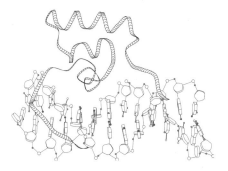
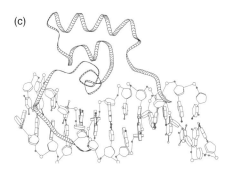

(d)

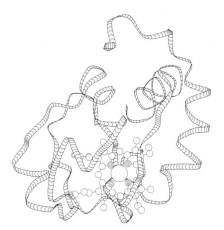
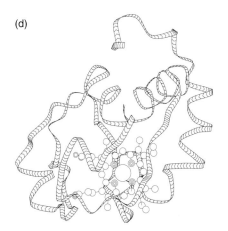

Fig. 1.10 (*See overleaf*)

(e)

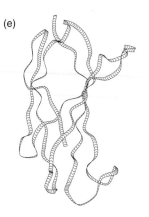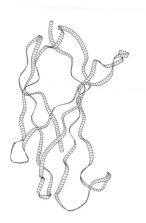

(f)

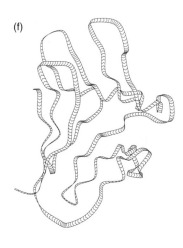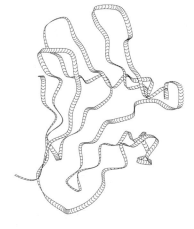

(g)

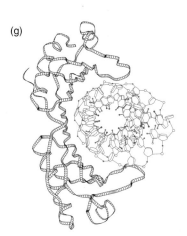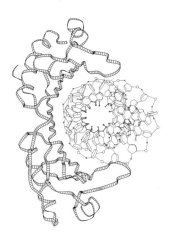

Fig. 1.10

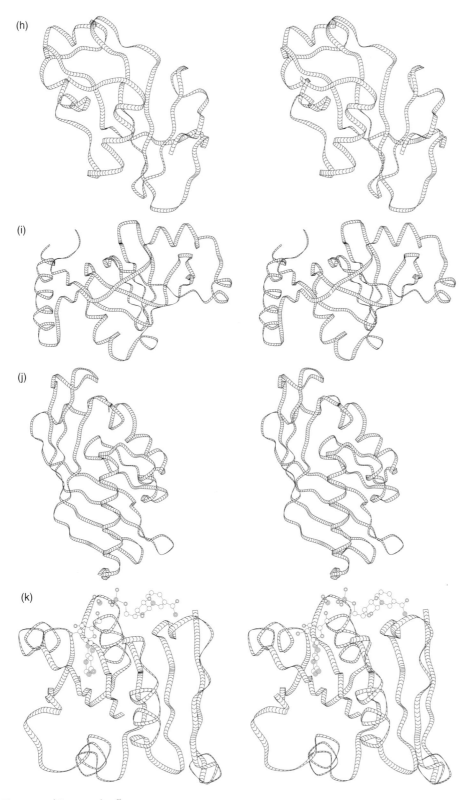

Fig. 1.10 (*See overleaf*)

(l)

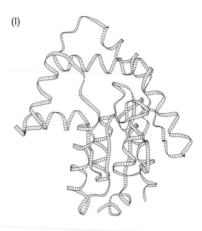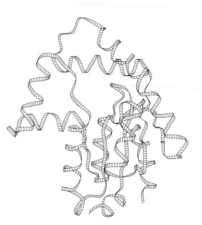

(m)

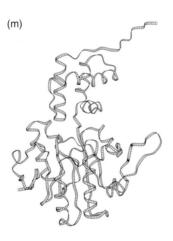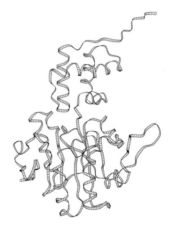

(n)

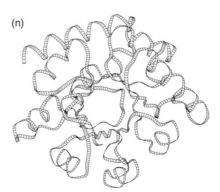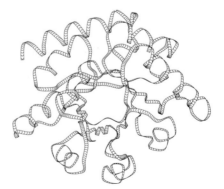

Fig. 1.10

(o)

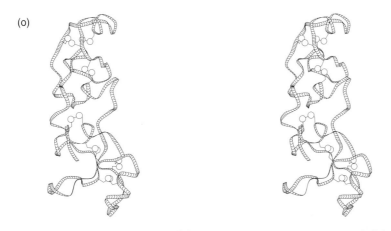

Fig. 1.10 An album of protein structures. (a) Engrailed homeodomain [1ENH], (b) utrophin calmodulin homology domain [1BHD], (c) HIN recombinase, DNA-binding domain [1HCR], (d) rice embryo cytochrome *c* [1CCR], (e) fibronectin cell-adhesion module type III-10 [1FNA], (f) mannose-specific agglutinin (lectin) [1NPL], (g) TATA-box-binding protein core domain [1CDW], (h) barnase [1BRN], (i) lysyl-tRNA synthetase [1BBW], (j) scytalone dehydratase [3STD], (k) alcohol dehydrogenase, NAD-binding domain [1EE2], (l) adenylate kinase [3ADK], (m) chemotaxis receptor methyltransferase [1AF7], (n) thiamin phosphate synthase [2TPS], (o) porcine pancreatic spasmolytic polypeptide [2PSP].

Web resources Macromolecular structures

The World Wide PDB (wwPDB) is a collaboration between three primary archival projects to integrate the archiving and distribution of biological macromolecular structures:

online resource centre

- The **Research Collaboratory for Structural Bioinformatics (RCSB)** (USA)

- The **Macromolecular Structure Database (MSD)** (at the European Bioinformatics Institute (EBI), Hinxton, UK)

- **The Protein Data Bank/Japan** (Osaka, Japan)

The wwPDB sites accept depositions, process new entries and maintain the archives.

Other databanks reorganize, and provide access to the data, including:

- **Structural Classification of Proteins (SCOP)** is a carefully curated database of all protein domains, classified according to structure, function and evolution.

- **The Molecular Modeling DataBase (MMDB)** is the project within the US National Center for Biotechnology Information (NCBI) ENTREZ system, treating experimentally determined macromolecular structures.

Naturally, there is considerable overlap between the sites. Each has its own strengths, based in many cases on the research interests of the contributing scientists. For instance, the Macromolecular Structure Database at the European

$\longrightarrow$

→

Web resources (*continued*)

Bioinformatics Institute maintains the Protein Quaternary Structure site, which gives the probable state of assembly of multichain proteins in their biologically active forms. Different sites differ also in their 'look and feel', and users will discover their own preferences.

These and many other sites provide search facilities to identify structures of interest. For instance, to locate a protein of interest in SCOP, the user can traverse the structural hierarchy, or search via keywords, such as protein name, PDB code, function (including Enzyme Commission number), name of fold (for instance, barrel). For each structure, SCOP provides textual information (including the full text of the entry), pictures and links to other databases.

Protein structure prediction and engineering

The amino acid sequence of a protein dictates its three-dimensional structure. In a medium of suitable solvent and temperature conditions, such as provided by a cell interior, proteins fold spontaneously to their active states. Chaperones help proteins to fold properly, but they catalyse the process rather than direct it.

If amino acid sequences contain sufficient information to specify three-dimensional structures of proteins, it should be possible to devise an algorithm to predict protein structure from amino acid sequence. This has proved elusive, although recent progress has been impressive. In consequence, in addition to pursuing the fundamental problem of *a priori* prediction of protein structure from amino acid sequence, scientists have defined less-ambitious goals:

1. *Secondary structure prediction* Which segments of the sequence form helices and which form strands of sheet?

2. *Fold recognition* Given a library of known protein structures and their amino acid sequences, and the amino acid sequence of a protein of unknown structure, can we find the structure in the library that is most likely to have a folding pattern similar to that of the protein of unknown structure?

3. *Homology modelling* Suppose a target protein, of known amino acid sequence but unknown structure, is homologous to one or more proteins of known structure. Then we expect that much of the structure of the target protein will resemble that of the known protein, and it can serve as a basis for a model of the target structure. The completeness and quality of the result depend crucially on how similar the sequences are. As a rule of thumb, if the sequences of two related proteins have 50% or more identical residues in an optimal alignment, the structures are likely to have similar conformations over >90% of the model. (This is a conservative estimate, as the following illustration shows.)

Here are the aligned sequences, and superposed structures, of two related proteins, hen egg white lysozyme (black) and baboon α-lactalbumin (blue). The sequences are closely related (37% identical residues in the aligned sequences), and the structures are very similar. Each protein could serve as a good model for the other, at least as far as the course of the mainchain is concerned.

Chicken lysozyme   `KVFGRCELAAAMKRHGLDNYRGYSLGNWVCAAKFESNFNTQATNRNTDGS`
Baboon α-lactalbumin   `KQFTKCELSQNLY--DIDGYGRIALPELICTMFHTSGYDTQAIVEND-ES`

Chicken lysozyme   `TDYGILQINSRWWCNDGRTPGSRNLCNIPCSALLSSDITASVNCAKKIVS`
Baboon α-lactalbumin   `TEYGLFQISNALWCKSSQSPQSRNICDITCDKFLDDDITDDIMCAKKILD`

Chicken lysozyme   `DGN-GMNAWVAWRNRCKGTDVQA-WIRGCRL-`
Baboon α-lactalbumin   `I--KGIDYWIAHKALC-TEKL-EQWL--CE-K`

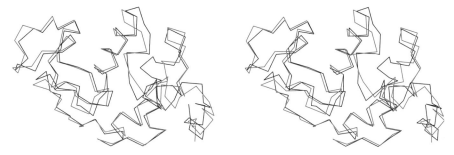

Critical Assessment of Structure Prediction (CASP)

Judging of techniques for predicting protein structures requires blind tests. To this end, J. Moult initiated biennial CASP (Critical Assessment of Structure Prediction) programmes. Crystallographers and NMR spectroscopists in the process of determining a protein structure are invited to (1) publish the amino acid sequence several months before the expected date of completion of their experiment, and (2) commit themselves to keeping the results secret until an agreed date. Predictors submit models, which are held until the deadline for release of the experimental structure. Then the predictions and experiments are compared—to the delight of a few and the chagrin of most.

The results of CASP evaluations record progress in the effectiveness of predictions, which has occurred partly because of the growth of the databanks but also because of improvements in the methods. We shall discuss protein structure prediction in Chapter 6.

Protein engineering

Molecular biologists used to be like astronomers—we could observe our subjects but not modify them. This is no longer true. In the laboratory we can modify nucleic acids and proteins at will. We can probe them by exhaustive mutation to see the effects on function. We can endow old proteins with new functions, as in the development of catalytic antibodies. We can even create new ones.

Many rules about protein structure were derived from observations of natural proteins. These rules do not *necessarily* apply to engineered proteins. Natural proteins have features required by general principles of physical chemistry, and by the mechanism of protein evolution. Engineered proteins must obey the laws of physical chemistry but not the constraints of evolution. Engineered proteins can explore new territory.

Proteomics

The proteome, in analogy with the genome, is the set of proteins of an organism. Proteomics combines the census, distribution, interactions, dynamics and expression patterns of the proteins within living systems. It is a data-intensive subject, depending on high-throughput measurements. These include DNA microarrays, and mass spectrometry.

DNA microarrays

See Box:
Applications of DNA microarrays

DNA microarrays, or DNA chips, are devices for checking a sample *simultaneously* for the presence of many sequences. DNA microarrays can be used (1) to determine expression patterns of different proteins by detection of mRNAs, or (2) for genotyping, by detection of different variant gene sequences, including but not limited to single-nucleotide polymorphisms (SNPs). It is possible to measure simple presence or absence, or to quantitate relative abundance.

From the point of view of bioinformatics, DNA arrays are yet another prolific stream of data creation. They demand effective design of archives and information retrieval systems. One advantage is that the data are all so new that the field is not encumbered with data structures and formats based on older generations of hardware and programs.

Applications of DNA microarrays

- *Identifying genetic individuality in tissues or organisms, or genotyping* Detection of SNPs is one example. In humans and animals, this permits correlation of genotype with susceptibility to disease. In bacteria, this permits identifying mechanisms of development of drug resistance by pathogens.

- *Investigating cellular states and processes* Patterns of expression that change with cellular state or growth conditions can give clues to the mechanisms of processes such as sporulation, or the change from aerobic to anaerobic metabolism.

- *Diagnosis of genetic disease* Testing for the presence of mutations can confirm the diagnosis of a suspected genetic disease. Detection of carriers can help in counselling prospective parents.

- *Diagnosis of infectious disease* Microarrays can detect viruses or other pathogens in blood samples. It may be possible to recognize strains resistant to certain antibiotics, guiding optimal treatment and patient isolation protocols.

$\longrightarrow$

Applications of DNA microarrays *(continued)*

- *Specialized diagnosis of disease* Different types of leukaemia, for example, can be identified by different patterns of gene expression. Knowing the exact type of the disease is important for prognosis, and for selecting treatment. More generally, expression profiling of tumours permits analysis of development and progression of the disease.

- *Genetic warning signs* Some diseases are not determined entirely and irrevocably by genotype, but the probability of their development is correlated with genes or their expression patterns. A person aware of an enhanced risk of developing a condition can in some cases improve his or her prospects by adjustments in lifestyle, or in some cases even by prophylactic surgery.

- *Drug selection* Detection of genetic factors that govern responses to drugs, that in some patients render treatment ineffective and in others cause unusual serious adverse reactions.

- *Target selection for drug design* Proteins showing enhanced transcription in particular disease states might be candidates for attempts at pharmacological intervention. Detection of genes expressed in pathogens are useful for identification of the pathogen, and for choosing targets for drug design.

- *Pathogen resistance* Comparisons of genotypes or expression patterns, between bacterial strains susceptible and resistant to an antibiotic, point to the proteins involved in the mechanism of resistance.

- *Measuring temporal variations in protein expression* permits timing the course of many interesting processes, including: (1) responses to pathogen infection, (2) responses to environmental change and (3) changes during the cell cycle.

Mass spectrometry

Mass spectrometry is a physical technique that characterizes molecules by measurements of the masses of their ions. Applications to proteomics include:

- Rapid identification of the components of a complex mixture of proteins.

- Partial sequencing of proteins and nucleic acids.

- Analysis of post-translational modifications, or substitutions relative to an expected sequence.

- Measuring extents of hydrogen–deuterium exchange, to reveal the solvent exposure of individual sites. This provides information about static conformation, dynamics—including folding, and aggregation—and interactions.

Systems biology

The watchword of systems biology is integration.

Integration has two aspects. One is the study of patterns *within a cell or an organism*: patterns of protein–protein and protein–nucleic acid interactions, patterns of metabolic pathways and control cascades, and patterns of protein expression. Patterns have both static and dynamic aspects. Identification of pairs of proteins that bind to each other, and the assembly of pairwise interactions into a network, produces a static pattern. The flow of metabolites through a network of enzymes, or the flow of information down a control cascade, is a dynamic pattern.

The other aspect of integration is the comparison of occurrence, activities and interactions of genes and proteins *across different species.* The reason why the comparative approach is so powerful in biology is that the systems we are trying to understand arose through processes of evolution. Different species illuminate one another. To understand what it means to be human, we must appreciate both what we have in common with other species and how we differ.

High-throughput methods of genomics and proteomics provide data about sequences, expression patterns and interactions. From genome sequences we can infer the amino acid sequences of an organism's complement of proteins. Proteomics tells us how expression patterns of these proteins vary within the organism, how they change during development or in response to changes in conditions, and how they cooperate with one another. Systems biology takes these data as pieces of a jigsaw puzzle that extends in both space and time. To understand the complex and delicate instrument that is the living cell, we must fit the pieces into their frame.

Clinical implications

There is consensus that the sequencing of the human and other genomes will lead to improvements in the health of mankind. Even discounting some of the more outrageous claims—hype springs eternal—categories of applications include:

1. **Diagnosis of disease and disease risks** DNA sequencing can detect the absence of a particular gene, or a mutation. Identification of specific gene sequences associated with diseases will permit fast and reliable diagnosis of conditions (a) when a patient presents with symptoms, (b) in advance of appearance of symptoms, as in tests for inherited late-onset conditions such as Huntington disease (see Box), (c) for *in utero* diagnosis of potential abnormalities such as cystic fibrosis and (d) for genetic counselling of couples contemplating having children.

'Genetics loads the gun and environment pulls the trigger' —J. Stern

In many cases our genes do not irrevocably condemn us to contract a disease, but raise the probability that we will. An example of a risk factor detectable at the genetic level involves α_1-antitrypsin, a protein that normally functions to inhibit elastase in the alveoli of the lung. People homozygous for the Z mutant of α_1-antitrypsin (342Glu→Lys) express only a dysfunctional protein. They are at risk of emphysema, because of damage to the lungs from endogenous elastase unchecked by normal inhibitory activity; and also of liver disease, because of accumulation of a polymeric form of the mutant α_1-antitrypsin in hepatocytes where it is synthesized. Smoking makes the development of emphysema all but certain. In these cases, the disease is brought on by a *combination* of genetic and environmental factors.

Often the relationship between genotype and disease risk is much more difficult to pin down. Some diseases such as asthma depend on interactions of many genes, as well as on environmental factors. In other cases, a gene may be all present and correct, but a mutation elsewhere may alter its level of expression or distribution among tissues. Such abnormalities must be detected by measurements of protein activity. Analysis of protein expression patterns is also an important way to measure response to treatment.

Huntington disease

Huntington disease is an inherited neurodegenerative disorder affecting approximately 30 000 people in the USA. Its symptoms are quite severe, including uncontrollable dance-like (choreatic) movements, mental disturbance, personality changes and intellectual impairment. Death usually follows within 10–15 years after the onset of symptoms. The gene arrived in New England during the colonial period, in the seventeenth century. It may have been responsible for some accusations of witchcraft. The gene has not been eliminated from the population, because the age of onset — 30–50 years — is after the typical reproductive period.

Formerly, members of affected families had no alternative but to face the uncertainty and fear, during youth and early adulthood, of not knowing whether they had inherited the disease. The discovery of the gene for Huntington disease in 1993 made it possible to identify affected individuals. The gene contains expanded repeats of the trinucleotide CAG, corresponding to polyglutamine blocks in the corresponding protein, *huntingtin*. (Huntington disease is one of a family of neurodegenerative conditions resulting from trinucleotide repeats.) The larger the block of CAGs, the earlier the onset and more severe the symptoms. The normal gene contains 11–28 CAG repeats. People with 29–34 repeats are unlikely to develop the disease, and those with 35–41 repeats may develop only relatively mild symptoms. However, people with >41 repeats are almost certain to suffer full Huntington disease.

The inheritance is marked by a phenomenon called *anticipation*: the repeats grow longer in successive generations, progressively increasing the severity of the disease and reducing the age of onset. For some reason this effect is greater in paternal than in maternal genes. Therefore, even people in the borderline region, who might bear a gene containing 29–41 repeats, should be counselled about the risks to their offspring.

---

2. **Genetics of reponses to therapy—customized treatment** Because people differ in their ability to metabolize drugs, different patients with the same condition may

require different treatment. Sequence analysis permits selecting drugs and dosages optimal for individual patients, a fast-growing field called ***pharmacogenomics***. Physicians can thereby avoid experimenting with different therapies, a procedure that is dangerous in terms of side effects — often even fatal — and in any case expensive. Treatment of patients for adverse reactions to prescribed drugs consumes billions of dollars in health care costs.

For example, the very toxic drug 6-mercaptopurine is used in the treatment of childhood leukaemia. A small fraction of patients used to die from the treatment, because they lack the enzyme thiopurine methyltransferase, needed to metabolize the drug. Testing of patients for this enzyme identifies those at risk.

Conversely, it may become possible to use drugs that are safe and effective in a minority of patients, but which have been rejected before or during clinical trials because of inefficacy or severe side effects in the majority.

3. **Identification of drug targets** Ideally, a drug will affect the symptoms or underlying causes of a disease by interaction with a specific protein to alter its function. This protein is the *target* of the drug discovery process. The specificity of the interaction is important: interaction of the drug with other proteins may lead to unacceptable side effects. Identification of a target provides the focus for subsequent steps in the drug design process. Among drugs now in use, the targets of about half are receptors, about a quarter enzymes and about a quarter hormones. Approximately 7% act on unknown targets.

The growth in bacterial resistance to antibiotics is creating a crisis in disease control. There is a very real possibility that our descendants will look back at the second half of the twentieth century as a narrow window during which bacterial infections could be controlled, and before and after which they could not.

The urgency of finding new drugs is mitigated by the increasing availability of data on which to base their development. Genomics can suggest targets. ***Differential genomics***, and comparison of protein expression patterns, between drug-sensitive and drug-resistant strains of pathogenic bacteria, can pinpoint the proteins responsible for drug resistance. The study of genetic variation between tumour and normal cells can identify differentially expressed proteins as potential targets for anticancer drugs.

4. **Gene therapy** If a gene is missing or defective, we'd like to replace it or at least supply its product. If a gene is overactive, we'd like to turn it off.

Direct supply of proteins is possible for many diseases, of which insulin replacement for diabetes and Factor VIII for a common form of haemophilia are perhaps the best known.

Gene transfer has succeeded in animals, for production of human proteins in the milk of sheep and cows. In human patients, gene replacement therapy for cystic fibrosis using adenovirus has shown encouraging results.

One approach to blocking genes is called 'antisense therapy.' The idea is to introduce a short stretch of DNA or RNA that binds in a sequence-specific manner to a region of a gene. Binding to endogenous DNA can interfere with transcription; binding to mRNA can interfere with translation. Antisense therapy has shown some efficacy against cytomegalovirus and Crohn disease.

Antisense therapy is very attractive, because going directly from target sequence to blocker short-circuits many stages of the drug design process.

The future

This century will see a revolution in healthcare development and delivery. Barriers between 'blue sky' research and clinical practice are tumbling down. It is possible that a reader of this book will discover a cure for a disease that would otherwise kill him or her. It is extremely likely that Szent-Gyorgi's quip, 'Cancer supports more people than it kills' will come true. One hopes that this will happen because the research establishment succeeds in developing therapeutic or preventative measures against tumours rather than merely by imitating their uncontrolled growth.

Recommended reading

A glimpse of the future?

Blumberg, B.S. (1996). Medical research for the next millennium. *The Cambridge Review*, **117**, 3–8. [A fascinating prediction of things to come, some of which are already here.]

The intellectual setting

Mayr, E. (2004). *What Makes Biology Unique? Considerations on the Autonomy of a Scientific Discipline*. Cambridge University Press, Cambridge. [Perspectives, from a self-described 'dirty-fingernails biologist' with an unequalled clarity of mind.]

The overall biological context

Doolittle, W.F. (2000). Uprooting the tree of life. *Scientific American*, **282(2)**, 90–95. [Implications of analysis of sequences for our understanding of the relationships between living things.]

Genomic sequence determination

Green, E.D. (2001). Strategies for systematic sequencing of complex organisms. *Nature Reviews Genetics*, **2**, 573–583. [A clear discussion of possible approaches to large-scale sequencing projects. Includes list of, and links to, ongoing projects for sequencing of multicellular organisms.]

Sulston, J. and Ferry, G. (2002). *The Common Thread: A Story of Science, Politics, Ethics and the Human Genome*. Bantam, New York. [A first-hand account.]

Discussions of databases and information retrieval

Altschul, S.F., Madden, T.L., Schaffer, A.A., Zhang, J., Zhang, Z., Miller, W. and Lipman, D.J. (1997). Gapped BLAST and PSI-BLAST: a new generation of protein database search programs. *Nucleic Acids Research*, **25**, 3389–3402.

Frishman, D., Heumann, K., Lesk, A. and Mewes, H.-W. (1998). Comprehensive, comprehensible, distributed and intelligent databases: current status. *Bioinformatics*, **14**, 551–561.

Wheeler, D.L., Barrett, T., Benson, D.A., Bryant, S.H., Canese, K., Chetvernin, V., Church, D.M., DiCuccio, M., Edgar, R., Federhen, S., Geer, L.Y., Kapustin, Y., Khovayko, O., Landsman, D., Lipman, D.J., Madden, T.L., Maglott, D.R., Ostell, J., Miller, V., Pruitt, K.D., Schuler, G.D., Sequeira, E., Sherry, S.T., Sirotkin, K., Souvorov, A., Starchenko, G., Tatusov, R.L., Tatusova, T.A., Wagner, L. and Yaschenko, E. (2007). Database resources of the National Center for Biotechnology Information. *Nucleic Acids Research*, **35**, D5–D12.

Ouzounis, C.A. and Karp, P.D. (2002). The past, present and future of genome-wide re-annotation. *Genome Biology*, comment2001.1–comment2001.6. [Discussion of how to improve the quality of annotation in databanks, making use of the growth in relevant data.]

Valencia, A. (2002). Search and retrieve. *EMBO Reports*, **3**, 396–400. [Examination of what information-retrieval tools we have and what tools we will need.]

Proteins

Branden, C. I. and Tooze, J. (1999). *Introduction to Protein Structure*, 2nd edn. Garland, New York. [A fine introductory text.]

Lesk, A. M. (2001). *Introduction to Protein Architecture: The Structural Biology of Proteins*. Oxford University Press, Oxford.

Lesk, A. M. (2004). *Introduction to Protein Science: Architecture, Function and Genomics.* Oxford University Press, Oxford. [Companion volumes to *Introduction to Bioinformatics*, with a focus on protein structure, function and evolution.]

The transition to electronic publishing

Berners-Lee, T. and Hendler, J. (2001). Publishing on the semantic web. *Nature*, **410**, 1023–1024. [From the inventor of the Web.]

Legal aspects

Human Genome Project Information Website: Genetics and Patenting
http://www.ornl.gov/hgmis/elsi/patents.html

Bobrow, M. and Thomas, S. (2002). Patenting DNA. *Current Opinion in Molecular Therapy*, **4**, 542–547.

Kieff, F.S., ed. (2003). Perspectives on properties of the Human Genome Project. *Advances in Genetics*, vol. **50**. [A collection of articles, discussing legal aspects of genomics and bioinformatics.]

online resource centre

Web resources

For general background
D. Casey of Oak Ridge National Laboratory has written extremely useful *compact* introductions to molecular biology providing essential background for bioinformatics:
Primer on Molecular Genetics (1992).
Washington, DC: Human Genome Program, US Department of Energy.
http://www.ornl.gov/sci/techresources/Human_Genome/publicat/primer2001/index.shtml

Genome statistics
http://bioinformatics.weizmann.ac.il/mb/statistics.html

Taxonomy sites
Species 2000—a comprehensive index of all known plants, animals, fungi and microorganisms:
http://www.sp2000.org
Tree of life—phylogeny and biodiversity:
http://phylogeny.arizona.edu/tree

Databases of genetics of disease
http://www.ncbi.nlm.nih.gov/omim/
http://www.geneclinics.org/profiles/all.html

→

→

Web resources (*continued*)

Lists of databases
http://www.infobiogen.fr/services/dbcat/
http://www.ebi.ac.uk/biocat/

List of tools for analysis
http://www.ebi.ac.uk/Tools/index.html

Debate on electronic access to the scientific literature
http://www.nature.com/nature/debates/e-access/)

Exercises, Problems and Weblems

Exercises

Exercise 1.1 (a) The Sloan Digital Sky Survey is a mapping of the Northern sky over a 5-year period. The data in release 5 amount to about 15 terabytes (1 byte = 1 character; 1 Tb = 10^{12} bytes). To how many human genome equivalents does this correspond? (b) The Earth Observing System/Data Information System (EOS/DIS)—a series of long-term global observations of the Earth—is estimated to require 15 petabytes of storage (1 petabyte = 10^{15} bytes). To how many human genome equivalents will this correspond? (c) Compare the data storage required for EOS/DIS with that required to store the complete DNA sequences of every inhabitant of the USA. (Ignore savings available using various kinds of storage compression techniques. Assume that each person's DNA sequence requires 1 byte/nucleotide.)

Exercise 1.2 (a) How many CDs would be required to store the entire human genome? (c) How many DVDs would be required to store the entire human genome? (In all cases assume that the sequence is stored as 1 byte/per character, uncompressed.)

Exercise 1.3 Suppose you were going to prepare the Box on Huntington disease (page 57) for a web site. For which words or phrases would you provide links?

Exercise 1.4 The end of the human β-haemoglobin gene has the nucleotide sequence:
``` ... ctg gcc cac aag tat cac taa ```
(a) What is the translation of this sequence into an amino acid sequence? (b) Write the nucleotide sequence of a single base change producing a silent mutation in this region. (A silent mutation is one that leaves the amino acid sequence unchanged.) (c) Write the nucleotide sequence, and the translation to an amino acid sequence, of a single base change producing a missense mutation in this region. (d) Write the nucleotide sequence, and the translation to an amino acid sequence, of a single base change producing a mutation in this region that would lead to premature truncation of the protein. (e) Write the nucleotide sequence of a single base change producing a
```

mutation in this region that would lead to improper chain termination resulting in extension of the protein.

Exercise 1.5 On a photocopy of the Box, *Complete pairwise sequence alignment of human PAX-6 protein and* Drosophila melanogaster *eyeless*, indicate with a highlighter the regions aligned by PSI-BLAST.

Exercise 1.6 On a photocopy of the Box, *Complete pairwise sequence alignment of human PAX-6 protein and* Drosophila melanogaster *eyeless*, highlight the regions in the human PAX-6 protein aligned to the *Drosophila* circadian clock protein (page 39).

Exercise 1.7 (a) What cut-off value of E would you use in a PSI-BLAST search if all you want to know is whether your sequence is already in a databank? (b) What cut-off value of E would you use in a PSI-BLAST search if you want to locate distant homologues of your sequence?

Exercise 1.8 In designing an antisense sequence, estimate the minimum length required to avoid exact complementarity to many random regions of the human genome.

Exercise 1.9 It is suggested that all living humans are descended from a common ancestor called Eve, who lived approximately 140 000–200 000 years ago. (a) Assuming six generations per century, how many generations have there been between Eve and the present? (b) If a bacterial cell divides every 20 minutes, how long would be required for the bacterium to go through that number of generations?

Exercise 1.10 Name an amino acid that has physicochemical properties similar to (a) leucine, (b) aspartic acid, (c) threonine. We expect that such substitutions would in most cases have relatively little effect on the structure and function of a protein. Name an amino acid that has physicochemical properties very different from (d) leucine, (e) aspartic acid, (f) threonine. Such substitutions might have severe effects on the structure and function of a protein, especially if they occur in the interior of the protein structure.

Exercise 1.11 In Fig. 1.7(a), does the direction of the chain from the N-terminus to the C-terminus point up the page or down the page? In Fig. 1.7(b), do the directions of the chain from the N-terminus to the C-terminus point up the page or down the page?

Exercise 1.12 From inspection of Fig. 1.9, how many times does the chain pass between the domains of *M. jannaschii* ribosomal protein L1?

Exercise 1.13 On a photocopy of Fig. 1.10(k and l), indicate with highlighter the helices (in red) and strands of sheet (in blue). On a photocopy of Fig. 1.10(g and m), divide the protein into domains.

Exercise 1.14 On a photocopy of the superposition of chicken lysozyme and baboon α-lactalbumin structures, indicate with a highlighter two regions in which the conformation of the mainchain is different.

Exercise 1.15 Which of the structures shown in Fig. 1.10 contains the following domain?

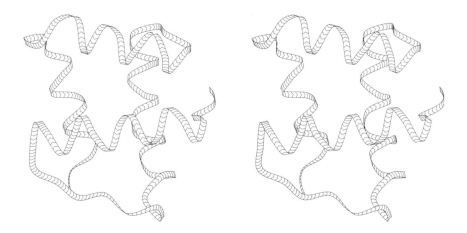

Exercise 1.16 In the PERL program on page 18, estimate the fraction of the text of the program that contains comment material. (Count full lines and half lines.)

Exercise 1.17 Modify the PERL program that extracts species names from PSI-BLAST output so that it would also accept names given in the form [D. melanogaster]

Exercise 1.18 Modify the PERL program that extracts species names from PSI-BLAST output so that it would count the number of sequences from each species occurring in the list.

Exercise 1.19 What is the nucleotide sequence of the molecule shown in Plate I?

Problems

Problem 1.1 The following table contains a multiple alignment of partial sequences from a family of proteins called ETS domains. Each line corresponds to the amino acid sequence from one protein, specified as a sequence of letters each specifying one amino acid. Looking down any column shows the amino acids that appear at that position in each of the proteins in the family. In this way patterns of preference are made visible.

```
TYLWEFLLKLLQDR.EYCPRFIKWTNREKGVFKLV..DSKAVSRLWGMHKN.KPD
VQLWQFLLEILTD..CEHTDVIEWVG.TEGEFKLT..DPDRVARLWGEKKN.KPA
IQLWQFLLELLTD..KDARDCISWVG.DEGEFKLN..QPELVAQKWGQRKN.KPT
IQLWQFLLELLSD..SSNSSCITWEG.TNGEFKMT..DPDEVARRWGERKS.KPN
IQLWQFLLELLTD..KSCQSFISWTG.DGWEFKLS..DPDEVARRWGKRKN.KPK
IQLWQFLLELLQD..GARSSCIRWTG.NSREFQLC..DPKEVARLWGERKR.KPG
IQLWHFILELLQK..EEFRHVIAWQQGEYGEFVIK..DPDEVARLWGRRKC.KPQ
VTLWQFLLQLLRE..QGNGHIISWTSRDGGEFKLV..DAEEVARLWGLRKN.KTN
ITLWQFLLHLLLD..QKHEHLICWTS.NDGEFKLL..KAEEVAKLWGLRKN.KTN
LQLWQFLVALLDD..PTNAHFIAWTG.RGMEFKLI..EPEEVARLWGIQKN.RPA
IHLWQFLKELLASP.QVNGTAIRWIDRSKGIFKIE..DSVRVAKLWGRRKN.RPA
RLLWDFLQQLLNDRNQKYSDLIAWKCRDTGVFKIV..DPAGLAKLWGIQKN.HLS
RLLWDYVYQLLSD..SRYENFIRWEDKESKIFRIV..DPNGLARLWGNHKN.RTN
IRLYQFLLDLLRS..GDMKDSIWWVDKDKGTFQFSSKHKEALAHRWGIQKGNRKK
LRLYQFLLGLLTR..GDMRECVWWVEPGAGVFQFSSKHKELLARRWGQQKGNRKR
```

On a photocopy of this page:

(a) Using a coloured highlighter, mark, in each sequence, the residues in different classes in different colours:

small residues:	G A S T
medium-sized nonpolar residues:	C P V I L
large non-polar residues:	F Y M W
polar residues:	H N Q
positively charged residues:	K R
negatively charged residues:	D E

(b) For each position containing the same amino acid in every sequence, write the letter symbolizing the common residue in upper case below the column. For each position containing the same amino acid in all but one of the sequences, write the letter symbolizing the preferred residue in lower case below the column.

(c) What patterns of periodicity of conserved residues suggest themselves?

(d) What secondary structure do these patterns suggest in certain regions?

(e) What distribution of conservation of charged residues do you observe? Propose a reasonable guess about what kind of molecule these domains interact with.

Problem 1.2 Classify the structures appearing in Fig. 1.10 in the following categories: α-helical, β-sheet, $\alpha + \beta$, α/β linear, α/β-barrels, little or no secondary structure.

Problem 1.3 Generalize the PERL program on page 18 to print the translations of a DNA sequence in all six possible reading frames.

Problem 1.4 Write a PERL program to read a CLUSTAL-W alignment, such as the alignment of mitochondrial cytochromes b from elephants and mammoths, and to count the number of sequence mismatches between all pairs of proteins.

Problem 1.5 For which of the following sets of fragment strings does the PERL program on pages 20–1 work correctly?

(a) Would it correctly recover:

Kate, when France is mine and I am
yours, then yours is France and you are mine.

from:

Kate, when France
France is mine
is mine and
and I am\nyours
yours then
then yours is France
France and you are mine\n

(b) Would it correctly recover:

One woman is fair, yet I am well; another is wise, yet I am well; another virtuous, yet I am well; but till all graces be in one woman, one woman shall not come in my grace.

from:

One woman is
woman is fair,
is fair, yet I am
yet I am well;
I am well; another
another is wise, yet I am well;
yet I am well; another virtuous,
another virtuous, yet I am well;
well; but till all
all graces be
be in one woman,
one woman, one
one woman shall
shall not come in my grace.

(c) Would it correctly recover:

That he is mad, 'tis true: 'tis true 'tis pity;
And pity 'tis 'tis true.

from:

That he is
is mad, 'tis
'tis true
true: 'tis true 'tis
true 'tis
'tis pity;\n
pity;\nAnd pity
pity 'tis
'tis 'tis
'tis true.\n

In (c), would it work if you deleted all punctuation marks from these strings?

Problem 1.6 Generalize the PERL program on pages 20–1 so that it will correctly assemble all the fragments in the previous problem. (Warning—this is not an easy problem.)

Problem 1.7 Write a PERL program to find motif matches as illustrated in the Box on page 26. (a) Demand exact matches. (b) Allowing one mismatch, but no insertions or deletions.

Problem 1.8 PERL is capable of great concision. Here is an alternative version of the program to assemble overlapping fragments (see pages 20–1):

```perl
#!/usr/bin/perl

$/ = "";
@fragments = split("\n",<DATA>);

foreach (@fragments) { $firstfragment{$_} = $_; }

foreach $i (@fragments) {
  foreach $j (@fragments) { unless ($i eq $j) {
    ($combine = $i . "XXX" . $j) =~ /([\S ]{2,})XXX\1/;
    (length($1) <= length($successor{$i})) || { $successor{$i} = $j };
  }                       }
  undef $firstfragment{$successor{$i}};
}

$test = $outstring = join "", values(%firstfragment);
while ($test = $successor{$test}) { ($outstring .= "XXX" . $test) =~ s/([\S ]+)XXX\1/\1/; }

$outstring =~ s/\\n/\n/g; print "$outstring\n";

__END__
the men and women merely players;\n
one man in his time
All the world's
their entrances,\nand one man
stage,\nAnd all the men and women
They have their exits and their entrances,\n
world's a stage,\nAnd all
their entrances,\nand one man
in his time plays many parts.
merely players;\nThey have
```

This is a good example of what to avoid. Anyone who produces code like this should be fired immediately. The absence of comments, and the tricky coding and useless brevity, make it difficult to understand what the program is doing. A program written in this way is difficult to debug and virtually impossible to maintain. Someday you may succeed someone in a job and be presented with such a program to work on. You will have my sympathy.

(a) Photocopy the concise program listed in this problem and the original version on pages 20–1 so that they appear side-by-side on a page. Wherever possible, map each line of the concise program into the corresponding set of lines of the long one.

(b) Prepare a version of the concise program with enough comments to clarify what it is doing (for this you could consider adapting the comments from the original program) and how it is doing it. Do not change any of the executable statements (back to the original version or to anything else); just add comments.

Weblems

Weblem 1.1 Identify the sources of all quotes from Shakespeare's plays in the Box on alignment (page 26).

Weblem 1.2 Identify web sites that give *elementary* tutorial explanations and/or on-line demonstrations of (a) the polymerase chain reaction (PCR), (b) Southern blotting,

(c) restriction maps, (d) suffix tree, (e) heapsort. Write a one-paragraph explanation of these terms based on these sites.

Weblem 1.3 To which phyla do the following species belong? (a) starfish, (b) lamprey, (c) tapeworm, (d) ginko tree, (e) scorpion, (f) jellyfish, (g) sea anemone.

Weblem 1.4 What are the common names of the following species? (a) *Acer rubrum*, (b) *Orycteropus afer*, (c) *Beta vulgaris*, (d) *Pyractomena borealis*, (e) *Macrocystis pyrifera*.

Weblem 1.5 A typical British breakfast consists of: eggs (from chickens) fried in lard, bacon, kippered herrings, grilled cup mushrooms, fried potatoes, grilled tomatoes, baked beans, toast, and tea with milk. Write the complete taxonomic classification of the organisms from which these are derived.

Weblem 1.6 Recover and align the mitochondrial cytochrome *b* sequences from horse, whale and kangaroo. (a) Compare the degree of similarity of each pair of sequences with the results from comparison of the pancreatic ribonuclease sequences from these species in the case study on page 27. Are the conclusions from the analysis of mitochondrial cytochrome *b* sequences consistent with those from analysis of the pancreatic ribonucleases? (b) Compare the *relative* similarity of these sequences with the results from comparison of the pancreatic ribonuclease sequences from these species in the case study. Are the conclusions from the analysis of mitochondrial cytochrome *b* sequences consistent with those from analysis of the pancreatic ribonucleases?

Weblem 1.7 Recover and align the pancreatic ribonuclease sequences from sperm whale, horse and hippopotamus. Are the results consistent with the relationships shown by the SINES?

Weblem 1.8 We observed that the amino acid sequences of cytochrome *b* from elephants and the mammoth are very similar. One hypothesis to explain this observation is that a functional cytochrome *b* might *require* so many conserved residues that cytochromes *b* from all animals are as similar to one another as the elephant and mammoth proteins are. Test this hypothesis by retrieving cytochrome *b* sequences from other mammalian species, and check whether the cytochrome *b* amino acid sequences from more distantly related species are as similar as the elephant and mammoth sequences.

Weblem 1.9 Recover and align the cytochrome *c* sequences of human, rattlesnake and monitor lizard. Which pair appears to be the most closely related? Is this surprising to you? Why or why not?

Weblem 1.10 Send the sequences of pancreatic ribonucleases from horse, minke whale and red kangaroo (see case study on page 27) to the T-coffee multiple-alignment server: http://tcoffee.vital-it.ch/cgi-bin/Tcoffee/tcoffee_cgi/index.cgi Is the resulting alignment the same as that shown in the case study, produced by CLUSTAL-W?

Weblem 1.11 Create a multiple sequence alignment of the mitochondrial cytochrome *b* genes from Indian and African elephants, and Siberian mammoths. How many mismatches are there between each pair of sequences? Is the result consistent with the suggestion from comparison of protein sequences that the mammoth is more closely related to the Indian than to the African elephant?

Weblem 1.12 Linnaeus divided the animal kingdom into six classes: mammals, birds, amphibia (including reptiles), fishes, insects and worms. This implies, for instance, that he considered crocodiles and salamanders more closely related than crocodiles and birds. Thomas Huxley, on the other hand, in the nineteenth century, grouped reptiles and birds together. For three suitable proteins with homologues in crocodiles, salamanders and birds, determine the similarity between the homologous sequences. Which pair of animal groups appears most closely related? Who was right, Linnaeus or Huxley?

Weblem 1.13 When was the last new species of primate discovered?

Weblem 1.14 In how many species have PAX-6 homologues been discovered? To which phyla do they belong?

Weblem 1.15 What are the classifications in SCOP of the proteins in Fig. 1.10(a), (e), (g), (i), (l) and (o)?

Weblem 1.16 Identify three modular proteins in addition to fibronectin itself that contain fibronectin III domains.

Weblem 1.17 Find six examples of diseases other than diabetes and haemophilia that are treatable by administering missing protein directly. In each case, what protein is administered?

Weblem 1.18 To what late-onset disease are carriers of a variant apolipoprotein E gene at unusually high risk? What variant carries the highest risk? What is known about the mechanism by which this variant influences the development of the disease?

Weblem 1.19 For approximately 10% of Europeans, the painkiller codeine is ineffective because the patients lack the enzyme that converts codeine into the active molecule, morphine. What is the most common mutation that causes this condition?

Weblem 1.20 Find the page in SCOP headed by Protein: Thermopin from *Thermobifida fusca*. (a) What is the PDB code of this protein? (b) To what superfamily does this protein belong? (c) For what homologue of thermopin in the chicken is the structure known? (d) By clicking on embedded links, get to a page that displays the abstract of the article describing the structure determination. How many clicks are there in the shortest path that you found? What were the URLs of the intermediate sites on this path?

Weblem 1.21 Monotremes, of which the best known is probably the platypus (*Ornithorhynchus anatinus*), form an order in the class *mammalia*. Only a few species of monotremes are known. (a) Which if any of them are critically endangered? (b) Where are they found in the wild? (c) What is the nature of the current threat to their continued survival? (d) What gene sequences, if any, are known from endangered monotreme species? (e) Suppose you wanted to sequence a gene from an endangered monotreme species. Identify a zoo that harbours a specimen.

Genome organization and evolution

Learning goals

1. Knowing the basic sizes and organizing principles of simple and complex genomes.

2. Understanding how genomes are analysed, and the relationship of gene sequences to phenotypic features, including inherited diseases.

3. Recognizing the importance and the difficulty of deriving from a complete genome sequence the amino acid sequences of the proteins encoded, and assigning functions to these proteins.

4. Understanding how to find genes associated with inherited diseases, and how the availability of the complete human genome has changed such investigations.

5. Knowing the general ideas of the contents of particular genomes, and how the genomes of prokaryotes and eukarya differ systematically, and to appreciate the implications of general surveys of genomes of different organisms.

6. Realizing that published genomes record the characteristics of only a single individual; and that there is considerable variation within populations, and great variation between separated populations of organisms belonging to the same species.

7. Appreciating the power of DNA sequences in studying human history, including inference of human migration patterns, and as records of plant and animal domestication.

8. Recognizing the power of DNA sequencing for personal identification, its application in paternity and criminal cases, and the questions of social policy it raises.

9. Appreciating the power of comparative genomics to identify features responsible for differences between species—for instance, what is it that makes us human?

Genomes and proteomes

The genome of a typical bacterium comes as a single DNA molecule of about 5 million characters, about the same as a fairly large book. If extended, the bacterial genome would be about 2 mm long. (It has to fit into a cell of diameter of about 0.001 mm). The DNA of higher organisms is organized into chromosomes—normal human cells contain 23 chromosome pairs. The total amount of genetic information per cell—the sequence of nucleotides of DNA—is very nearly constant for all members of a species, but varies widely between species (see Box across for longer list):

The numbers of characters in the *E. coli* genome and in the First Folio edition of Shakespeare's plays differ by less than 0.1%.

Organism	Genome size (base pairs)
Epstein–Barr virus	0.172×10^6
Bacterium (*E. coli*)	4.6×10^6
Yeast (*S. cerevisiae*)	12.5×10^6
Nematode worm (*C. elegans*)	100.3×10^6
Thale cress (*A. thaliana*)	115.4×10^6
Fruit fly (*D. melanogaster*)	128.3×10^6
Human (*H. sapiens*)	3223×10^6

Different patterns of proteins also characterize different cells. The relationship between DNA content and protein content is not direct. Not all DNA codes for proteins. Conversely, some genes exist in multiple copies. In eukarya, many genes produce several different proteins by alternative splicing. Therefore, the amount of protein sequence information in a cell, much less the number and pattern of different proteins expressed, cannot easily be estimated from the genome size.

Genes

A single gene coding for a particular protein corresponds to a sequence of nucleotides along one or more regions of a molecule of DNA. The DNA sequence is collinear with the protein sequence. In species for which the genetic material is double-stranded DNA, genes may appear on either strand. Bacterial genes are continuous regions of DNA. Therefore, the functional protein-coding unit of genetic sequence information from a bacterium is a string of $3N$ nucleotides encoding a string of N amino acids, or a string of N nucleotides encoding a structural RNA molecule. Such a string, equipped with annotations, would form a typical entry in an archive of genetic sequences.

Genome sizes

Organism	Number of base pairs	Number of genes	Comment
φX-174	5386	10	virus infecting *E. coli*
Human–mitochondrion	16 569	37	subcellular organelle
Epstein-Barr virus (EBV)	172 282	80	cause of glandular fever
Mycoplasma pneumoniae	816 394	680	cause of cyclic pneumonia epidemics
Rickettsia prowazekii	1 111 523	878	bacterium cause of epidemic typhus
Treponema pallidum	1 138 011	1039	bacterium cause of syphilis
Borrelia burgdorferi	1 471 725	1738	bacterium cause of Lyme disease
Aquifex aeolicus	1 551 335	1749	bacterium from hot spring
Thermoplasma acidophilum	1 564 905	1509	archaeal prokaryote lacks cell wall
Campylobacter jejuni	1 641 481	1708	frequent cause of food poisoning
Methanococcus jannaschii	1 664 970	1783	archaeal prokaryote thermophile
Helicobacter pylori	1 667 867	1589	chief cause of stomach ulcers
Haemophilus influenzae	1 830 138	1738	bacterium cause of middle ear infections
Thermotoga maritima	1 860 725	1879	marine bacterium
Archaeoglobus fulgidus	2 178 400	2437	another archaeon
Deinococcus radiodurans	3 284 156	3187	radiation-resistant bacterium
Synechocystis	3 573 470	4003	cyanobacterium 'blue-green alga'
Vibrio cholerae	4 033 460	3890	cause of cholera
Mycobacterium tuberculosis	4 411 529	4275	cause of tuberculosis
Bacillus subtilis	4 214 814	4779	popular in molecular biology
Escherichia coli	4 639 221	4406	molecular biologists' all-time favourite
Pseudomonas aeruginosa	6 264 403	5570	largest prokaryote sequenced as yet
Saccharomyces cerevisiae	12.1×10^6	6172	yeast, first eukaryotic genome sequenced
Caenorhabditis elegans	95.5×10^6	19 099	the worm
Arabidopsis thaliana	1.17×10^8	25 498	flowering plant (angiosperm)
Drosophila melanogaster	1.8×10^8	13 601	the fruit fly
Takifugu rubripes	3.9×10^8	30 000	puffer fish (fugu fish)
Human	3.2×10^9	20 500	
Wheat	16×10^9	30 000	
Salamander	10^{11}	?	
Psilotum nudum	10^{11}	?	whisk fern — a simple plant

In eukarya the nucleotide sequences that encode the amino acid sequences of individual proteins are organized in a more complex manner. Frequently one gene appears split into separated segments in the genomic DNA. An *exon* is a stretch of DNA retained in the mature messenger RNA (mRNA) that the ribosome translates into protein. An *intron* is an **int**ervening **region** between two exons. Cellular machinery splices together the proper segments of initial RNA transcripts, based on signal sequences flanking the exons in the sequences themselves. Many introns are very long—in some cases substantially longer than the exons.

Control information organizes the expression of genes. Control mechanisms may turn genes on and off, or regulate gene expression more finely. Cascades of controls respond to concentrations of nutrients, or to stress. Regulatory networks orchestrate complex programmes of development during the lifetime of the organism.

Many control regions of DNA lie near the segments coding for proteins. They contain *signal sequences* that serve as binding sites for the molecules that transcribe the DNA sequence, or sequences that bind regulatory molecules that can *block* transcription. Bacterial genomes contain examples of contiguous genes, coding for several proteins that catalyse successive steps in an integrated sequence of reactions, all under the control of the same regulatory sequence. F. Jacob, J. Monod and E. Wollman named these *operons*. One can readily understand the utility of a parallel control mechanism.

Eukaryotic chromosomes contain complexes of DNA with histones. Chromatin re-modelling is an important mechanism of transcriptional control. Reversible chemical modification of histones, by a variety of reactions including deacetylation, methylation, decarboxylation, phosphorylation, ubiquitinylation and sumoylation, leads to alterations of the DNA–histone interactions that render transcription initiation sites more or less accessible.

In animals, methylation of DNA provides the signals for tissue-specific expression of developmentally regulated genes. DNA methylation is stable during tissue differentiation, surviving cell division. When a cell divides, enzymes copy the methylation patterns, preserving the settings of the regulatory switches.

Products of certain genes cause cells to commit suicide—a process called *apoptosis*. Defects in the apoptotic mechanism leading to uncontrolled growth are observed in some cancers, and stimulation of these mechanisms is a general approach to cancer therapy.

The conclusion is that to reduce genetic data to individual coding sequences is to disguise the very complex nature of the inter-relationships among them, and to ignore the historical and integrative aspects of the genome. Robbins has expressed the situation unimprovably:

'...Consider the 3.2 gigabytes of a human genome as equivalent to 3.2 gigabytes of files on the mass-storage device of some computer system of unknown design. Obtaining the sequence is equivalent to obtaining an image of the contents of that mass-storage device. Understanding the sequence is equivalent to reverse engineering that unknown computer system (both the hardware and the 3.2 gigabytes of software) all the way back to a full set of design and maintenance specifications.

...

> The genes that code for proteins, and for structural RNA molecules, present only the *static* picture of the genome.

'Reverse engineering the sequence is complicated by the fact that the resulting image of the mass-storage device will not be a file-by-file copy, but rather a streaming dump of the bytes in the order they were entered into the device. Furthermore, the files are known to be fragmented. In addition, some of the device contains erased files or other garbage. Once the garbage has been recognized and discarded and the fragmented files reassembled, the reverse engineering of the codes can be undertaken with only a partial, and sometimes incorrect, understanding of the CPU on which the codes run. In fact, deducing the structure and function of the CPU is part of the project, since some of the 3.2 gigabytes are the binary specifications for the computer-assisted-manufacturing process that fabricates the CPU. In addition, one must also consider that the huge database also contains code generated from the result of literally millions of maintenance revisions performed by the worst possible set of kludge-using, spaghetti-coding, opportunistic hackers who delight in clever tricks like writing self-modifying code and relying upon undocumented system quirks.'

—Robbins, R.J. (1992). Challenges in the human genome project. *IEEE Engineering in Medicine and Biology*, **11**, 25–34. (©1992 IEEE).

Proteomics

An organism's genome gives a complete but static set of specifications of the potential life of that individual. The state of development of the organism, and its activity at the molecular level at any moment, depend primarily on the amounts and distribution of its proteins. The **proteome project** deals in an integral way with patterns of expression of proteins in biological systems, in ways that complement and extend genome projects.

What kinds of data would we like to measure, and what mature experimental techniques exist to determine them? The basic goal is a spatio-temporal description of the deployment of proteins in the organism. The rates of synthesis of different proteins vary among different tissues and different cell types and states of activity. Methods are available for efficient analysis of transcription patterns of multiple genes. However, because proteins 'turn over' at different rates, it is also necessary to measure proteins directly. High-resolution two-dimensional polyacrylamide gel electrophoresis (2D PAGE) shows the pattern of protein content in a sample. Mass-spectrometric techniques identify the proteins into which the sample has been separated. We shall develop these topics in Chapter 7.

Application of these methods provides a picture of the protein-based activity of an organism, as the genome provides a complete set of potential proteins. R. Simpson has drawn an analogy: if the genome is a list of the instruments in an orchestra, the proteome is the orchestra in the process of playing a symphony.

Historically the chemical problem of determining amino acid sequences of proteins directly was solved before the genetic code was established and before methods for determination of nucleotide sequences of DNA were developed. However, the amino acid sequences of an organism's proteins are inherent in its genome sequence, by virtue of the genetic code. Indeed, new protein sequence data are now being determined by translation of DNA sequences, rather than by direct sequencing of proteins.

Should any distinction be made between amino acid sequences determined directly from proteins and those determined by translation from DNA? First, we must assume that it is possible correctly to identify within DNA sequences the regions that encode

F. Sanger's sequencing of insulin in 1955 first proved that proteins had definite amino acid sequences, a proposition that until then was hypothetical.

proteins. The pattern-recognition programs that address this question are subject to three types of errors: a genuine protein sequence may be missed entirely, or an incomplete protein may be reported, or a gene may be incorrectly spliced. Several variations on the theme add to the complexity: genes for different proteins may overlap, or genes may be assembled from exons in different ways, in different tissues or even in individual cells. Conversely, some genetic sequences that appear to code for proteins may in fact be defective or not expressed. *A protein inferred from a genome sequence is a hypothetical object until an experiment verifies its existence.*

In many cases the expression of a gene produces a molecule that must be modified within a cell, to make a *mature* protein that differs significantly from the one suggested by translation of the gene sequence. In many cases the missing details of **post-translational modifications**—the molecular analogues of body piercing—are quite important. Post-translational modifications include addition of ligands (for instance the covalently bound haem group of cytochrome *c*), glycosylation, methylation, excision of peptides and many others. Patterns of disulphide bridges—primary chemical bonds between cysteine residues—cannot be deduced directly from the amino acid sequence. In some cases, mRNA is edited before translation, creating changes in amino acid sequences that are not inferrable from the genes.

Cleavage of a peptide is a common post-translational modification. In some cases, cleavage converts an inactive form of a protein to an active one. The proteases active in digestion of our food are examples. In other cases, the effect is to promote correct folding. For instance, insulin is synthesized as a single-chain precursor that folds properly, after which excision of a peptide produces the mature oligomeric form.

Most post-translational cleavage reactions are carried out by proteases. Alternatively, **inteins** are proteins that have a 'self-splicing' activity. They autocatalytically excise internal peptides and join the ends. (In contrast, peptide excision from proinsulin leaves two chains that are *not* joined by a peptide bond.)

Eavesdropping on the transmission of genetic information

How hereditary information is stored, passed on and implemented is perhaps *the* fundamental problem of biology. Three types of maps have been essential (see Box):

1. Linkage maps of genes

2. Banding patterns of chromosomes

3. DNA sequences

These maps represent three very different types of data. Genes, as discovered by Mendel, were entirely abstract entities. Chromosomes are physical objects, banding patterns their visible landmarks. Only with DNA sequences are we dealing directly with stored hereditary information in its physical form.

It was the very great achievement of biology during the last century to forge connections between these three types of data. The first steps—and giant strides they were indeed—proved that, for any chromosome, the maps are one-dimensional arrays, and indeed that they are collinear. Any schoolchild now knows that genes are strung out along chromosomes, and that each gene corresponds to a DNA sequence. But the proofs of these statements earned a large number of Nobel prizes.

Splitting a long molecule of DNA—for example, the DNA in an entire chromosome—into fragments of convenient size for cloning and sequencing requires additional maps to report the order of the fragments, to facilitate assembly of the entire sequence from the sequences of the fragments. A *restriction endonuclease* is an enzyme that cuts DNA at a specific sequence, usually about 6 bp long. Cutting DNA with several restriction enzymes with different specificities produces sets of overlapping fragments. From the sizes of the fragments it is possible to construct a *restriction map*, stating the order and distance between the restriction endonuclease cleavage sites. A mutation in one of these cleavage sites will change the sizes of the fragments produced by the corresponding enzyme, allowing the mutation to be located in the map.

Restriction enzymes can produce fairly large pieces of DNA. Cutting the DNA into smaller pieces, which are cloned and ordered by sequence overlaps (as in the example using text on pages 19–21) produces a finer dissection of the DNA called a *contig map*.

Gene maps, chromosome maps and sequence maps

1. A *gene map* is classically determined by observed patterns of heredity. Linkage groups and recombination frequencies can detect whether genes are on the same or different chromosomes, and, for genes on the same chromosome, how far apart they are. The principle is that the farther apart two linked genes are, the more likely they are to recombine, by crossing over during meiosis. Indeed, two genes on the same chromosome but very far apart will appear to be unlinked. The unit of length in a gene map is the Morgan, defined by the relationship that 1 cM corresponds to a 1% recombination frequency. (We now know that 1 cM $\sim 1 \times 10^6$ bp in humans, but it varies with the location in the genome and with the distance between genes.)

2. *Chromosome banding pattern maps* Chromosomes are physical objects. Banding patterns are visible features on them. The nomenclature is as follows: in many organisms, chromosomes are numbered in order of size, 1 being the largest. The two arms of chromosomes, separated by the centromere, are called the p (petite = short) arm and q (queue) arm. Regions within the chromosome are numbered p1, p2, . . . and q1, q2 . . . outward from the centromere. Subsequent digits indicate subdivisions of bands. For example, certain bands on the q arm of human chromosome 15 are labelled 15q11.1, 15q11.2, 15q12. Originally bands 15q11 and 15q12 were defined; subsequently 15q11 was divided into 15q11.1 and 15q11.2.
   Deletions of substantial segments of DNA are observable in changes in banding patterns. (Smaller deletions are observable by fluorescent *in-situ* hybridization,

$\longrightarrow$

⟶ ──

Gene maps, chromosome maps and sequence maps *(continued)*

or FISH; see page 77.) The observation of banding patterns was crucial to the identification of chromosomes as the vessels of heredity (see *Introduction to Genomics*, Chapter 1.)

Many deletions are associated with inherited diseases. For instance, deletion in the 15q region in the human are associated with Prader–Willi and Angleman syndromes. These syndromes have the interesting feature that the alternative clinical consequences depend on whether the affected chromosome is paternal or maternal. This observation of **genomic imprinting** shows that the genetic information in a fertilized egg is not simply the bare DNA sequences contributed by the parents. Chromosomes of paternal and maternal origin have different states of methylation, signals for differential expression of their genes. The process of modifying the DNA which takes place during differentiation in development is already present in the zygote.

3. ***The DNA sequence itself*** Physically a sequence of nucleotides in the molecule, computationally a string of characters A, T, G and C. Genes are regions of the sequence, in many cases interrupted by non-coding regions.

Identification of genes associated with inherited diseases

In the past, the connections between chromosomes, genes and DNA sequences have been essential for identifying the molecular deficits underlying inherited diseases, such as Huntington disease or cystic fibrosis. Sequencing of the human genome has changed the situation radically.

Given a disease attributable to a defective protein:

- If we know the protein involved, we can pursue rational approaches to therapy.

- If we know the gene involved, we can devise tests to identify sufferers or carriers.

- In many cases, knowledge of the chromosomal location of the gene is unnecessary for either therapy or detection; it is required only for identifying the gene, providing a bridge between the patterns of inheritance and the DNA sequence. (This is not true of diseases arising from chromosome abnormalities.)

For instance, in the case of sickle-cell anaemia, we know the protein involved. The disease arises from a single point mutation in haemoglobin. We can proceed directly to drug design. We need the DNA sequence only for genetic testing and counselling. In contrast, if we know neither the protein nor the gene, we must somehow retrace the steps back to the gene from the phenotype, a process called **positional cloning** or **reverse genetics**. Positional cloning used to involve a kind of 'Tinker to Evers to Chance' cascade from the gene map to the chromosome map to the DNA sequence. Later we shall see how recent developments have short-circuited this process.

Patterns of inheritance identify the type of genetic defect responsible for a condition. Simple Mendelian inheritance patterns show, for example, that Huntington disease and cystic fibrosis are caused by single genes. To find the gene associated with cystic fibrosis it was necessary to begin with the gene map, using linkage patterns of heredity in affected families to localize the affected gene to a particular region of a particular chromosome. Knowing the general region of the chromosome, it was then possible to search the DNA of that region to identify candidate genes, and finally to pinpoint the particular gene responsible and sequence it (see Boxes, *Identification of the cystic fibrosis gene,* and *Positional cloning: finding the cystic fibrosis gene*). In contrast, many diseases do not show simple inheritance, or, even if only a single gene is involved, heredity creates only a predisposition, the clinical consequences of which depend on environmental factors. The full human genome sequence, and measurements of expression patterns, will be essential to identify the genetic components of these more complex cases.

Mappings between the maps

A gene linkage map can be calibrated to chromosome banding patterns through observation of individuals with deletions or translocations of parts of chromosomes. The genes responsible for phenotypic changes associated with a deletion must lie within the deletion. Translocations are correlated with altered patterns of linkage and recombination.

There have been several approaches to coordinating chromosome banding patterns with individual DNA sequences of genes:

- In **fluorescent *in situ* hybridization** (FISH) a probe sequence is labelled with fluorescent dye. The probe is hybridized with the chromosomes, and the chromosomal location where the probe is bound shows up directly in a photograph (see Plate II). Typical resolution is $\sim 10^5$ bp, but specialized new techniques can achieve high resolution, down to 1 kbp. Simultaneous FISH with two probes can detect linkage and even estimate genetic distances. This is important in species for which the generation time is long enough to make standard genetic approaches inconvenient. FISH can also detect chromosomal abnormalities.

- *Somatic cell hybrids* are rodent cells containing few, one, or even partial human chromosomes. (Chromosome fragments are produced by irradiating the human cells prior to fusion. Such lines are called **radiation hybrids**.) Hybridization of a probe sequence with a panel of somatic cell hybrids, detected by fluorescence, can identify which chromosome contains the probe. This approach has been superseded by use of clones of yeast, bacteria or phage containing fragments of human DNA in artificial chromosomes (YACs, BACs and PACs, i.e. yeast, bacterial and bacteriophage P1 artificial chromosomes).

Of course, with sequences from the human or other organisms for which the complete genome sequence is known, these methods are obsolete. Given a DNA sequence, one would just look it up—but mind the gaps!

Identification of the cystic fibrosis gene

Cystic fibrosis, a disease known to folklore since at least the Middle Ages and to science for about 500 years, is an inherited recessive autosomal condition. Its symptoms include intestinal obstruction; reduced fertility including anatomical abnormalities (especially in males); and recurrent clogging and infection of lungs—the primary cause of death now that there are effective treatments for the gastrointestinal symptoms. Approximately half the sufferers die before age 25 years, and few survive beyond 50. Cystic fibrosis affects 1/2500 individuals in the American and European populations. Approximately 1/25 Caucasians carry a mutant gene, and 1/65 African-Americans. The protein that is defective in cystic fibrosis also acts as a receptor for uptake of *Salmonella typhi*, the pathogen that causes typhoid fever. Increased resistance to typhoid in heterozygotes—who do not develop cystic fibrosis itself but are carriers of the mutant gene—probably explains why the gene has not been eliminated from the population.

The pattern of inheritance showed that cystic fibrosis was the effect of a single gene. However, the actual protein involved was unknown. It had to be found via the gene.

Clinical observations provided the gene hunters with useful clues. It was known that the problem had to do with Cl^- transport in epithelial tissues. People had long recognized that children with excessive salt in their sweat—tasteable when kissing an infant on the forehead—were short-lived. Modern physiological studies showed that epithelial tissues of cystic fibrosis patients cannot reabsorb chloride. When closing in on the gene, the expected distribution, among tissues, of its expression, and of the type of protein implicated, were useful guides.

In 1989 the gene for cystic fibrosis was isolated and sequenced. This gene—called *CFTR* (cystic fibrosis transmembrane conductance regulator)—codes for a 1480 amino acid protein that normally forms a cyclic AMP-regulated epithelial Cl^- channel. The gene, comprising 24 exons, spans a 250 kb region. For 70% of mutant alleles, the mutation is a 3 bp deletion, deleting the residue Phe508 from the protein. This mutation is denoted del508. The effect of the deletion is defective translocation of the protein, which is degraded in the endoplasmic reticulum rather than transported to the cell membrane.

An *in utero* test for cystic fibrosis is based on recovery of foetal DNA. A PCR primer is designed to give a 154 bp product from the normal allele and a 151 bp product from the del508 allele.

Clinicians have taken advantage of the fact that the affected tissues of the airways are easily accessible, to experiment with gene therapy. Unfortunately, the results have so far been disappointing. Use of genetically engineered adenovirus sprayed into the respiratory passages to deliver the correct gene to epithelial tissues caused inflammation. An alternative approach now being pursued is to deliver the normal *CFTR* gene via liposomes.

Positional cloning: finding the cystic fibrosis gene

The process by which the gene responsible for cystic fibrosis was found has been called *positional cloning* or *reverse genetics*.

♦ A search in family pedigrees for a linked marker showed that the cystic fibrosis gene was close to a known variable number tandem repeat (VNTR), DOCR-917. Somatic cell hybrids placed this on chromosome 7, band q3.

♦ Other markers found were linked more tightly to the target gene. It was thereby bracketed by a VNTR in the *MET* oncogene and a second VNTR, D7S8. The target gene lies 1.3 cM from *MET* and 0.9 cM from D7S8—localizing it to a region of approximately 1–2 million bp. A region this long could well contain 100–200 genes.

♦ The inheritance patterns of additional markers from within this region localized the target more sharply to within 500 kb. A technique called *chromosome jumping* made the exploration of the region more efficient.

♦ A 300 kb region at the correct distance from the markers was cloned. Probes were isolated from the region, to look for active genes, characterized by an upstream CCGG sequence. (The restriction endonuclease *Hpa*II is useful for this step; it cuts DNA at this sequence, but only when the second C is not methylated, i.e. when the gene is active.)

♦ Identification of genes in this region by sequencing.

♦ Checking in animals for genes similar to the candidate genes turned up four likely possibilities. Checking these possibilities against a complementary DNA (cDNA) library from sweat glands of cystic fibrosis patients and healthy controls identified one probe with the right tissue distribution for the expected expression pattern of the gene responsible for cystic fibrosis. One long coding segment had the right properties, and indeed corresponded to an exon of the cystic fibrosis gene. Most cystic fibrosis patients have a common alteration in the sequence of this gene—a 3 bp deletion, deleting the residue Phe508 from the protein.

Proof that the gene was correctly identified included:

♦ 70% of cystic fibrosis alleles have the deletion. It is not found in people who are neither sufferers nor likely to be carriers.

♦ Expression of the wild-type gene in cells isolated from patients restores normal Cl^- transport.

♦ Knockout of the homologous gene in mice produces the cystic fibrosis phenotype.

♦ The pattern of gene expression matches the organs in which it is expected.

♦ The protein encoded by the gene would contain a transmembrane domain, consistent with involvement in transport.

High-resolution maps

Formerly, genes were the only visible portions of genomes. Now, markers are no longer limited to genes with phenotypically observable effects, which are anyway too sparse for an adequately high-resolution map of the human genome. Now that we can interrogate DNA sequences directly, any features of DNA that vary among individuals can serve as markers, including the following.

- *Variable number tandem repeats* (**VNTRs**), also called minisatellites. VNTRs contain regions 10–100 bp long, repeated a variable number of times—same sequence, different number of repeats. In any individual, VNTRs based on the same repeat motif may appear only once in the genome; or several times, with different lengths on different chromosomes. The distribution of the sizes of the repeats is the marker. Inheritance of VNTRs can be followed in a family and mapped to a disease phenotype like any other trait. VNTRs were the first genetic sequence data used for personal identification—genetic fingerprints—in paternity and in criminal cases.

Formerly, VNTRs were observed by producing *restriction fragment length polymorphisms* (**RFLPs**) from them. VNTRs are generally flanked by recognition sites for the same restriction enzyme, which will neatly excise them. The results can be spread out on a gel, and the distribution of their lengths detected by Southern blotting. However, it is much easier and more efficient to measure the sizes of VNTRs by amplifying them with PCR, and this method has replaced the use of restriction enzymes.

- *Short tandem repeat polymorphisms* (**STRPs**), also called microsatellites. STRPs are regions of only 2–5 bp but repeated many times; typically 10–30 consecutive copies. They have several advantages as markers over VNTRs, one of which is a more even distribution over the human genome.

There is no reason why these markers need lie within expressed genes, and usually they do not. (The CAG repeats in the gene for huntingtin and certain other disease genes are exceptions.)

Panels of microsatellite markers greatly simplify the identification of genes. It is interesting to compare a recent project to identify a disease gene now that the human genome sequence is available, with such classic studies as the identification of the gene for cystic fibrosis (see Box: *Identification of a gene for Berardinelli–Seip syndrome*).

Distinguish: VNTRs are characteristics of genome sequences; RFLPs are artificial mixtures of short stretches of DNA created in the laboratory in order to identify VNTRs.

Identification of a gene for Berardinelli–Seip syndrome

Berardinelli–Seip syndrome (congenital generalized lipodystrophy) is an autosomal recessive disease. Its symptoms include absence of body fat, insulin-resistant diabetes and enhanced rate of skeletal growth.

To determine the gene involved, a group led by J. Magré subjected DNA from members of affected families to linkage analysis and homozygosity mapping with

$\longrightarrow$

Identification of a gene for Berardinelli–Seip syndrome *(continued)*

a genome-wide panel containing ~400 microsatellite markers of known genetic location, with an average spacing of ~10 cM. In this procedure, a fixed panel of primers specific for the amplification and analysis of each marker is used to compare whole DNA of affected individuals with that of unaffected relatives. The measurements reveal the lengths of the repeats associated with each microsatellite. For every microsatellite, each observed length is an allele. Identifying microsatellite markers that are closely linked to the phenotype localizes the desired gene. The measurements are done efficiently and in parallel using commercial primer sets and instrumentation.

Two markers in chromosome band 11q13—D11S4191 and D11S987—segregated with the disease, and some affected individuals born from consanguineous families were homozygous for them. Finer probing, mapping with additional markers, localized the gene on chromosome 11 to a region of about 2.5 Mb.

There are 27 genes in the implicated 2.5 Mb region and its vicinity. Sequencing these genes in a set of patients identified a deletion of three exons in one of them. It was proved to be the disease gene by comparing its sequences in members of the families studied, and demonstrating a correlation between the presence of the syndrome and abnormalities in the gene. None of the other 26 genes in the suspect interval showed such correlated alterations.

Previous studies had identified a different gene, *BSCL1,* at 9q34, in other families with the same syndrome. The gene *BSCL1* has not yet been identified. It is possible that abnormalities in these two genes produce the same effect because their products participate in a common pathway which can be blocked by dysfunction of either.

The gene on chromosome 11, *BSCL2*, contains 11 exons spanning 14 kb. It encodes a 398-residue protein, named seipin. Observed alterations in the gene include large and small deletions, and single amino acid substitutions. The effects are consistent with loss of functional protein, either by causing frameshifts or truncation, or a missense mutation Ala212→Pro that credibly interferes with the stability of a helix or sheet in the protein structure.

Seipin has homologues in mouse and *Drosophila*. There are no clues to the function of any of the homologues, although they are predicted to contain transmembrane helices. What does provide suggestions about the aetiology of some aspects of the syndrome is the expression pattern, highest in brain and testis. This might be consistent with earlier endocrinological studies of Berardinelli–Seip syndrome that identified a problem in the regulation of release of pituitary hormones by the hypothalamus. Discovery of a protein of unknown function involved in the syndrome opens the way to investigation of what may well be a new biological pathway.

Additional mapping techniques deal more directly with the DNA sequences, and can short-circuit the process of gene identification:

- A *contig* or *contiguous clone map*, is a series of overlapping DNA clones of known order along a chromosome from an organism of interest—for instance, human—stored in yeast or bacterial cells as YACs or BACs. A contig map can produce a very fine mapping of a genome. In a YAC, human DNA is stably integrated into a small extra chromosome in a yeast cell. A YAC can contain up to 10^6 bp. In principle, the entire human genome could be represented in 10 000 YAC clones. In a BAC, human DNA is inserted into a plasmid in an *E. coli* cell. (A plasmid is a small piece of double-stranded DNA found in addition to the main genome, usually but not always circular.) A BAC can carry about 250 000 bp. Despite their smaller capacities, BACs are preferred to YACs because of their greater stability and ease of handling.

◆ A *sequence tagged site* (STS) is a short, sequenced region of DNA, typically 200–600 bp long, that appears in a unique location in the genome. It need not be polymorphic. An STS can be mapped into the genome by using PCR to test for the presence of the sequence in the cells containing a contig map.

One type of STS arises from an *expressed sequence tag* (EST), a piece of cDNA (complementary DNA, i.e. a DNA sequence derived from the mRNA of an expressed gene). The sequence contains only the exons of the gene, spliced together to form the sequence that encodes the protein. cDNA sequences can be mapped to chromosomes using FISH, or located within contig maps.

How do contig maps and STS facilitate identifying genes? If you are working with an organism for which the full genome sequence is not known, but for which full contig maps are available for all chromosomes, you would identify STS markers tightly linked to your gene, then locate these markers in the contig maps.

Picking out genes in genomes

Computer programs for genome analysis identify *open reading frames* or *ORFs*. An ORF is a region of DNA sequence that begins with an initiation codon (ATG) and ends with a stop codon. An ORF is a potential protein-coding region.

Approaches to identifying protein coding regions choose from or combine two possible approaches:

1. **Detection of regions similar to known coding regions from other organisms.** These regions may encode amino acid sequences similar to known proteins, or may be similar to ESTs. Because ESTs are derived from mRNA, they correspond to genes known to be transcribed. It is necessary to sequence only a few hundred initial bases of cDNA to give enough information to identify a gene: characterization of genes by ESTs is like indexing poems or songs by their first lines.

2. *Ab initio* **methods, that seek to identify genes from the properties of the DNA sequences themselves.** Computer-assisted annotation of genomes is more complete and accurate for bacteria than for eukarya. Bacterial genes are relatively easy to identify because they are contiguous—they lack the introns characteristic of eukaryotic genomes—and the intergene spaces are small. In higher organisms, identifying genes is harder. Identification of exons is one problem, assembling them is another. Alternative splicing patterns present a particular difficulty.

A framework for *ab initio* gene identification in eukaryotic genomes includes the following features:

- The initial (5′) exon starts with a transcription start point, preceded by a core promotor site such as the TATA box typically ~30 bp upstream. It is free of in-frame stop codons, and ends immediately before a dinucleotide GT splice signal. (Occasionally a non-coding exon precedes the exon that contains the initiator codon.)

- Internal exons, like initial exons, are free of in-frame stop codons. They begin immediately after an AG splice signal and end immediately before a GT splice signal.

- The final (3′) exon starts immediately after an AG splice signal and ends with a stop codon, followed by a polyadenylation signal sequence. (Occasionally a non-coding exon follows the exon that contains the stop codon.)

All coding regions have non-random sequence characteristics, based partly on codon usage preferences. Empirically, it is found that statistics of hexanucleotides perform best in distinguishing coding from non-coding regions. Starting from a set of known genes from an organism as a training set, pattern-recognition programs can be tuned to particular genomes.

Accurate gene detection is a crucial component of genome sequence analysis. This problem is an important focus of current research.

Genome sequencing projects

A list of active genome consortia and centres appears on a web page of the US National Library of Medicine (see Box). Of the 459 groups, a few are major players, and most are specialized to only a few projects. Thirteen institutions participate in more than 20 genome sequencing projects. Four hundred groups, distributed around the world, work on four or fewer genomes. As sequencing becomes less expensive, and the equipment more compact, more individual institutions are likely to acquire instruments and do at least their prokaryotic sequencing 'in-house'.

Major genome sequencing centres associated with 20 or more projects

Institution	Number of projects
US Department of Energy Joint Genome Institute	435
J. Craig Venter Institute	302
The Institute for Genomic Research (TIGR)	296
Washington University	184
Broad Institute	157

$\longrightarrow$

Major genome sequencing centres associated with 20 or more projects *(continued)*

Institution	Number of projects
Gordon and Betty Moore Foundation Marine Microbiology Initiative	114
Wellcome Trust Sanger Institute	96
Genoscope	70
Baylor College of Medicine	47
Institut Pasteur	34
University of Tokyo	33
Integrated Genomics	24

Source: http://www.ncbi.nlm.nih.gov/genomes/static/lcenters.html

Genomes on the Web

Completely sequenced genomes currently include several hundred bacteria, over 20 archaea, many viruses and organelles, and over 30 eukarya (see Table below). Almost all the results are freely available on the Web. Many others are in progress (not counting assemblies from metagenomics sequencing projects):

A sample of completed eukaryotic genomes

Mammals

Human	*Homo sapiens*
Chimpanzee	*Pan troglodytes*
Macaque	*Macaca mulatta*
Mouse	*Mus musculus*
Norway or brown rat	*Rattus norvegicus*
Dog	*Canis familiaris*
Cow	*Bos taurus*
African elephant	*Loxodonta africana*
Opossum	*Monodelphis domestica*

Other chordates

Chicken	*Gallus gallus*
Frog	*Xenopus tropicalis*
Zebrafish	*Danio rerio*
Fugu fish	*Takifugu rubripes*
Pufferfish	*Tetraodon nigroviridis*
Sea squirt (tunicate)	*Ciona intestinalis*
Tunicate	*Ciona savignyi*

A sample of completed eukaryotic genomes *(continued)*

Higher plants

Thale cress	*Arabidopsis thaliana*
Rice	*Oryza sativa*
Maize (corn)	*Zea mays*
Lotus	*Lotus japonicus*
Barrel medic	*Medicago truncatula*
Tomato	*Lycopersicon esculentum*
Black cottonwood	*Populus trichocarpa*

Other eukarya

Fruit fly	*Drosophila melanogaster*
Anopheles mosquito	*Anopheles gambiae*
Dengue mosquito	*Aedes aegypti*
Honeybee	*Apis mellifera*
Nematode worm	*Caenorhabditis elegans*
Baker's yeast	*Saccharomyces cerevisiae*
Fission yeast	*Schizosaccharomyces pombe*
Fungus	*Candida glabrata* CBS138
Fungus	*Debaryomyces hansenii* CBS767
Microsporidian	*Encephalitozoon cuniculi*

Sequencing of the genomes of many other organisms is in progress. The site http://www.ebi.ac.uk/2can/genomes/genomes.html gives brief descriptions of the species represented, and their scientific, clinical and/or practical (e.g. Baker's yeast) significance. A more complete description of the current status of genome projects appears at the site http://www.genomesonline.org/:

Current status of genome projects

Total completed and published	676
Prokaryotic completed or in progress	3000
Eukaryotic completed or in progress	1000

as of 1 January 2008

Groups involved in many full-genome sequencing projects create and maintain databases focused on individual species. Scientists with specialized expertise assume responsibility for curation and annotation of the data. The analysis includes identification of genes, and assignment of function to their products. The results embed the genome in the context of other information about the individual species, arising from other data streams such as proteomics.

For instance, the Comprehensive Yeast Genome Database (CYGD), based at the Munich Information Center for Protein Sequences (MIPS), organizes and presents information on sequence, structure, function and molecular interactions in *Saccharomyces cerevisiae* (http://mips.gsf.de/genre/proj/yeast/). The MIPS group, one of the leading bioinformatics groups in Europe, has provided the nexus of computational support for numerous collaborative sequencing projects, including yeast and *Arabidopsis thaliana*.

Several groups, including MIPS, have developed tools specialized for information retrieval and comparative analysis of genomes. Others include the Ensembl (at the Wellcome Trust Sanger Centre, Hinxton, UK) and University of California at Santa Cruz genome browsers (http://www.ensembl.org, http://genome.ucsc.edu).

online resource centre

Web resources Genome databases

Lists of completed genomes:
http://www.ncbi.nlm.nih.gov/PMGifs/Genomes/allorg.html

http://www.ebi.ac.uk/genomes/mot/index.html

http://pir.georgetown.edu/pirwww/search/genome.html

Organism-specific databases:
http://www.unl.edu/stc-95/ResTools/biotools/biotools10.html

http://www-fp.mcs.anl.gov/~gaasterland/genomes.html

http://www.hgmp.mrc.ac.uk/GenomeWeb/genome-db.html

http://www.bioinformatik.de/cgi-bin/browse/Catalog/Databases/Genome_Projects/

Genomes of prokaryotes

The genetic material of most prokaryotic cells takes the form of a large single circular piece of double-stranded DNA, usually less than 5 Mb long. In addition, cells may contain plasmids.

The protein-coding regions of bacterial genomes do not contain introns. In many prokaryotic genomes the protein-coding regions are partially organized into *operons*—tandem genes transcribed into a single mRNA molecule, under common transcriptional control. In bacteria, the genes of many operons code for proteins with related functions. For instance, successive genes in the *trp* operon of *E. coli* code for proteins that catalyse successive steps in the biosynthesis of tryptophan (see Fig. 2.1). In archaea, a metabolic relationship between genes in operons is less frequently observed.

The typical prokaryotic genome contains only a relatively small amount of non-coding DNA (in comparison with eukarya), distributed throughout the sequence. In *E. coli*, only ~11% of the DNA is non-coding.

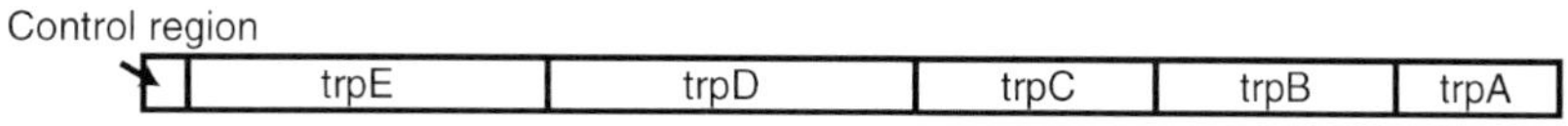

Control region

	trpE	trpD	trpC	trpB	trpA

Fig. 2.1 The *trp* operon in *E. coli* begins with a control region containing promoter, operator and leader sequences. Five structural genes encode proteins that catalyse successive steps in the synthesis of the amino acid tryptophan from its precursor chorismate:

$$\text{chorismate} \rightarrow \underset{(1)}{\underset{\text{nilate}}{\text{anthra-}}} \rightarrow \underset{(2)}{\underset{\text{anthranilate}}{\text{phosphoribosyl-}}} \rightarrow \underset{(3)}{\underset{\text{phosphate}}{\text{indoleglycerol-}}} \rightarrow \underset{(4)}{\text{indole}} \rightarrow \underset{(5)}{\underset{\text{phan}}{\text{trypto-}}}$$

Reaction step (1): *trpE* and *trpD* encode two components of anthranilate synthase. This tetrameric enzyme, comprising two copies of each subunit, catalyses the conversion of chorismate to anthranilate. Reaction step (2): the protein encoded by *trpD* also catalyses the subsequent phosphoribosylation of anthranilate. Reaction step (3): *trpC* encodes another bifunctional enzyme, phosphoribosylanthranilate isomerase — indoleglycerolphosphate synthase. It converts phosphoribosyl anthranilate to indoleglycerolphosphate, through the intermediate, carboxyphenylaminodeoxyribulose phosphate. Reaction steps (4) and (5): *trpB* and *trpA* encode the β and α subunits, respectively, of a third bifunctional enzyme, tryptophan synthase (an $\alpha_2\beta_2$ tetramer). A tunnel within the structure of this enzyme delivers, without release to the solvent, the intermediate produced by the α subunit — indoleglycerolphosphate $\rightarrow$ indole — to the active site of the β subunit — which converts indole $\rightarrow$ tryptophan.

A separate gene, *trpR*, not closely linked to this operon, codes for the trp repressor. The repressor can bind to the operator sequence in the DNA (within the control region) only when binding tryptophan. Binding of repressor blocks access of RNA polymerase to the promoter, turning the pathway off when tryptophan is abundant. Further control of transcription in response to tryptophan levels is exerted by the attenuator element in the mRNA, within the leader sequence. The attenuator region (a) contains two tandem Trp codons and (b) can adopt alternative secondary structures, one of which terminates transcription. Levels of tryptophan govern levels of Trp-tRNAs, which govern the rate of progress of the tandem Trp codons through the ribosome. Stalling on the ribosome at the tandem Trp codons in response to low tryptophan levels reduces the formation of the mRNA secondary structure that terminates transcription.

The genome of the bacterium *Escherichia coli*

Escherichia coli, strain K-12, has long been the workhorse of molecular biology, The genome of strain MG1655, published in 1997 by the group of F. Blattner at the University of Wisconsin, contains 4 639 221 bp in a single circular DNA molecule, with no plastids. Approximately 89% of the sequence codes for proteins or structural RNAs. An inventory reveals:

- 4284 protein-coding genes
- 122 structural RNA genes
- non-coding repeat sequences
- regulatory elements
- transcription/translation guides

- transposases

- prophage remnants

- insertion sequence elements

- patches of unusual composition, likely to be foreign elements introduced by horizontal transfer.

Analysis of the genome sequence required identification and annotation of protein-coding genes and other functional regions. Many *E. coli* proteins were known before the sequencing was complete, from many years of intensive investigation: 1853 proteins had been described before publication of the genome sequence. Other genes could be assigned functions from identification of homologues by searching in sequence databanks. The narrower the range of specificity of the function of the homologues, the more precise could be the assignment. Currently, over 60% of proteins can be assigned at least a general function (see Box). Other regions of the genome are recognized as regulatory sites, or mobile genetic elements, also on the basis of similarity to homologous sequences known in other organisms.

Introduction to Genomics contains several examples.

We visualize the contents of bacterial and organelle genomes as concentric circular diagrams, looking vaguely like 'tie-dyed' patterns. Complex patterns of colour coding serve as a visual 'feature table.' The site http://wishart.biology.ualberta.ca/BacMap/index.html contains an atlas of bacterial genome diagrams.[1]

Distribution of *E. coli* proteins among 22 functional groups

Functional class	Number	%
Regulatory function	45	1.05
Putative regulatory proteins	133	3.10
Cell structure	182	4.24
Putative membrane proteins	13	0.30
Putative structural proteins	42	0.98
Phage, transposons and plasmids	87	2.03
Transport and binding proteins	281	6.55
Putative transport proteins	146	3.40
Energy metabolism	243	5.67
DNA replication, recombination, modification and repair	115	2.68
Transcription, RNA synthesis, metabolism and modification	55	1.28

→

1 Stothard, P., Van Domselaar, G., Shrivastava, S., Guo, A., O'Neill, B., Cruz, J., Ellison, M. and Wishart, D.S. (2005). BacMap: an interactive picture atlas of annotated bacterial genomes. *Nucleic Acids Research*, **33**, D317–D320.

Distribution of *E. coli* proteins among 22 functional groups *(continued)*

Functional class	Number	%
Translation, post-translational protein modification	182	4.24
Cell processes (including adaptation, protection)	188	4.38
Biosynthesis of cofactors, prosthetic groups and carriers	103	2.40
Putative chaperones	9	0.21
Nucleotide biosynthesis and metabolism	58	1.35
Amino acid biosynthesis and metabolism	131	3.06
Fatty acid and phospholipid metabolism	48	1.12
Carbon compound catabolism	130	3.03
Central intermediary metabolism	188	4.38
Putative enzymes	251	5.85
Other known genes (gene product or phenotype known)	26	0.61
Hypothetical, unclassified, unknown	1632	38.06

Source: F.R. Blattner *et al.* (1997). The complete genome sequence of *Escherichia coli* K12. *Science*, **277**, 1453–1462.

The distribution of protein-coding genes over the genome of *E. coli* does not seem to follow any simple rules, either along the DNA or on different strands. Indeed, comparison of strains suggests that the genes are mobile.

The *E. coli* genome is relatively gene dense. Genes coding for proteins or structural RNAs occupy ~89% of the sequence. The average size of an ORF is 317 amino acids. If the genes were evenly distributed, the average intergenic region would be 130 bp; the observed average distance between genes is 118 bp. However, the sizes of intergenic regions vary considerably. Some intergenic regions are large. These contain sites of regulatory function, and repeated sequences. The longest intergenic region, 1730 bp, contains non-coding repeat sequences.

Approximately three-quarters of the transcribed units contain only one gene; the rest contain several consecutive genes, or operons. It is estimated that the *E. coli* genome contains 630–700 operons. Operons vary in size, although few contain more than five genes. The genes in operons tend to have related functions.

In some cases, the same DNA sequence encodes parts of more than one polypeptide chain. One gene codes for both the τ and γ subunits of DNA polymerase III. Translation of the entire gene forms the τ subunit. The γ subunit corresponds approximately to the N-terminal two-thirds of the τ subunit. A frameshift on the ribosome at this point leads to chain termination 50% of the time, causing a 1:1 ratio of expressed τ and γ

subunits. There do not appear to be any overlapping genes in which different reading frames both code for expressed proteins.

In other cases, the same polypeptide chain appears in more than one enzyme. A protein that functions on its own as lipoate dehydrogenase is also an essential subunit of pyruvate dehydrogenase, 2-oxoglutarate dehydrogenase and the glycine cleavage complex.

Having the complete genome, we can examine the protein repertoire of E. coli. The largest class of proteins are the enzymes—appoximately 30% of the total genes. Many enzymatic functions are shared by more than one protein. Some of these sets of functionally similar enzymes are very closely related, and appear to have arisen by duplication, either in E. coli itself, or in an ancestor or gene donor species. Other sets of functionally similar enzymes have very dissimilar sequences, and differ in specificity, regulation, or intracellular location.

Several features of E. coli's generous endowment of enzymes give it a versatile metabolic competence, which allows it to grow and compete under varying conditions:

- It can synthesize all components of proteins and nucleic acids (amino acids and nucleotides), and cofactors.

- It has metabolic flexibility: Both aerobic and anaerobic growth are possible, utilizing different pathways of energy capture. It can grow on many different carbon sources. Not all metabolic pathways are active at any given time; the alternatives allow response to changes in conditions.

- Even for specific metabolic reactions there are many cases of multiple enzymes. These provide redundancy, and contribute to an ability to tune metabolism to varying conditions, through complementary control mechanisms.

- However, E. coli does not possess a complete range of enzymatic capacity. It cannot fix CO_2 or N_2.

We have described here some of the *static* features of the E. coli genome and its protein repertoire. Current research has elucidated dynamic aspects, including the mechanisms that govern protein expression patterns in time and space.

The genome of the archaeon *Methanococcus jannaschii*

S. Luria once suggested that to determine common features of all life one should not try to survey everything, but rather identify the organism most different from us and see what we have in common with it. The assumption was that the way to do this would be to find an organism adapted to the most different environment.

Deep-sea exploration has revealed environments as far from the familiar as those portrayed in science fiction. Hydrothermal vents are underwater volcanoes emitting hot lava and gases through cracks in the ocean floor. They create niches for communities of living things disconnected from the surface, which depend on the minerals exuded from the vent as inorganic nutrients. They support living communities of microorganisms that are the only known forms of life not dependent on sunlight, directly or indirectly, for their energy source.

The microorganism *Methanococcus jannaschii* was collected from a hydrothermal vent 2600 m deep off the coast of Baja California, Mexico, in 1983. It is a thermophilic organism, surviving at temperatures from 48 to 94°C, with an optimum at 85°C. It is a strict anaerobe, capable of self-reproduction from inorganic components. Its overall metabolic equation is to synthesize methane from H_2 and CO_2.

Methanococcus jannaschii belongs to the archaea, one of the three major divisions of life along with the bacteria and eukarya (see Fig. 1.2). The archaea comprise groups of prokaryotes, including organisms adapted to extreme environmental conditions such as high temperature and pressure, or high salt concentration. However, many archaea are not extremophiles.

The genome of *M. jannaschii* was sequenced in 1996 by The Institute for Genomic Research (TIGR). It was the first archaeal genome sequenced. It contains a large chromosome containing a circular double-stranded DNA molecule 1 664 976 bp long, and two extrachromosomal elements of 58 407 and 16 550 bp. There are 1784 predicted protein-coding regions, of which 1728 are on the chromosome, and 44 and 12 on the large and small extrachromosomal elements, respectively. Some RNA genes contain introns. As in other prokaryotic genomes there is little non-coding DNA.

Methanococcus jannaschii would appear to satisfy Luria's goal of finding our most distant extant relative. Comparison of its genome sequence with others shows that it is distantly related to other forms of life. Only 42% of the genes have been assigned a function. However, to everyone's great surprise, archaea are in some ways more closely related to eukarya than to bacteria! They are a complex mixture. In archaea, proteins involved in transcription, translation and regulation are more similar to those of eukarya. Archaeal proteins involved in metabolism are more similar to those of bacteria.

The genome of one of the simplest organisms: *Mycoplasma genitalium*

Mycoplasma genitalium is an infectious bacterium, the cause of nongonococcal uretheitis. Its genome was sequenced in 1995 by a collaboration of groups at TIGR, The Johns Hopkins University and The University of North Carolina. The genome is a single DNA molecule containing 580 070 bp. At the time, this was the smallest cellular genome yet sequenced. So far, *M. genitalium* is the closest we have to a **minimal organism,** the smallest capable of independent life. (Viruses, in contrast, require the cellular machinery of their hosts.)

The genome is dense in coding regions. A total of 468 genes have been identified as expressed proteins. Some regions of the sequence are gene rich, others gene poor, but overall 85% of the sequence is coding. The average length of a coding region is 1040 bp. As in other bacterial genomes, the coding regions do not contain introns. Further compression of the genome is achieved by overlapping genes. It appears that many of these have arisen through loss of stop codons.

The gene repertoire of *M. genitalium* includes some that encode proteins essential for independent reproduction, such as those involved in DNA replication, transcription and translation, plus ribosomal and transfer RNAs (rRNAs and tRNAs). Other genes are specific for the infectious activity, including adhesins that mediate binding to infected cells, other molecules for defence against the host's immune system, and a large number of transport proteins. As an adaptation to the parasitic lifestyle of the organism, there has been widespread loss of metabolic enzymes, including those responsible for amino acid biosynthesis—indeed, one of the 20 amino acids is absent from all *M. genitalium* proteins (see Weblem 2.7).

Genomes of eukarya

It is rare in science to encounter a completely new world containing phenomena entirely unsuspected. The complexity of the eukaryotic genome is such a world (see Box).

Inventory of a eukaryotic genome

Moderately repetitive DNA

◆ Functional

 – dispersed gene families
   ● e.g. actin, globin
 – tandem gene family arrays
   ● rRNA genes (250 copies)
   ● tRNA genes (50 sites with 10–100 copies each in human)
   ● histone genes in many species

◆ Without known function

 – short interspersed elements (SINEs)
   ● Alu is an example
   ● 200–300 bp long
   ● 100 000s of copies (300 000 Alu)
   ● scattered locations (not in tandem repeats)
 – long interspersed elements (LINEs)
   ● 1–5 kb long
   ● 10–10 000 copies per genome
 – pseudogenes

Highly repetitive DNA

◆ Minisatellites

 – composed of repeats of 14–500 bp segments
 – 1–5 kb long
 – many different ones
 – scattered throughout the genome

◆ Microsatellites

 – composed of repeats of up to 13 bp
 – ~100s of kb long
 – ~10^6 copies/genome
 – most of the heterochromatin around the centromere

$\longrightarrow$

→

Inventory of a eukaryotic genome *(continued)*

◆ Telomeres
  - contain a short repeat unit (typically 6 bp: TTAGGG in human genome, TTGGGG in *Paramecium,* TAGGG in trypanosomes, TTTAGGG in *Arabidopsis*)
  - 250–1000 repeats at the end of each chromosome

In eukaryotic cells, the majority of DNA is in the nucleus, separated into bundles of nucleoprotein, the chromosomes. Each chromosome contains a single double-stranded DNA molecule. Smaller amounts of DNA appear in organelles—mitochondria and chloroplasts. The organelles originated as intracellular parasites. Organelle genomes usually have the form of circular double-stranded DNA, but are sometimes linear and sometimes appear as multiple circles. The genetic code by which organelle genes are translated differs from that of nuclear genes.

Nuclear genomes of different species vary widely in size (see page 71). The correlation between genome size and complexity of the organism is very rough. It certainly does not support any preconception that humans stand on a pinnacle. In many cases differences in genome size reflect different amounts of simple repetitive sequences, often referred to as 'junk DNA'.

In addition to variation in DNA content, eukaryotic species vary in the number of chromosomes and distribution of genes among them. Some differences in the distribution of genes among chromosomes involve translocations, or chromosome fragmentations or joinings. For instance, humans have 23 pairs of chromosomes; chimpanzees have 24. Human chromosome 2 is equivalent to a fusion of chimpanzee chromosomes 12 and 13 (see Fig. 2.2). The difficulty of chromosome pairing during mitosis in a zygote after such an event can contribute to the reproductive isolation associated with species separation.

Other differences in chromosome complement reflect duplication or hybridization events. The wheat first used in agriculture, in the Middle East at least 10 000–15 000 years ago, is a diploid called einkorn (*Triticum monococcum),* containing 14 pairs of chromosomes. Emmer wheat (*T. turgidum* ssp. *dicoccum*), also cultivated since Palaeolithic times, and durum wheat (*T. turgidum* ssp. *durum*), are merged hybrids of relatives of einkorn with other wild grasses, to form tetraploid species. Additional hybridizations, with different wild grasses, gave hexaploid forms, including spelt (*T. aestivum* ssp. *spelta*), and modern common wheat *T. aestivum* ssp. *aestivum*. Triticale, a robust crop developed in modern agriculture and currently used primarily for animal feed, is an artificial genus arising from crossing durum wheat (*T. turgidum* ssp. *durum*) and rye (*Secale cereale*). Most triticale varieties are hexaploids.

Variety of wheat	Classification	Chromosome complement
Einkorn	*Triticum monococcum*	AA
Emmer wheat	*Triticum turgidum* ssp. *dicoccum*	AABB
Durum wheat	*Triticum turgidum* ssp. *durum*	AABB

Variety of wheat	Classification	Chromosome complement
Spelt	*Triticum aestivum* ssp. *spelta*	AABBDD
Common wheat	*Triticum aestivum* ssp. *aestivum*	AABBDD
Triticale	*Triticosecale*	AABBRR

A = genome of original diploid wheat or a relative, B = genome of a wild grass *Aegilops speltoides* or *Triticum speltoides*, or a relative, D = genome of another wild grass, *Triticum tauschii* or a relative, R = genome of rye *Secale cereale*.

All these species are still cultivated—some to only minor extents—and have their individual uses in cooking. Spelt, or *farro* in Italian, is the basis of a well-known soup; pasta is made from durum wheat; and bread from *T. aestivum* ssp. *aestivum*.

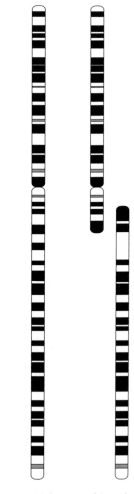

Fig. 2.2 Left: human chromosome 2. Right: matching chromosomes from chimpanzee.

Recent investigations of the history of wheat go beyond simple chromosome counts, to studies of relationships between species and subspecies at the genomic level. General results have measured the decay of synteny between orthologous regions after polyploidization, and mapping of insertions and deletions. Particular results include identification of mutations that confer properties favourable for agriculture. These properties include survival under stressful climate or soil conditions; and firmer attachment of grains to spikes, preserving them for harvesting against dispersal by wind.

A species that undergoes a revolutionary genomic change such as polyploidization is threatened with a penalty in the form of loss of genetic diversity. For the change must have occurred initially in only one or a few individuals, founders of new populations. Evidence for gene flow between domestic and wild forms of wheat suggests a mechanism for recovery and maintenance of genetic diversity.

> For a corresponding discussion of maize domestication see *Introduction to Genomics*, chapter 3.

Gene families

In addition to duplications of entire chromosomes, duplications of individual genes are common, as a result of unequal crossing-over. As a result, gene families within single chromosomes are common in eukarya.

Some family members are **paralogues**—related genes that have diverged to provide separate functions in the same species. (**Orthologues**, in contrast, are homologues that perform the same function in different species. For instance, human α and β globin are paralogues, and human and horse myoglobin are orthologues.) Other related sequences may be pseudogenes, which may have arisen by duplication, or by retrotransposition from mRNA, followed by the accumulation of mutations to the point of loss of function. The human globin gene cluster is a good example (see Box).

The globin gene cluster

Human haemoglobin genes and pseudogenes appear in clusters on chromosomes 11 and 16. The normal adult human synthesizes primarily three types of globin chains: α and β chains, which assemble into haemoglobin $\alpha_2\beta_2$ tetramers; and myoglobin, a monomeric protein found in muscle. Other forms of haemoglobin, encoded by different genes, are synthesized in the embryonic and foetal stages of life. Other globins are unlinked; they arose long before this cluster diverged.

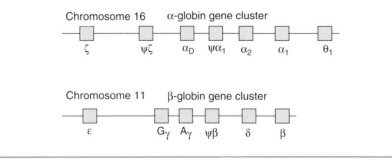

The globin gene cluster *(continued)*

The α gene cluster on chromosome 16 extends over 28 kbp. It contains three functional genes: ζ, and two α genes identical in their coding regions, α_1 and α_2; three pseudogenes, ψζ, $\psi\alpha_1$ and $\psi\alpha_2$; and another homologous gene the function of which is obscure, θ_1. The β gene cluster on chromosome 11 extends over 50 kbp. It includes five functional genes: ε; two γ genes (G_γ and A_γ), which differ in one amino acid, δ; β; and one pseudogene, ψβ. The genes for myoglobin, neuroglobin and cytoglobin are unlinked from both of these clusters.

All human haemoglobin and myoglobin genes have the same intron/exon structure. They contain three exons separated by two introns:

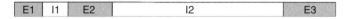

Here E = exon and I = intron. The lengths of the regions in this figure reflect the human β-globin gene. This exon/intron pattern is conserved in most expressed vertebrate globin genes, including haemoglobin α and β chains and myoglobin. In contrast, the genes for plant globins have an additional intron, genes for *Paramecium* globins have one fewer intron, and genes for insect globins contain none. The gene for human neuroglobin, a homologue expressed at low levels in the brain, contains three introns, like plant globin genes.

The distribution of haemoglobin genes and pseudogenes on the chromosomes appears to reflect their evolution via duplication and divergence.

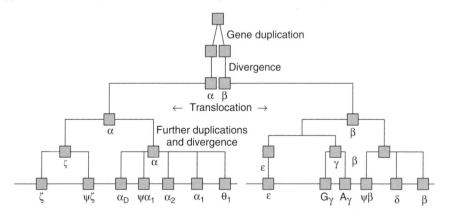

The expression of these genes follows a strict developmental pattern. In the embryo (up to 6 weeks after conception) two haemoglobin chains are primarily synthesized—ζ and ε—which form a $\zeta_2\varepsilon_2$ tetramer. Between 6 weeks after conception until about 8 weeks after birth, foetal haemoglobin—$\alpha_2\gamma_2$—is the predominant species. This is succeeded by adult haemoglobin—$\alpha_2\beta_2$.

Thalassaemias are genetic diseases associated with defective or deleted haemoglobin genes. Most Caucasians have four genes for the α chain of normal adult haemoglobin, two alleles of each of the two tandem genes α_1 and α_2. Therefore α-thalassaemias can present clinically in different degrees of severity, depending on

→
The globin gene cluster *(continued)*

how many genes express normal α chains. Only deletions leaving fewer than two active genes present as symptomatic under normal conditions. Observed genetic defects include deletions of both genes (a process made more likely by the tandem gene arrangement and repetitive sequences which make crossing-over more likely); and loss of chain termination leading to transcriptional 'readthrough', creating extended polypeptide chains which are unstable.

β-Thalassaemias are usually point mutations, including missense mutations (amino acid substitutions), or nonsense mutations (changes from a triplet coding for an amino acid to a stop codon) leading to premature termination and a truncated protein, mutations in splice sites, or mutations in regulatory regions. Certain deletions including the normal termination codon and the intergenic region between δ and β genes create δ–β fusion proteins.

The genome of *Saccharomyces cerevisiae* (Baker's yeast)

Yeast is one of the simplest known eukaryotic organisms. Its cells, like our own, contain a nucleus and other specialized intracellular compartments. The sequencing of its genome, by an unusually effective international consortium involving ~100 laboratories, was completed in 1992. The yeast genome contains 12 057 500 bp of nuclear DNA, distributed over 16 chromosomes. The chromosomes range in size over an order of magnitude, from the 1352 kbp chromosome IV to the 230 kbp chromosome I.

The yeast genome contains 6172 predicted protein-coding genes, ~140 genes for rRNAs, 40 genes for small nuclear RNAs (snRNAs) and 275 tRNA genes. In two respects, the yeast genome is denser in coding regions than the known genomes of the more complex eukarya *Caenorhabditis elegans, Drosophila melanogaster* and human: (1) Introns are relatively rare, and relatively small. Only 231 genes in yeast contain introns. (2) There are fewer repeat sequences compared with more complex eukarya.

A duplication of the entire yeast genome appears to have occurred ~150 Mya. This was followed by translocations of pieces of the duplicated DNA and loss of one of the copies of most (~92%) of the genes.

Of the 6172 protein-coding genes, 4778 correspond to molecules to which a function can be assigned. About 1000 more contain some similarity to known proteins in other species. Another ~800 are similar to ORFs in other genomes that correspond to unknown proteins. Many of these homologues appear in prokaryotes. Only about one-third of yeast proteins have identifiable homologues in the human genome.

In taking censuses of genes, it has been useful to classify their functions into broad categories. The following classification of yeast protein functions is taken from http://mips.gsf.de/genre/proj/yeast/Search/Catalogs/catalog.jsp

Functional category	Number of proteins
Metabolism	1514
Energy	367
Cell cycle and DNA processing	1007
Transcription	1078
Protein synthesis	480
Protein fate (folding, modification, destination)	1154
Protein with binding function or cofactor requirement (structural or catalytic)	1048
Regulation of metabolism and protein function	249
Cellular transport, transport facilities and transport routes	1038
Cellular communication/signal transduction mechanism	234
Cell rescue, defence and virulence	554
Interaction with the environment	463
Transposable elements, viral and plasmid proteins	120
Cell fate	273
Development (systemic)	69
Biogenesis of cellular components	862
Cell type differentiation	452
Total functionally classified proteins	4778
Functionally unclassified proteins	1394

Yeast is a testbed for development of methods to assign functions to gene products. The search for homologues has been exhaustive and continues. Collections of mutants exist containing a knockout of every gene. (A unique sequence 'bar code' introduced into each mutant facilitates identification of the ones that grow under selected conditions.) Cellular localization and expression patterns are being investigated. Several types of measurements, including those based on activation of transcription by pairs of proteins that can form dimers, are producing catalogues of interprotein interactions.

The genome of *Caenorhabditis elegans*

The nematode worm *C. elegans* entered biological research in the 1960s, at the express invitation of Sydney Brenner. He recognized its potential as an organism sufficiently complex to be interesting, yet simple enough to permit complete analysis, at the cellular level, of its development and neural circuitry.

The *C. elegans* genome, completed in 1998, provided the first full DNA sequence of a multicellular organism. The *C. elegans* genome contains ~97 Mbp of DNA distributed on paired chromosomes I, II, III, IV, V and X (see Table). There is no Y chromosome. Different genders in *C. elegans* appear in the XX genotype, a self-fertilizing hermaphrodite; and the XO genotype, a male.

Distribution of *C. elegans* genes

Chromosome	Size (Mb)	Number of protein genes	Density of protein genes (kb/gene)	Number of tRNA genes
I	7.9	2803	5.06	13
II	8.5	3259	3.65	6
III	7.6	2508	5.40	9
IV	9.2	3094	5.17	7
V	9.8	4082	4.15	5
X	10.1	2631	6.54	3

The *C. elegans* genome is about eight times larger than that of yeast, and its 19 099 predicted genes are approximately three times the number of yeast. The gene density is relatively high for a eukaryote, with ~1 gene/5 kb of DNA. Exons cover ~27% of the genome; the genes contain an average of five introns each. Approximately 25% of the genes are in clusters of related genes.

Many *C. elegans* proteins are common to other life forms. Others are apparently specific to nematodes: 42% of proteins have homologues outside the phylum; 34% are homologous to proteins of other nematodes; and 24% have no known homologues outside *C. elegans* itself. Many of the proteins have been classified according to structure and function (see Box).

C. elegans: 20 most common protein domains

Type of domain	Number
Seven-transmembrane spanning chemoreceptor	650
Eukaryotic protein kinase domain	410
Two-domain, C4 type zinc finger	240
Collagen	170

$\longrightarrow$

→

C. elegans: 20 most common protein domains *(continued)*

Type of domain	Number
Seven-transmembrane spanning receptor (rhodopsin-family)	140
C_2H_2-type zinc finger	130
C-type lectin	120
RNA recognition motif	100
C_3HC_4-type (RING finger) zinc fingers	90
Protein tyrosine phosphatase	90
Ankyrin repeat	90
WD domain G-β repeat	90
Homeobox domain	80
Neurotransmitter-gated ion channel	80
Cytochrome P450	80
Conserved C-terminal helicase	80
Short chain and alcohol dehydrogenases	80
UDP-glucoronosyl and UDP-glucosyl transferases	70
EGF-like domain	70
Immunoglobulin superfamily	70

Source: *C. elegans* genome consortium paper in the 11 December 1998 issue of *Science*.

Several kinds of RNA genes have been identified. The *C. elegans* genome contains 659 genes for tRNA, almost half of them (44%) on the X chromosome. Spliceosomal RNAs appear in dispersed copies, often identical. (**Spliceosomes** are the organelles that convert pre-mRNA transcripts to mature mRNA by excising introns, and stitching the exons together.) RNAs appear in a long tandem array at the end of chromosome I. 5S RNAs appear in a tandem array on chromosome V. Some RNA genes appear in introns of protein-coding genes.

The *C. elegans* genome contains many repeat sequences. Approximately 2.6% of the genome consists of tandem repeats. Approximately 3.6% of the genome contains inverted repeats; these appear preferentially within introns, rather than between genes. Repeats of the hexamer sequence TTAGGC appear in many places. There are also simple duplications, involving hundreds to tens of thousands of kilobases.

The genome of *Drosophila melanogaster*

Drosophila melanogaster, the fruit fly, has been the subject of detailed studies of genetics and development for almost a century. Its genome sequence, the product of a collaboration between Celera Genomics and the Berkeley Drosophila Genome Project, was announced in 1999.

The chromosomes of *D. melanogaster* are nucleoprotein complexes, with variation in their structure along their lengths. Approximately one-third of the genome is contained in heterochromatin, highly coiled and compact (and therefore densely staining) regions flanking the centromeres. The other two-thirds is euchromatin, a relatively uncoiled, less compact form. Most of the active genes are in the euchromatin. The heterochromatin in *D. melanogaster* contains many tandem repeats of the sequence AATAACATAG, and relatively few genes.

The total chromosomal DNA of *D. melanogaster* contains ~180 Mbp. The sequence released in 1999 consists of the euchromatic portion, ~120 Mbp. In 2007 an additional ~15 Mbp of heterochromatin sequence was assembled.

The genome is distributed over five chromosomes: three large autosomes, a Y chromosome and a fifth tiny chromosome containing only ~1 Mbp of euchromatin. The fly's 13 601 genes are approximately double the number in yeast, but are fewer than in *C. elegans*, perhaps a surprise. The average density of genes in the euchromatin sequence is 1 gene/8 kb, much lower than the typical 1 gene/kb densities of prokaryotic genomes.

The heterochromatin contains at least ~250 protein-coding genes. They differ from typical euchromatic protein-coding genes by containing longer introns. Most of the intron sequences are repetitive, predominantly fragmented transposable elements.

Despite the fact that insects are not very closely related to mammals, the fly genome is useful in the study of human disease. It contains homologues of 289 human genes implicated in various diseases, including cancer, and cardiovascular, neurological, endocrinological, renal, metabolic and haematological diseases. Some of these homologues have different functions in humans and flies. Other human disease-associated genes can be introduced into, and studied in, the fly. For instance, the gene for human spinocerebellar ataxia type 3, when expressed in the fly, produces similar neuronal cell degeneration. There are now fly models for Parkinson disease and malaria.

The non-coding regions of the *D. melanogaster* genome must contain regions controlling spatiotemporal patterns of development. The developmental biology of the fly has been studied very intensively. It is therefore an organism in which the study of the genomics of development should prove extremely informative.

The compact genome was one reason for adopting *Arabidopsis* by the research community. *A. thaliana* is called 'the fruit fly of botany'.

The genome of *Arabidopsis thaliana*

As a flowering plant, *A. thaliana* is a very distant relative of the other higher eukaryotic organisms for which genome sequences are available. It invites comparative analysis to identify common and specialized features.

Arabidopsis thaliana has a relatively small genome—146 Mb—distributed over five chromosomes. (The maize genome is almost 20 times larger.) The *Arabidopsis* Genome

Initiative reported 115.4 Mbp of genomic DNA sequence in 2000. There are five pairs of chromosomes, containing 25 498 predicted genes. The genome is relatively compact, with 1 gene/4.6 kb on the average. This figure is intermediate between prokaryotes and *Drosophila,* and roughly similar to *C. elegans.* The genes of *Arabidopsis* are relatively small. Exons are typically 250 bp long, and introns relatively small, with mean length 170 bp. Typical of plant genes is an enrichment of coding regions in GC content.

The *Arabidopsis thaliana* genome

| | Chromosome | | | | | |
	1	2	3	4	5	Total
Length (bp)	29 105 111	19 646 945	23 172 617	17 549 867	25 353 409	115 409 949
Number of genes	6543	4036	5220	3825	5874	25 498
Density (kb/gene)	4.0	4.9	4.5	4.6	4.4	
Mean gene length	2078	1949	1925	2138	1974	

Most *Arabidopsis* proteins have homologues in animals, but some systems are unique, among higher organisms, to plants. These include cell wall production and photosynthesis. It might be expected that these need special proteins that might not be shared with animals. Many proteins shared with animals have diverged widely since the last common ancestor. Typical of another difference between plants and animals, 25% of the nuclear genes have signal sequences governing their transport into organelles — mitochondria and chloroplasts — compared with 5% of mitochondrial-targeted nuclear genes in animals.

The *Arabidopsis* nuclear genome is relatively compact. Protein-coding genes contain an average of 5.4 exons, of average length 276 bp, separated by relatively short introns ~165 bp long. The intergenic spacing is also short, ~4.6 kb. A feature of plant genes is that the G+C content of exons (44%) is higher than that of introns (32%).

The structure of the *A. thaliana* genome reveals both local and genome-wide duplications. There were probably *three* polyploidizations, estimates of the dates of which vary widely. The ranges: 225–300 million years ago for the first, 150–170 million years ago for the second, and 25–40 million years ago for the most recent, have been suggested. In addition, local duplications have affected ~17% of genes. Close relatives, such as cabbage and cauliflower, have undergone additional polyploidizations during the 12 million years since they diverged from *Arabidopsis.*

Higher plants must integrate the effects of three genomes — nuclear, chloroplast and mitochondrial. The organelle genomes are much smaller:

Gene distribution in *A. thaliana* between nucleus and organelles

	Nucleus	Chloroplast	Mitochondrion
Genome size (kb)	125 100	154	367
Protein genes	25 498	79	58
Density (kb/protein gene)	4.5	1.2	6.25

Many genes for proteins synthesized by nuclear genes and transported to organelles appear to have originated in the organelles and been transferred to the nucleus.

Genome analysis must address questions of divisions of labour. Relative to animal cells, organelles in plant cells bear a greater metabolic burden, if only because of the activities of chloroplasts. Chloroplast genomes are relatively gene dense, with preserved gene order. In plant mitochondria, genes are more widely spaced, and recombination is more common. Mitochondrial and chloroplast genes contain fewer introns:

Genome:	Nucleus	Chloroplast	Mitochondrial
Genes containing introns:	80%	18.4%	12%

The *Arabidopsis* proteome contains many proteins specific to plants, including those involved in photosynthesis, and metabolism of components of cell walls:

- Plants have many special metabolic pathways, for photosynthesis; and for metabolism of cell wall components, alkaloids and growth regulators such as auxins and gibberellins. Complex metabolism requires the genome to encode a large and varied set of enzymes.

- Plants are threatened by pathogens, and have evolved defence mechanisms dissimilar from our immune system. One weapon against pathogens involves production of reactive oxygen species. Plants synthesize some defence molecules against animals, and others that attract pollinators. These have provided useful sources of flavours, fragrances, and drugs, encompassing traditional 'herbal medicine' and modern pharmacology.

- In keeping with the essential role of light in plant life, *Arabidopsis* has many light sensors, that regulate development and circadian responses.

- *Arabidopsis* is rich in genes that encode water-transporting channels, peptide-hormone transporters, metabolic and biosynthetic enzymes, and proteins involved in defence, detoxification and environmental sensing.

Comparing the proteins encoded in the nuclear genome of *Arabidopsis* with human proteins, the fraction of homologues observed varies with functional category. For protein synthesis, 60% of nuclear-encoded *Arabidopsis* genes have human homologues.

For transcription regulation, the figure is only 30%. It is not that transcription is poorly represented in plant genomes; it is just that plants do it differently. In fact, plants have several times as many transcription factors as the fruit fly. Although many components of the signal transduction pathways familiar from animals are absent in plants, plants have developed specific transcription factor families unknown in animals.

Many *Arabidopsis* genes are homologous to human genes implicated in disease. For instance, plants and animals have similar DNA repair systems, and *Arabidopsis* has a homologue of BRCA2. For some human disease-associated genes, the plant homologue is more similar to the human protein than those from fruit fly or *C. elegans*. Study of the function of the plant homologues will be illuminating, even though it is unlikely that *Arabidopsis* will be suitable for clinical trials of drugs intended for human use!

The genome of *Homo sapiens* (The Human Genome)

Notice

PERSONS attempting to find a motive in this narrative will be prosecuted; persons attempting to find a moral in it will be banished; persons attempting to find a plot in it will be shot.

Mark Twain, Preface to *The Adventures of Huckleberry Finn*

In February 2001, the International Human Genome Sequencing Consortium and Celera Genomics published, separately, drafts of the human genome. On April 14, 2003 the finishing of the genome was announced—reduced error rate, closure of most gaps. This date was within a few days of the 50th anniversary of the publication of the Watson–Crick model for the structure of DNA.

The sequence amounts to $\sim3.2 \times 10^9$ bp, 30 times larger than the genomes of *C. elegans* or *D. melanogaster*. One reason for this disparity in size is that coding sequences form less than 5% of the human genome; repeat sequences over 50%. Perhaps the most surprising feature was the small number of genes identified. The finding of only about 20 000–25 000 genes suggests that alternative splicing patterns make a very significant contribution to our protein repertoire. It is estimated that $\sim35\%$ of genes have alternative splicing patterns.

The human genome is distributed over 22 chromosome pairs plus the X and Y chromosomes. The DNA contents of the autosomes range from 279 Mbp down to 48 Mbp. The X chromosome contains 163 Mbp and the Y chromosome only 51 Mbp.

The exons of human protein-coding genes are relatively small compared with those in other known eukaryotic genomes. The introns are relatively long. As a result many protein-coding genes span long stretches of DNA. For instance, the dystrophin gene, coding for a 3685 amino acid protein, is >2.4 Mbp long.

Protein-coding genes

Analysis of the human protein repertoire implied by the genome sequence has proved difficult because of the problems in reliably detecting genes, and because of alternative splicing patterns. Of the estimated 20 000–25 000 genes, the top categories in a functional classification are:

Function	Number	% of genome
Nucleic acid binding	2207	14.0%
DNA binding	1656	10.5%
DNA repair protein	45	0.2%
DNA replication factor	7	0.0%
Transcription factor	986	6.2%
RNA binding	380	2.4%
Structural protein of ribosome	137	0.8%
Translation factor	44	0.2%
Transcription factor binding	6	0.0%
Cell cycle regulator	75	0.4%
Chaperone	154	0.9%
Motor	85	0.5%
Actin binding	129	0.8%
Defence/immunity protein	603	3.8%
Enzyme	3242	20.6%
Peptidase	457	2.9%
Endopeptidase	403	2.5%
Protein kinase	839	5.3%
Protein phosphatase	295	1.8%
Enzyme activator	3	0.0%
Enzyme inhibitor	132	0.8%
Apoptosis inhibitor	28	0.1%
Signal transduction	1790	11.4%
Receptor	1318	8.4%
Transmembrane receptor	1202	7.6%
G-protein-linked receptor	489	3.1%
Olfactory receptor	71	0.4%
Storage protein	7	0.0%
Cell adhesion	189	1.2%
Structural protein	714	4.5%
Cytoskeletal structural protein	145	0.9%
Transporter	682	4.3%
Ion channel	269	1.7%
Neurotransmitter transporter	19	0.1%
Ligand binding or carrier	1536	9.7%
Electron transfer	33	0.2%
Cytochrome P450	50	0.3%

→

Function	Number	% of genome
Tumour suppressor	5	0.0%
Unclassified	4813	30.6%
Total	15 683	100.0%

Source: http://www.ebi.ac.uk/proteome/
[under Functional classification of *H. sapiens* using Gene Ontology (GO):
General statistics (InterPro proteins with GO hits)]

A classification based on structure revealed the most common types:

Protein	Number
Immunoglobulin and major histocompatibility complex domain	591
Zinc finger, C2H2 type	499
Eukaryotic protein kinase	459
Rhodopsin-like GPCR superfamily	346
Serine/Threonine protein kinase family active site	285
EGF-like domain	259
RNA-binding region RNP-1 (RNA recognition motif)	214
G-protein beta WD-40 repeats	196
Src homology 3 (SH3) domain	194
Pleckstrin homology (PH) domain	188
EF-hand family	185
Homeobox domain	179
Tyrosine kinase catalytic domain	173
Immunoglobulin V-type	163
RING finger	159
Proline rich extensin	156
Fibronectin type III domain	151
Ankyrin-repeat	135
KRAB box	133
Immunoglobulin subtype	128
Cadherin domain	118
PDZ domain (also known as DHR or GLGF)	117

→

Protein	Number
Leucine-rich repeat	113
Serine proteases, trypsin family	108
Ras GTPase superfamily	103
Src homology 2 (SH2) domain	100
BTB/POZ domain	99
TPR repeat	92
AAA ATPase superfamily	92
Aspartic acid and asparagine hydroxylation site	91

Source: http://www.ebi.ac.uk/proteome/

Repeat sequences

Repeat sequences comprise >50% of the genome:

♦ Transposable elements, or interspersed repeats—almost half the entire genome! These include the LINEs and SINEs (see Box).

Type of transposable elements in the human genome

Element	Size (bp)	Copy number	Fraction of genome
Short interspersed nuclear elements (SINEs)	100–300	1 500 000	13%
Long interspersed nuclear elements (LINEs)	6000–8000	850 000	21%
Long terminal repeats	15 000–110 000	450 000	8%
DNA transposon fossils	80–3000	300 000	3%

♦ Retroposed pseudogenes

♦ Simple 'stutters'—repeats of short oligomers. These include the minisatellites and microsatellites. Trinucleotide repeats such as CAG, corresponding to glutamine repeats in the corresponding protein, are implicated in numerous diseases.

♦ Segmental duplications, of blocks of ~10–300 kb. Interchromosomal duplications appear on non-homologous chromosomes, sometimes at multiple sites. Some intra-chromosomal duplications include closely spaced duplicated regions many kilobases long of very similar sequence implicated in genetic diseases; for example Charcot–Marie–Tooth syndrome type 1A, a progressive peripheral neuropathy resulting from duplication of a region containing the gene for peripheral myelin protein 22.

♦ Blocks of tandem repeats, including gene families

RNA

RNA genes in the human genome include:

1. A total of 497 RNA genes. One large cluster contains 140 tRNA genes within a 4 Mbp region on chromosome 6.

2. Genes for 28S and 5.8S rRNAs appear in a 44 kb tandem repeat unit of 150–200 copies. 5S RNA genes also appear in tandem arrays containing 200–300 genes, the largest of which is on chromosome 1.

3. Small nucleolar RNAs include two families of molecules that cleave and process rRNAs.

4. Spliceosomal snRNAs, including the U1, U2, U4, U5 and U6 snRNAs, many of which appear in clusters of tandem repeats of nearly identical sequences, or inverted repeats.

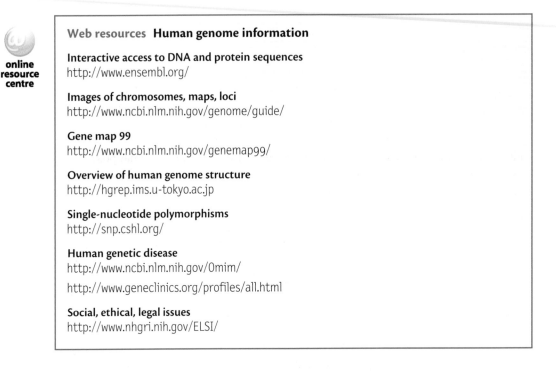

online resource centre

Web resources Human genome information

Interactive access to DNA and protein sequences
http://www.ensembl.org/

Images of chromosomes, maps, loci
http://www.ncbi.nlm.nih.gov/genome/guide/

Gene map 99
http://www.ncbi.nlm.nih.gov/genemap99/

Overview of human genome structure
http://hgrep.ims.u-tokyo.ac.jp

Single-nucleotide polymorphisms
http://snp.cshl.org/

Human genetic disease
http://www.ncbi.nlm.nih.gov/Omim/
http://www.geneclinics.org/profiles/all.html

Social, ethical, legal issues
http://www.nhgri.nih.gov/ELSI/

Single-nucleotide polymorphisms (SNPs) and haplotypes

All people, except identical siblings, have unique DNA sequences. Comparisons between unrelated individuals reveal overall differences between whole-genome sequences of ~0.1%. Many of the differences between individuals have the form of *Single-nucleotide polymorphisms*, or *SNPs*. There are also many short deletions.

A SNP (pronounced 'snip') is a genetic variation between individuals, limited to a single base pair which can be substituted, inserted or deleted. Sickle-cell anaemia is an example of a disease caused by a specific SNP: an A→T mutation in the β-globin gene changes a Glu→Val, creating a sticky surface on the haemoglobin molecule that leads to polymerization of the deoxy form.

Not all SNPs are linked to diseases. Many are not within functional regions (although the density of SNPs is higher than the average in regions containing genes). Some SNPs that occur within exons are mutations to synonymous codons, or cause substitutions that do not significantly affect protein function. Other types of SNPs can cause more than local perturbation to a protein: (1) A mutation from a sense codon to a stop codon, or vice versa, will cause either premature truncation of protein synthesis or 'readthrough'. (2) A deletion or insertion will cause a phase shift in translation.

The A, B and O alleles of the genes for blood groups illustrate these possibilities. A and B alleles differ by four SNP substitutions. They code for related proteins that add different saccharide units to an antigen on the surface of red blood cells.

Allele	Sequence	Saccharide
A	...gctggtgaccctt...	n-Acetylgalactosamine
B	...gctcgtcaccgcta...	Galactose
O	...cgtggt-acccctt...	—

The O allele has undergone a mutation causing a phase shift, and produces no active enzyme. The red blood cells of type O individuals contain neither the A nor the B antigen. This is why people with type O blood are universal donors in blood transfusions. The loss of activity of the protein does not seem to carry any adverse consequences. Indeed, individuals of blood types B and O have greater resistance to smallpox.

Stong correlation of a disease with a specific SNP is advantageous in clinical work, because it is relatively easy to test for affected people or carriers. But if a disease arises from dysfunction of a specific protein, there ought to be many sites of mutations that could cause inactivation. However, a particular site may predominate if (1) all bearers of the gene are descendants of a single individual in whom the mutation occurred, and/or (2) the disease results from a *gain* rather than loss of a specific property, such as in the ability of sickle-cell haemoglobin to polymerize, and/or (3) the mutation rate at a particular site is unusually high, as in the Gly380→Arg mutation in the fibroblast growth receptor gene *FGFR3*, associated with achondroplasia (a syndrome including short stature).

In contrast, many independent mutations have been detected in the *BRCA1* and *BRCA2* genes, loci associated with increased disposition to early-onset breast and ovarian cancer. The normal gene products function as tumour suppressors. Insertion or deletion mutants causing phase shifts generally produce a missing or inactive protein. But it cannot be deduced *a priori* whether a novel *substitution* mutant in *BRCA1* or *BRCA2* confers increased risk or not.

Treatments of diseases caused by defective or absent proteins include:

1. **Providing normal protein** We have mentioned insulin for diabetes, and Factor VIII for the most common type of haemophilia. Another example is the administration of human growth hormone in patients with an absence or severe reduction in normal levels. Use of recombinant proteins eliminates the risk of transmission of AIDS through blood transfusions or of Creutzfeld–Jakob disease from growth hormone isolated from crude pituitary extracts.

2. *Lifestyle adjustments that make the function unnecessary* Phenylketonuria (PKU) is a genetic disease caused by deficiency in phenylalanine hydroxylase, the enzyme that converts phenylalanine to tyrosine. Accumulation of high levels of phenylalanine causes developmental defects, including mental retardation. The symptoms can be avoided by a phenylalanine-free diet. Screening of newborns for high blood phenylalanine levels is legally required in the USA and many other countries.

3. *Gene therapy* to replace absent proteins is an active field of research.

Other clinical applications of SNPs reflect correlations between genotype and reaction to therapy (pharmacogenomics). For example, a SNP in the gene for *N*-acetyl transferase (*NAT-2*) is correlated with peripheral neuropathy—weakness, numbness and pain in the arms, legs, hands or feet—as a side effect of treatment with isoniazid (isonicotinic acid hydrazide), a common treatment for tuberculosis. Patients who test positive for this SNP are given alternative treatment.

SNPs are distributed throughout the genome, occurring on average every 2000 bp. Although they arose by mutation, many positions containing SNPs have low mutation rates, and provide stable markers for mapping genes.

Each of us bears an accumulated collection of SNPs reflecting mutations that occurred in our ancestors. Some constellations of SNPs are co-inherited as blocks. Others are not: mutations in *different* DNA molecules of diploid chromosomes become separated within a single generation, by assortment. Mutations on the *same* chromosome become separated more slowly, by recombination. Haploid sequences, such as most of the human Y chromosome or mitochondrial DNA, are not subject to recombination. Mutations in these sequences remain together.

Mutations in the same DNA molecule in diploid chromosomes will become unlinked by recombination events that occur between their loci. The greater the separation between two sites, the greater the frequency of recombination. However, recombination rates vary widely along the genome, by several orders of magnitude. SNPs on opposite sides of recombinational 'hot spots' are more likely to be separated in any generation. SNPs lying within recombination-poor ('cold') regions will tend to stay together.

In humans, many 100 kb regions tend to remain intact. They show the expected number of SNPs, but relatively few of the possible combinations. An average SNP density of 0.1%, or 1 SNP/kb, suggests ~100 SNPs per 100 kb. The genome of any individual may possess, or may lack, each of them, giving a very large number (2^{100}) of possible combinations. However, many 100 kb regions show fewer than five combinations of SNPs. These discrete combinations of SNPs in recombination-poor regions define an individual's **haplotype**, or '*haplo*id geno*type*' (see Box).

Haplotypes are local combinations of genetic polymorphisms that tend to be co-inherited.

Haplotype distributions

Our individual genomes are characterized by a distribution of genetic markers. SNPs are convenient features to observe, and to study within and across populations. Although the overall density of SNPs in our genomes is ~1 SNP/5 kb,

$\longrightarrow$

→

Haplotype distributions *(continued)*

many 100 kb regions show only a few (typically 2–4) of the possible combinations of SNPs, suggesting that recombination is rare within the region. These segments, which remain intact, are separated by intervals in which recombination is more frequent.

The few discrete combinations of SNPs define the *haplotype* of an individual. The International HapMap project collects and curates haplotype distributions from several human populations.

Haplotypes are difficult to measure, because it is essential to determine which SNPs appear in the *same* DNA strand. Clearly, study of mixed samples from several individuals can determine the frequencies of individual SNPs but not their correlation into haplotypes. Even a sample containing both chromosomes from a diploid cell mixes the contributions of both copies of the region. However, mass spectral studies of amplified single-copy DNA molecules, produced by dilution, can identify the *combination* of SNPs appearing together on the same chromosome, allowing unambiguous haplotyping.

Haplotypes provide a very economical characterization of entire genomes. They simplify the search for genes responsible for diseases—or any other phenotype–genotype correlations. For field biologists, including anthropologists, haplotypes permit detection of migratory and interbreeding patterns in populations.

In looking for genes responsible for diseases or other phenotypic traits, haplotypes provide a magnifying glass. The goal is to correlate phenotype with genetic sequence. The target may be to identify one base out of 3×10^9. By correlating phenotype with *haplotype,* much less sequencing data must be collected to localize the site to within the typical length of a haplotype block, perhaps ~100 kb, containing only a few genes. Another way to look at it is to regard boundaries between haplotype blocks as like the grooves in a bar of chocolate that permit it to be broken easily into bite-size fragments.

Systematic measurements and collections of SNPs

Variations in human genomes are the subject of several large-scale projects.

The SNP consortium (http://snp.cshl.org) collects human SNPs. Its database currently contains nearly 5 million SNPs.

The International HapMap project collects and curates haplotype distributions from several human populations. SNPs are its raw material, from which it identifies the correlations among them. Phase I of the project, published in October 2005, had the goal of measuring the distributions of at least one SNP every 5 kb across the whole human genome. Blood samples were provided by 269 individuals from four continents (see Box). Over 1 million SNPs of significant frequency (>5%) were documented. In addition, 10 selected 500 kb regions were fully sequenced from 48 of the samples. Phase II will extend the analysis of the samples to determine an additional 4.6 million SNPs from the same individuals.

Origin of samples for the International HapMap Project

Population origin	Location	Number of individuals	Relationships
Yoruba	Ibadan, Nigeria	90	30 parent–offspring trios
Northern and western European descent	Utah, USA	90	30 parent–offspring trios
Han Chinese	Beijing, China	45	
Japanese	Tokyo, Japan	44	

Why the choice of parent–offspring combinations? A difficulty in determining haplotypes in heterozygous regions of diploid chromosomes, is how to determine which SNPs lie on the *same* DNA. Comparison of parental and child sequences can sort the observed SNPs into haploid contributions.

Source: The International HapMap Consortium (2005). A haplotype map of the human genome. *Nature*, **437**, 1299–1320.

The work of the International HapMap Consortium, together with other studies, show that:

♦ Most of the variations appear in all populations sampled. Some of the inter–population differences reflect different relative amounts of the same SNPs.

♦ However, a very few SNPs are unique to particular populations. For example, out of over 1 million SNPs, only 11 are consistently different between all individuals of European origin in the sample studied, and all individuals of Chinese or Japanese origin in the sample studied.

♦ The genomes of individuals from Japan and China are very similar, suggesting more recent common ancestry than other population pairs in the study.

♦ The X chromosome varies more between different populations than other chromosomes. This may arise from the fact that males contain only one X chromosome, the genes on which are therefore more subject to selective pressure. Recombinations of X chromosomes can occur, but only in females.

♦ Lengths of haplotype blocks vary among the different sources of samples. They tend to be shorter among populations from Africa, consistent with the idea of African origin of the human species.

The International HapMap Consortium paid due attention to ethical, legal and social issues. Informed consent of the donors preceded collection of samples. The procedure for informed consent involved not only individual agreement, but also community engagement, including interactive explanation of the project. Samples were labelled anonymously. In fact, more samples were collected than used (similar in some ways to the principle of issuing blank cartridges to a firing squad). Nevertheless, characteristics of a population constitutes personal information, the release of which may affect all individuals in the population, including those who were never asked to contribute a sample, and even those who refused. For this reason, the HapMap Consortium did not collect medical information about the sample contributors, even under the protections of consent and anonymity.

Genetic diversity in anthropology

SNP data are of great utility in anthropology, giving clues to historical variations in population size, and migration patterns.

Degrees of genetic diversity are interpretable in terms of the size of the founding population. Founders are the original set of individuals from whom an entire population is descended. Founders can be either original colonists, such as the Polynesians who first settled New Zealand, or merely the survivors of a near-extinction. Cheetahs show the effects of a population bottleneck, estimated to have occurred 10 000 years ago. All living cheetahs are as closely related to one another as siblings. Extrapolations of mitochondrial DNA variation in contemporary humans suggest a single maternal ancestor who lived 140 000–200 000 years ago. Calling her Eve suggests that she was the first woman. But fossil evidence for human ancestors goes back much longer. Mitochondrial Eve was the founder of a surviving population following a near-extinction.

There is now consensus that our species, *Homo sapiens,* arose in Africa approximately 100 000–150 000 years ago. The evidence for human origins in Africa is that contemporary genetic diversity is highest there. The mitochondrial DNA haplogroup L1 (see Box), believed to be the oldest haplotype that survives, is found in the KhoiSan of the Kalahari Desert in southern Africa, and in the Biaka pygmies of the central African rainforest.

Human mitochondrial DNA haplogroups

Human mitochondrial DNA is a double-stranded closed circular molecule 16 569 bp long. It is inherited almost exclusively through maternal lines. A fertilized egg contains the mother's mitochondria. Although sperm contain mitochondria—essential to provide energy for their motility—the few paternal mitochondria that enter the egg are selectively eliminated. As a haploid entity, mitochondrial DNA is therefore not subject to recombination, and changes only by mutation.

Mitochondrial DNA is estimated to adopt 1 mutation every 25 000 years. This gives a reasonable rate of divergence to trace human migration patterns. [Nuclear DNA mutates ∼10 times more slowly than mitochondrial DNA because (1) histones protect it, (2) active repair mechanisms edit out some mutations and (3) the activity of mitochondria in respiration exposes the DNA to mutagenic oxygen radicals.]

Human mitochondrial DNA contains genes for 22 tRNAs, two rRNAs and 13 proteins. The major non-coding region is the control region, or D-loop, involved in regulation and initiation of replication. This region is ∼1 kb long. It shows a higher rate of substitution than the rest of the mitochondrial genome, by a factor of about 4.

$\longrightarrow$

→

Human mitochondrial DNA haplogroups *(continued)*

Different mitochondrial DNA sequences are associated with different populations. Mutations are referred to the first human mitochondrial DNA sequence determined, called the Cambridge Reference Sequence. Groups of related sequences are called haplogroups. (The distribution of the number of sequence differences between different individuals has peaks at ~70 for Africans and ~30 for non-Africans.) The original classification of sequence variants depended on changes in restriction sites (see Fig. 2.3). This was followed by explicit sequencing of the control region, focusing on its two highly polymorphic segments. For finest resolution, contemporary studies are now more frequently determining full mitochondrial DNA sequences, except in cases of ancient DNA where the best recoverable material may be fragmentary.

Several databases focus on human mitochondrial genomes, including MITOMAP (http://www.mitomap.org) and mtDB (http://www.genpat.uu.se/mtDB).

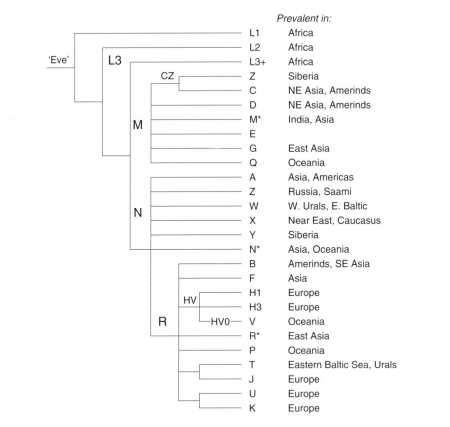

Fig. 2.3 Phylogenetic tree of major mitochondrial haplogroups. The nomenclature began with a study of Native Americans, or Amerinds, and the letters A, B, C and D were assigned to them. Other letters were introduced, and (as more detailed sequencing data appeared) were subdivided as needed (HV0 was formerly called pre-V).

Migrations beginning approximately 60 000 years ago took our ancestors around the world, and continue to do so. Unlike modern population flows, documented in historical records, we depend on archaeological relics, contemporary genomics and linguistics to infer the timing, the routes, the numbers of individuals and even perhaps the motivation, of ancient migrations.

Population-specific SNPs are informative about migrations. Mitochondrial sequences provide information about female ancestors, and Y chromosome sequences provide information about male ancestors. For example, it has been suggested that the population of Iceland—first inhabited about 1100 years ago—is descended from Scandinavian males, and from females from both Scandinavia and the British Isles. Mediaeval Icelandic writings refer to raids on settlements in the British Isles.

See *Introduction to Genomics,* chapter 3.

Other crucial transitions in human social organization, such as turning from hunting to agriculture, can be seen in domestications of other species such as maize and dog. Genomic data are joining classical archaeological evidence to illuminate times and places of domestications (see Box).

Genetic analysis of cattle domestication

Animal resources are an integral and essential aspect of human culture. Analysis of DNA sequences sheds light on their historical development and on the genetic variety characterizing modern breeding populations.

Contemporary domestic cattle include those familiar in Western Europe and North America, *Bos taurus;* and the zebu of Africa and India, *Bos indicus.* The most obvious difference in external appearance is the humpback of the zebu. It has been widely believed that the domestication of cattle occurred once, about 8000–10 000 years ago, and that the two species subsequently diverged.

Analyses of mitochondrial DNA sequences from European, African and Asian cattle suggest, however, that (1) all European and African breeds are more closely related to each other than either is to Indian breeds, and (2) the two groups diverged about 200 000 years ago, implying recent independent domestications of different species. The similarity in physical appearance of the African and Indian zebu (and other similarities at the molecular level; for instance, VNTR markers in nuclear DNA) must then be attributable to importation of cattle from India to East Africa.

A fascinating relationship between human DNA sequences and language families has been investigated by L.L. Cavalli-Sforza and colleagues. These studies have proved useful in working out inter-relationships among American Indian languages. They confirm that the Basques, known to be a linguistically isolated population, have also been genetically isolated.

With the study of isolated populations, anthropological genetics provides data useful in medicine, as mapping disease genes is easier if the background variation is low. Genetically isolated populations in Europe include, in addition to the Basques, the Finns, Icelanders, Welsh and Lapps.

Genetic diversity and personal identification

Variations in our DNA sequences give us individual fingerprints, useful for identification and for establishment of relationships, including but not limited to questions of paternity. The use of DNA analysis as evidence in criminal trials is now well established.

Genetic fingerprinting techniques were originally based on patterns of VNTRs, but have been extended to include analysis of other features including mitochondrial DNA sequences.

For most of us, all our mitochondria are genetically identical, a condition called homoplasmy. However, some individuals have mitochondria with different DNA sequences; called heteroplasmy. Such sequence variation in a disease gene can complicate the observed inheritance pattern of the disease.

The most famous case of heteroplasmy involved Tsar Nicholas II of Russia. After the revolution in 1917 the Tsar and his family were taken to exile in Yekaterinburg in Central Russia. During the night of 16–17 July 1918, the Tsar, Tsarina Alexandra, at least three of their five children, plus their physician and three servants who had accompanied the family, were killed, and their bodies buried in a secret grave. When the remains were rediscovered, assembly of the bones and examination of the dental work suggested—and sequence analysis confirmed—that the remains included an expected family group. The identity of the remains of the Tsarina were proved by matching the mitochondrial DNA sequence with that of a maternal relative, Prince Philip, Chancellor of the University of Cambridge, Duke of Edinburgh and grandnephew of the Tsarina.

However, comparisons of mitochondrial DNA sequences of the putative remains of Nicholas II with those of two maternal relatives revealed a difference at base 16 169: the Tsar had a C and the relatives a T. Extreme political and even religious sensitivities mandated that no doubts were tolerable. Further tests showed that the Tsar was heteroplasmic; T was a minor component of his mitochondrial DNA at position 16 169. To confirm the identity beyond any reasonable question, the body of Grand Duke Georgij, brother of the Tsar, was exhumed, and was shown to have the same rare heteroplasmy.

Evolution of genomes

The availability of complete information about genomic sequences has redirected research (see Box). A general challenge in analysis of genomes is to identify 'interesting events'. A background mutation rate in coding sequences is reflected in *synonymous* nucleotide substitutions: changes in codons that do not alter the amino acid. With this as a baseline, one can search for instances in which there are significantly higher rates of *non-synonymous* nucleotide substitutions: changes in codons that cause mutations in the corresponding protein. (Note, however, that synonymous changes are not necessarily selectively neutral.)

Distribution of genome projects

Organism	Complete	Draft assembly	In progress	Total
Prokaryotes	622	453	457	1532
Archaea	49	4	29	82
Bacteria	573	449	428	1450
Eukarya	22	128	174	324
Animals	4	53	82	139
Mammals	2	20	23	45
Birds		1	2	3
Fishes		3	6	9
Insects	1	19	17	37
Flatworms	1	1	3	4
Roundworms	1	4	12	17
Amphibians			2	2
Reptiles			1	1
Other animals		6	19	25
Plants		6	32	40
Land plants	2	4	25	31
Green Algae		2	7	9
Fungi	10	49	29	88
Ascomycetes	8	41	20	69
Basidiomycetes	1	6	4	11
Other fungi	1	2	5	8
Protists	6	18	27	51
Apicomplexans	1	9	7	17
Kinetoplasts	1	2	5	8
Other protists	4	7	14	25
Total:	644	581	631	1856

These projects have the goal of a high-quality full-genome sequence for a new species. Other high-throughput sequencing efforts involve: (1) Resequencing: sequencing a new individual of a species for which a full reference sequence is available. J.D. Watson's genome is an example. Related projects probe for specific variations, e.g. testing for mutations in BRCA1 and BRCA2 to assess risk of breast or ovarian cancer. (2) Comparative genomics: selection of interesting regions and sequencing homologous segments from many species, e.g. the ENCODE project (see p. 123). (3) Metagenomics, including Environmental Shotgun Sequencing, to measure variation in the biota at a point in space and time (see p. 128).

Source: http://www.ncbi.nlm.nih.gov/genomes/static/gpstat.html

Given two aligned gene sequences, we can calculate K_s = the number of synonymous substitutions, and K_a = the number of non-synonymous substitutions. (The calculation involves more than simple counting because of the need to estimate and correct for possible multiple changes.) A high ratio of K_s/K_a identifies pairs of sequences apparently showing positive selection, possibly even functional changes.

The new field of comparative genomics treats questions that can only now be addressed, such as:

- What *genes* do different phyla share? What genes are unique to different phyla? Do the arrangements of these genes in the genome vary from phylum to phylum?

- What homologous **proteins** do different phyla share? What proteins are unique to different phyla? Does the integration of the activities of these proteins vary from phylum to phylum? Do the mechanisms of control of expression patterns of these proteins vary from phylum to phylum?

- What **biochemical functions** do different phyla share? What biochemical functions are unique to different phyla? Does the integration of these biochemical functions vary from phylum to phylum? If two phyla share a function, and the protein that carries out this function in one phylum has a homologue in the other, does the homologous protein carry out the same function?

The same questions could be asked about different species within each phylum.

M.A. Andrade, C. Ouzounis, C. Sander, J. Tamames and A. Valencia compared the protein repertoire of species from the three major domains of life: *Haemophilus influenzae* represented the bacteria, *Methanococcus jannaschii* the archaea, and *Saccharomyces cerevisiae* (yeast) the eukarya. Their classification of protein functions contained as major categories processes involving energy, information, and communication and regulation:

General functional classes

- Energy

  - Biosynthesis of cofactors, amino acids
  - Central and intermediary metabolism
  - Energy metabolism
  - Fatty acids and phospholipids
  - Nucleotide biosynthesis
  - Transport

- Information

  - Replication
  - Transcription
  - Translation

- Communication and regulation

  - Regulatory functions
  - Cell envelope/cell wall
  - Cellular processes

The number of genes in the three species, known at the time of that study, are:

Species	Number of genes
Haemophilus influenzae	1680
Methanococcus jannaschii	1735
Saccharomyces cerevisiae	6278

Are there, among these, shared proteins for shared functions? In the category of energy, proteins are shared across the three domains. In the category of communication, proteins are unique to each domain. In the categories of regulation and information, archaea share some proteins with bacteria and others with eukarya.

Analysis of shared functions among all domains of life has led people to ask whether it might be possible to define a *minimal organism*—that is, an organism with the smallest gene complement consistent with independent life based on the central DNA→RNA→protein dogma (i.e. excluding protein-free life forms based solely on RNA). The minimal organism must have the ability to reproduce, but not be required to compete in growth and reproductive rate with other organisms. One may reasonably assume a generous nutrient medium, relieving the organism of biosynthetic responsibility, and dispense with stress reponse functions including DNA repair.

The smallest known independent organism is *Mycoplasma genitalium,* with 468 pre-dicted protein sequences. In 1996, A.R. Mushegian and E.V. Koonin compared the genomes of *M. genitalium* and *H. influenzae.* (At the time, these were the only completely sequenced bacterial genomes.) The last common ancestor of these widely diverged bacteria lived about 2 billion years ago. Of 1703 protein-coding genes of *H. influenzae*, 240 are homologues of proteins in *M. genitalium.* Mushegian and Koonin reasoned that all of these must be essential, but might not be sufficient for autonomous life—because some essential functions might be carried out by unrelated proteins in the two organisms. For instance, the common set of 240 proteins left gaps in essential pathways, which could be filled by adding 22 enzymes from *M. genitalium.* Finally, removing functional redundancy and parasite-specific genes gave a list of 256 genes as the proposed necessary *and* sufficient minimal set.

What is in the proposed minimal genome? Functional classes include:

- Translation, including protein synthesis

- DNA replication

- Recombination and repair—a second function of essential proteins involved in DNA replication.

- Transcription apparatus

- Chaperone-like proteins

- Intermediary metabolism—the glycolytic pathway

- No nucleotide, amino acid or fatty acid biosynthesis

◆ Protein export machinery

◆ Limited repertoire of metabolite transport proteins

It should be emphasized that the viability of an organism with these proteins has not been proven. Moreover, even if experiments proved that some minimal gene content—the proposed set or some other set—is necessary and sufficient, this does not answer the related question of identifying the gene complement of the common ancestor of *M. genitalium* and *H. influenzae*, or of the earliest cellular forms of life. Only 71% of the proposed set of 256 proteins have recognizable homologues among eukaryotic *or* archaeal proteins.

Nevertheless, identification of functions necessarily common to all forms of life allows us to investigate the extent to which different forms of life accomplish these functions in the same ways. Are similar reactions catalysed in different species by homologous proteins? Genome analysis has revealed families of proteins with homologues in archaea, bacteria and eukarya. The assumption is that these have evolved from an individual ancestral gene through a series of speciation and duplication events, although some may be the effects of horizontal transfer. The challenge is to map common functions and common proteins.

Several thousand protein families have been identified with homologues in archaea, bacteria and eukarya. Different species contain different amounts of these common families: in bacteria, the range is from *Aquifex aeolicus*, 83% of the proteins of which have archaeal and eukaryotic homologues, to *Borrelia burgdorferi*, in which only 52% of the proteins have archaeal and eukaryotic homologues. Archaeal genomes have somewhat higher percentages (62–71%) of proteins with bacterial and eukaryotic homologues. But only 35% of the proteins of yeast have bacterial and archaeal homologues.

Does the common set of proteins carry out the common set of functions? Among the proteins of the minimal set identified from *M. genitalium*, only ~30% have homologues in all known genomes. Other essential functions must be carried out by unrelated proteins, or possibly by unrecognized homologues. The protein families for which homologues carry out common functions in archaea, bacteria and eukarya are enriched in those involved in translation and biosynthesis:

Protein functional class	Number of families appearing in all known genomes
Translation, including ribosome structure	53
Transcription	4
Replication, recombination, repair	5
Metabolism	9
Cellular processes: (chaperones, secretion, cell division, cell wall biosynthesis)	9

The picture is emerging that evolution has explored the vast potential of proteins to different extents for different types of functions. It has been most conservative in the area of protein synthesis.

Web resources Databases of aligned gene families

Pfam: Protein families database
http://www.sanger.ac.uk/Software/Pfam/

COG: Clusters of Orthologous Groups
http://www.ncbi.nlm.nih.gov/COG/

HOBACGEN: Homologous Bacterial Genes Database
http://pbil.univ-lyon1.fr/databases/hobacgen.html

HOVERGEN: Homologous Vertebrate Genes Database
http://pbil.univ-lyon1.fr/databases/hovergen.html

TAED: The Adaptive Evolution Database
http://www.sbc.su.se/ liberles/TAED.html

Many other sites contain data on individual families.

Please pass the genes: horizontal gene transfer

Learning that *Streptomyces griseus* trypsin is more closely related to bovine trypsin than to other microbial proteinases, Brian Hartley commented in 1970 that, '. . . the bacterium must have been infected by a cow'. It was a clear case of lateral or horizontal gene transfer—a bacterium picking up a gene from the soil in which it was growing, which an organism of another species had deposited there. The classic experiments on bacterial transformation by Griffith and by O. Avery, C. MacLeod and M. McCarthy that identified DNA as the genetic material are another example. In general, horizontal gene transfer is the acquisition of genetic material by one organism from another, by natural rather than laboratory procedures, through some means other than descent from a parent during replication or mating. Several mechanisms of horizonal gene transfer are known, including direct uptake, as in the pneumococcal transformation experiments, or via a viral carrier.

Analysis of genome sequences has shown that horizontal gene transfer is not a rare event, but has affected most genes in microorganisms. It requires a change in our thinking from ordinary 'clonal' or parental models of heredity. Evidence for horizontal transfer includes (1) discrepancies among evolutionary trees constructed from different genes, and (2) direct sequence comparisons between genes from different species:

◆ In *E. coli*, 755 ORFs (a total of 547.8 kb, ~18% of the genome) appear to have entered the genome by horizontal transfer after divergence from the *Salmonella* lineage 100 Mya.

◆ In microbial evolution, horizontal gene transfer is more prevalent among operational genes—those reponsible for 'housekeeping' activities such as biosynthesis—than among informational genes—those responsible for organizational activities such as transcription and translation. For example: *Bradyrhizobium japonicum,* a nitrogen-fixing bacterium symbiotic with higher plants, has two glutamine synthetase genes. One is similar to those of its bacterial relatives; the other 50% identical to those of higher plants. Rubisco (ribulose-1,5-bisphosphate carboxylase/oxygenase), the enzyme that first fixes carbon dioxide at the entry to the Calvin cycle of photosynthesis, has been passed around between bacteria, mitochondria and algal plastids, as well as undergoing gene duplication. Many phage genes appearing in the *E. coli* genome provide further examples and point to a mechanism of transfer.

Nor is the phenomenon of horizontal gene transfer limited to prokaryotes. Both eukarya and prokaryotes are chimeras. Eukarya derive their informational genes primarily from an organism related to *Methanococcus*, and their operational genes primarily from proteobacteria, with some contributions from cyanobacteria and methanogens. Almost all informational genes from *Methanococcus* itself are similar to those in yeast. Nor is gene transfer limited to ancient ancestors. The human genome revealed hundreds of bacterial proteins among our genes. Conversely, at least eight human genes appeared in the *M. tuberculosis* genome.

The observations hint at the model of a 'global organism', a genetic common market, or even a World Wide DNA Web from which organisms download genes at will! How can this be reconciled with the fact that the discreteness of species has been maintained? The conventional explanation is that the living world contains ecological 'niches' to which individual species are adapted. It is the discreteness of niches that explains the discreteness of species. But this explanation depends on the stability of normal heredity to maintain the fitness of the species. Why wouldn't the global organism break down the lines of demarcation between species, just as global access to pop culture threatens to break down lines of demarcation between national and ethnic cultural heritages? Perhaps the answer is that it is the informational genes, which appear to be less subject to horizontal transfer, that determine the identity of the species. Metagenomic sampling is illuminating these questions (see page 128).

It is interesting that although evidence for the importance of horizontal gene transfer is overwhelming, it was dismissed for a long time as rare and unimportant. The source of the intellectual discomfort is clear: parent-to-child transmission of genes is at the heart of the Darwinian model of biological evolution whereby selection (differential reproduction) of parental phenotypes alters gene frequencies in the next generation. For offspring to gain genes from elsewhere than their parents smacks of Lamarck and other discredited alternatives to the paradigm. The evolutionary tree as an organizing principle of biological relationship is a deeply ingrained concept: scientists display an environmentalist-like fervour in their commitment to trees, even when trees are not an appropriate model of a network of relationships (see Chapter 5). Perhaps it is well to recall that Darwin knew nothing of genes, and the mechanism that generated the variation on which selection could operate was a mystery to him. Maybe he would have accepted horizontal gene transfer more easily than his followers!

Comparative genomics of eukarya

A comparison of the genomes of yeast, fly, worm and human revealed 1308 groups of proteins that appear in all four. These form a conserved core of proteins for basic functions, including metabolism, DNA replication and repair, and translation.

These proteins are made up of individual protein domains, including single-domain proteins, oligomeric proteins, and modular proteins containing many domains (the biggest, the muscle protein titin, contains 250–300 domains.) The proteins of the worm and fly are built from a structural repertoire containing about three times as many domains as the proteins of yeast. Human proteins are built from about twice as many as those of the worm and fly. Most of these domains also appear in bacteria and archaea, but some are specific to (probably, invented by) vertebrates (see Table). These include proteins that mediate activities unique to vertebrates, such as defence and immunity proteins, and proteins in the nervous system; only one of them is an enzyme, a ribonuclease.

Distribution of probable homologues of predicted human proteins

Vertebrates only	22%
Vertebrates and other animals	24%
Animals and other eukarya	32%
Eukarya and prokaryotes	21%
No homologues in animals	1%
Prokaryotes only	1%

To create new proteins, inventing new domains is an unusual event. It is far more common to create different combinations of existing domains in increasingly complex ways. A common mechanism is by accretion of domains at the ends of modular proteins (see Fig. 2.4). This process can occur independently, and take different courses, in different phyla.

Gene duplication followed by divergence is a mechanism for creating protein families. For instance, there are 906 genes + pseudogenes for olfactory receptors in the human genome. These are estimated to bind ~10 000 odour molecules. Homologues have been demonstrated in yeast and other fungi (some comparisions *are* odorous), but it is the need of vertebrates for a highly developed sense of smell that multiplied and specialized the family to such a great extent. Eighty percent of the human olfactory receptor genes are in clusters. Compare the small size of the globin gene cluster (pages 95–6), which did not require such great variety.

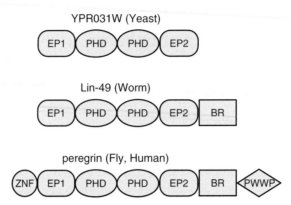

Fig. 2.4 Evolution by accretion of domains, of molecules related to perigrin, a human protein that probably functions in transcription regulation. The *C. elegans* homologue, lin-49, is essential for normal development of the worm. The yeast homologue is involved in histone acetylation. The proteins contain these domains: ZNF = C_2H_2-type zinc finger (not to be confused with acetylene; C and H stand for cysteine and histidine); EP1 and EP2 = Enhancer of polycomb 1 and 2, PHD = plant homeodomain, a repressor domain containing the $C_4H_3C_3$ type of zinc finger, BR = bromo domain; PWWP = domain containing sequence motif Pro-Trp-Trp-Pro.

The ENCODE project

The ENCODE project (ENCyclopedia of DNA Elements) has the ultimate goal of developing methods for comprehensive identification of functional regions of the human genome, including coding and regulatory regions. A selected portion of the human genome—1%, about 30 Mb—has been the initial focus. The basic approaches have been comparative genomics and expression profiling, and involved both laboratory and computational analysis.

Regions corresponding to the selected human genome segments from 28 vertebrates have been sequenced (see Table and Fig. 2.5). These data illuminate each other. The ENCODE project will apply, improve, and develop as necessary, a variety of experimental and computational methods. Lessons learned from work with the selected subset will guide the scaling up of successful methods to analysis of entire genomes (see http://www.genome.gov/10005107).

Species targeted by the ENCODE project			
	Quality of sequencing		
	High	**Medium**	**Unfinished**
Class:			
Actinopterygii	Zebrafish		
Actinopterygii	Fugu fish		
Amphibia	Frog		
Aves	Chicken		

→

→

Species targeted by the ENCODE project *(continued)*

| | Quality of sequencing | | |
	High	Medium	Unfinished
Class *Mammalia*			
Order: **Suborder:**			
Monotremata			Platypus
Marsupialia		Opossum	
Proboscidia			African elephant
Insectivora			Tenrec
Xenarthra			Armadillo
Insectivora			Hedgehog
Insectivora			Shrew
Chiroptera			Bat
Artiodactyla	Cow		
Carnivora	Dog		
Carnivora			Cat
Rodentia	Mouse		
Rodentia	Rat		
Lagomorpha			Rabbit
Primates *Prosimii*			Galago
Primates *Prosimii*			Mouse lemur
Primates *Platyrrhini*			Duski titi
Primates *Platyrrhini*			Owl monkey
Primates *Platyrrhini*			Marmoset
Primates *Catarrhini*			Colobus
Primates *Catarrhini*	Macaque		
Primates *Catarrhini*			Baboon
Primates *Hominidae*	Chimpanzee		
Primates *Hominidae*	Human		

High-quality sequence will be finished to state-of-the-art standards, including resolving difficult regions. Medium-quality sequence will have >8-fold coverage, with manual refinement of assembly. Unfinished sequences are whole-genome shotguns; the coverage may vary and assembly may be incomplete.

Coordinating with ENCODE, the HapMap project focuses on variations among humans in 10 of the ENCODE regions. Sequences from 48 individuals from different geographic origins, yielded 30 000 SNPs.

Analysis of function involves two steps: deciding whether a segment has functional significance, and, if so, identifying what it does. Approximately 5% of the human genome is conserved with respect to mouse and rat sequences. The original idea was to use conservation as a filter to identify functional regions. We shall see that this was only a partial success, which itself is interesting. Only about a third of this 5% is predicted to encode protein. Analysis of function will require treatment of both protein-coding and non-protein-coding regions.

Accordingly, the criteria for selection of regions for the ENCODE project included choosing regions with ranges of gene density, and of non-exonic conservation with

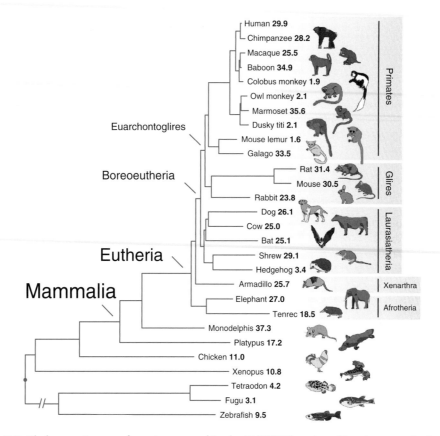

Fig. 2.5 Phylogenetic tree of species treated in the ENCODE project. Numbers are the length of sequence examined, in Mb. (Reproduced and modified with permission from Margulies E.H. *et al.* (2007). Analyses of deep mammalian sequence alignments and constraint predictions for 1% of the human genome. *Genome Research*, **17**, 760–774. Copyright 2007. Cold Spring Harbor Laboratory Press).

respect to the mouse sequence. The result is a set of 44 discrete regions, spread around different human chromosomes and the syntenic regions in other species. These include well-studied regions such as the α- and β-globin loci, and the region containing the gene for the cystic fibrosis transmembrane conductance regulator (CFTR), for which sequence information from different species is known.

Chromosome	Approximate sizes of ENCODE regions in Mb (gene of interest)
1	0.5
2	0.5, 0.5, 0.5, 0.5
4	0.5
5	0.5, 0.5, 1.0 (interleukin)

Chromosome	Approximate sizes of ENCODE regions in Mb (gene of interest)
6	0.5, 0.5, 0.5, 0.5
7	0.5, 1.0, 1.1, 1.2, 1.9 (CFTR)
8	0.5
9	0.5
10	0.5
11	0.5, 0.5, 0.6, 0.5 (Apo cluster), 1.0 (β-globin)
12	0.5
13	0.5, 0.5
14	0.5, 0.5
15	0.5
16	0.5, 0.5, 0.5 (α-globin)
18	0.5, 0.5
19	1.0
20	0.5
21	0.5, 1.7
22	1.7
X	0.5, 1.2

The results of the ENCODE measurements confirmed many expectations, but turned up a number of surprises:

♦ The correlation between conservation and function of segments was imperfect: 60% of the sequences that appear to be under evolutionary constraint map to functional elements. Will functions ultimately be discovered for the remaining 40%? Conversely, many regions of known function appear *not* to be conserved, among the different vertebrate genomes and among different human individuals.

♦ Almost all the human genome is transcribed. Transcripts detected form an overlapping cover of nearly the entire genome. Of course most of this involves regions that do not code for proteins, or for known structural RNAs. Some of these transcripts function in regulation of gene expression, for instance small interfering RNAs (siRNAs). But most of the transcripts do not have a known function.

♦ The relationship between chromatin structure and regulation of transcription and replication has been exposed in greater detail. In local regions of the sequence, there is a good correlation between chromatin accessibility and histone modification patterns and the locations and activity of transcriptional start sites. On a larger scale—on the order of 1 Mb—there is a relationship between histone modifications and the temporal pattern of DNA replication.

Metagenomics: the collection of genomes in a coherent environmental sample

Classically, microbiologists studied prokaryotes by growing them in culture, producing pure strains for detailed study. Powerful as the methods were, and useful as they were for clinical applications and research, they were also blinders that hindered full appreciation of the variety and interactions of species in natural environments. Microbiologists have described and named only a very small fraction of bacterial species. Viruses present even greater variety, by far.

DNA sequencing has made it possible to:

- Clarify evolutionary relationships through comparison of full genomes.

- Use high-throughput sequencing methods to study a cross-section of the life in a natural sample.

- Study the majority of strains that are difficult to grow in culture.

- Understand the relationships and interactions among different species that share an ecosystem. The ecosystem could be a region of the ocean, a sample of acid mine drainage, or the gut of a person or animal.

- Compare the populations that occupy neighbouring and distant locations. For example, studying the variation of microbial populations with depth at some location in the ocean, or surveying how populations vary with latitude and longitude across a region of ocean or land.

- Reveal unsuspected levels of complexity—of both individual species and their interactions—for example, the very great variety of viruses inhabiting the ocean.

A millilitre of ocean water may contain 100–200 species. A gram of soil may contain 4000. How many strains? Don't ask!

From natural samples containing complex mixtures, it is possible to amplify and determine sequences directly, without culturing individual organisms. This is called *environmental shotgun sequencing*. With modern equipment, it is possible to generate very large amounts of sequence data. It is correspondingly difficult to assemble the data into coherent genomes, a problem aggravated by horizontal gene transfer. In fact, a lot of horizontal gene transfer threatens to break down the idea of discrete species. This would be a real revolution in biology.[2]

In a set of shotgun sequences of DNA from a soil sample, <1% of the reads showed overlaps.

Alternatively, one can focus on a few selected molecules, and survey organisms for these sequences. This would offer some understanding of the diversity and distribution of organisms in the sample. The molecule of choice continues to be 15S rRNA. This is partly because of its traditional role as a molecule that varies at the appropriate rate

2 See: D. M. Ward *et al.* (2008). Genomics, environmental genomics and the issue of microbial species. *Heredity*, **99**, 1–13.

to distinguish phylogenetic branching patterns. In addition, rRNA is not very prone to horizontal gene transfer. It thereby preserves the distinctions between taxa. In fact, it may well disguise the mixing that has taken place by horizontal transfer of other genes.

Characterizing an environmental sample by its rRNAs has other serious limitations:

◆ *Ribosomal RNA does not reveal any details of the metabolism or other adaptations of the species or strains* K-12 MG1655 and O157:H7 are strains of *E. coli*.

Strain	Genome size size (bp)	Number of genes
K-12 MG1655	4 639 221	4406
O157:H7	5 594 477	5416

E. coli O157:H7 is a virulent strain, responsible for outbreaks of disease.

The two strains conserve a common 4.1 Mb sequence, with 98.3% base identity, with 2027 gaps. This is the core of the species genome. The remaining 1.4 Mb of O157:H7 is rich in foreign DNAs acquired by horizontal transfer, including 24 prophages and prophage-like elements.

Despite the high similarity of much of these genomes, there is substantial divergence in the proteome. Strain O157:H7 encodes 1632 proteins and 20 tRNAs not present in K-12 MG1655. This is not exclusively from genes in the extra 1.4 Mb: K-12 MG1655 encodes 528 genes not present in O157:H7. These proteomes differ much more than intraspecific variations observed in higher animals.

◆ **Viruses do not contain ribosomes and therefore are invisible to probes for 16S rRNA** Viruses appear to be the 'dark matter' of nature: they exist in unsuspected numbers and variety. Many viral proteins are very different from the sets of molecules familiar in cellular organisms. Anyone with the ambition of deriving a catalogue of protein folding patterns, on the basis of the results of current structural genomics projects, should live in dire fear of what the combined viral proteome will reveal.

Perhaps the most ambitious harvesting of metagenomics data came from the *Sorcerer II* Global Ocean Sampling Expedition.[3] During a round-the-world trip between 8 August 2003 and 22 May 2004, samples were collected at ~320 km intervals along a >8000 km route that started in Halifax, Nova Scotia, Canada, along the East Coast of the USA, the Gulf of Mexico, the Galapagos Islands, across the Pacific Ocean to Australia, through the Indian Ocean, South Africa and back across the Atlantic to the USA.

The expedition was inspired by the *HMS Challenger* expedition of 1872–1876, a survey of ocean geology, climate and biology.

Selected fractions with cells of size 0.1–0.8 μm were filtered to focus on bacteria. Of 7.7 million sequencing reads from these samples, amounting in total to 6.3×10^9 bp, there remained, upon counting for overlaps, almost 6 Gb of unique sequence.

---

3 S. Yooseph *et al.* (2007). The Sorcerer II Global Ocean Sampling Expedition: expanding the universe of protein families. *PLoS Biology*, **13**, e16; Rusch, D.B. *et al.* (2007). The Sorcerer II Global Ocean Sampling Expedition: Northwest Atlantic through eastern tropical Pacific. *PLoS Biology*, **13**, e77.

Over half the reads were unique; that is, they had ≤98% sequence similarity to previously reported sequences. Contigs were assemblable into >3 million whole-genome scaffolds.

The 16S RNA data from the shotgun sequencing revealed 811 distinct sequence types (below 97% identity). Over half represented putative novel species. Note the absence of Archaea from the most highly represented taxa.

Phylum or class	Fraction
α-Proteobacteria	0.32
Unclassified proteobacteria	0.155
γ-Proteobacteria	0.132
Bacteriodetes	0.13
Cyanobacteria	0.079
Firmicutes	0.075
Actinobacteria	0.046
Marine group A	0.022
β-Proteobacteria	0.017
OP11	0.008
Unclassified bacteria	0.008
δ-Proteobacteria	0.005
Planctomycetes	0.002
ε-Proteobacteria	0.001

Source: Rusch, D.B. *et al.* (2007). The Sorcerer II Global Ocean Sampling Expedition: Northwest Atlantic through eastern tropical Pacific. PLoS *Biology*, **13**, e77.

Translations of gene sequences identified over 6 million bacterial and viral proteins. Of these, 1700 have no detectable similarity to previously known proteins. If all are indeed novel—remember that sequence-based tools do not always successfully identify structural similarity in distantly related proteins—the results will almost double the number of known protein families.

The data flow from metagenomics, already large, will only increase by leaps and bounds, as sequencing methods become even more powerful. The challenges to bioinformatics will be not only in the quantity of the data but in their novelty and variety. In our understanding of the organization of life, the paradigm will shift from the hierarchical model to which since Linnaeus we have been accustomed, to a higher dimension of complexity. Coming to terms with this will require conceptual as well as computational breakthroughs.

Recommended reading

C. elegans Sequencing Consortium (1999). How the worm was won. The *C. elegans* genome sequencing project. *Trends in Genetics*, **15**, 51–58. [Description of the project in which high-throughput DNA sequencing was originally developed, and its results, the first metazoan genome to be sequenced.]

Ashburner, M. (2006). *Won for All: How the* Drosophila *Genome Was Sequenced.* Cold Spring Harbor Laboratory Press, Cold Spring Harbor, NY. [A personal description—a 'blow-by-blow' account—of the fly genome project.]

Ureta-Vidal, A., Ettwiller, L. and Birney, E. (2003). Comparative genomics: genomewide analysis in metazoan eukarya. *Nature Reviews Genetics*, **4**, 251–262. [Introduction to the repertoire of solved genomes and how genomes from different species illuminate one another.]

Zhang, R. and Zhang, C.-T. (2006). The impact of comparative genomics on infectious disease research. *Microbes and Infection*, **8**, 1613–1622. [How comparative genomics of prokaryotes is elucidating infectious disease, including identification of virulence determinants, drug targets, candidate vaccines and diagnostic markers.]

Doolittle, W.F. (1999). Lateral genomics. *Trends in Cell Biology*, **9**, M5–M8. [How the discovery of horizontal gene transfer has upset traditional views of evolution.]

Koonin, E.V. (2000). How many genes can make a cell?: the minimal-gene-set concept. *Annual Review of Genomics and Human Genetics*, **1**, 99–116. [A summary of work on comparative genomics.]

Kwok, P.-Y. and Gu, Z. (1999). SNP libraries: why and how are we building them? *Molecular Medicine Today* **5**, 538–543. [Progress and rationale for databases of single-nucleotide polymorphisms.]

Southan, C. (2004). Has the yo–yo stopped? An assessment of human protein-coding gene number. *Proteomics*, **4**, 1712–1726. [How many genes are there in the human genome? Estimates have been changing.]

Bentley, D.R. (2004). Genomes for medicine. *Nature*, **429**, 440–445. [Summary and discussion of the results of genomic sequencing and their applications.]

Dubcovsky, J. and Dvorak, J. (2007). Genome plasticity a key factor in the success of polyploid wheat under domestication. *Science* **316**, 1862–1866.

Publications of the drafts of the human genome sequences appeared in special issues of *Nature*, **15** February 2001, containing the results of the publicly supported Human Genome Project, and *Science*, **16** February 2001, containing the results produced by Celera Genomics. These are landmark issues.

The May 2001 issue of *Genome Research*, Volume 11, Number 5, is devoted to the human genome.

The completion of the human genome in 2003 was announced in various press releases, and described in issues of *Nature* and *Science* magazines:

Collins, F.S., Green, E.D., Guttmacher, A.E. and Guyer, M.S. (2003). A vision for the future of genomics research. *Nature*, **422**, 835–847.

Building on the DNA Revolution. (2003). Special section, *Science*, **300**, 11 April 2003, pp. 277ff.

Xu, J. (2006). Microbial ecology in the age of genomics and metagenomics: concepts, tools and recent advances. *Molecular Ecology*, **15**, 1713–1731.

Eisen, J.A. (2007) Environmental shotgun sequencing: its potential and challenges for studying the hidden world of microbes. *PLoS Biology*, **5**, e82 [Two papers discussing the current status and conceptual problems of coming to terms with metagenomics.]

Exercises, Problems and Weblems

Exercises

Exercise 2.1 The overall base composition of the *E. coli* genome is $A = T = 49.2\%$, $G = C = 50.8\%$. In a random sequence of 4 639 221 nucleotides with these proportions, what is the expected number of occurrences of the sequence CTAG?

Exercise 2.2 The *E. coli* genome contains a number of pairs of enzymes that catalyse the same reaction. How would this affect the use of knockout experiments (deletion or inactivation of individual genes) to try to discern function?

Exercise 2.3 Which of the categories used to classify the functions of yeast proteins (see page 98) would be appropriate for classifying proteins from a prokaryotic genome?

Exercise 2.4 Which occurred first, a man landing on the moon or the discovery of deep-sea hydrothermal vents? Guess first, then look it up.

Exercise 2.5 Gardner syndrome is a condition in which large numbers of polyps develop in the lower gastrointestinal tract, leading inevitably to cancer if untreated. In every observed case, one of the parents is also a sufferer. What is the mode of the inheritance of this condition?

Exercise 2.6 The gene for retinoblastoma is transmitted along with a gene for esterase D to which it is closely linked. However, either of the two alleles for esterase D can be transmitted with either allele for retinoblastoma. How do you know that retinoblastoma is not the direct effect of the esterase D genotype?

Exercise 2.7 If all somatic cells of an organism have the same DNA sequence, why is it necessary to have cDNA libraries from different tissues?

Exercise 2.8 Suppose you are trying to identify a gene causing a human disease. You find a genetic marker 0.75 cM from the disease gene. To within approximately how many base pairs have you localized the gene you are looking for? Approximately how many genes is this region likely to contain?

Exercise 2.9 Leber hereditary optic neuropathy (LHON) is an inherited condition that can cause loss of central vision, resulting from mutations in mitochondrial DNA. You are asked to counsel a woman who has normal mitochondrial DNA and a man with LHON, who are contemplating marriage. What advice would you give them about the risk to their offspring of developing LHON?

Exercise 2.10 Glucose-6-phosphate dehydrogenase deficiency is a single-gene recessive X-linked genetic defect affecting hundreds of millions of people. Clinical consequences include haemolytic anaemia and persistent neonatal jaundice. The gene has not been eliminated from the population because it confers resistance to malaria. In this case, general knowledge of metabolic pathways identified the protein causing the defect. Given the amino acid sequence of the protein, how would you determine the chromosomal location(s) of the corresponding gene?

Exercise 2.11 Before DNA was recognized as the genetic material, the nature of a gene—in detailed biochemical terms—was obscure. In the 1940s, G. Beadle and E. Tatum observed that single mutations could knock out individual steps in biochemical pathways. On this basis they proposed the *one gene—one enzyme* hypothesis. On a photocopy of Fig. 2.1, draw lines linking genes in the figure to numbered steps in the sequence of reactions in the pathway. To what extent do the genes of the *trp* operon satisfy the one gene–one enzyme hypothesis and to what extent do they present exceptions?

Exercise 2.12 The figure shows human chromosome 5 (left) and the matching chromosome from a chimpanzee. On a photocopy of this figure indicate which regions show an inversion of the banding pattern.

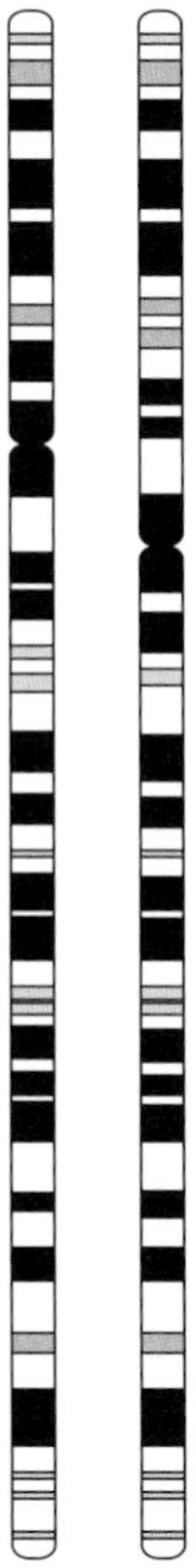

Exercise 2.13 Describe in general terms how the FISH picture in Plate II would appear if the affected region of chromosome 20 were not deleted but translocated to another chromosome.

Exercise 2.14 755 open reading frames entered the *E. coli* genome by horizontal transfer in the 14.4 million years since divergence from *Salmonella*. What is the average rate of horizontal transfer in kb/year? To how many typical proteins (~300 amino acids) would this correspond? What percentage of known genes entered the *E. coli* genome via horizontal transfer?

Exercise 2.15 To what extent is a living genome like a database? Which of the following properties are shared by living genomes and computer databases? Which are properties of living genomes but not databases? Which are properties of databases but not living genomes?

(a) Serve as repositories of information.

(b) Are self-interpreting.

(c) Different copies are not identical.

(d) Scientists can detect errors.

(e) Scientists can correct errors.

(f) There is planned and organized responsibility for assembling and disseminating the information.

Problems

Problem 2.1 Summarize the experimental evidence that shows that the genetic linkage map on any single chromosome is linearly ordered.

Problem 2.2 For *M. genitalium* and *H. influenzae,* what are the values of (a) gene density in genes/kb, (b) average gene size in bp, (c) number of genes. Which factor contributes most to the reduction of genome size in *M. genitalium* relative to *H. influenzae*?

Problem 2.3 It is estimated that the human immune system can produce 10^{15} antibodies. Would it be feasible for such a large number of proteins each to be encoded entirely by a separate gene, the diversity arising from gene duplication and divergence? A typical gene for an IgG molecule is ~2000 bp long.

Weblems

Weblem 2.1 On photocopies of Figs. 1.2, 1.3 and 1.4, indicate the positions of species for which full-genome sequences are known. (http://www.ebi.ac.uk/genomes/)

Weblem 2.2 What are the differences between the standard genetic code and the vertebrate mitochondrial genetic code?

Weblem 2.3 What is the chromosomal location of the human myoglobin gene?

Weblem 2.4 Find examples of additional completely sequenced eukaryotic genomes not listed in the table on pages 84–5. Find at least two in each of the categories: Mammals, Other chordates, Higher plants, Other eukarya.

Weblem 2.5 The table on page 117 contains the statistics of current status of genome projects, as of 1 January 2008. What are the current numbers?

Weblem 2.6 What is the number of occurrences of the tetrapeptide CTAG in the *E. coli* genome? Is it over-represented or under-represented relative to the expectation for a random sequence of the same length and base composition as the *E. coli* genome? (See Exercise 2.1.)

Weblem 2.7 Plot a histogram of the cumulative number of completed genome sequences in each year since 1995.

Weblem 2.8 (a) How many predicted ORFs are there on *Saccharomyces cerevisiae* chromosome X? (b) How many tRNA genes?

Weblem 2.9 Which amino acid is entirely lacking in the proteins of *M. genitalium*? How does the genetic code of *M. genitalium* differ from the standard one?

Weblem 2.10 In the human, 1 cM $\sim 10^6$ bp. In yeast, approximately how many base pairs correspond to 1 cM?

Weblem 2.11 Sperm cells are active swimmers, and contain mitochondria. At fertilization the entire contents of a sperm cell enters the egg. How is it therefore that mitochondrial DNA is inherited from the mother only?

Weblem 2.12 The box on pages 95–6 show the duplications and divergences leading to the current human α- and β- globin gene clusters. (a) In which species, closely related to ancestors of humans, did these divergences take place? (b) In which species related to ancestors of humans did the developmental pattern of expression pattern ($\zeta_2\epsilon_2$ = embryonic; $\alpha_2\gamma_2$ = foetal; $\alpha_2\gamma_2$ = adult) emerge?

Weblem 2.13 Are language groups more closely correlated with variations in human mitochondrial DNA or Y chromosome sequences? Suggest an explanation for the observed result.

Weblem 2.14 The mutation causing sickle-cell anaemia is a single base change A→T, causing the change Glu→Val at position 6 of the β chain of haemoglobin. The base change occurs in the sequence 5'-GTGAG-3' (normal) → GTGTG (mutant). What restriction enzyme is used to distinguish between these sequences, to detect carriers? What is the specificity of this enzyme?

Weblem 2.15 What mutation is the most common cause of phenylketonuria (PKU)?

Weblem 2.16 Find three examples of mutations in the CFTR gene (associated with cystic fibrosis) that produce reduced but not entirely absent chloride channel function. What are the clinical symptoms of these mutations?

Weblem 2.17 Find an example of a genetic disease that is: (a) autosomal dominant, (b) autosomal recessive (other than cystic fibrosis), (c) X-linked dominant, (d) X-linked recessive, (e) Y-linked, (f) the result of abnormal mitochondrial DNA (other than Leber's hereditary optic neuropathy).

Weblem 2.18 (a) Identify a state of the USA in which newborn infants are routinely tested for homocystinuria. (b) Identify a state of the USA in which newborn infants are *not* routinely tested for homocystinuria. (c) Identify a state of the USA in which newborn infants are routinely tested for biotinidiase. (d) Identify a state of the USA in which newborn infants are *not* routinely tested for biotinidiase. (e) What are the clinical consequences of failure to detect homocystinuria, or biotinidiase deficiency?

Weblem 2.19 (a) What is the normal function of the protein that is defective in Menke disease? (b) Is there a homologue of this gene in the *A. thaliana* genome? (c) If so, what is the function of this gene in *A. thaliana*?

Weblem 2.20 *Duchenne muscular dystrophy (DMD)* is an X-linked inherited disease causing progressive muscle weakness. DMD sufferers usually lose the ability to walk by the age of 12, and life expectancy is no more than about 20–25 years. *Becker muscular dystrophy (BMD)* is a less severe condition involving the same gene. Both conditions are usually caused by deletions in a single gene, dystrophin. In DMD there is complete absence of functional protein; in BMD there is a truncated protein retaining some function. Some of the deletions in cases of BMD are longer than others that produce DMD. What distinguishes the two classes of deletions causing these two conditions?

Weblem 2.21 What chromosome of the cow contains a region homologous to human chromosome region 8q21.12?

CHAPTER 3

Scientific publications and archives: media, content and access

Learning goals

1. Understanding the trajectory of the development of the scientific literature, and how it has affected the practice of science.

2. Recognizing the differences in accessibility and convenience between paper and computer access to scientific journals; to appreciate the differences between traditional and digital libraries.

3. Appreciating how economic trends are affecting the scientific publishing industry.

4. Becoming expert in identifying articles relevant to any subject of interest.

5. Understanding the rights and responsibilities associated with open access.

6. Coming to terms with the explosion of scientific information, and how to find modes of keeping track of what you need to know.

7. Being aware of large-scale efforts at digitization of library materials and how the results will be made available.

8. Appreciating some of the general problems associated with curating and distributing information in a high-quality scientific database.

9. Understanding the problems involved in the processing of queries that require appeal to multiple databases.

10. Knowing the principles of machine learning.

$\longrightarrow$

> →
>
> Learning goals (*continued*)
>
> 11. Being able to distinguish different types of computer languages, to understand their relative strengths, and to know how to decide which to use for different purposes.
>
> 12. Appreciating the power and limitations of natural language processing by computer.

The scientific literature

J.R. Oppenheimer said: 'The best way to send information is to wrap it up in a person.'

Scientific publication began as interpersonal communication. Schools—known as far back as Pythagoras—and lectures, seminars and discussions all involve *oral* communication, often supplemented by demonstrations (see Box) or audiovisual material. Only now is this changing: computers are becoming ever more important mediators of human-to-human communication. Computer-to-computer communication is also playing an essential role in research, and in society as a whole.

Robert Hooke at the Royal Society

In 1662 Robert Hooke became Curator of Experiments at the newly formed Royal Society of London. His duties included a demonstration of novel experimental results *every week* at the Society's regular meetings. He fulfilled this responsibility for 41 years! Hooke's wide-ranging interests included the discovery of the cell, in April 1683.

Formal *written* articles or books, sometimes based on transcriptions of talks, constitute the conventional scientific literature. Classically, scientists presented their major results as full-length books. Euclid, Copernicus and Newton are well-known examples. The first scientific journal, the *Proceedings of the Royal Society*, began publication in 1800, as a collection of abstracts of longer publications. A few scientists continued to present their results as books, notably Darwin's *On the Origin of Species by Means of Natural Selection* and Freud's *The Interpretation of Dreams*. However, monographs and articles became the preferred form of publication of scientific results.

Today, in addition to journals, formats of scientific publication include: presentations at meetings; books, or chapters contributed to books; material on the Web; films; radio and television programmes; and podcasts. Formal academic publications must pass the test of 'peer review', an imperfect but nevertheless valuable criterion of quality.

The Web provides an alternative to paper as a mechanism of distribution of the regular scientific literature. It has created novel media and forms of publication. These include bulletin boards, 'blogs, course notes, presentations from scientific meetings or courses, and organized compendia such as the Wikipedia.

Web-based publications uninhibited by peer review are of highly variable quality. In addition, web sites are volatile. Anyone who browses encounters many pointers to vanished links. Indeed, the *mean* lifetime of a wibsite is ~3 months. However, unlike the organic world, the Web has no efficient mechanism for decay and turnover of dead matter.

> The mean lifetime of a web site is comparable to the mean lifetime of a human erythrocyte.

Before the internet, the scientific literature appeared on paper, primarily in the form of journals. Today, journals appear electronically as well as on paper. Many scientists with adequate internet access now only rarely visit a library to read journals. Indeed, the major reason for the survival of paper copies at all is the need of publishers and subscribers to work out a mutually satisfactory and secure economic model to charge for access (see next section). The emergence of digital libraries has (at least) two important implications:

1. The delivery of the literature is delocalized, from a repository of paper copies at one or more fixed sites, to any point with Web access and suitable authorization (such as a personal or institutional password).

2. Computational methods of information retrieval help to identify relevant articles from the vast amount of information available. Searching has replaced browsing. Does this bid goodbye to serendipity? The standard retort: 'Why should you have to buy a full bottle of wine if all you want is a glass?' seems to me to miss the point that the difference between searching and browsing is not merely one of quantity, but of variety. The web site BananaSLUG (mascot of the University of California at Santa Cruz) throws in a random word to a set of search terms, with the goal of rounding up more than the usual suspects. (Try http://www.bananaslug.com/) Readers can experiment with it and draw their own conclusions about its effectiveness.

Economic factors governing access to scholarly publications

In the traditional economic model of scientific journals, a scientific organization or a commercial publisher produces, and distributes at regular intervals, a paper-bound 'issue' containing one or more articles. Many journals include ancillary material such as book reviews or meeting announcements. *Costs of production* include (see Fig. 3.1):

- Activities of an editorial office: receiving submissions, organizing peer review, deciding on acceptance or rejection (sometimes a long, drawn-out process).

- Preparation of accepted manuscripts: copy-editing, layout, etc.

- Printing and distribution of the journal issue as a physical object.

Journals' *sources of support* include:

- Sales—mostly by subscription.

- In many cases, page charges to authors.

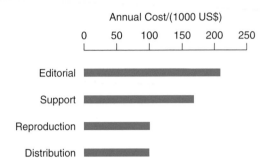

Fig. 3.1 Estimates of annual costs of production and distribution of a typical scientific journal. Assumptions of average journal characteristics: 8.3 issues per volume; 123 articles per volume, selected from 205 manuscripts submitted; per volume: 1439 article pages, 260 special graphics pages, 1728 total pages; 5800 subscriptions.

Editorial costs include manuscript handling, identification and communication with referees, copy-editing and formatting, indexing, preparation of table of contents.

Support costs include marketing, questions involving rights and permissions, administration and financial management.

Reproduction costs include printing, collating, binding, preparation of offprints.

Distribution costs of paper versions include wrapping, labelling, mailing and maintenance of subscription lists.

A few journals earn considerably more from commercial advertising than from subscriptions, but these are exceptions.

(Data from: King, D.W. and Tenopir, C. (2004). Scholarly journal and digital database pricing: threat or opportunity? Available at: http://web.utk.edu/~tenopir/pub/chapters.html).

- Donation of time by editorial boards and referees, who are usually employees of universities, research institutes or related industrial installations, and *not* paid by the publisher.

- Fees for permissions to reproduce material (a minor component).

- For some journals, subsidy from scientific societies, or from foundations.

- Meeting and employment announcements

- A few journals publish commercial advertisements.

Until relatively recently, demand was fairly inelastic. Academic libraries accepted the responsibility of taking virtually all reputable journals in the teaching and research fields of their institutions. Often a university would buy several copies of a journal—for a central library and one or more specialized departmental libraries. No longer! Large increases in costs have broken the budgets of academic libraries, leading to cutbacks in subscriptions.

Several trends have buffetted the system:

- More papers are being published. As a result, many existing journals are publishing more pages each year: larger issues and/or higher publication frequency. This is one of the factors driving up costs.

◆ The larger volume of publications puts libraries under financial pressure: to pay the increased costs of purchasing greater volumes of material, to bind it and to provide space—in principle in perpetuity—to house the purchased journals and make them available. Ultimately the material will require conservation and repair. To save money, libraries reduce purchases, leading to lower print runs and higher prices per copy—a vicious cycle.

◆ Electronic facilities reduce the costs of paper, printing and distribution, and also editorial costs. For instance, electronic typesetting and desktop publishing programs greatly reduce the cost of including equations and graphics in a technical article.

◆ Electronic distribution extends the potential format of journal articles, which may more liberally include colour, movies, sound clips and web links.

◆ Nevertheless, journal costs to libraries, *even for journals distributed electronically*, are increasing. Mergers of publishing companies present libraries with monolithic suppliers, at liberty to increase fees. In defence, libraries are joining to gain collective strength. OhioLINK, a consortium of college and university libraries, and the State Library, in Ohio (central USA) was a pioneer. The combined library system of the University of California campuses can get into the ring with even the largest publishers. (See http://libraries.universityofcalifornia.edu/scholarly/)

◆ Much of the user community supports open access (see next section).

The main tension between readers and publishers is economic. However, scientists depend crucially on journals in another, purely academic, respect: peer review. Today, any research group could simply post their results on the Web. What the journals provide, through the review process, is a (putative) guarantee of trustworthiness and quality in the articles published. Despite a few well-publicized exceptions, the system works fairly well. The peer-review process also acts as an (again, putative) guarantee and ranking of the quality of *scientists*. A major component of judging and comparing scientists, for career milestones of hiring and promotion, depends on the record of *peer-reviewed* publication. This too works well—not without exceptions—but it is a very expensive solution of the problem, and there does not seem to be any logical link between the quality of a scientist's achievements and the medium in which they are reported. A cynic has quipped: 'The day Harvard or Stanford gives tenure to someone for Web postings, 90% of scientific journals will disappear overnight'.

A European Commission report notes that: 'Scientific journals fulfill a double role of *certification* and *dissemination* of knowledge. Dewatripont, M. *et al.* (2006) (see Recommended reading).

Open access

Open access is a redefinition of the author/publisher/reader relationship. Open access retains the peer-review process as the criterion for publication. Then:

◆ Accepted articles are placed on the Web, with free access, immediately upon publication. (Many commercial journals impose a delay—typically 6 months—between paper publication and electronic posting.)

◆ Authors retain copyright, rather than assigning it to the publisher as most journals require. However, not only are articles accessible, the content is freely usable by

For a directory of open-access journals see: http://www. doaj.org readers, subject to giving proper credit to authors. For example, it is permissible to reproduce and distribute articles in a set of teaching materials. Such use of material not subject to open access would require consent of the author or publisher as copyright holder, or, in the case of older material, expiration of copyright protection.

◆ Costs of publication are transferred from readers to authors.

Open access has wide popular appeal in the scientific community. Supporters include both individual scientists in their dual roles of producers and consumers of the literature, and funding agencies. The Wellcome Trust in the UK, and the US National Institutes of Health, require open access to publications reporting work that they have supported.

Some countries have tried to mandate open access of scientific literature by legislation, but no laws have come into force. On 2 May 2006, US Senators John Cornyn (Texas) and Joseph Lieberman (Connecticut) proposed the Federal Research Public Access Act (S. 2695). Passage and presidential approval of this act would have required public availability, via the internet, of *all* journal articles resulting from work supported by funding from US Government agencies involved in large-scale research support. This bill never came to a vote, and awaits reintroduction.

The Public Library of Science (PLoS)

The PLoS was started in October 2000 by H.E. Varmus, P.O. Brown and M.B. Eisen, to put the principles of open access into practice. It is a non-profit organization of professional scientists, not commercial publishers. Its goals include:

◆ Public access to the scientific literature.

◆ To organize the scientific literature so that it is computer searchable.

◆ To encourage developments of innovative approaches to information retrieval from the scientific literature.

Rights and responsibilities of open access

Guidelines for usage of material from PLoS journals appear in the Creative Commons Attribution License (see http://www.plos.org/journals/license.html or http://creativecommons.org/licenses/by/2.5/ for summaries, or http://creativecommons.org/licenses/by/2.5/legalcode for the full details).

With funding from the Gordon and Betty Moore Foundation and other sources, the PLoS has launched a series of journals, which will permit exploration of different relationships—including but not limited to economic ones—between authors, publishers and readers.

Traditional and digital libraries

On most university campuses, a large, fortress-like building occupies a prominent and central site. The functions of the traditional library include:

♦ Archiving, cataloguing and curating printed material.

♦ Providing to readers facilities for convenient access to the literature.

♦ Helping readers find what they need. This involves both help in *identification* of suitable sources of information, and help in gaining *access* to them within the library.

♦ Organizing the reservation, borrowing and return, and photocopying of material. To enforce security—primarily against loss, theft or damage. Regular surveillance for mis-shelved books is essential: a mis-shelved book is worse than a lost one—because it continues to occupy space.

♦ Providing a place to read, study, meet and communicate.

♦ Organizing exhibits and lectures.

♦ Interacting with other libraries to enhance user services; such as organizing inter-library loans.

♦ Speaking up in readers' interests. The American Library Association organized opposition to the loss of readers' rights to privacy imposed by the USA Patriot Act of 2001. Of course librarians also mediate between their readers and publishers, in areas such as open access, and in negotiating subscription prices.

In a digital library, in contrast:

♦ The archives are in electronic form, not on paper.

♦ Provision of access is at a computer screen, rather than off a shelf. This detaches the point of access from the point of repository. The archives can be anywhere. The access point can be anywhere. Most airports and even some airplanes provide internet access. Several cities—including Minneapolis and Philadelphia in the USA; Aarhus, in Denmark; and Taipei, in Taiwan—have effected complete saturation of wireless access over their entire areas. Many other cities are discussing the idea.

♦ Librarian assistance to readers must be done by e-mail or telephone, not requiring a person stationed at the access point.

♦ Security concerns more typically take the form of password control rather than protection against physical theft or damage. Damage to a library computer is damage to library *facilities*, not to library *contents*.

♦ Exhibits and lectures can be made available over the Web. The contents of exhibits are not limited by available wall or display cabinet space.

♦ The digital analogue of the place to study, meet and communicate is perhaps the computer café. Or perhaps collaborative study, meeting and communication will all be entirely on-line. Physical proximity may be unnecessary for intellectual intercourse.

◆ Developing enhanced user services, and representing readers' interests, remain viable goals.

The detachment of archive and access points raises the question of whether the central library building is necessary. At urban universities especially, the footprint of the building occludes a large chunk of prime real estate. Why not go out into the country somewhere—many miles away if necessary—dig a big hole, install computer equipment and connections—keeping at most a skeleton staff —and establish high-speed electronic links to the user community 'downtown'? Then replace the central library with a sports facility. The most important loss in such a scenario is the provision of quiet, distraction-free rooms in which readers can concentrate. Some people can find such a place in a home or office. Many people who can't, don't realize what they are missing.

How to populate a digital library

Many publishers distribute material in electronic form. Although there are no real technical obstacles, many economic questions involving the renegotiation of traditional financial relationships among authors, publishers, subscribers and readers remain unsolved. Older publications that exist only on paper can be scanned in and redistributed electronically. Scanning initially produces a page image, readable by a human but not easily intelligible to a computer. At this stage it is not easy to search the material for some particular text. Optical character recognition can convert the page images to searchable text. However, even the best programs, operating on the cleanest material, require labour-intensive hand-editing because of errors in character recognition.

An obstacle to creating large-scale digital libraries by scanning is the restriction imposed by copyright. In the UK, copyright law began in 1709 during the reign of Queen Anne, shortly after the unification of England and Scotland which prompted legal reconciliations of certain aspects of the disparate systems. Under current UK law, printed material remains under copyright for 70 years from the end of the calendar year of the death of the last surviving author.

The University of Michigan Library holdings amount to 2 380 000 000 pages, containing 743 750 000 000 words.

Legal impediments notwithstanding, the Google Library Project has organized the large-scale scanning of material from over a dozen academic libraries, including those of Harvard, Princeton, Stanford, the Universities of Michigan and California, and The New York Public Library in the USA, the Oxford University Library in the UK, and libraries in Germany and Spain. Different libraries have adopted different stances with respect to material remaining in copyright. For instance, the agreement between Google and Princeton University envisages digitizing only material in the public domain, a selection of about 1 million books, out of a total University collection of 6 million printed works and 5 million manuscripts. In contrast, the agreement between Google and the University of Michigan involves comprehensive scanning, but limited *release* of the product. All results will be searchable. However, in the case of material restricted by copyright, only a small amount of material will be accessible, amounting to a few sentences around the instance of the search term. (For an example see the document 'Project Overview' at http://www.lib.umich.edu/mdp/)

The information explosion

Recalling 'The Sorcerer's Apprentice', efficient delivery can be a mixed blessing. The growth in number of practising scientists has led to an increase in the quantity of publications, by expansion of existing journals and proliferation of new ones. The literature has passed a threshold beyond which it is impossible for anyone to read all the literature in any given field (http://www.eyeforpharma.com/print.asp?news=30550).

It is extremely difficult to command reasonably comprehensive knowledge of even a relatively narrow speciality. Outside one's own particular field, it is impossible. This has stimulated the growth of secondary journals aimed at non-specialists, containing reviews or tutorial presentations. For instance, the Laboratory of Molecular Biology in Cambridge contains a small library focused on the research themes of the institute. It subscribes to 300 journals on paper (in addition to many of these and other journals available electronically). Approximately 14% of the journals on the shelves have titles such as *Reviews of . . .* , or *Trends in . . .* or *Current Opinion in . . .* , etc.

The conclusion is that a reader *must* be selective. Search engines play an essential role in helping to pick out articles, on the basis of combinations of keywords, that match a specific topic. From one relevant publication, links to 'related articles' can identify others. Even this procedure threatens to produce reading lists with more material than one can assimilate, without ruthlessly imposing the narrowest possible focus.

> It is estimated that the scientific literature increases by 2000 pages *every minute,* and that it would take five years for anyone to read the new scientific literature produced each day.

The Web — higher dimensions

The Web does more than provide a convenient channel for distribution of information. Hypertext, by supporting links among different sites, changes fundamentally the *dimensionality* of our access to information.

The presentation of material in a typical traditional book or article is *linear:* you are intended to read successive lines on successive pages. Footnotes are an exception. In some cases, footnotes contain lengthy commentary on or extension of the main text, but often they are merely citations. And most people tend to skip sections of technical detail, at least on a first reading.

Hypertext makes a difference. Web sites contain embedded links. Internal links take you to other portions of the text of a current document, or to associated images, movies or sounds. External links take you, in the first instance, to sites containing related or supplementary information. Following links from those sites will take you everywhere, without limit.

Even the traditional scientific literature formed a network of inter-related data and ideas. Hypertext facilitates fluid navigation of this network. A few readers may remember days in which one read a journal in the library of a biology department, found a reference to a journal located in the chemistry department library, crossed campus to read the cited article, found there a reference to an article in the mathematics library . . .

New media — video, sound

The internet allows the extension of media of presentation, from the traditional text with interspersed pictures, to incorporation of material in many other formats. Extensive sets of coloured pictures, movies or audio are in many cases the best way to present scientific results. For instance, it is very difficult to capture the important details of a complex macromolecular structure as a 'still' picture; a movie affords a much better perception of three dimensions.

The intrinsic importance of sound in some fields — musicology for example — is obvious. In biology, the study of songs of birds, and of whales, is also obviously based on sounds. But there are applications in biomedicine also: the University of Pennsylvania Medical School has distributed sound clips of healthy and abnormal cardiac activity as MP3 files that physicians in training download onto their iPods.

The internet even enhances the distribution of text. For example, the author of a computer program can make it available for downloading. If the program were lengthy, publishing it on paper only would be extremely inconvenient and require the user to retype it.

Searching the literature

The component of the scientific literature that is available in electronic form constitutes a database. To search this database is to specify a set of criteria — in the form of combinations of keywords. The result of the search will be a list of documents that match the specifications. A 'follow-up' question is a subsequent, related search, with modification of the criteria. For example, if a search returned a large number of articles published over the course of many years, one might wish to search again, limiting the results to recent articles, published during the past two years.

For readers interested in the biomedical literature, the standard access route is the specialized database PubMed, from the US National Library of Medicine. PubMed is the bibliographic component of the composite database ENTREZ, maintained by the US National Center for Biotechnology Information.

Most searching over the Web is text searching. However, pattern matching of pictorial information has been very important in interpretation of aerial and satellite pictures for battlefield reconnaisance, for forecasting of agricultural yields, in detecting rare events in bubble chamber photographs, and in analysing shoppers' habits in supermarkets.

General search engines such as Google are very powerful, and will return lists of web sites including but not limited to articles in the scientific literature. These are the most comprehensive results. Google Scholar is a specialized offshoot of the general Google search engine that covers academic publications: primarily books and journal articles. Its advantages include simplicity of use, some access to material in commercial journals not generally publicly available, and the facility to list publications that cite a selected one. (The Science Citation Index also contains this information.) Between the *backward* references cited in the paper, and the *forward* references in which the paper is cited, any particular work takes its place as a node in a network of publications about its subject.

Bibliography management

Staying on top of a subject

Suppose you have used search engines, literature databases, etc. to gain a rounded appreciation of the state of knowledge about some topic. How can you effectively assimilate continuing developments? One way would be to revisit the same general sources, and sort out recent additions from the ones you have seen before. Many libraries have 'New-titles' shelves.

A more efficient approach is to ask the sources to take the initiative of informing you when new information appears. This takes the form of 'current awareness' features in search engines and databases. For instance, many publishers will alert you to publications that cite selected articles. Or you can specify a protein sequence and receive alerts of new sequences related to your 'standing orders' by similarity of either sequence or keyword. More generally, you can specify general search parameters, and receive updated results of an automatic periodic repetition of the search (see, for example: My NCBI, described at http://www.ncbi.nlm.nih.gov/books/bv.fcgi?rid=helppubmed.section. pubmedhelp.My_NCBI, or K. Wolfe's PubCrawler software, http://pubcrawler.gen.tcd.ie).

An inconvenience of many alerting agents is that they inform you by e-mail. An alternative is any of a number of systems called RSS (Really Simple Syndication). These require that: (1) the sites, from which you want to receive alerts, broadcast updates of their information onto the Web, and (2) you run a program, called an *aggregator*, that collects information from the feeds, filters it according to some keyword specification and displays what remains in a window on your screen.

Organizing and sharing the harvest

What is the electronic analogue of a stack of reprints cluttering up your desk? You can download selected articles, and save them in a directory. Browsers allow saving lists of 'bookmarks' or 'favourites'. These lists have the disadvantages of (1) often containing only short, cryptic, information; and (2) residing locally on a disk, making it an effort to synchronize saved URLs if you use more than one computer (not uncommonly a desktop at work and a laptop elsewhere). Several programs allow you to collect literature references *online*—in a form accessible from any internet-connected computer, and which is operating system independent. For example, the program Connotea (http://www.connotea.org/), developed by the Nature Publishing Group, stores URLs, with keywords, or 'tags' to serve as mnemonics if necessary; if the URL corresponds to an article in the scientific literature, the reference will automatically be associated with the entry. A comments field allows more extended annotation. The references are exportable into articles in preparation.

In addition to organizing any individual's sites of interest, the results are shareable. Colleagues with overlapping interests can browse one another's files. Several individuals contributing to the comments field effectively create a mini-blog about each site. Teachers can prepare material for classes—including traditional 'reading lists', and collections of sources of data or other reference material.

The social aspects of sharing annotated URL lists may well be more popular than the purely bibliographic ones. The site http://del.icio.us has some similarities to Connotea, but emphasizes the social aspects of sharing URL annotations. An example of a list primarily intended for *other* people to read would be an online bridal registry, with shared 'write-access' to allow flagging of items as purchased.

Databases

A database is an organized collection of information, in computer-readable form. Defining characteristics of databases include:

- the contents
- the *ontology:* the list of valid terms and their definitions
- the logical structure, or the expression of the interrelationships among the data, called the *schema*
- the format of the data
- the routes for selective retrieval of data, and presentation of the results, or passing them on to a program for analysis
- databases include links to other information resources—other databases, references to original publication of data, tutorial background, etc.

Any database project must assemble all of these. In addition, many independent avenues of retrieval are possible: Anyone can write his or her own 'front end' to any distributed or web-accessible database. Usually, but not always, only the institution that maintains the archives takes responsibility for curation and annotation of the data.

Database contents

Most databases limit themselves to a circumscribed subject. Of course different databases have horizons of different breadths. But most have a unifying theme.

For example, the EMBL Nucleotide Sequence Database, also known as EMBLBank (in partnership with GenBank and the DNA Databank of Japan), collects, curates and annotates nucleotide sequences, including genome sequences. FlyBase is a database containing *Drosophila* genes and genomes, plus material of interest to the community of scientists carrying out research on *D. melanogaster* and its near relatives. FlyBase includes a bulletin board showing schedules of meetings and courses, an atlas of pictures and movies, and links to other relevant sites.

The EMBL Nucleotide Sequence Database and FlyBase contain overlapping material, presented from different points of view, and set within different contexts of additional material, facilities and links.

It is tempting to regard certain databases as *primary*—those that originally gather the data, and are responsible for curation (applying quality control, standardizing format and providing annotation) and archiving. Staff at primary databases have expertise in the experimental techniques that produced the data. *Secondary* databases, or derived databases, would then be those that take data from the primary databases, recombine them, reannotate and reformat them, re-present them and provide different informational environments, different facilities and different links. However, the distinction between primary and secondary databases is not as clear-cut as it used to be.

The literature as a database

Medline® (Medical Literature Analysis and Retrieval System Online) is the bibliographic database of the US National Library of Medicine. Medline has been integrated into PubMed, the bibliographical component of the US National Center for Biotchnology Information database ENTREZ.

Medline covers the scientific literature of fields related to research, teaching and delivery of healthcare. It includes relevant areas of fundamental science. Medline is not primarily patient-oriented. MedlinePlus® is a less technical information resource about healthcare. For instance, a search in MedlinePlus for allopurinol, a drug used in treating gout and related diseases (see pages 436–7), returns a page linking to descriptions of the diseases in which this drug can be used, recommendations for use, side effects, etc. A search for allopurinol in MedLine (via PubMed) returns ~7000 technical articles, most of which do not deal directly with the prescription of allopurinol in current clinical practice.

Organization

The internal structure of a database must reflect the inter-relationships of the contents, in a way that facilitates answering queries. Types of database organization in common use include:

- In a *hierarchical structure*, items are classified, and clustered, at multiple levels. The original Linnaean taxonomy and its many descendants, including the Tree of Life (http://www.tolweb.org/tree/), are examples. The databases SCOP and CATH present hierarchies of protein structures based on evolutionary relationship and structural similarity. We shall see that the markup language XML provides a natural format for databases of information with natural hierarchical structures.

- In a famous 1970 paper, E.F. Codd of IBM Corporation described the *relational database*. The basic unit of a relational database is a set of correspondences between different features of the database contents, called *tables*. Codd showed how set-theoretic operations (union, intersection, difference, Cartesian product) on tables facilitate processing of logically complex queries. Mature software is available, both open source and commercial, for managing relational databases, and for processing queries.

A relational database of amino acids might have one table that associates with each amino acid its name, its three- and one-letter codes, its volume, its accessible surface area and the chemical nature of the distal atoms in its sidechain (see Box). This organization makes it easy to answer queries of the form: what are the three-letter codes of all amino acids the sidechains of which have distal carboxyl groups? The operation required is to select, from the first table, all rows in which the *distal group* has the value carboxyl, and report the three-letter code column from those rows. Or, it would be easy to extract a subtable showing the surface areas of all amino acids with distal methyl groups. This is called a *view*. A compound query might take the form: what are the three-letter codes of all amino acids that have volumes greater than 120 $Å^3$ with distal carboxyl or amide groups?

Two tables from a relational database of properties of amino acids

Amino acid	3-letter code	1-letter code	Volume ($Å^3$)	Surface area ($Å^2$)	Distal group
Alanine	Ala	A	88.6	115	Methyl
Arginine	Arg	R	173.4	225	Guanidinium
Asparagine	Asn	N	111.1	150	Amide
Aspartic acid	Asp	D	114.1	160	Carboxyl
Cysteine	Cys	C	108.5	135	Sulphydryl
Glutamic acid	Glu	E	138.4	190	Carboxyl
Glutamine	Gln	Q	143.8	180	Amide
Glycine	Gly	G	60.1	75	Hydrogen
Histidine	His	H	153.2	195	Imidazole
Isoleucine	Ile	I	166.7	175	Methyl
Leucine	Leu	L	166.7	170	Methyl
Lysine	Lys	K	168.6	200	Amino
Methionine	Met	M	162.9	185	Methyl
Phenylalanine	Phe	F	189.9	210	Phenyl
Proline	Pro	P	112.7	145	Pyrrolidine
Serine	Ser	S	89.0	115	Hydroxyl
Threonine	Thr	T	116.1	140	Hydroxyl
Tryptophan	Trp	W	227.8	255	Indole
Tyrosine	Tyr	Y	193.6	230	Phenol
Valine	Val	V	140.0	155	Methyl

→

Two tables from a relational database of properties of amino acids *(continued)*

Distal group	H-bond donor	H-bond acceptor
Amide	yes	yes
Amino	yes	no
Carboxyl	no	yes
Guanidinium	yes	yes
Hydrogen	no	no
Hydroxyl	yes	yes
Indole	yes	yes
Methyl	no	no
Phenol	yes	yes
Phenyl	no	no
Pyrrolidine	yes	no
Sulphydryl	yes	no

The second table associates with each distal atom grouping the hydrogen-bonding potential. The common column—the distal atom grouping—allows queries that reflect correlations of these properties, for instance: what are the three-letter codes of all amino acids, the sidechains of which can serve as hydrogen-bond donors? To combine the information in both tables would involve, in Codd's terms, a *join* of the two tables.

The general join operation amounts to forming the Cartesian product of the two tables. (The Cartesian product of two sets is the set of ordered pairs of elements, one from each set. If the sets contain n and m elements, respectively, the Cartesian product will contain nm elements.) There are 20 entries in the first table and 12 in the second. The Cartesian product would contain 240 rows, including:

Parts of a join from a relational database of properties of amino acids

From first table: From second table:

Alanine	Ala	A	88.6	115	Methyl	Amide	Yes	Yes
Alanine	Ala	A	88.6	115	Methyl	Amino	Yes	Yes

. . .

| Alanine | Ala | A | 88.6 | 115 | Methyl | Methyl | No | No |

. . .

Aspartic acid	Asp	D	114.1	160	Carboxyl ‖	Amide	Yes	Yes

. . .

Aspartic acid	Asp	D	114.1	160	Carboxyl ‖	Carboxyl	No	Yes

. . .

To report the three-letter codes of amino acids that have sidechains that could serve as hydrogen-bond acceptors, we perform what is called a *natural join*. This retains from the combined tables those rows in which columns 6 and 7 are equal. Of the three rows shown, one contains methyl in both columns 6 and 7. The others would be rejected. The survivors form a new table containing columns specifying the hydrogen-bonding potential associated with each amino acid. We could then select from this combined table the rows containing hydrogen-bond acceptors, and report their three-letter codes. One such row would appear as follows, after merging the two equal distal group columns following the natural join:

Aspartic acid	Asp	D	114.1	160	Carboxyl	No	Yes

From this row, we could extract the three-letter code Asp.

The relational database organization lends itself naturally to processing complex queries constructed as logical compositions of simpler queries. A somewhat artificial example: What are the three-letter codes of amino acids with volumes between 100 and 120 AND [(that can serve as hydrogen bond donors AND NOT serve as hydrogen-bond acceptors) OR (that have surface areas greater than 120 Å^2 AND have distal methyl groups)].

The Structured Query Language (SQL) is a fairly well standardized syntax for probing relational databases with queries of this type. Complex queries containing logical connectives are translatable into Codd's set operations on tables. For a flavour of SQL, the query in the preceding paragraph would appear:

```
SELECT <3_letter_code> FROM <amino_acid_table>
WHERE (sidechain_volume between 100 AND 120)
AND
(H-bond_donor = "yes" AND H-bond_acceptor = "no")
OR
(surface_area > 100 AND distal_group = "methyl"))
```

Annotation

A typical entry in a database in molecular biology might contain the sequence of a gene. However, the entry will contain more than the bare nucleotide sequence. It will also contain:

◆ **Reference information:** citations of the publications that served as the source of the entry, the history of the entry in the database and accession information assigned by the database.

◆ **Interpretative information:** for example, the limits of exons within the sequence.

◆ **Links to other information:** perhaps a protein sequence database containing information about product encoded and the function attributed to that product, or other entries in the same or other databases describing homologous genes.

When databases were more thematically focused and isolated, there was a comfortable and clear distinction between the primary data and the annotations. Annotations tended to be free-form comments, some expressed more casually than others. Recently, many database mergers have occurred, in response to the need to assemble a wide spectrum of information about gene sequences (and many other items). As a result of mergers, and of the importance of ontologies and computer-interpretable formats, entries in databases have taken more formal structures. It is growing more difficult to draw as sharp a distinction between data and annotation.

Some of the information in entries is more reliable than others. Nucleic acid sequences, determined by modern techniques with generous coverage, are quite accurate. On the other hand, assignment of function to gene products in the absence of direct experimental information is an important challenge in database annotation. It is a common practice to transfer functional annotation from a previously annotated homologous protein. This approach relies on the assumptions that (1) because homologous proteins have similar sequences and structures, they have similar functions, and (2) the annotation of the homologue is correct. Often, but certainly not always, these assumptions are valid. Because of the phenomenon of 'recruitment', proteins very similar or even identical in sequence can adopt different functions (see page 360). This can lead to misannotation.

Database quality control

If errors do enter databases—either in data or in annotations —they tend to propagate into other databases, and are very difficult to extirpate.

In principle there are two approaches to improving database quality: keeping errors out in the first place, and removing them when they have been detected.

As part of the 'get it right the first time' approach, database curation and annotation has emerged as a new profession. Curators bring to their activities a specialized complement of skills and attitudes. The quality of their work translates directly into the quality of the databases.

Nevertheless, the high volume and diversity of subjects of scientific papers makes it difficult for database staff to keep up adequately with the workload. An alternative approach is to involve the scientists who publish papers in the harvesting of database entries based on their results. For example, the Protein Data Bank accepts from authors a virtually complete entry corresponding to the structure deposited.

Databank staff carry out validation procedures, but rarely add significant amounts of material.

Nevertheless, despite the professionalism of the curators, and the assiduity of their checking, errors will appear. One problem is to identify them and the second is to remove them. One approach to identifying them is to enlist experts as external curators to examine databank entries in their own specialities. Often, database users call attention to errors.

Once identified, errors can be corrected in a 'master copy' of a database, particularly if the database management is in the hands of a single institution or a close-coupled partnership.

However, correction at source is not enough, because:

> The proliferation of divergent copies, of an object that is continually changing anyway, makes it difficult to reproduce a published investigation.

1. Many users create local versions of databases. These copies will contain the errors that appeared at the time of downloading. The dissemination of any corrections is at the mercy of the frequency of updating of the downloaded versions.

2. Many other databanks assimilate, reintegrate and redisseminate data, processes which may shield errors from correction, especially if items are not carefully tagged with their site and date of origin.

One attractive idea is to create 'knowbots'—robot programs that sweep the web checking for errors. Knowbots are a delocalized form of UNIX 'daemons'. However, security issues would block them from most sites.

What is possible are programs that offer 'health checks' of versions of databases. Two examples are:

- The PDBREPORT database[1] contains the results of validation software, WHAT_CHECK, applied to each entry in the Protein Data Bank. The program tests the validity and consistency of the *format*, and also analyses the structures, detecting outliers in stereochemical properties, such as bond lengths or angles, and looking for inconsistencies in hydrogen-bonding patterns. It has been pointed out by crystallographers—very very emphatically—that outliers do not necessarily signal errors in the structure determination. (Of course, non-outliers also may or may not be errors.)

> Gene Ontology (GO) is a classification scheme for protein function (see page 207).

- GOChase provides Web-based utilities to detect errors in GO-based annotations, arising from updates in GO itself that are not correctly propagated.[2]

GOChase offers four facilities:

1. **Tracking the history of redefinitions of any GO identification number** The Box shows the return from a query about GO identification number GO:0006489 in the Biological Process component of GO.

---

1 See http://swift.cmbi.ru.nl/gv/pdbreport/ and Hooft, R.W.W., Vriend, G., Sander, C. and Abola, E.E. (1996). Errors in protein structures. *Nature*, **381**, 272.

2 Park, Y.R., Park, C.H. and Kim, J.H. (2005). GOChase: correcting errors from Gene Ontology-based annotations for gene products. *Bioinformatics*, **21**, 829–831.

History of gene ontology ID GO:0006489, reported by GOChase

GOChase-HistoryResolver

Your input: GO:0006489

dolichyl diphosphate biosynthesis (GO:0006489): the formation from simpler components of dolichyl diphosphate, a diphosphorylated dolichol derivative.

GO:0019408: dolichol biosynthesis
GO:0006488: dolichol-linked oligosaccharide biosynthesis
GO:0046465: dolichyl diphosphate metabolism
GO:0006489: dolichyl diphosphate biosynthesis

Date	Action	GO History
Mar 01, 2001	Move to under	metabolism (GO:0008152)
	Move to under	biosynthesis (GO:0009058)
Oct 01, 2001	Move out from	metabolism (GO:0008152)
	Move to under	lipid metabolism (GO:0006629)
	Move to under	catabolism (GO:0009056)
	Move to under	protein metabolism (GO:0019538)
Aug 01, 2002	Move out from	biosynthesis (GO:0009058)
	Move to under	protein biosynthesis (GO:0006412)
Oct 01, 2002	Move out from	catabolism (GO:0009056)
	Move to under	biosynthesis (GO:0009058)
	New definition	GO:0006489 (dolichyl diphosphate biosynthesis)
	Term name change	dolichyl diphosphate biosynthesis (GO:0006489)
		changed from dolichyl-diphosphate biosynthesis (GO:0006489)
Jul 01, 2003	Move out from	protein biosynthesis (GO:0006412)
	Move out from	protein modification (GO:0006464)
	Move out from	protein metabolism (GO:0019538)
Aug 01, 2003	Move to under	protein biosynthesis (GO:0006412)
	Move to under	protein modification (GO:0006464)
	Move to under	protein metabolism (GO:0019538)
Jul 01, 2004	Move to under	metabolism (GO:0008152)

2. **Correction of obsolete terms** For any query term which has been merged into another term, or which has become obsolete for any other reason, the program returns the new term that should replace it.

3. GOChase will examine a file containg GO ID numbers, and **report required updates**.

4. Given a GO ID number, GOChase will **probe** a selected set of databases for items annotated with the term.

Database access

Most databases in molecular biology permit general, free-of-charge, public access to the data (see Box). In general, users can read the data, but almost never make changes. 'Reading' the data usually means seeing a presentation of the data through some program running in a browser. Many 'front ends' may exist for the same database, with individual appearances and different sets of links.

Some but not all databases permit users to download raw data in bulk. For this to be worthwhile, the data must be in a generally accessible format. To this end, some databases maintain a version in which each entry appears as plain text (called a *flat file*). This is not necessarily the most useful internal format but facilitates general data exchange. Other collections are maintained using widely available database management systems. These are easily distributable among installations running equivalent software. The relational database format is an example. XML, discussed later in this chapter, is another.

All databases must carefully impose controls on permission to modify their contents. Databases in molecular biology are generally maintained by specific institutions, or by limited partnerships of specific institutions. External users can submit information and suggest corrections or other changes, but not modify the database directly. To the extent that external specialists may be invited to curate data about particular topics, the databases will have to consider mechanisms of extending modification rights to these external curators.

Open and free access to articles in journals, and open and free access to the data the articles report, are related but separate issues.

Public access to scientific data

Scientists in the academic world who determine novel data, such as gene sequences or protein structures, are expected to deposit the data in publicly accessible databanks. To do this is, at least potentially, to sacrifice commercial rights, or the intellectual advantages of unshared knowledge in a competitive field of research. The commercial sector of research in molecular biology—prominently including but not limited to the pharmaceutical and biotechnology industries—generally regards as proprietary the results that its scientists generate.

Even within the academic world, this is not a new conflict. Early in the eighteenth century, Isaac Newton demanded access to data collected by the Astronomer Royal, John Flamsteed, in order to prepare a new edition of his

$\longrightarrow$

→

Public access to scientific data *(continued)*

Principia. Flamsteed refused, claiming ownership of the data despite its having been collected while he occupied an official government post.

Today, journals and granting agencies require deposition of data. Journals will not accept papers without confirmation of receipt from an appropriate databank. Although these rules now have general acceptance, their establishment was controversial.

Science made an exception to its mandatory deposition policy in publishing the Human Genome Draft Sequence by J.C. Venter and co-workers in 2001. For criticism of this waiver, see Powledge[*]. A similar waiver applied to the publication of the genome of one of the strains of rice, eliciting similar criticisms[†]. Conversely, *Science* did require deposition in publicly accessible databanks of the sequence of the strain of influenza virus active in the 1918–19 pandemic. R. Kurzweil and W. Joy criticized the *non-withholding* of this sequence, on the grounds that terrorists might use the information to recreate the virus and use it as a weapon[‡]

[*] Powledge, T.M. (2001). Changing the rules? EMBO *Reports*, **2**, 171–172.
[†] Petsko, G.A. (2002). Grain of truth. *Genome Biology* **3**, comment 1007.1–comment1007.2.
[‡] *The New York Times*, September 17, 2005.

Links

The utility of a database depends on the quality of its links as well as on its contents. Internal links allow navigation around the database itself. External links make connection to other databases, including but not limited to literature databases containing references.

Figure 3.2 shows the SWISS-PROT entry for crambin, a protein of unknown function found in the seeds of the Abyssinian kale *Crambe abyssinica*. The terms highlighted in blue contain links. These include:

- Relevant reference information, some specific to the entry (for instance, bibliographical information about papers reporting the sequence and structure) or relevant but not specific to the entry (for instance, information about the taxonomic classification of the source organism).

- Links to other databases, including InterPro, Gene3D, Pfam, PRINTS, PROSITE, ProDOM and BLOCKS.

- The *feature table*, indicating annotations of structural roles of different residues, including the assignments of secondary structure: helices and strands of sheet.

The actual sequence is a very small portion of the entry!

Another important type of link launches a calculation, to analyse selected data. Consider the retrieval of amino acid sequences from the UniProt database. Searching for serpins in *Caenorhabditis elegans*, insisting on proteins not labelled as hypothetical, retrieved 32 entries. Figure 3.3 shows the top of the screen, containing the first eight 'hits'. You can select all 32 hits, or any selection, by checking boxes at the left. You can then pass the chosen sequences directly to a multiple sequence alignment program or to BLAST, with a single click. It is not necessary to save the sequences, or to cut-and-paste them into a different window.

UniProtKB/Swiss-Prot entry P01542 [CRAM_CRAAB] C...   http://ca.expasy.org/uniprot/P01542

| 🏠 **ExPASy Home page** | **Site Map** | **Search ExPASy** | **Contact us** | **Swiss-Prot** |

Search | Swiss-Prot/TrEMBL | for | crambin | Go | Clear |

UniProtKB/Swiss-Prot entry P01542

| Printer-friendly view |
| Submit update |
| Quick BlastP search |
| Entry history |

[Entry info] [Name and origin] [References] [Comments] [Cross-references] [Keywords] [Features] [Sequence] [Tools]

Note: most headings are clickable, even if they don't appear as links. They link to the user manual or other documents.

Entry information

Entry name	**CRAM_CRAAB**
Primary accession number	**P01542**
Secondary accession numbers	None
Integrated into Swiss-Prot on	July 21, 1986
Sequence was last modified on	May 30, 2000 (Sequence version 2)
Annotations were last modified on	March 20, 2007 (Entry version 64)

Name and origin of the protein

Protein name	**Crambin**
Synonyms	None
Gene name	**Name: THI2**
From	Crambe abyssinica (Abyssinian [TaxID: crambe) (Abyssinian kale) 3721]
Taxonomy	Eukaryota; Viridiplantae; Streptophyta; Embryophyta; Tracheophyta; Spermatophyta; Magnoliophyta; eudicotyledons; core eudicotyledons; rosids; eurosids II; Brassicales; Brassicaceae; Crambe.

References

[1] PROTEIN SEQUENCE.
   DOI=10.1021/bi00522a013; PubMed=6895315 [NCBI, ExPASy, EBI, Israel, Japan]
   Teeter M.M., Mazer J.A., L'Italien J.J.;
   "Primary structure of the hydrophobic plant protein crambin.";
   Biochemistry 20:5437-5443(1981).

[2] X-RAY CRYSTALLOGRAPHY (1.5 ANGSTROMS), AND DISULFIDE BONDS.
   Hendrickson W.A., Teeter M.M.;
   "Structure of the hydrophobic protein crambin determined directly from the anomalous scattering of sulphur.";
   Nature 290:107-113(1981).

[3] X-RAY CRYSTALLOGRAPHY (1.05 ANGSTROMS).
   PubMed=8188676 [NCBI, ExPASy, EBI, Israel, Japan]
   Yamano A., Teeter M.M.;
   "Correlated disorder of the pure Pro22/Leu25 form of crambin at 150 K refined to 1.05-A resolution.";
   J. Biol. Chem. 269:13956-13965(1994).

[4] X-RAY CRYSTALLOGRAPHY (0.89 ANGSTROMS).
   DOI=10.1074/jbc.272.15.9597; PubMed=9092482 [NCBI, ExPASy, EBI, Israel, Japan]

Fig. 3.2

Yamano A., Heo N.-H., Teeter M.M.;
"Crystal structure of Ser-22/Ile-25 form crambin confirms solvent, side chain substate correlations.";
J. Biol. Chem. 272:9597-9600(1997).
[5] STRUCTURE BY NMR.
PubMed=3338468 [NCBI, ExPASy, EBI, Israel, Japan]
Lamerichs R.M.J.N., Berliner L.J., Boelens R., de Marco A., Llinas M., Kaptein R.;
"Secondary structure and hydrogen bonding of crambin in solution. A two-dimensional NMR study.";
Eur. J. Biochem. 171:307-312(1988).

Comments
- **FUNCTION**: The function of this hydrophobic plant seed protein is not known.
- **SUBCELLULAR LOCATION**: Secreted protein.
- **MISCELLANEOUS**: Two isoforms exists, a major form PL (shown here) and a minor form SI.
- **SIMILARITY**: Belongs to the plant thionin (TC 1.C.44) family [view classification].

Copyright
Copyrighted by the UniProt Consortium, see http://www.uniprot.org/terms. Distributed under the Creative Commons Attribution-NoDerivs License.

Cross-references

Sequence databases

| PIR | A01805; KECX. |

3D structure databases

PDB	1AB1; X-ray; @=1-46. [ExPASy / RCSB / EBI]
	1CBN; X-ray; @=1-46. [ExPASy / RCSB / EBI]
	1CCM; NMR; @=1-46. [ExPASy / RCSB / EBI]
	1CCN; NMR; @=1-46. [ExPASy / RCSB / EBI]
	1CNR; X-ray; @=1-46. [ExPASy / RCSB / EBI]
	1CRN; X-ray; @=1-46. [ExPASy / RCSB / EBI]
	1CXR; NMR; A=1-46. [ExPASy / RCSB / EBI]
	1EJG; X-ray; A=1-46. [ExPASy / RCSB / EBI]
	1JXT; X-ray; A=1-46. [ExPASy / RCSB / EBI]
	1JXU; X-ray; A=1-46. [ExPASy / RCSB / EBI]
	1JXW; X-ray; A=1-46. [ExPASy / RCSB / EBI]
	1JXX; X-ray; A=1-46. [ExPASy / RCSB / EBI]
	1JXY; X-ray; A=1-46. [ExPASy / RCSB / EBI]
	1YV8; NMR; A=1-46. [ExPASy / RCSB / EBI]
	1YVA; NMR; A=1-46. [ExPASy / RCSB / EBI]
	2EYA; NMR; A=1-46. [ExPASy / RCSB / EBI]
	2EYB; NMR; A=1-46. [ExPASy / RCSB / EBI]
	2EYC; NMR; A=1-46. [ExPASy / RCSB / EBI]
	2EYD; NMR; A=1-46. [ExPASy / RCSB / EBI]
	Detailed list of linked structures.
ModBase	P01542.

Family and domain databases

InterPro	IPR001010; Thionin.
	Graphical view of domain structure.
Gene3D	G3DSA:3.30.70.10; Thionin; 1.
Pfam	PF00321; Thionin; 1.
	Pfam graphical view of domain structure.
PRINTS	PR00287; THIONIN.
PROSITE	PS00271; THIONIN; 1.
ProDom	[Domain structure / List of seq. sharing at least 1 domain]
BLOCKS	P01542.

Fig. 3.2 (*See Overleaf*)

Database interoperability

How can we deal with questions that require appeal to multiple databases at once? There are two general approaches:

- Merge several databases into a single one with the combined contents of the contributors.

- Develop methods for intercommunication between databases, that allow dissection and distribution of queries, and recombination of responses.

Other
SWISS-3DIMAGE P01542.
LinkHub   P01542; -.
ProtoNet   P01542.
UniRef   View cluster of proteins with at least 50% / 90% / 100% identity.

Keywords
3D-structure; **Direct protein sequencing**; **Plant defense**.

Features

Feature table viewer

Key	From	To	Length	Description	FTId
CHAIN	1	46	46	Crambin.	PRO_0000221479
DISULFID	3	40			
DISULFID	4	32			
DISULFID	16	26			
VARIANT	22	22	1	P -> S (in isoform SI).	
VARIANT	25	25	1	L -> I (in isoform SI).	
STRAND	2	6	5		
HELIX	7	17	11		
TURN	18	20	3		
HELIX	23	30	8		
STRAND	36	38	3		

Sequence information
Length: **46 AA** [This is the length of the unprocessed precursor]   Molecular weight: **4736 Da** [This is the MW of the unprocessed precursor]   CRC64: **919E68AF159EF722** [This is a checksum on the sequence]

```
         10         20         30         40
TTCCPSIVAR SNFNVCRLPG TPEALCATYT GCIIIPGATC PGDYAN
```

         P01542 in FASTA format

View entry in original UniProtKB/Swiss-Prot format
View entry in raw text format (no links)
Report form for errors/updates in this UniProtKB/Swiss-Prot entry

BLAST   BLAST submission on ExPASy/SIB or at NCBI (USA)   Sequence analysis tools: ProtParam, ProtScale, Compute pI/Mw, PeptideMass, PeptideCutter, Dotlet (Java)

ScanProsite, MotifScan   Submit a homology modeling request to SWISS-MODEL

NPSA Sequence analysis tools

ExPASy Home page   **Site Map**   **Search ExPASy**   **Contact us**   **Swiss-Prot**
Hosted by ▐◀▌ CBR Canada   Mirror sites: Australia Brazil China Korea Switzerland

Fig. 3.2 SWISS-PROT entry for crambin. Reproduced by permission.

Historically, there were good reasons why databases kept a pretty sharp focus on a selected topic. Database projects reflected the interests and expertise of small groups of dedicated individuals. The data representation and organization flowed from the natural properties of the information. Moreover, in the early days, funding levels remained relatively small. With no earmarked categories of funding, databases had to compete with—and often were obliged to disguise (really not too strong a word) themselves as—research projects. This was another factor promoting specialization.

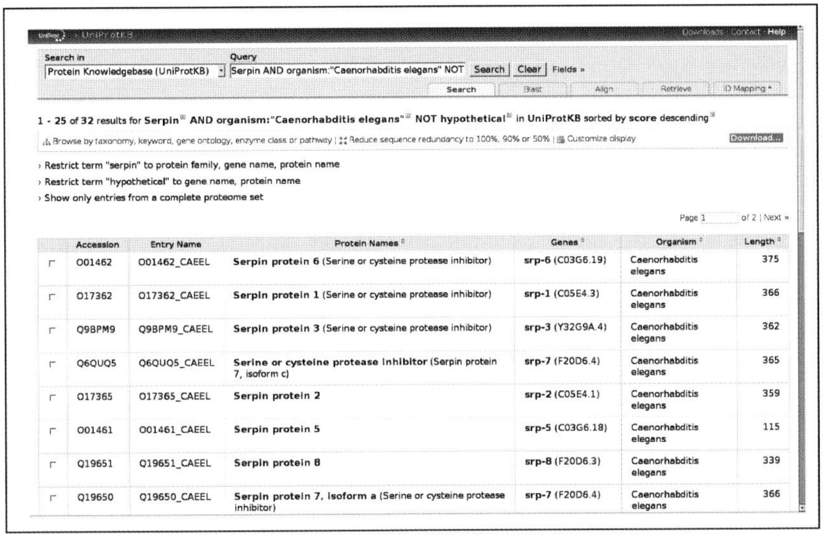

Fig. 3.3 Results of search in UniProt, for serpins from *C. elegans*, demanding no hypothetical molecules. Only the first eight 'hits' are shown. The software permits selection of any or all sequences by checking boxes at the left, launching a BLAST search or submission to a multiple sequence alignment program directly. [The UniProt Consortium (2007). The Universal Protein Resource (UniProt). *Nucleic Acids Research*, **35**, D193–D197. www.uniprot.org].

The overall growth, and consolidation of effort, in recent years, of genome sequencing and associated bioinformatics and databanking activities, has given a natural impetus to merging of information resources. Some, for instance UniProt, have assimilated a number of separate databases into 'the universal protein resource', as they describe themselves. ENTREZ, the databases maintained at the National Center for Biotechnology Information in the USA, close-couples 25 databanks, with facilities for simultaneous searching. Of course the common managerial superstructure facilitates the integration of these databanks.

The alternative approach is to leave individual databanks separate, and to layer a query system on top of them. This system would:

1. Dissect information retrieval requests into partial questions that would be farmed out to different databases, and then

2. Merge the responses into a coherent conclusion.

This is an active area of current research. Most people would not consider it a solved problem.

Common to all approaches is the goal of facile interaction among different databases. This involves both a careful specification of the ontology and schema of each database, so that the outside world can correctly interpret its contents, and mechanisms for handling queries within a framework free of commitment to any specific database organization. CORBA®—Common Object Request Broker Architecture—is such a system, which has many adherents in the bioinformatics community.

One obvious component of integration is consistency checking and reconciling of disagreements in data or annotation.

Data mining

The examples we have discussed of information retrieval from databases have involved the framing by a user of a specific set of criteria, and the return of relevant entries, selected according to the criteria. Consider alternatively a scientific field in the exploratory phase, where a large amount of data has become available, and the challenge is to understand what underlying patterns exist. The first step is to generate hypotheses about those patterns. Perhaps experts might guess what to look for. Testing and refining the experts' hypotheses then requires computer programs that probe information archives with sets of queries, seeking relationships and correlations in the data. This is the traditional way that science has made progress.

Now, the power of programs permits them to take the initiative in data exploration, to some extent. For example, programs can be adapted to assign data to classes on the basis of 'training' with examples, even if it is not possible explicitly to specify the rules that define the classes. It is even possible for a program to suggest hypotheses about patterns implicit in our data. This amounts to a partial automation of scientific research.

Machine learning is a computational approach to data analysis in which, through analysis of relevant information resources, computer programs achieve the ability to infer properties of data. Two complementary aspects are:

1. **knowledge discovery**—descriptions, or even *explanations*, of regularities in the data, and

2. **successful forecasting**, or predictive modelling.

Sophisticated numerical methods applied to data analysis include:

- **Statistical techniques**, including clustering and classification algorithms, and principal component analysis (identification of a small number of possibly composite parameters that account for most of the variation in a set of data). Hidden Markov Models are the most powerful methods for detecting homologous sequences.

- **Artificial neural networks** (see Chapter 6). Neural networks are the method of choice for prediction of secondary structures of proteins.

- **Support vector machines** are algorithms for classification that outperform neural networks in a number of applications.

Both artificial neural networks and support vector machines are data structures and algorithms for *supervised learning*. In supervised learning, the general framework of a program is constructed, but the details depend on choices of parameters. By exposing the program to a number of objects of known classification, and telling the program whether its prediction was correct or not (the supervision phase), the program can tune its parameters to give the optimal performance on unknowns.

The computer programs that implement some machine-learning techniques, including artificial neural networks, have complex internal structures. Large numbers of variable parameters give them versatility; optimization of the parameters by training can achieve impressive accuracy in classifying input data. A disappointing aspect is that it is usually impossible to 'pick apart' a trained network, to harvest any insights into the structure of the data, that are expressible in a simple, understandable, form.

R. Hamming wrote, 'The goal of computing is insight, not numbers'. Most people today want both.

Some statistical methods do provide such insight, at least by identifying which are the important variables, or combinations of variables.

An example of a program that achieves *unsupervised learning* is T. Kohonen's self-organizing map (SOM). A two-dimensional SOM is a neural network that clusters similar items of high-dimensional data and projects the relationships into a space of lower dimension, such as a plane (see Box). Reduction to two dimensions is most convenient because the results are easy to visualize; however, this is not a limitation of the SOM technique.

Application of self-organizing maps to analyse olfactory perception space

Odours are an important component of our perceptual environment, and play crucial roles in the sensory lives of many mammals. From the molecular point of view, a set of receptor protein molecules mediates recognition and distinction of odours. At the psychological level, humans can distinguish ~10 000 odours. However, it is difficult to classify odours: there is no natural distance measure, or 'metric', that would allow us to say, of the odours of banana, apple and strawberry for example, which pair is the most similar. Moreover, judgements of smell have a component that varies with cultural background, and may be influenced by drugs or disease. Loss of acuity of smell is an early symptom of Alzheimer's disease.

Ultimately, we should like to define mappings among (1) perceptual odour space, (2) the molecular structures of the active principles and (3) the combinatorial code by which differential binding of ~10 000 molecules to the panel of ~1000 odorant-receptor proteins creates sensation.

Madany Mamlouk, Chee-Ruiter, Hofmann and Bower have applied T. Kohonen's SOMs to classification of odours*. The Aldrich Flavor and Fragrance Catalog† contains data for 851 chemicals, which are assigned profiles according to 278 odour descriptors—a high-dimensional space indeed. The characterization of each chemical is not numerical but rather a record of which perceptual properties it possessed or lacked. Here is a small fragment:

Odorant	Fruity	Pineapple	Sweet	Apple	Coconut	Nutty
Hexyl butyrate	Yes	Yes	Yes	No	No	No
Methyl-2-methylbutyrate	Yes	No	Yes	Yes	No	No
6-Amyl-α-pyrone	No	No	Yes	No	Yes	Yes

Source: Madany Mamlouk, A. (2002). Quantifying olfactory perception. Diploma Thesis, University of Lübeck, Germany.

To each of the 851 chemicals corresponds a string of 278 bits. The Hamming distances between pairs of such profiles—the number of positions at which the two profiles disagree—is the most obvious way to create a dissimilarity matrix.

→

Typically mammals express ~1000 homologous odorant-receptor proteins.

164

Application of self-organizing maps to analyse olfactory perception space *(continued)*

Applied to this matrix, the statistical technique of multidimensional scaling reduced the space to 32 dimensions but not farther.

The SOM neural network classified and clustered the data and projected it into two dimensions (see Fig. 3.4). Not surprisingly, citrus fruits form a class. A less obvious example of odours considered similar are caramel and vanilla. Moreover, as the map is a projection from many dimensions, orange and refreshing are also neighbours.

Do the clusters reflect similarities of chemical structure? Flavour and fragrance chemists have tried very hard to determine predictive rules for odours, based on molecular shape and spectroscopic properties. Success has proved elusive. At the level of general chemical composition, Madany Mamlouk *et al.* mapped the

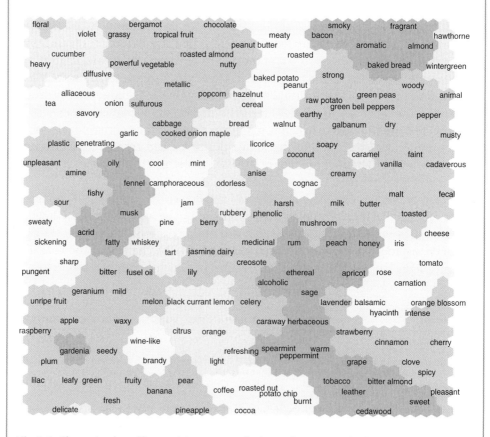

Fig 3.4 Clustering by self-organizing map technique of perceptual odorant space. The 851 chemicals cluster into 37 groups. Reprinted, with modification, from Madany Mamlouk, A., Chee-Ruiter, C., Hofmann, U.G. and Bower, J.M. (2003). Quantifying olfactory perception: mapping olfactory perception space by using multidimensional scaling and self-organizing maps. *Neurocomputing*, **52–54**, 591–597. Copyright 2003, with permission from Elsevier.

→ --

Application of self-organizing maps to analyse olfactory perception space *(continued)*

nitrogen- and sulphur-containing compounds from their data set onto the clusters and found that they segregate into separated groups.

[*] Madany Mamlouk, A., Chee-Ruiter, C., Hofmann, U.G. and Bower, J.M. (2003). Quantifying olfactory perception: mapping olfactory perception space by using multidimensional scaling and self-organizing maps. *Neurocomputing*, **52–54**, 591–597.
[†] Sigma Aldrich Chemicals Company, Milwaukee, WI, USA, 1996.

Programming languages and tools

A computer program is a set of orders that a computer will execute. At the moment of execution, the orders must be specified in a form that can activate the computer; that is, the orders must be in a form that corresponds to the computer's limited repertoire of basic operations. Human beings would like to specify the orders in a human language. This has led to the development of 'pidgin' languages that allow people to write computer programs in languages as close as possible to natural mathematical discourse, but followed by translation into the computer's operation set. FORTRAN was the first of these.

Programming languages differ from natural human languages in many respects, including a restricted horizon of possibility of expression, and very strict intolerance to error.

A similar intolerance to error affects the preparation and formatting of data to be read by computer programs. To serve as input to a program, data must be (1) presented according to specific rules—for example, terms restricted to a controlled vocabulary, and (2) properly formatted. There is a tension between user-friendliness and program-friendliness in the requirements.

Another distinction, which is not as sharp as it used to be, classifies programs into *systems programs* and *applications programs*. Applications programs are generally specific to one or more users; they solve a particular problem in a particular field, and they are active in a computer for limited times, after which they report an answer and disappear. In contrast, systems programs govern the overall workflow of the computer, are common to all users and are consistent with the use of the computer to solve a wide variety of problems (by means of individual applications programs). For instance, a program to superpose two or more protein structures would be an application program. The programs that create the general operating environment—for instance UNIX or Microsoft Windows—are systems programs.

Operating systems offer many specific facilities in addition to their overall 'house-keeping' functions. To create lists of orders invoking the facilities of the operating system is to write a program called a *script*.

The boundaries between systems and applications program are becoming fluid. All the features of the editor with which I am typing this paragraph are specific to the problem of accepting and editing text. However, many people use it, it was distributed with the operating system, and it remains active even when I am finished with this passage. Conversely, many programmers who put together large and powerful packages that address a variety of problems—for example, retrieval of genetic sequences from a databank—boast of having written 'program systems' (rather than applications programs, but not systems programs).

Traditional programming languages

Notable computer languages include FORTRAN, C and C++. A separate program called the *compiler* translates a program in these languages into the appropriate set of computer instructions. The maturity of compiler technology, together with the understanding of algorithms provided by computer scientists, and the experience and skill of the community of programmers, combine to make these languages most suitable for large-scale computations which push the available resources to the limit.

Another advantage of not writing in native machine language is code *portability:* the ability of one program, written in FORTRAN, C, or C++, to run on a large variety of platforms. It is true that each target machine language requires its own compiler. But writing a compiler needs be done only once per machine, and there is mature software which facilitates compiler construction. Then an entire literature of programs becomes executable.

Scripting languages

Many extremely useful tasks require only minimal computer resources. For instance, the translation of a gene sequence into an amino acid sequence requires only a straightforward table-look-up of each codon (see Chapter 1). For these, a simple program achieves adequate throughput: what is important is to save programmer time. The computer time required is often negligible.

Indeed, there has been a steady trend in the relative costs of hardware and software. The balance is tipping, steeply, in the direction of high costs of creating software relative to purchasing and maintaining hardware. Programming practice has reacted with tools and languages that streamline the effort required to write code that works correctly, even at some cost in efficiency of execution.

Several languages provide such facilities, including PERL, PYTHON and RUBY. At least in their initial versions they were *interpreted* languages. This means that the systems program that carried out the commands skipped the step of compilation to machine language, but simulated the stated operation on a line-by-line basis. In principle this makes for less efficient execution. In any case, it is a legitimate price to pay for the ease of writing the program and the sharp curtailment of the 'debugging' phase. Often the difference in execution time is unnoticeable.

Some languages can be run in either interpretive or compiled mode, for instance LISP. Demonstration by a new interpreted language such as PERL of significant advantages and popular appeal will elicit writing of a compiler, or at least a more efficient interpreter. (A superficially attractive but ultimately ineffective idea is to write a translation program that will convert the scripting language into a language which can be compiled. This will often not speed up execution significantly if the original interpreter calls upon programs written in the compiled language anyway.)

Program libraries specialized for molecular biology

Programmers usually construct new programs by combining well-established components. For instance, an algorithm may contain a step that requires sorting a list, or solving linear equations. Subprograms for these steps are widely available. All programs depend on standard libraries for input/output. Almost never does one write a program completely 'from scratch.'

In addition to standard libraries for numerical analysis and text processing, there are libraries specialized for molecular biology. Different libraries are associated with different programming languages. For example, BioPERL (http://www.bioperl.org) contains modules that implement common computational tasks in bioinformatics, written in PERL. Typical tasks in the repertoire include translation of nucleic acid sequences to protein sequences, or sequence alignments. Modules can be integrated smoothly into a new program.

Java — computing over the Web

The Java language has a syntax with many similarities to C and C++. Its operating environment is designed to address the following problem:

Suppose the creator of a web site wants to provide a program which users can run interactively from a browser. If the program is run on a computer at the web site, and if many users simultaneously avail themselves of the facilities, the hardware on which the web site is running will come under pressure. An example of this mode is the NCBI BLAST server, which in a typical month fields ~6.5 million enquiries, and runs them on a 300-CPU cluster.

An alternative is to ask each user to provide the computer power. Without leaving the web site, the browser will dynamically download programs (called *applets*). The programs will be run on the user's computer.

This, in turn, creates a security problem: the user must give the web site access to resources on the user's computer. A web site that can download executable code and gain access to the local files can do considerable damage, including crashing the computer, snooping around the file system to steal or damage confidential information, or carrying out unwanted invasive activity such as displaying unsolicited advertising material.

The basic idea of the way to protect the user is as follows: the downloaded Java program is not run directly by the user's operating system, but involves an intermediate agent. The user's system simulates an internal computer—called a virtual machine—which runs the Java program. (Each *actual* operating system requires its own Java virtual machine to provide the executable environment for programs written in Java. Automatic portability of Java programs is a concomitant.) The virtual machine carefully restricts the resources to which the Java program running under its auspices has access. The local virtual machine imposes the rules; the distant web site programmer must follow them.

Java is a *compiled* language. Although usually executed from a browser, Java programs can stand alone. In contrast, programs in JavaScript are *interpreted* by a browser.

Markup languages

. . . .
Algorithms +
Data
Structures =
Programs.
— N. Wirth
. . . .

Markup languages implement data structures, as essential a component of programs as executable instructions. Data structures are the organization of the information on which a program acts. Choice of the proper data structure is a crucial aspect of programming.

The term markup originally described editors' annotations to manuscripts, that control the appearance of the final published text without explicitly appearing in it. An example would be designation of certain words to appear in italics. Computer-typesetting systems include formatting commands: the UNIX facilities of the 'roff' family are an early example, and D. Knuth's TeX system is an 'all possible bells and whistles' development. HTML (hypertext markup language) is primarily a presentational markup language.

The utility of the close coupling of annotation with contents extends, beyond presentation markup, to organization of data in files. Such a structure provides an alternative to traditional *positional formatting*. Positional formatting is specifying how to interpret an item in a file through rigid rules specifying *where* the item appears. Typical examples of positional formatting are: 'the number of bases in the sequence appears in columns 10–16'. or 'Items, separated by white space, appear in the order: gene name, source organism, number of bases, sequence'. The markup approach achieves greater flexibility by surrounding each item with a local descriptor. The line:

<number of bases>5386</number of bases>

could appear *anywhere* in a file. A program, or a human reader, would recognize what the number 5386 signified. The syntax *<descriptor>*value*</descriptor>* is common to many markup languages, including HTML. The descriptor is called a *tag*. The material enclosed by the beginning and end of the tag is called the *element*. Standardization of the syntax simplifies the construction of the software to interpret it.

Tag-element combinations provide *self-describing* data. Moreover, the data description is *local*; that is, contiguous with individual data items. In contrast, the learning goals that appear at the beginnings of the chapters of this book are descriptions of contents that are *not* local to the sections to which they refer.

Flexibility of format comes at a price, most obviously in a rather cumbersome and bloated appearance of the files. Nor is adult supervision entirely unnecessary: the ontology of the data must specify acceptable ranges of values. Programs could not be asked to swallow:

<number of bases>Tuesday</number of bases>

Therefore, any file in a markup language requires a *schema*: a list of allowed element and attribute names, and allowed ranges of values. This permits validating a file for proper formatting and consistency. A Document Type Definition, itself written in a standardized language, specifies the schema. Note that <number of bases>Tuesday</number of bases> is valid *syntax* but invalid with respect to any reasonable *schema*.

There are many markup languages, specialized for different types of data. One of the most general is XML (extensible markup language) used in many databases and information-retrieval systems. XML assumes a tree-based, or hierarchical, structure of the material. Lower-level tags and elements can appear within higher-level ones.

An XML database of mammalian species might contain the following:

```
<mammals>
    <genus>Homo
        <species>sapiens common_name='human'</species>
        <species>neanderthalis common_name='neanderthal man'</species>
    </genus>
</mammals>
```

Note the three nested levels of tags: mammals, genus, species. The species elements include the common name as an *attribute*. In an alternative schema, the common name might be a separate tag within the species.

It would be more difficult to construct an XML database of information that is non-hierarchical. Consider a database of information about movies. It would be possible to define an XML schema in which the movie title was at a higher level in the hierarchy than the list of performers. Then it would be easy to probe the database with a movie title, and retrieve the cast. It would, in contrast, be more difficult to retrieve all the movies in which Peter Sellers acted. In an alternative schema, the performers could be at a higher level than the movies, making it easy to search for an actor or actress, but then it would be difficult to probe with a title and retrieve the cast. A relational database would be a more natural way to organize the data if one wanted to be able to query with *either* movie title or performer. However, facilities for such queries are not completely incompatible with XML. Even in a database structured hierarchically with an XML schema, it is possible to index it in different ways to support versatile approaches to retrieval, including non-hierarchical ones.

XML, unlike HTML, is not directly concerned with appearance or presentation. On the other hand, it is perfectly possible to write formatting programs that control the presentation of the contents of an XML file such as the mammal–genus–species example. Such a program could follow convention to display genus and species names in italics. Different programs could impose independent decisions about how to display common names. One program could display common names in boldface, another in plain Roman type.

In contrast, in an HTML file, the decision to display common names in boldface type would be *irrevocably* implemented by tags: neanderthal man. Moreover, it is impossible to make up novel tags for HTML format (without approval by an international commission). In other words, the schema of HTML has been fixed. This has the advantage of complete portability and the disadvantage of inflexibility.

Markup languages in general, and XML and HTML in particular, are becoming standard in database construction and distribution, for:

- **Archiving and curating data** XML provides a general and flexible structure compatible with organizing information from many different fields and applications.

Data validation—checking that the values of the elements are consistent with the schema—is straightforward. The results provide a format for data interchange, facilitating database interoperability.

- **Providing data to programs** Insertion of a parser between an XML data file and an application program can simplify the input phase of a calculation.

- **Ease of data extraction and presentation** Selection of data and formatting into an HTML file can be a natural and fluent mapping that facilitates conversion of data into a form that is both human-friendly and distributable over the Web. Other markup languages provide facilities for describing graphics. These are profoundly involved with *both* data structure and presentation.

Natural language processing

Biomedical research depends crucially on the quality of the data and annotations in databanks. Some annotations are generated automatically from the data; others are extracted by human readers from articles in the scientific literature. Extraction from the literature is a labour-intensive activity that will not be able to keep up with the increasing rate of published articles. Will it be possible for computers to take over this task? Unlike most input prepared for computers in strictly defined formats, the literature, aimed primarily at human–human communication, appears in a natural language, although of course many articles contain equations and tables. Much of the contemporary scientific literature is written in English.

Natural language refers to the verbal—oral and/or textual—forms of human–human communication. (Few people think it a realistic goal for computers to deal with the grunt-and-gesture communications especially common in certain cities.) **Natural language processing** by computer means: *at least* the analysis of a stream of spoken or written words that a human could interpret, and *at best* an appropriate reaction—acting on a command or providing a suitable response in the natural language.

Natural language processing has been a goal of computing for decades. Early hopes, during the 1950s and 1960s, for achieving automatic language translation were unfulfilled (see Box).

Automatic translation?

An apocryphal story about automatic translation concerns a program that converted English to Russian and back. From the input 'The spirit is willing but the flesh is weak' came back 'The vodka is fine but the meat is rotten'. (That this occurred in a computer system is an urban myth. The first traceable publication of this joke actually is in a newspaper over a century ago: *The Decatur*, Illinois, USA Herald,

$\longrightarrow$

→

Automatic translation? *(continued)*

20 January, 1903, p. 5.) A true computer translation howler was the rendering of '...la Cour de Justice considère la création d'un sixième poste d'avocat général' as '...the Court of Justice is considering the creation of a sixth general avocado station'*.

* Wheeler, P.J. and Lawson, V. (March, 1982). Computing ahead of the linguists. *Ambassador International*, 21–22.

A major difficulty in natural language processing is the ambiguity of words and even phrases. If a man married to a lawyer asks his wife to 'press his suit', does he want sartorial or forensic action? Human beings extract the meaning from such phrases by using contextual clues to resolve ambiguities. No reader would interpret the third line of Keats's 'Ode to a Nightingale':

My heart aches, and a drowsy numbness pains
My sense, as though of hemlock I had drunk,
Or emptied some dull opiate to the drains
One minute past, and Lethe-wards had sunk

as signifying that the poet had just poured his opiate down his kitchen sink. Keats was deliberately using archaic senses of words.

Computer programs have access neither to life experience nor to context-related clues—is the lawyer's husband holding a garment or a folder of papers? Therefore, ambiguities are difficult to circumvent. A simplification is to restrict the field of discourse. For instance, an early natural language processing system provided an interface to a database of information about the Boston Red Sox (a local sports team).

Natural language processing in bioinformatics has set as goals the extraction of information from the relevant scientific literature and databases. Applications of textual analysis of databases of biomedical literature include:

Identifying keywords and combinations of keywords

For instance, given a list of names of genes and a list of names of diseases, it should be possible to identify papers that contain references to combinations of genes and diseases, and to produce a list of gene-disease combinations based on co-occurrences in one or more papers. Several aspects of this problem make it more challenging than simple keyword search. Many biological entities have multiple synonyms. Conversely, many terms appear in several technical categories and are also used in colloquial senses. As an extreme example, consider: 'common cold', 'cold sore', 'cold shock protein', 'kept in a cold room', 'cold finger', 'paroxysmal cold haemoglobinuria', 'cold turkey', 'cold compresses', 'colicigonenic plasmid Cold-CA23' and 'Cold Spring Harbor Laboratory', all of which appear in technical articles. Even in the restricted sphere of the biomedical literature, disambiguation challenges abound.

Bioinformaticians have applied synonym dictionaries, syntactic analysers that parse sentences to assign parts of speech to words—cold is a noun in only two of the phrases in the preceding paragraph—and a variety of machine learning models that try to assemble context information by analysing the groups of terms that accompany each potential meaning of a word.

Knowledge extraction: protein–protein interactions

There are several approaches to deriving a list of interacting proteins, some experimental and some theoretical. One is to extract information automatically from the scientific literature. For instance, an article entitled: 'Calnuc binds to Alzheimer's β-amyloid precursor protein and affects its biogenesis' recently appeared in the *Journal of Neurochemistry*.[3] (Of course, it makes no difference whether the sentence is in the title or the body of the article.) A human reader could harvest for a protein interaction database the pair: calnuc and Alzheimer's β-amyloid precursor protein.

To extract this information automatically, it would help to have a list of protein names. The challenge is to write a program that can identify, within processed text, patterns of the form:

<protein name> ... <bind or some equivalent verb> ... <protein name>

The ... allows for various kinds of intervening material. For instance, another recent article has the title: 'Ubiquitin binds to and regulates a subset of SH3 domains'.[4] The program should recognize the verb 'binds' and ignore 'to and regulates'. Alternatively, if one were trying to deduce regulatory networks, then a different verb would form part of the pattern.

The title of this paper is relatively general with respect to the proteins that bind ubiquitin. A sentence in the abstract of that paper: 'the yeast endocytic protein Sla1, as well as the mammalian proteins CIN85 and amphiphysin, carry ubiquitin-binding SH3 domains', would, if properly parsed, permit extraction of three specific SH3 domains that bind ubiquitin.

One word within the ... that the pattern should not ignore is 'not': the sentence 'Auxin-binding protein 1 does not bind auxin within the endoplasmic reticulum despite this being the predominant subcellular location for this hormone receptor'[5] satisfies the pattern but is a false positive. It is not enough to check for the presence of 'not'; consider: 'The human anti-apoptotic proteins cIAP1 and cIAP2 bind but do not inhibit caspases.'[6]

A valuable tool for language processing is a *syntactic analyser*; a program that parses natural language text. It identifies nouns, verbs, etc. It specifies relationships among words; for instance, which noun or noun phrase is the subject of which verb (see Box).

3 Lin, P. *et al.* (2007). *Journal of Neurochemistry*, **100**, 1505–1514.
4 Stamenova, S.D. *et al.* (2007) *Molecular Cell*, **25**, 273–284.
5 Tian, H., Klämbt, D. and Jones, A.M. (1995). *Journal of Biological Chemistry*, **270**, 26962–26969.
6 Eckelman, B.P. and Salveson, G.S. (2006). *Journal of Biological Chemistry*, **281**, 3253–3260.

Syntactic analysis: parsing of English text

Applied to the sentence:

Mutations change the base sequence of DNA.

a syntactic analyser would return:

```
[ROOT
  [S
    [NP [NNS Mutations]
    [VP [VBP alter
      [NP
        [NP [DT the [JJ nucleotide] [NN sequence]]
        [PP [IN of
          [NP [NNP DNA]]]]]]
      [. .]]]
```

which could be displayed as a tree structure:

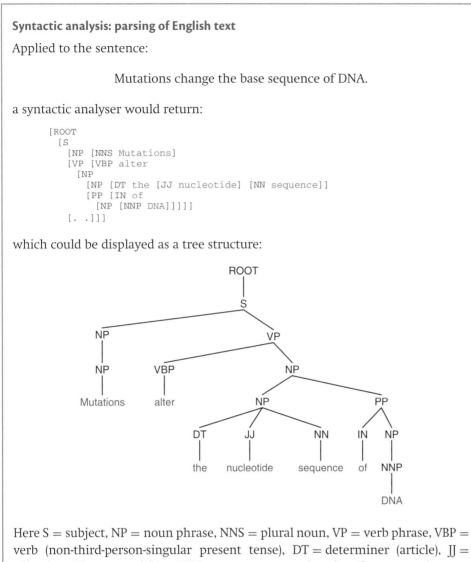

Here S = subject, NP = noun phrase, NNS = plural noun, VP = verb phrase, VBP = verb (non-third-person-singular present tense), DT = determiner (article), JJ = adjective, IN = preposition, NNP = proper noun, singular (for a complete set of definitions see: http://www.computing.dcu.ie/~acahill/tagset.html).

Automatic text-mining software does not work perfectly (see Problem 3.3). Some people believe that there are fundamental limitations that will never be overcome. Nevertheless, for extracting information from the literature to create the complete and high-quality annotations in the databases on which research crucially depends, what else is there? Annotation by human action is labour-intensive and error-prone. Databanks cannot augment their staffs by sufficient numbers of well-trained annotation experts to do the job. The only real alternative to successful natural language processing is distributed annotation: Authors of journal articles distil database annotations from their own results.

Applications of text mining

Computational analysis of texts of articles in the biomedical literature offers a series of challenges. The results have been successful in supporting the identification of relevant information for collection into databases, and even in generating useful suggestions for treatments of diseases.

One goal is to identify papers that contain targeted types of information. For example, the protein sequence databank SWISS-PROT stores information about protein function, and protein post-translational modifications. BIND is a database of protein–protein interactions. Identification of papers containing relevant information supports the work of the curators of these databases. Because the set of terms that might be relevant is so diffuse, simple keyword searches do not suffice. For instance, to identify post-translational modifications, a search for 'phosphorylation' would pick up papers describing not only the phosphorylation of proteins—which are relevant—but also the phosphorylation of glucose or fructose—which are not.

Selection of papers is already a useful result, even if a human curator must read them. The next step would be automatic extraction of the information from the paper. This is a challenge and focus of current research. CASP-like evaluations track progress.

The most basic task in computer analysis of an article is to identify the names that appear: names of genes, proteins, metabolites, drugs and diseases (or, more generally, phenotypes). Name identification depends heavily on dictionaries, but natural language processing contributes semantic information helpful in both recognizing names themselves and recognizing modifiers of names.

The next level is to identify associations and interactions. Examples include attempts to correlate genes or proteins with diseases, or, more generally, to assign function to genes or proteins. To extract associated terms, the minimal pattern must include at least two names and one interaction, the interaction specified by a word or a phrase. We have already seen examples of the combination:

<protein name> ... binds ... <protein name>

There are many other protein–protein interactions, such as

<protein name> ... regulates ... <protein name>.

More complex combinations are very important: a correlation between a set of interacting proteins and two or more apparently unrelated diseases can show a hidden relationship in the mechanism underlying the diseases.

Identification of references to individual genes and proteins

A basic task is to identify in a body of text the names of the relevant objects, such as genes and proteins. The difficulty is the wide range and ambiguity of names, and the use of common words as parts of gene names. The problem of identifying the species from which a gene arises is very difficult, as many genes have equivalent names in different mammalian species.

Chang, Schütze and Altman have developed a program called GAPSCORE that identifies gene and protein names within submitted text.[7] One might think that simply creating a dictionary and looking for its entries would suffice. Dictionaries are of course at the core of any identification procedure. But many gene names have other meanings. For instance, 'ring' (which stands for 'Really Interesting New Gene') can also appear in articles in the biomedical literature in the context of chemical structure: 'histidine ring', or in histology: 'signet-ring cell'. Even the common colloquial sense of the word, ring, as an item of jewellery, appears in the scientific literature in connection with metal-elicited contact dermatitis. Also, a dictionary should include a thesaurus, specifying, for example, that PTEN and MMAC1 are synonyms. (PTEN stands for Phosphatase and TEnsin Homolog. MMAC1 stands for Mutated in Multiple Advanced Cancers 1.)

GAPSCORE scores terms according to a statistical model based on:

- **Dictionary lookup** A table of known gene names.

- **Appearance** Many gene names have the form NAT1; other gene or protein names end in 'in'. Many enzyme names end in -ase.

- **Variations** The title of a recent paper included the phrase 'conformational changes of apo- and holocalmodulin'; the prefixes apo and holo are used only for proteins.

- **Syntax/context** The name of a protein or gene must be a noun. It is likely to be associated with certain other words. such as 'expression' or 'mutated', or even 'gene' itself. To utilize such word combinations as effectively as possible requires syntactic analysis.

- **Word morphology**: The derivation and formation of terms. For example, any short term that begins cdk . . . is likely to be a cyclin-dependent protein kinase.

Submitting to GAPSCORE only the title of a paper:[8] 'Neuroprotection by transforming growth factor-beta1 involves activation of nuclear factor-kappaB through phosphatidylinositol-3-OH kinase/Akt and mitogen-activated protein kinase-extracellular-signal regulated kinase signaling pathways.' returned the following:

	Gene or protein name	Quality (score)
1	Mitogen-activated protein kinase	Excellent (1.00)
2	Phosphatidylinositol-3-OH kinase	Excellent (1.00)
3	Transforming growth factor-beta1	Excellent (1.00)
4	Nuclear factor-kappaB	Good (0.60)
5	Activation	Poor (0.07)
6	Neuroprotection	Poor (0.04)

Note that the Greek letters β and κ are spelt out in full.

It is very important to recognize species differences in correlations between genes and drug activities. Tamoxifen, used widely against breast cancer, was originally developed as a birth-control pill. It is a fine contraceptive for rats but *promotes* ovulation in women.

7 Chang, J.T., Schütze, H. and Altman, R.B. (2004). GAPSCORE: finding gene and protein names one word at a time. *Bioinformatics*, **20**, 216–225.

8 Zhu, Y., Culmsee, C., Klumpp, S. and Krieglstein, J. (2004). *Neuroscience* **123**, 897–906. http://bionlp. stanford.edu/gapscore/

Identification of interactions

R. Hofmann and A. Valencia developed a system for datamining PubMed by natural language processing to identify genes, proteins and their interactions. Their results are available in a database named iHOP, or Information Hyperlinked Over Proteins (http://www.ihop-net.org/UniPub/iHOP/).

The basic item of iHOP data is a sentence from an abstract of an article appearing in PubMed. Appearances of any gene name, or synonym, in two different sentences provide a link. Currently the system contains 12 000 000 sentences, referring to 80 000 genes, from 1500 organisms.

An example of iHOP and its navigation facilities appears in Fig. 3.5.

Interaction networks and diseases

Some genetic diseases show simple Mendelian inheritance. They are the effect of a single gene. Other genetic diseases may arise from mutations of any of several genes. This suggests the involvement of a pathway or network, that has several vulnerable points. Still more complex are several diseases that appear to share a common protein interaction network.

Sam, Liu, Li, Friedman and Lussier applied data mining techniques based on natural language processing to identify relationships between diseases through sharing of components of a protein interaction network. They combined two sets of data:

1. **Relationships between proteins and diseases**—This data set associated 154 diseases with 1931 proteins.

2. **A protein interaction network**—A set of relationships among proteins, including binary interactions and direct complex formation. This data set included 20 317 interaction pairs from 1140 proteins.

For each pair of diseases, the associated proteins were checked for identity or interaction; that is, one protein might be associated with both diseases. Or, one protein associated with one disease might be paired in the interaction network with another protein associated with the other disease. Either contributes to a link between the two diseases.

A pair of diseases that share both common proteins and interactions is xeroderma pigmentosum and Cockayne syndrome (see Box, and Fig. 3.6). Both diseases involve defects in DNA repair systems. Of the proteins shared by both diseases, some mutations in XPB lead to the combined syndrome called the XP/CS complex, with both sets of symptoms. Mutations in ERCC6 are associated with Cockayne syndrome. The tumour antigen p53—which does not interact with any of the other proteins—is likely to be, not the primary lesion, but the subject of unrepaired damage leading to enhanced cancer susceptibility.

At the time of this work, the close connection between xeroderma pigmentosum and Cockayne syndrome, both effects of repair dysfunction, was already known. What was and still is not well understood is what, beyond the known functional defects, produces the differences in phenotype associated with the two diseases. In this respect, the mutations that produce the combined symptoms—the XP/CS complex—may be the ones that provide the clues.

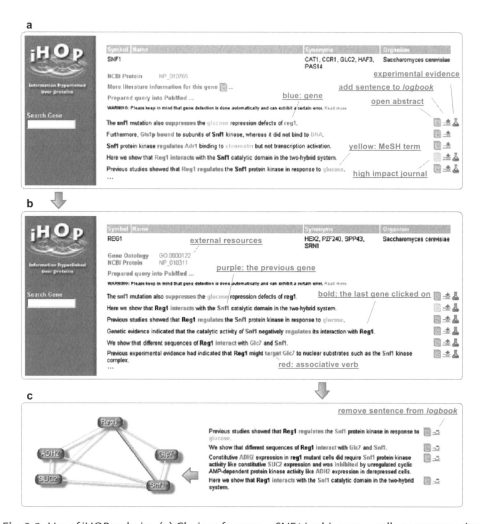

Fig. 3.5 Use of iHOP web site. (a) Choice of a gene — SNF1 in this case — calls up presentation of information about that gene and its interactions. Frame (a) contains five sentences describing SNF1 (many others are omitted). Each sentence describes an interaction and/or function of SNF1. On the right is a link to the full abstract in which the sentence appeared. The top sentence links the current gene of focus, SNF1, with another, reg1. Clicking on any mention of reg1 will shift the focus to it, opening another window. (b) The corresponding window for REG1. Note that the top sentences in this frame contain SNF1 as well as REG1. Information about the predecessor governs the ranking and ordering of the sentences in the new window. (c) In the course of navigation through iHOP, relationships can be collected into a logbook or gene model. The interaction network relating the selected proteins appears as a graph in a separate window. (From: Hoffmann, R. and Valencia, A. (2005). Implementing the iHOP concept for navigation of biomedical literature. *Bioinformatics*, **21** Suppl. 2, ii252–ii258). Reproduced by permission of Oxford University Press.

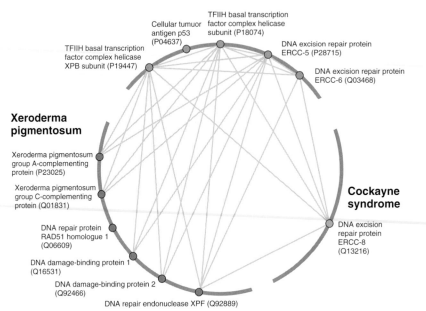

Fig. 3.6 Interaction network of proteins associated with xeroderma pigmentosum and Cockayne syndrome. Arc at lower left: proteins associated with xeroderma pigmentosum. Arc at lower right: proteins associated with Cockayne syndrome. Arc at top: proteins associated with both. Lines indicate interaction pairs. Note that there is only one direct interaction between a protein associated with xeroderma pigmentosum only and another associated with Cockayne syndrome only. (From: Sam, L., Liu, Y., Li, J., Friedman, C. and Lussier, Y.A. (2007). Discovery of protein interaction networks shared by diseases. *Pacific Symposium on Biocomputing*, **12**, 76–87.)

Xeroderma pigmentosum and Cockayne syndrome: two diseases of DNA repair

- *Xeroderma pigmentosum* is a genetic disorder involving a defect in the ability to repair damage caused by ultraviolet light. This leads most obviously to great sensitivity to sunlight, including the tendency, upon even short exposure, to sunburn, blisters and freckles. More devastating is the predisposition to development of malignant tumours, presumably arising from unrepaired damage to tumour suppressor genes.

- *Cockayne syndrome* shares with xeroderma pigmentosum a sensitivity to sunlight, but involves other symptoms including abnormal growth and development leading to short stature, retinal and other neurological degeneration, and premature ageing. Risk of skin cancer is normal, not elevated as in xeroderma pigmentosum.

$\longrightarrow$

> Xeroderma pigmentosum and Cockayne syndrome: two diseases of DNA repair *(continued)*
>
> ◆ A small number of cases of the *xeroderma pigmentosum/Cockayne complex* syndrome are known. These patients show symptoms of both diseases.
>
> | Disease | Genes in which mutations appear include: |
> | --- | --- |
> | Xeroderma pigmentosum | *XPA, XPB (ERCC3), XPC, XPD (ERCC2), XPE (DDB2), XPF (ERCC4), XPG (RAD2, ERCC5), XPV (POLH)* |
> | Cockayne syndrome | *CSB ERCC6 (CSB), ERCC8 (CSA)* |
> | XP/CS complex | *XPB (ERCC3), XPD (ERCC2), XPG (ERCC5)* |

Hypothesis generation

The literature implicitly contains many unsuspected relationships. D.R. Swanson read papers that connected magnesium and epilipsy, and papers that connected epilepsy and migraine headaches. Taken together, these suggested to him that there should be a relationship between magnesium and migrane. Subsequent research confirmed such a link. Swanson had other successes, including the suggestion that fish oil would benefit patients with Raynaud syndrome (a disorder affecting blood vessels of the extremities). Subsequent research confirmed this suggestion also.

Automation of Swanson's approach is an obvious goal; implementation of effective methods is not so easy.

P. Srinivasan and B. Libbus developed software to apply Swanson's approach. They searched for applications of turmeric, a spice from the rhizomes of the plant *Curcuma longa*, containing the active compound curcumin.[9] In Asia, turmeric is in common use in cooking. Its medicinal properties are also well known. It is an analgesic and an antiseptic, used for treatment of burns, stomach ulcers, skin diseases and the common cold.

A PubMed search for turmeric OR curcumin OR curcuma returned 1175 documents. From these, using natural language processing, Srinivasan and Libbus extracted terms with names of genes or genomes; enzymes; and amino acids, peptides or proteins, and ranked these terms by how frequently they turned up in the articles identified. They then reprobed PubMed using these results as search terms, and extracted from the results, and ranked, terms referring to Diseases or Syndromes; Neoplastic Processes (= terms referring to cancer).

The idea is that this procedure would link turmeric with certain diseases, *through* the medium of genes, genomes, enzymes or proteins (see Fig. 3.7). The results embody suggestions that turmeric would have some relationship with the diseases, and perhaps even be useful in their treatment.

---

9 Srinivasan, P and Libbus, B. (2004). Mining MEDLINE for implicit links between dietary substances and diseases. *Bioinformatics*, **20**, Supplement 1, i290–i296.

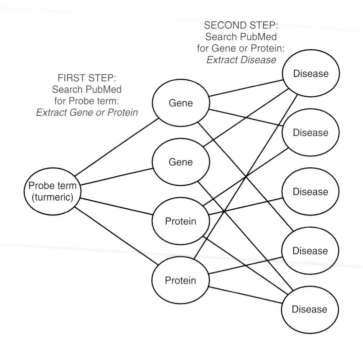

Fig. 3.7 The goal is to link a probe term, such as turmeric, with a set of diseases. In a two-stage procedure, first probe PubMed with the probe term, and recover names of genes, genomes, enzymes and proteins. These links from turmeric to molecules have a 'strength' proportional to the number of times the term appears in the articles that PubMed identifies as related to turmeric. A second stage probes PubMed again, separately, with each of the molecules identified in the first stage. This time analysis of the articles extracts names of diseases. Again the ranking of the molecule–disease link is proportional to the number of times the disease term appears in the articles that PubMed identified in the second stage. A connection between turmeric and a disease, through *two* strong links, is suggestive of a relationship between turmeric and the disease.

Srinivasan and Libbus discussed three diseases:

◆ Retinal diseases, including diabetic retinopathy, inflammation and glaucoma.

◆ Crohn disease.

◆ Disorders related to the spinal core, including inflammation following injury, and an autoimmune disease resembling multiple sclerosis.

A common feature of all these diseases is inflammation. A common set of proteins linking turmeric with the disease includes TNF-α, MAPK, NF-*k*B, COX-2, and other cytokines and interleukins. Knowing the molecules involved in the links between turmeric and diseases means that scientists can understand the mechanism by which turmeric might be expected to act. The result is not merely a correlation, but provides a rationale of the relevance of turmeric to the disease. This in turn usefully guides design of experiments to evaluate and elucidate the connection, and the clinical utility of the probe substance, turmeric.

Recommended reading

The transition to electronic publishing

Lesk, M. (2004). *Understanding Digital Libraries*, 2nd edn. Morgan Kaufmann, San Francisco. [Introduction to the transition from traditional libraries to information provision by computer.]

Berners-Lee, T. (with Mark Fischetti) (2000). *Weaving the Web: The Original Design and Ultimate Destiny of the World Wide Web*. Harper Business, New York.

Berners-Lee, T. and Hendler, J. (2001). Publishing on the semantic web. *Nature*, **410**, 1023–1024. [From the inventor of the Web.]

Gorder, P.F. (2006). Digital libraries come of age. *Computing in Science and Engineering*, **8(5)**, 6–10.

Butler, D. and Campbell, P. (2001). Future e-access to the primary literature, *Nature Web Debates*, 5 April 2001. http://www.nature.com/nature/debates/e-access/introduction.html [Introduction to a continuing discussion, about the web, on the web.]

Malakoff, D. (2003). Scientific publishing: opening the books on open access. *Science*, 302, 550–554. [Description of the journals published by the Public Library of Science.]

Spedding, V. (2003). Great data, but will it last? *Research Information*, Spring 2003, 16–20. [Problems of preservation of digital information. This journal has many articles of interest to scientists whose research depends on the quality and computer accessibility of data.]

Garson, L.R. (2004). Communicating original research in chemistry and related sciences. *Accounts in Chemical Research*. 37, 141–148.

Winograd, S. and Zare, R.N. (1995). 'Wired' science or whither the printed pages. *Science*, **269**, 615. [The authors, among the most distinguished of contemporary scientists, raise questions that are still not answered after more than 10 years.]

Discussions of developments in access and pricing in scientific journals

Van Orsdel, L.C. and Born, K. (2006). Journals in the time of Google. *Library Journal*, **131** (7), 39–44.

SQW Ltd (2004). *Costs and Business Models in Scientific Research Publishing*. The Wellcome Trust, London.

King, D.W. and Tenopir, C. (2004). Scholarly journal and digital database pricing: threat or opportunity? Available at: http://web.utk.edu/~tenopir/pub/chapters.html

King, D.W. (2007). The cost of journal publishing: a literature review and commentary. *Learned Publishing*, **20**, 85–106.

Dewatripont, M., Ginsburgh, V., Legros, P., Walckiers, A., Devroey, J.-P., Dujardin, M., Vadooren, F., Dubois, P., Foncel, J., Ivaldi, M. and Heusse, M.-D. (2006). *Study on the Economic and Technical Evolution of the Scientific Publication Markets in Europe*. European Commission, Directorate-General for Research, Brussels. [A recent, and thorough exposition of the issues, and some recommendations.]

Reviews of the achievements, challenges and resources for applications of natural language processing in bioinformatics

Krallinger, M. and Valencia, A. (2005). Text-mining and information-retrieval services for molecular biology. *Genome Biology*, **6**, 224.

Cohen, A.M. and Hersh, W.R. (2005). A survey of current work in biomedical text mining. *Briefings in Bioinformatics*, **6**, 57–71.

Shatkay, H. (2005). Hairpins in bookstacks: information retrieval from biomedical text. *Briefings in Bioinformatics*, **6**, 222–238.

Bosak, J. and Bray, T. (1999). XML and the second-generation web. *Scientific American* **280** (5), 89–93. [An introduction to XML, including descriptions of the problems that motivated its development, and the solutions it provides.]

Exercises, Problems and Weblems

Exercises

Exercise 3.1 Suppose a university library is considering purchase of electronic access to a very broad spectrum of scientific journals. Information about usage patterns of different journals are recordable at the *Publishers'* web sites. (a) How could a university librarian make use of this information to help make difficult choices in the face of budgetary pressure? (b) Is it to a publisher's financial advantage to make this information available to university librarians?

Exercise 3.2 Consider a database of audio clips (e.g. recordings of broadcasts of speeches by Winston Churchill). You want to create software to make this database searchable by computer, using *spoken* English sentences as search objects. (a) Suppose you had software that would perform accurate speech recognition; that is, conversion of speech to text. How could you use this to solve the problem. (b) How, in general terms, *might* you try to solve the problem *without* using speech → text conversion?

Exercise 3.3 According to the data on pages 150–1, which amino acids satisfy the compound query shown on page 152?

Exercise 3.4 For what types of data are the following markup languages specialized? (a) VRML, (b) CML, (c) BSML, (d) LOGML.

Exercise 3.5 Rewrite the XML fragment containing a database of mammals on page 169 converting the common name from an attribute to a tag.

Exercise 3.6 The sentence 'Time flies like an arrow' is ambiguous. (a) Explain three potential meanings of this sentence, treating Time as (1) a noun, (2) a verb, (3) an adjective (modifying flies). (b) Could you reject any of these meanings because they do not correctly obey rules of grammar? (c) Could you reject any of these meanings because they are not consistent with ordinary experience?

Exercise 3.7 Compose a search pattern to detect interacting proteins analogous to <protein name> ... <binds or some equivalent verb> ... <protein name> based on the noun association instead of the verb binds.

Exercise 3.8 A simple way to try to find enzyme names in text is to search for words that end in -ase. Think of 10 English words ending in -ase that are NOT names of enzymes. What is the longest such word ending in -ase that you can find? Of the words you suggest, would any of them be likely to appear in an article in the biomedical literature? One obvious word ending in -ase, not an enzyme name, that does appear in the biomedical literature is *disease*. (To turn this exercise into a Weblem, look for an online rhyming dictionary.)

Problems

Problem 3.1 From the data in Fig. 3.1, (a) for sales of 5800 subscriptions, what price per subscription would give a 5% profit over costs? (b) How many subscriptions would be required to make a 5% profit while charging half the cost of subscription found in (a)? Assume for simplicity that the cost of reproduction does *not* increase, but that the cost of distribution is linearly proportional to the number of copies distributed. (c) What would have to be charged for an electronic subscription (no paper version produced) to make a 5% profit if there are still 5800 subscribers. Assume for simplicity zero reproduction and distribution costs.

Problem 3.2 Consider the query: what are the three-letter codes of all amino acids that have volumes greater than 120 Å^2 with distal carboxyl or amide groups? Draw a Venn diagram showing, separately, the distributions of three-letter codes of side-chains, according to distal functional groups and volumes. Show the overlaps of the distributions and indicate the residues that satisfy the query.

Problem 3.3 If the text: 'under a spreading chestnut tree the village smithy stands' is submitted to a syntactic analyser, the result is:

```
[ROOT
  [S
    [PP [IN Under]
      [NP [DT a] [JJ spreading] [NN chestnut] [NN tree]]]
    [NP [DT the] [NN village]]
    [ADVP [RB smithy]]
    [VP [VBZ stands]]
```

or, in tree form:

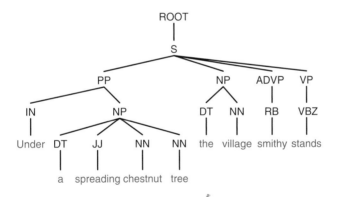

Here S = subject, PP = prepositional phrase, IN = preposition, NP = noun phrase, DT = determiner (article), JJ = adjective, NN = singular noun, ADVP = adverbial phrase, RB = adverb, VP = verb phrase, VBZ = verb (third person singular present tense)

(a) Parse this text yourself, and derive a graph comparable with the figure. (b) In what ways does your analysis differ from that of the computer program? (c) Suppose the syntactic analyser were coupled to a standard dictionary. Would it be reasonable to expect the program to recognize that village is an adjective and smithy a noun? (d) Can you think of other examples in English in which an adverb is constructed as a noun + suffix y? (e) What clues in the punctuation—or, more precisely, in the absence of punctuation—might permit the program to conclude that spreading modifies tree rather than chestnut?

Weblems

Weblem 3.1 From PubMed, determine how many papers related to breast cancer were published in 2007. Read the abstract of one of them. How long did it take you? Multiply the time required by the number of papers. If you wanted to be an expert on the subject of breast cancer, how much time would it take you to read all the abstracts of all the papers on the subject published in 2007?

Weblem 3.2 How many open access journals are there categorized as in (a) the Health Sciences, (b) in Biology and Life Sciences, and (c) in Agriculture and Food Sciences? (see the Directory of Open Access Journals: http://www.doaj.org/)

Weblem 3.3 When will the works of William Butler Yeats go out of copyright: (a) in the UK? (b) in the US?

Weblem 3.4 The home page of the RCSB includes links to query facilities that permit identification and retrieval of particular macromolecular structures from its contents, by specifying desired features. Find at least two other sites that permit identification of structures from the Protein Data Bank. There will be substantial overlap in the features of these query systems. What if any unique facilities does each have?

Weblem 3.5 (a) Give examples of three types of information about fruit flies that appear in *both* the EMBL Nucleotide Sequence Database and FlyBase. (b) Give examples of three types of information about fruit flies that appear in FlyBase but *not* in the EMBL Nucleotide Sequence Database.

Weblem 3.6 Find, in the Protein Data Bank, two structures of the same protein, determined by X-ray crystallography at different resolution: one at low resolution (≥ 2.9 Å) and the other at high resolution (≤ 1.9 Å). Remember: the lower the number specifying the resolution, the higher the potential quality of the result. Check the PDBREPORT entries for the two entries, in http://swift.cmbi.ru.nl/gv/pdbreport/ Compare the evaluations of the two structures by the software underlying the reports.

Weblem 3.7 Fig. 3.3 shows the results of a search for serpins in *C. elegans.* The search also returned several molecules called serpentine receptors. (a) What is the most obvious reason why these results are surprising? (b) Are serpentine receptors members of the serpin family?

Weblem 3.8 (a) Search PubMed for the article: Levitt, M. and Chothia, C. (1976). Structural patterns in globular proteins. *Nature*, **261**, 552–558. Can you get the full text of the article via PubMed? Print a copy of the page that provides the most detailed information about this article that is accessible to you from PubMed.
(b) Do a Google search for this article. (It is enough to enter search terms Levitt Chothia Structural Patterns). Try the first two or three links returned. Did any of those give you access to the full text of the article?
(c) Search Google Scholar for the same article. Under this entry, select Web Search. Can you easily find a site containing a complete text of the article? If so, print the page containing the link to the site that provides access to the full text, and highlight the link to the full-text site.

(d) Compare the ease of getting to the full text through PubMed, Google and Google scholar.

(e) What conclusions can you draw?

(f) Now suppose you knew or suspected that if the full text were available on the Web it would be in pdf format. Do a Google search for keywords:

Levitt Chothia structural patterns pdf

Do the results suggest that you should reconsider your conclusions?

Weblem 3.9 On 2 May 2006, US Senators John Cornyn (Texas) and Joseph Lieberman (Connecticut) proposed the Federal Research Public Access Act (S. 2695). (a) Summarize the salient provisions of the legislation (Senator Lieberman's web site is at: http://lieberman.senate.gov Select Education in the pull-down menu under Issues & Legislation, then enter S.2695 in the Bill number frame.) (b) What groups support this legislation, and what groups oppose it?

Weblem 3.10 Submit to GAPSCORE (http://bionlp.stanford.edu/gapscore/) (a) the title + abstract and (b) the entire text of the following paper: Zhu, Y., Culmsee, C., Klumpp, S. and Krieglstein, J. (2004). Neuroprotection by transforming growth factor-beta1 involves activation of nuclear factor-kappaB through phosphatidylinositol-3-OH kinase/Akt and mitogen-activated protein kinase-extracellular-signal regulated kinase signaling pathways. *Neuroscience*, **123**, 897–906. What gene and protein names does GAPSCORE find in the title + abstract that were not found in the title? (see page 175). What gene and protein names does GAPSCORE find in the full paper that were not found in the title + abstract?

Weblem 3.11 CbiT is a protein the function of which was originally hypothesized, on the basis of homology to proteins of known function, to be a precorrin-8w decarboxylase in the biosynthesis of vitamin B12. The crystal structure, solved in 2002, showed it to be a methyltransferase.[10] In many databases, the annotations were corrected. How many entries with incorrect annotation as a precorrin-8w decarboxylase can you still find?

---

10 Keller, J.P., Smith, P.M., Benach, J., Christendat, D., deTitta, G.T., and Hunt, J.F. (2002). The crystal structure of MT0146/CbiT suggests that the putative precorrin-8w decarboxylase is a methyltransferase. *Structure*, **10**, 1475–1487.

Archives and information retrieval

Learning goals

1. Understanding the general kinds of data describing the molecules and processes of life that are assembled in the databanks that support research and applications in biology, medicine, agriculture and technology.

2. Knowing the basic infrastructure of bioinformatics, in terms of the sites and responsibilities of the major archival projects.

3. Understanding the basic concepts of information retrieval, including how to frame queries.

4. Gaining facility with general search engines on the Web, and with specific sites for bioinformatics.

5. Knowing how to search for specific information about sequences, structures, metabolic pathways and relationships to disease, and how to launch analyses of the data retrieved.

This chapter introduces the specialized information retrieval skills that will allow you to make effective use of the databanks in molecular biology. The goal is to give you familiarity with basic operations. It will then be easy to improve and develop your technique, and to learn in more detail the facilities, and inter-relationships and interactions of resources available on the Web. Convenient sources of training materials include the tutorials embedded in many databanks. An example is the ENTREZ tutorial site at the US National Center for Biotechnology Information: http://www.ncbi.nlm.nih.gov/entrez/tutor.html

Database indexing and specification of search terms

An index is a set of pointers to information in a database. You have explored the entire World Wide Web with a general search engine such as Google, and have visited specialized databases in molecular biology. You proposed one or more search terms, and the retrieval program checked for them in its tables of indices. The model is that the database is composed of *entries*—discrete coherent parcels of information. The software identified entries with contents relevant to your interest. An example of the simplest paradigm is that you submit the term 'horse' and the program returns a list of entries that contain the term horse.

A full search of the Web would turn up information about many different aspects of horses—molecular biology, breeding, racing, poems about horses—most of which you don't want to see. For a successful search, it is not enough to mention what you *do* want—you must specialize your search to ensure that your desired responses don't get buried in a mass of extraneous rubbish. (Of course rubbish is merely whatever *other* people are interested in.)

To focus the results, information-retrieval programs accept multiple query terms or keywords. A search for 'horse liver alcohol dehydrogenase' would produce responses specialized to this enzyme. The search would, most probably, identify entries that contain all four keywords that you submitted: horse AND liver AND alcohol AND dehydrogenase. Poems about horses would not appear among its top hits (except in the unlikely event that a poem contained all four keywords).

It is possible to ask for other logical combinations of indexing terms. For instance, if a search engine didn't know about transatlantic spelling differences, it would be useful to be able to search for 'hemoglobin OR haemoglobin.' Note that a search for 'hemoglobin haemoglobin' would probably be interpreted as 'hemoglobin AND haemoglobin' which would pick up documents written by international committees or orthographically challenged expatriates. (Some web sites deliberately include both spellings.) Similar considerations apply to sulfur/sulphur, etc.

If you wanted to know about other dehydrogenases, you could ask for dehydrogenase NOT alcohol. This would retrieve entries that contain the term dehydrogenase but did NOT contain the word alcohol. You would find entries about lactate dehydrogenase, malate dehydrogenase, etc. You would miss references to review articles that compared alcohol dehydrogenases with other dehydrogenases, or alignments of the sequences of many dehydrogenases including alcohol dehydrogenase. You might regret missing these.

Many database search engines will allow complex logical expressions such as (haemoglobin OR hemoglobin) AND (dehydrogenase NOT alcohol). Construction of such expressions is an exercise in set theory. Drawing Venn diagrams helps in formulating the query. Although the logic of a search is independent of the software used to query a database, different programs demand different syntax to express the same conditions. For example, the query for dehydrogenase NOT alcohol might have to be entered as DEHYDROGENASE –ALCOHOL or DEHYDOGENASE !ALCOHOL.

Specialized databases, including those in molecular biology, impose a structure on the information, to separate different categories of information. This is essential. Active biomedical scientists include E(lisabetta) Coli, (John D.) Yeast, (Patrice) Rat, and a large number of Rabbits, as well as several Crystals and Blots. If you wanted to find papers published by these investigators, it would be naive to perform a general search of a molecular biology database with any of their surnames. Many databases provide separate indexing and searching of different categories of information. They permit searching for papers of which E. Coli is an AUTHOR.

Some categories, such as taxonomy, have controlled vocabularies. Often a query system presents the vocabulary terms to the user as choices from pull-down menus. The structure of taxonomic information is important in retrieval. To perform a search for 'globin NOT mammal', and pick out the relatively few entries about non-mammalian globins rather than the very many entries about globins, including human haemoglobins, that do not explicitly mention the term mammal, requires an information retrieval system that 'understands' the taxonomic hierarchy. Controlled vocabularies—limited, explicit and carefully defined sets of terms, known as *ontologies*—are also important for distributing queries among several databases.

A technical problem that frequently creates difficulty is how to enter terms containing non-standard characters such as accent marks or umlauts, cedillas, Greek letters and, as already mentioned, differences between US and British spelling. A specialized database such as NCBI's ENTREZ can handle the US–British spelling differences with a synonym dictionary. Programs that index the entire Web usually do not. Ignore the accent marks and hope for the best.

Follow-up questions

When searching in databases, it is rare that you will find exactly what you want on the first round of probing. Usually you have to modify the query, on the basis of the results initially returned. Most information-retrieval software permits consecutive, cumulative searches, with altered sets of search terms and/or logical relationships. Conversely, once you find what you were looking for, you will often want to extend your search to find related material. If you find a gene sequence, you might want to know about homologous genes in other organisms. Or whether a three-dimensional structure of the corresponding protein is available. Or you might want to read papers published about the sequence.

For these subsidiary queries, you need links between entries in the same or different databases. This is an example of the question of how one 'browses' in electronic libraries—a difficult problem, the subject of current research.

Suppose that you are interested in a particular gene. To find homologous genes you would like links to other items in the *same* database (a database of gene sequences). To find structures, or bibliographical references, related to that gene, you would like links *between* different databases (from the database of gene sequences to a database of three-dimensional structures, or to a bibliographical database). As the number of databases, and the variety of their contents, grows, intercommunication among them has become a high-priority goal. Indeed, the interactivity of the databases in molecular biology is growing more and more effective, so that these operations are fairly easy now—formerly one had to conduct separate searches on isolated databases. Most entries in molecular biology databases contain large numbers of embedded links. This is a generalization of the original model of a database as a closed set of independent entries that can be selected only by their indexed contents.

Database construction in bioinformatics involves activities that can be classified, to some extent, into **archiving**—with the major goals of conservation and curation of facts—and **interpreting and annotating**—the compilation of biological information in a form most useful to support research. Many archival databases specialize in different kinds of data—nucleic acid sequences, or protein sequences, or structures—for reasons in part historical and in part because of the different curatorial skills required. In many cases, archival and interpretative projects are carried out at the same institution and even by the same people. However, interpretative databases, independent of any archival project, are free to combine information from any available sources. Practical laboratory experience and expert knowledge of the experimental techniques used to generate the data are essential for curating an archival database, but are only extremely desirable for an interpretative database.

Two aspects of the recent development of bioinformatics databases stand out. One is the appearance of many projects that recombine the archived data in different ways. The other is the combination of many individual databases into larger and larger conglomerates. These processes overlap and often happen together. It is a question of whether the emphasis is on changes in the presentation of the data, or on changes in database contents. Most database unifications are outgrowths of prior collaborations, with varying degrees of intimacy in the result.

One must learn to think of the Web as a very high-dimensional space.

Embedded links have become a major component of database entries.

Analysis and processing of retrieved data

Sometimes as a result of a search you will want to launch a program, using the results retrieved as input. For instance, if you identify a protein sequence of interest, you might want to perform a PSI-BLAST search. This is somewhat different from a strictly keyword-based database entry retrieval problem. Formerly you would have to run one job to search for your data, store the results of your search and then run a separate, second, job, feeding the retrieved sequence to the application program by hand. However, like searches in multiple databases, several information-retrieval systems in molecular biology provide facilities for initiating such calculations. This makes for very much improved fluency in your sessions at the computer.

The archives

Although our knowledge of biological data is very far from complete, it is nevertheless of impressive size, and growing extremely rapidly. Many scientists are working to generate the data, and to carry out research projects analysing the results. There is a smooth and copious flow of results from the laboratory bench to databanking organizations; for archiving, curation and distribution to the research laboratory and the clinic.

Archiving of bioinformatics data was originally carried out by individual research groups motivated by an interest in the associated science. As the requirements for equipment and personnel grew—and the nature of the skills required changed, to include much more emphasis on computer science—national and, in most cases, international organizations have taken on the responsibility. To match the high volumes of data production, these projects have become very large-scale indeed. Anyone who has followed the entire history of the field cannot help being impressed by the replacement of tiny, low-profile and ill-funded projects carried out by a few dedicated individuals, to a multinational heavy industry subject to hostile takeovers and the scientific equivalent of leveraged buyouts.

Primary data collections related to biological macromolecules

- Nucleic acid sequences, including whole-genome projects
- Amino acid sequences of proteins
- Protein and nucleic acid structures
- Small-molecule crystal structures
- Protein functions
- Expression patterns of genes
- Networks: of metabolic pathways, of gene and protein interactions, and of control cascades
- Publications

Nucleic acid sequence databases

The world-wide nucleic acid sequence archive is a triple partnership of the National Center for Biotechnology Information (USA); the EMBL Nucleotide Database, or EMBLBank (European Bioinformatics Institute, UK); and the DNA Data Bank of Japan (National Institute of Genetics, Japan). These projects curate, archive and distribute

DNA and RNA sequences collected from genome projects, scientific publications and patent applications. The groups exchange data daily. As a result, the raw data are identical. However, the format in which they are presented, and the nature of the annotation, vary among these databanks. To ensure that these fundamental data are freely available, scientific journals require deposition of new nucleotide sequences, as a condition for publication of an article. Similar conditions apply to nucleic acid and protein structures.

The nucleic acid sequence databases, as distributed, are collections of entries. Each entry has the form of a text file containing data and annotations for a single contiguous sequence. Many entries are assembled from several published papers reporting overlapping fragments of a complete sequence.

Entries have a life history. Because of the desire on the part of the user community for rapid access to data, new entries are made available before completion of annotation and checking. Entries mature through the classes:

$$\text{Unannotated} \rightarrow \text{Preliminary} \rightarrow \text{Unreviewed} \rightarrow \text{Standard}$$

Rarely, an entry 'dies'—a few have been removed when they are determined to be erroneous.

A sample DNA sequence entry from the EMBL Nucleotide Database, including annotations as well as sequence data, is the *ATP7A* gene from the aardvark (see Box.) It encodes a protein involved in regulating copper levels. Mutations in the human homologue are implicated in Menkes syndrome, a progressive neurodegenerative disorder of copper metabolism.

The EMBL Nucleotide Database entry for *ATP7A* from the aardvark

```
ID    AAG47427; SV 1; linear; genomic DNA; STD; MAM; 675 BP.
XX
PA    AY011392.1
XX
DE    Orycteropus afer (aardvark) ATP7A
XX
OS    Orycteropus afer (aardvark)
OC    Eukaryota; Metazoa; Chordata; Craniata; Vertebrata; Euteleostomi; Mammalia;
OC    Eutheria; Afrotheria; Tubulidentata; Orycteropodidae; Orycteropus.
OX    NCBI_TaxID=9818;
XX
FH    Key             Location/Qualifiers
FH
FT    source          1..675
FT                    /organism="Orycteropus afer"
FT                    /mol_type="genomic DNA"
FT    CDS             AY011392.1:<1..>675
FT                    /codon_start=1
FT                    /gene="ATP7A"
FT                    /product="ATP7A"
FT                    /db_xref="GOA:Q9BFP6"
FT                    /db_xref="HSSP:Q04656"
FT                    /db_xref="InterPro:IPR001757"
FT                    /db_xref="InterPro:IPR006121"
FT                    /db_xref="UniProtKB/TrEMBL:Q9BFP6"
```

$\longrightarrow$

```
                The EMBL Nucleotide Database entry for ATP7A from the aardvark (continued)
FT                              /protein_id="AAG47427.1"
FT                              /translation="IVYQPHLITVEEIKKQIEAVGFPAFIKKQPKYLTLGAIDIERLKN
FT                              TSARSSEGSLQKSPSYTNDSTATFIIDGMHCKSCVSNIESALSTLQYVSSIAISLENRS
FT                              AIVKYNASSVTPETLRKAIEAVSPGQYTVSIISDVESIPNSPFSSSHQKIPLNIVSQPL
FT                              TQETVINISGMTCNSCVQSIEGVISKKAGVKSVQVSLADSSGVVEYDPLLTSPETLREE
FT                              IEN"
XX
SQ     Sequence 675 BP; 233 A; 136 C; 124 G; 182 T; 0 other; 264016655 CRC32;
       attgtttatc agcctcatct tatcacagta gaggaaataa aaaagcagat tgaagctgtg          60
       ggttttccag cattcatcaa aaaacagccc aagtacctta cattgggagc tattgacata         120
       gaacgtctaa agaacacatc tgccagatcc tcagaaggat cactgcaaaa gagtccatca         180
       tataccaatg attcaacagc cacttttatc atagatggca tgcattgtaa atcatgtgtg         240
       tcaaatattg aaagtgcttt atctacactc caatatgtaa gcagcatagc aatttcttta         300
       gagaataggt ctgccattgt aaaatataat gcaagctcag tcactccaga aaccctgaga         360
       aaggcaatag aggcagtatc accagggcaa tatactgtta gtattataag tgatgttgag         420
       agtatcccaa actctccttt tagctcatct catcaaaaaa tcccctttgaa catagtgagc         480
       cagcctctga ctcaagaaac tgtaataaac atcagtggca tgacttgtaa ttcttgtgta         540
       cagtctattg agggtgtcat atcaaaaaag gcaggtgtaa aatccgtaca agtctccctt         600
       gcagatagca gtggagttgt tgaatatgat cctctactaa cctctccaga aaccttgaga         660
       gaagaaatag aaaac                                                          675
//
```

A **feature table** (lines beginning FT) is a component of the annotation of an entry that reports properties of specific regions, for instance coding sequences (CDS). The aardvark *ATP7A* gene contains only one exon. Because feature tables are designed to be readable by computer programs—for example, to extract the amino acid sequence (see Exercise 4.4)—they have a more carefully controlled format and a more restricted vocabulary.

The feature table may indicate regions that:

◆ perform or affect function

◆ interact with other molecules

◆ affect replication

◆ are involved in recombination

◆ are a repeated unit

◆ have secondary or tertiary structure

◆ are revised or corrected

Genome databases and genome browsers

The general nucleic acid databases focus on collecting individual sequences. Associated with many full-genome sequences are Genome Browsers—databases bringing together all molecular information available about a particular species.

Ensembl

Ensembl (http://www.ensembl.org) is intended to be the universal information source for the human and other genomes. A goal is to collect and annotate all available information about genomic DNA sequences, link it to the master genome sequence and make it accessible to the many scientists who will approach the data with many different points of view and different requirements. To this end, in addition to collecting and organizing the information, very serious effort has gone into developing computational infrastructure, including establishment of suitable conventions of nomenclature. It is not trivial to devise a scheme for maintaining stable identifiers in the face of data that will be undergoing not only growth but revision. The most visible result of these efforts is the web site, very rich in facilities both for general browsing and for focusing in on details.

Ensembl is a joint project of the European Bioinformatics Institute and The Sanger Centre. However, Ensembl is organized as an open project, encouraging outside contributions. All but the most naive of readers must recognize the great demands that this will place on quality control procedures.

Data collected in Ensembl include genes, SNPs, repeats and homologies. Genes may be either known experimentally, or deduced from the sequence. Because the experimental support for annotation of the human genome is so variable, Ensembl records and presents the evidence for identification and annotation of every gene. Very extensive linking to other databases containing related information, such as Online Mendelian Inheritance in Man (OMIM^{TM}), or expression databases, extend the accessible information.

Ensembl and other genome browsers are structured around the sequences themselves. To focus on a desired region, users have available several avenues of selective entry into the system:

- Browsing—starting at the chromosome level then zooming in
- BLAST searches on a sequence or fragment
- Gene name
- Relationship to diseases, via OMIM
- Ensembl ID if the user knows it
- General text search

A text search in the Ensembl human genome browser for *BRCA1* produced the page displayed, showing the region around the *BRCA1* locus (see Plate III). The upper frame shows a megabase, mapped to the q21.2 and q21.31 bands of chromosome 17. It reports markers, and assigned genes. The bottom frame shows a more detailed view. Note the control panels between the two frames that permit navigation and 'zooming'. The bottom frame shows a 0.1 Mb region, reporting many more details, including the detailed structure of the *BRCA1* gene, and the SNPs observed.

Protein sequence databases

In 2002, three protein sequence databases—The Protein Information Resource (PIR), at the National Biomedical Research Foundation of the Georgetown University Medical Center in Washington, DC, USA; and SWISS-PROT and TrEMBL, from the Swiss Institute of Bioinformatics in Geneva and the European Bioinformatics Institute in Hinxton, UK—coordinated their efforts, to form the UniProt consortium. The partners in this enterprise share the database but continue to offer separate information-retrieval tools for access.

The PIR grew out of the very first sequence database, developed by Margaret O. Dayhoff—the pioneer of the field of bioinformatics. SWISS-PROT was developed at the Swiss Institute of Bioinformatics. TrEMBL contains the translations of genes identified within DNA sequences in the EMBL Nucleotide Database. TrEMBL entries are regarded as preliminary, and are converted—after curation and extended annotation—to mature SWISS-PROT entries.

Today, almost all amino acid sequence information arises from translation of gene sequences. However, even the amino acid sequence of a protein is *not* in general inferrable with confidence from the gene sequence. The main reason, in eukaryotes, is ambiguity in splicing. In addition, information about ligands, disulphide bridges, subunit associations, post-translational modifications, effects of mRNA editing, etc. is not available from DNA sequences. For instance, from genetic information alone one would not know that human insulin is a dimer linked by disulphide bridges. Protein sequence databanks collect this additional information from the literature and provide suitable annotations.

From UniProt, the entry for the amino acid sequence of the protein bovine pancreatic trypsin inhibitor, in SWISS-PROT format, is shown in the Box. Note that the sequence itself occupies only a relatively small amount of space in the entry.

Databases associated with SWISS-PROT

Two related databases closely associated with SWISS-PROT are the ENZYME DB, and PROSITE, a set of motifs.

The ENZYME DB stores the following information about enzymes:

- EC Number: a numerical identifier assigned by the Enzyme Commission (authorized by the International Union of Biochemistry and Molecular Biology; see http://www.chem.qmw.ac.uk/iubmb/enzyme/)

- Official name

- Alternative names, if any

- Catalytic activity

- Cofactors, if any

- General comments

- Links to SWISSPROT and other data banks

- Links to other enzymes with related activities

- Links to diseases associated with deficiency in the enzyme, if any known

One link, conspicuous by its absence, is to the Gene Ontology consortium classification of this enzyme (see page 207 and Weblem 4.23).

Figure 4.1 shows the ENZYME DB entry for peptidylglycine monooxygenase.

PROSITE contains common patterns of residues of sets of proteins. Such a pattern (or motif, or signature, or fingerprint, or template) is common to related proteins,

Amino acid sequence entry for bovine pancreatic trypsin inhibitor

05/17/04   #1

NiceProt View of Swiss-Prot: P00974

Entry information

Entry name	**BPT1_BOVIN**
Primary accession number	**P00974**
Secondary accession numbers	None
Entered in Swiss-Prot in	Release 01, July 1986
Sequence was last modified in	Release 10, March 1989
Annotations were last modified in	Release 44, June 2004

Name and origin of the protein

Protein name	**Pancreatic trypsin inhibitor [Precursor]**
Synonyms	**Basic protease inhibitor** **BPI** **BPTI** **Aprotinin**
Gene name	None
From	Bos taurus (Bovine) [TaxID: 9913]
Taxonomy	Eukaryota; Metazoa; Chordata; Craniata; Vertebrata; Euteleostomi; Mammalia; Eutheria; Cetartiodactyla; Ruminantia; Pecora; Bovidae; Bovinae; Bos.

References

[1] SEQUENCE FROM NUCLEIC ACID.
MEDLINE=87283904; PubMed=2441071; [NCBI, ExPASy, EBI, Israel, Japan]
Creighton T.E., Charles I.G.;
"Sequences of the genes and polypeptide precursors for two bovine protease inhibitors.";
J. Mol. Biol. 194:11-22(1987).
REFERENCES 2-13 DELETED

Comments

- **FUNCTION**: Inhibits trypsin, kallikrein, chymotrypsin, and plasmin.
- **SUBCELLULAR LOCATION**: Secreted.
- **PHARMACEUTICAL**: Available under the name Trasylol (Mile). Used for inhibiting coagulation so as to reduce blood loss during bypass surgery.
- **SIMILARITY**: Contains 1 BPTI/Kunitz inhibitor domain.
- **DATABASE**: NAME=Trasylol; NOTE=Clinical information on Trasylol; WWW="http://www.trasylol.com/".

Copyright

This Swiss-Prot entry is copyright. It is produced through a collaboration between the Swiss Institute of Bioinformatics and the EMBL outstation - the European Bioinformatics Institute.

Cross-references

EMBL	M20934; AAD13685.1; -.	[EMBL / GenBank / DDBJ] [CoDingSequence]
	5 ADDITIONAL CROSS-REFERENCES DELETED	
PIR	S00277; TIBO.	
PDB	1K09; 10-JUL-02.	[ExPASy / RCSB / EBI]
	46 ADDITIONAL STRUCTURES DELETED	
	Detailed list of linked structures.	
InterPro	IPR002223; Kunitz_BPTI.	
	Graphical view of domain structure.	
Pfam	PF00014; Kunitz_BPTI; 1.	

→

→ ───

Amino acid sequence entry for bovine pancreatic trypsin inhibitor *(continued)*

05/17/04   #2

	Pfam graphical view of domain structure.
PRINTS	PR00759; BASICPTASE.
ProDom	PD000222; Kunitz_BPTI; 1. [Domain structure / List of seq. sharing at least 1 domain]
SMART	SM00131; KU; 1.
PROSITE	PS00280; BPTI_KUNITZ_1; 1. PS50279; BPTI_KUNITZ_2; 1. PROSITE graphical view of domain structure.
HOVERGEN	[Family / Alignment / Tree]
BLOCKS	P00974.
ProtoNet	P00974.
ProtoMap	P00974.
PRESAGE	P00974.
DIP	P00974.
ModBase	P00974.
SMR	P00974; 6A778A4AD763FB19.
SWISS-2DPAGE	Get region on 2D PAGE.
UniRef	View cluster of proteins with at least 50% / 90% identity.

Keywords

Serine protease inhibitor; **Signal**; **Pharmaceutical**; **3D-structure**.

Features

Feature table viewer   Feature aligner

Key	From	To	Length	Description
SIGNAL	1	21	21	*Potential*.
PROPEP	22	35	14	
CHAIN	36	93	58	Pancreatic trypsin inhibitor.
PROPEP	94	100	7	
DOMAIN	40	90	51	BPTI/Kunitz inhibitor.
SITE	50	51	2	Reactive bond for trypsin.
DISULFID	40	90		
DISULFID	49	73		
DISULFID	65	86		
HELIX	38	41	4	
STRAND	53	59	7	
TURN	60	63	4	
STRAND	64	70	7	
STRAND	80	80	1	
HELIX	83	90	8	

Sequence information

Length: **100 AA** [This is the length   Molecular weight: **10903 Da** [This is the   CRC64: **6A778A4AD763FB19** [This
of the unprocessed precursor]   MW of the unprocessed precursor]   is a checksum on the sequence]

```
        10         20         30         40         50         60
        |          |          |          |          |          |
MKMSRLCLSV ALLVLLGTLA ASTPGCDTSN QAKAQRPDFC LEPPYTGPCK ARIIRYFYNA

        70         80         90        100
        |          |          |          |
KAGLCQTFVY GGCRAKRNNF KSAEDCMRTC GGAIGPWENL
```

P00974 in FASTA format

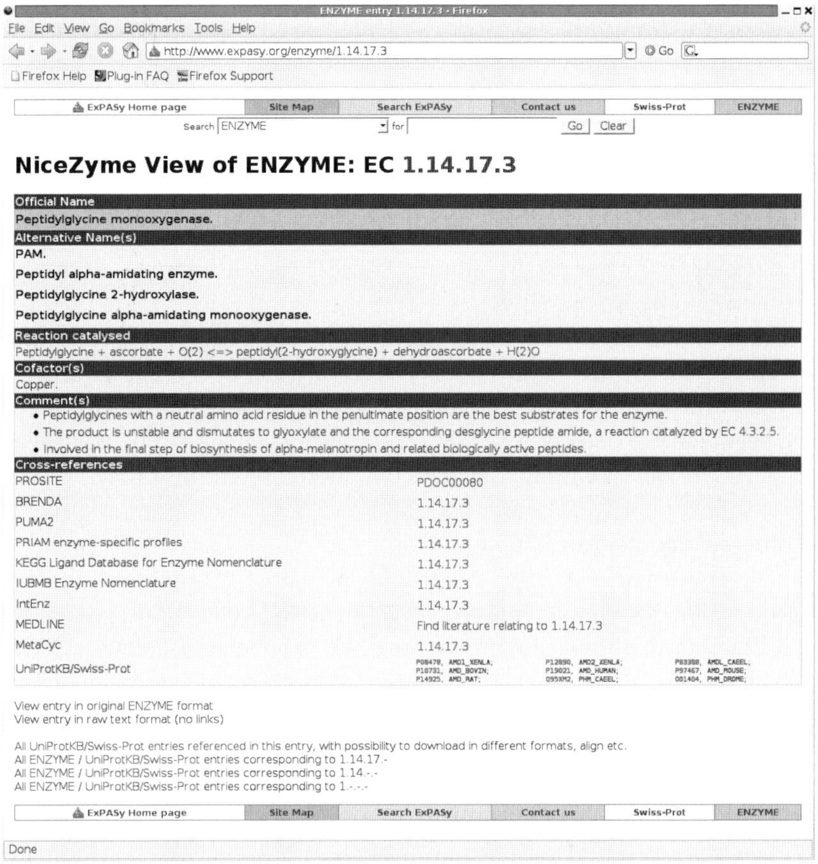

Fig. 4.1 The ENZYME DB view of peptidylglycine monooxygenase.

usually because of the requirements of binding sites that constrain the evolution of the protein family. For instance, the consensus pattern for inorganic pyrophosphatase is: D-[SGDN]-D-[PE]-[LIVMF]-D-[LIVMGAC]. The three conserved Ds bind divalent metal cations. Often, such a pattern identifies distant relationships not otherwise detectable by comparing sequences.

The Protein Information Resource (PIR) and associated databases

The PIR is one of the partners in UniProt. In addition, the PIR maintains several databases about proteins:

◆ PIRSF: the protein family classification system provides clustering of the sequences in UniProt according to their evolutionary relationships.

◆ iProClass, an integrated protein knowledgebase, is a gateway providing uniform access to over 90 biological databases, with flexible retrieval and navigation facilities.

◆ iProLINK (integrated Protein Literature, Information and Knowledge) is a gateway to the literature.

Databases of protein families

Evolutionary relationships are essential for making sense of biological data. Evolution provides the framework for an integrated appreciation of the properties of molecules and processes, and their similarities and difference in various species. Perhaps less obvious is that comparative studies illuminate, in an essential way, even individual molecules. Knowing only a single sequence, or structure, it is difficult to understand the significance of particular features. Patterns of conservation identify features that nature has found it necessary to retain. (PROSITE signatures are examples.) The challenge then is to figure out why.

Study of evolutionary patterns must begin with assembling a set of homologues. We again emphasize (1) the distinction between *homology* = descent from a common ancestor = a yes-or-no property, from *similarity* = some quantitative measure of the difference between two objects, and (2) that similarity can always be measured but it is rare to be able to observe homology directly. Therefore, in most cases, homology is an *inference* from similarity. (The methods and results of metagenomics may make it easier to observe homology directly.)

R. Doolittle suggested a general calibration of pairwise sequence similarity, for homology detection. Two full-length protein sequences (≥ 100 residues) that have $\geq 25\%$ identical residues in an optimal alignment are likely to be related. Below $\sim 15\%$ identical residues in an optimal alignment, we are mired in the noise. In this range of similarity, we have no reason to believe that the sequences are related — although they might be. Doolittle defined the range between 18 and 25% identity as 'the twilight zone', where there may be tantalizing suspicion of a relationship, but the evidence falls short of proof. In some cases the active site is better conserved than the bulk of the protein. In these cases, the appearance of a motif — such as the PROSITE consensus pattern for inorganic pyrophosphatase, D-[SGDN]-D-[PE]-[LIVMF]-D-[LIVMGAC] — can support the case for homology.

Multiple sequence alignments are much more powerful than pairwise sequence alignments. First, the additional data allow more accurate alignments. Secondly, the conservation patterns stand out far more sharply (see Problem 4.2).

Protein structure changes more conservatively than amino acid sequence. Therefore, inference of homology from structural similarity can link more distant relatives than sequence similarity can. In cases that lie in the twilight zone where sequence similarity is suggestive but not convincing, structural similarity is the court of last resort. In many cases, structural similarity can identify homologues even if no signal whatever — at least no signal detectable by current techniques — remains in the sequences.

It is common to refer to a group of related proteins as a family. Many databases classify proteins into families. These include sequence-oriented databases such as InterPro, Pfam and COG; and structure-oriented databases such as SCOP and CATH. The assignment of proteins to families is similar but not identical in various sources.

Most protein families contain many clusters of closer relatives. These form subfamilies. Conversely, two or more families can be grouped into superfamilies. Whereas the distinction between homologous and non-homologous proteins is objective (even if we cannot determine it with confidence in all cases), the clustering of homologues into subfamilies or superfamilies is partially a matter of convention or taste. Definition of subfamilies and superfamilies may legitimately differ among different databases.

Databases of structures

Structure databases archive, annotate and distribute sets of atomic coordinates. Started by the late Walter Hamilton at Brookhaven National Laboratories, Long Island, New York, USA in 1971, the major database for biological macromolecular structures is now the world-wide Protein Data Bank (wwPDB). It is a joint effort of the Research Collaboratory for Structural Bioinformatics (RCSB) (a distributed organization based at Rutgers University, in New Jersey; the San Diego Supercomputer Center, in California; and the University of Wisconsin, all in the USA); the Molecular Structure Database (at the European Bioinformatics Institute, in the UK); and the Protein Data Bank Japan (based at Osaka University). The wwPDB contains structures of proteins, nucleic acids and a few carbohydrates. The parent web site is http://www.wwpdb.org

The home pages of the wwPDB partners contain links to the data files themselves, to expository and tutorial material including short news items and the RCSB PDB Newsletter, to facilities for deposition of new entries, and to specialized search software for retrieving structures.

The Box shows part of a Protein Data Bank entry for a structure of spinach chloroplast thioredoxin.[1] The information contained includes:

◆ What protein is the subject of the entry, and what species it came from

◆ Who solved the structure, and literature references

◆ Experimental details about the structure determination, including information related to the general quality of the result, such as the resolution of an X-ray structure determination, and stereochemical statistics

◆ The amino acid sequence

◆ The atomic coordinates (lines beginning ATOM)

◆ What additional molecules appear in the structure, potentially including cofactors, inhibitors and water molecules (the keyword HETATM identifies the coordinates of these moieties)

◆ Assignments of secondary structure: helices, sheets

◆ Disulphide bridges

Protein Data Bank entry 1Faa, spinach chloroplast thioredoxin

```
HEADER     ELECTRON TRANSPORT                     13-JUL-00    1FAA
TITLE      CRYSTAL STRUCTURE OF THIOREDOXIN F FROM SPINACH CHLOROPLAST
```
                                                                    →

---

1 Capitani, G., Marković-Housley, Z., DelVal, G., Morris, M., Jansonius, J.N. and Schürmann, P. (2000). Crystal structures of two functionally different thioredoxins in spinach chloroplasts. *Journal of Molecular Biology*, **302**, 135–154.

Protein Data Bank entry 1FAA, spinach chloroplast thioredoxin (*continued*)

```
TITLE     2 (LONG FORM)
COMPND     MOL_ID: 1;
COMPND    2 MOLECULE: THIOREDOXIN F;
COMPND    3 CHAIN: A;
COMPND    4 FRAGMENT: LONG FORM;
COMPND    5 ENGINEERED: YES;
COMPND    6 MUTATION: YES
SOURCE     MOL_ID: 1;
SOURCE    2 ORGANISM_SCIENTIFIC: SPINACIA OLERACEA;
SOURCE    3 ORGANISM_COMMON: SPINACH;
SOURCE    4 CELLULAR_LOCATION: CHLOROPLAST;
SOURCE    5 EXPRESSION_SYSTEM: ESCHERICHIA COLI;
SOURCE    6 EXPRESSION_SYSTEM_COMMON: BACTERIA;
SOURCE    7 EXPRESSION_SYSTEM_PLASMID: PKK233-2 (MODIFIED)
KEYWDS     ELECTRON TRANSPORT
EXPDTA     X-RAY DIFFRACTION
AUTHOR     G.CAPITANI,Z.MARKOVIC-HOUSLEY,G.DELVAL,M.MORRIS,
AUTHOR    2 J.N.JANSONIUS,P.SCHURMANN
REVDAT    1    20-SEP-00 1FAA      0
JRNL        AUTH   G.CAPITANI,Z.MARKOVIC-HOUSLEY,G.DELVAL,M.MORRIS,
JRNL        AUTH 2 J.N.JANSONIUS,P.SCHURMANN
JRNL        TITL   CRYSTAL STRUCTURES OF TWO FUNCTIONALLY DIFFERENT
JRNL        TITL 2 THIOREDOXINS IN SPINACH CHLOROPLASTS
JRNL        REF    J.MOL.BIOL.                   V. 302   135 2000
JRNL        REFN   ASTM JMOBAK   UK ISSN 0022-2836
REMARK    1
REMARK    2
REMARK    2 RESOLUTION. 1.85 ANGSTROMS.
REMARK    3
REMARK    3 REFINEMENT.
REMARK    3    PROGRAM     : X-PLOR 3.851
REMARK    3    AUTHORS     : BRUNGER
REMARK    3

Additional information about details of solution of structure omitted ...

REMARK 900 RELATED ENTRIES
REMARK 900 RELATED ID: 1F9M    RELATED DB: PDB
REMARK 900 THIOREDOXIN F FROM SPINACH CHLOROPLAST (SHORT FORM)
REMARK 900 RELATED ID: 1FB0    RELATED DB: PDB
REMARK 900 THIOREDOXIN M FROM SPINACH CHLOROPLAST (REDUCED FORM)
REMARK 900 RELATED ID: 1FB6    RELATED DB: PDB
REMARK 900 THIOREDOXIN M FROM SPINACH CHLOROPLAST (OXIDIZED FORM)
DBREF  1FAA A    1   121  SWS    P09856   THIF_SPIOL       69     189
SEQADV 1FAA MET A   -2  SWS  P09856            CLONING ARTIFACT
SEQADV 1FAA TYR A   -1  SWS  P09856            CLONING ARTIFACT
SEQADV 1FAA TYR A    0  SWS  P09856            CLONING ARTIFACT
SEQADV 1FAA LEU A    1  SWS  P09856     MET    69 ENGINEERED
SEQADV 1FAA LEU A    3  SWS  P09856     GLN    71 ENGINEERED
SEQRES    1 A  124   MET TYR TYR LEU GLU LEU ALA LEU GLY THR GLN GLU MET
SEQRES    2 A  124   GLU ALA ILE VAL GLY LYS VAL THR GLU VAL ASN LYS ASP
SEQRES    3 A  124   THR PHE TRP PRO ILE VAL LYS ALA ALA GLY ASP LYS PRO
SEQRES    4 A  124   VAL VAL LEU ASP MET PHE THR GLN TRP CYS GLY PRO CYS
SEQRES    5 A  124   LYS ALA MET ALA PRO LYS TYR GLU LYS LEU ALA GLU GLU
SEQRES    6 A  124   TYR LEU ASP VAL ILE PHE LEU LYS LEU ASP CYS ASN GLN
SEQRES    7 A  124   GLU ASN LYS THR LEU ALA LYS GLU LEU GLY ILE ARG VAL
SEQRES    8 A  124   VAL PRO THR PHE LYS ILE LEU LYS GLU ASN SER VAL VAL
SEQRES    9 A  124   GLY GLU VAL THR GLY ALA LYS TYR ASP LYS LEU LEU GLU
SEQRES   10 A  124   ALA ILE GLN ALA ALA ARG SER
FORMUL    2   HOH   *34(H2 O1)
HELIX     1   1 GLY A    6  ALA A   12  1                               7
HELIX     2   2 THR A   24  ALA A   32  1                               9
HELIX     3   3 CYS A   46  TYR A   63  1                              18
HELIX     4   4 ASN A   77  GLY A   85  1                               9
HELIX     5   5 LYS A  108  ARG A  120  1                              13
```

Protein Data Bank entry 1FAA, spinach chloroplast thioredoxin *(continued)*

```
SHEET    1   A 5 VAL A  17  GLU A  19  0
SHEET    2   A 5 ILE A  67  ASP A  72  1  O  PHE A  68   N  THR A  18
SHEET    3   A 5 VAL A  37  PHE A  42  1  N  VAL A  38   O  ILE A  67
SHEET    4   A 5 THR A  91  LYS A  96 -1  O  THR A  91   N  MET A  41
SHEET    5   A 5 SER A  99  THR A 105 -1  O  SER A  99   N  LYS A  96
SSBOND   1 CYS A   46    CYS A   49
CISPEP   1 VAL A   89    PRO A   90            0           -0.06
CRYST1   30.600   63.100   31.600  90.00 110.70  90.00 P 1 21 1      2
ORIGX1      1.000000  0.000000  0.000000       0.00000
ORIGX2      0.000000  1.000000  0.000000       0.00000
ORIGX3      0.000000  0.000000  1.000000       0.00000
SCALE1      0.032680  0.000000  0.012349       0.00000
SCALE2      0.000000  0.015848  0.000000       0.00000
SCALE3      0.000000  0.000000  0.033829       0.00000
ATOM     1   N   LEU A   1      24.389  12.172  22.330  1.00 49.98          N
ATOM     2   CA  LEU A   1      23.617  11.064  22.997  1.00 51.12          C
ATOM     3   C   LEU A   1      22.228  10.829  22.381  1.00 51.55          C
ATOM     4   O   LEU A   1      21.316  11.634  22.547  1.00 50.88          O
ATOM     5   CB  LEU A   1      23.447  11.351  24.497  1.00 49.15          C
ATOM     6   CG  LEU A   1      24.373  10.670  25.513  1.00 47.15          C
ATOM     7   CD1 LEU A   1      23.831  10.905  26.924  1.00 45.33          C
ATOM     8   CD2 LEU A   1      24.488   9.185  25.215  1.00 44.91          C
ATOM     9   N   GLU A   2      22.076   9.713  21.674  1.00 53.58          N
ATOM    10   CA  GLU A   2      20.806   9.358  21.044  1.00 54.60          C
ATOM    11   C   GLU A   2      20.054   8.350  21.907  1.00 52.06          C
ATOM    12   O   GLU A   2      20.550   7.916  22.943  1.00 53.01          O
ATOM    13   CB  GLU A   2      21.043   8.762  19.650  1.00 60.19          C
ATOM    14   CG  GLU A   2      20.884   9.756  18.502  1.00 71.27          C
ATOM    15   CD  GLU A   2      19.687  10.694  18.678  1.00 77.29          C
ATOM    16   OE1 GLU A   2      19.521  11.277  19.779  1.00 79.51          O
ATOM    17   OE2 GLU A   2      18.910  10.850  17.705  1.00 81.98          O
ATOM    18   N   LEU A   3      18.857   7.975  21.477  1.00 48.15          N
ATOM    19   CA  LEU A   3      18.047   7.029  22.227  1.00 43.81          C
ATOM    20   C   LEU A   3      18.098   5.650  21.573  1.00 40.93          C
ATOM    21   O   LEU A   3      18.285   5.529  20.371  1.00 40.43          O
ATOM    22   CB  LEU A   3      16.598   7.531  22.280  1.00 44.07          C
ATOM    23   CG  LEU A   3      15.797   7.608  23.586  1.00 44.51          C
ATOM    24   CD1 LEU A   3      15.534   6.207  24.087  1.00 52.05          C
ATOM    25   CD2 LEU A   3      16.543   8.400  24.639  1.00 48.17          C

Coordinates of residues 4- 121 of protein omitted ...

ATOM   937   N   SER A 121       8.850 -16.291   7.411  1.00 55.56          N
ATOM   938   CA  SER A 121       9.537 -17.581   7.231  1.00 60.26          C
ATOM   939   C   SER A 121      10.923 -17.478   6.558  1.00 62.65          C
ATOM   940   O   SER A 121      11.870 -18.086   7.107  1.00 64.76          O
ATOM   941   CB  SER A 121       8.656 -18.556   6.432  1.00 60.35          C
ATOM   942   OG  SER A 121       7.280 -18.401   6.747  1.00 60.82          O
ATOM   943   OXT SER A 121      11.059 -16.809   5.503  1.00 63.39          O
TER    944       SER A 121
HETATM 945   O   HOH     1       2.260  -3.687  15.041  1.00 21.56          O
HETATM 946   O   HOH     2       0.884  -6.116  15.287  1.00 24.15          O
HETATM 947   O   HOH     3       0.912  13.888   2.773  1.00 30.06          O
HETATM 948   O   HOH     4      14.616  -4.966  11.156  1.00 26.65          O
HETATM 949   O   HOH     5       4.640  10.330   9.025  1.00 23.93          O
HETATM 950   O   HOH     6       3.040  -0.537  15.641  1.00 28.46          O
HETATM 951   O   HOH     7       6.246  -2.378  27.633  1.00 49.34          O

Coordinates of additional water molecules omitted ...

CONECT  358  375
CONECT  375  358
MASTER      211    0    0    5    5    0    0    6  977    1    2   10
END
```

The wwPDB overlaps several other databases. The Cambridge Crystallographic Data Centre archives the structures of small molecules; oligonucleotides appear in both the CCDC and the wwPDB. The combination of structural data from these sources is extremely useful in studies of conformations of the component units of biological macromolecules, and for investigations of macromolecule–ligand interactions, including but not limited to applications to drug design. The Nucleic Acid Structure Databank (NDB) at Rutgers University, New Jersey, USA also complements the wwPDB. The BioMagResBank, at the Department of Biochemistry, University of Wisconsin, USA—now a full partner in the RSCB—archives protein structures determined by nuclear magnetic resonance (NMR).

The archives collect not only the results of structure determination, but also the measurements on which they are based. The wwPDB keeps the data from X-ray structure determinations, and the BioMagResBank those from NMR.

The wwPDB assigns a four-character identifier to each structure deposited. The first character is a number from 1 to 9. Do not expect mnemonic significance. In many cases several entries correspond to one protein—solved in different states of ligation, or in different crystal forms, or re-solved using better crystals or more accurate data collection techniques. For instance, there have been at least four generations of sperm whale myoglobin crystal structures.

It is easy to retrieve a structure if you know its identifier. From the RCSB home page, entering a PDB ID and selecting Explore gives a one-page summary of the entry. Figure 4.2 shows the summary page for the spinach chloroplast thioredoxin structure, identifier 1FAA. Links from this page take you to:

- The publication in which the entry was described, via the bibliographic database PubMed

- Pictures of the structure (some of these may require that you install a viewing program on your computer)

- Access to the file containing the entry itself

- Lists of related structures, according to several different classifications of protein structures

- Stereochemical analysis—the distribution of bond lengths and angles, and conformational angles

- Sources of other information about this entry

- The sequence and secondary structure assignment

- Details about the crystal form and methods by which the crystals were produced

Searches for structures

Retrieval of a particular structure is easy, provided that you know its identifier. If not, how do you find it? A simple tool accessible from the RCSB home page permits a search for keywords. Entering SPINACH THIOREDOXIN returns eight entries, including 1FAA and other crystal structures, of the same molecule or mutants, in different

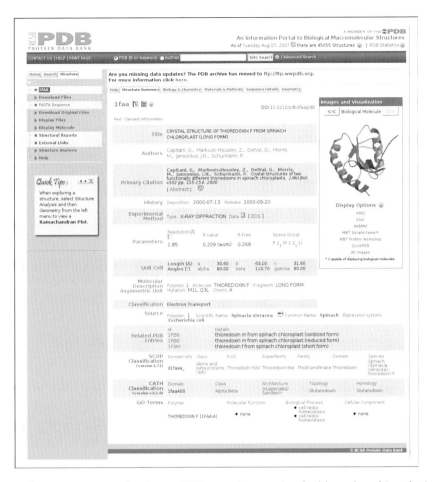

Fig. 4.2 The summary page for the wwPDB entry 1FAA, spinach chloroplast thioredoxin.

oxidation states. However, the search also returns several structures of glyceraldehyde 3-phosphate dehydrogenase. Why? Because embedded in the dehydrogenase structure entries is a reference to an article that contains the word thioredoxin in the title. Nevertheless, the information returned would easily permit you to choose structures to look at or analyse, according to your particular interest in this family of molecules. The RCSB site also offers more complex browsers. Using these, you could insist that the keywords appear in the molecule name. This would exclude the glyceraldehyde 3-phosphate dehydrogenase entries.

The Macromolecular Structure Database at the European Bioinformatics Institute (EBI) offers a useful list of facilities for searching and browsing the wwPDB, including OCA. OCA is a browser database for protein structure and function, integrating information from numerous databanks. Developed originally by J. Prilusky, OCA is supported by the EBI and is available there and at numerous mirror sites.

Another useful information source available at the EBI is the database of Probable Quaternary Structures (PQS) of the biologically active forms of proteins. Often the

The name OCA, in addition to being the Spanish word for goose, has the same relationship to PDB as A.C. Clarke's computer HAL in the movie *2001: A Space Odyssey* has to IBM.

asymmetric unit of the crystal structure, as deposited in the PDB entry, contains only part of the active unit, or alternatively multiple copies of the active unit. For many entries, it is not obvious how to go from information in the deposited entry to the active form.

Classifications of protein structures

Several web sites offer hierarchical classifications of all proteins of known structure according to their folding patterns:

- SCOP: Structural Classification of Proteins
- CATH: Class/Architecture/Topology/Homology
- DALI: based on extraction of similar structures from distance matrices
- CE: a database of structural alignments

These sites are useful general entry points to protein structural data. For instance, SCOP offers facilities for searching on keywords to identify structures, navigation up and down the hierarchy, generation of pictures, access to the annotation records in the PDB entries, and links to related databases (see Chapter 6).

Accuracy and precision of protein structure determinations

X-ray crystallography

X-ray crystallography produces estimates of the positions of the atoms in a molecule. It also produces estimates of their effective sizes, called **B-factors**. An important feature of the experimental data (the absolute values of the Fourier coefficients of the electron density) is that all atoms contribute to all observations. It is difficult to estimate errors in individual atomic positions. For protein crystal structures, B-factors are a useful index of the precision of the positions of the individual atoms.

Crystal structure determinations are at the mercy of the degree of order in different parts of the molecule. (Order is the extent to which different unit cells of the crystal are exact and static copies of one another.) The degree of order governs the available **resolution** of the experimental data. Resolution is an index of potential quality of an X-ray structure determination, measuring the ratio of the number of parameters to be determined to the number of observations. In structure determinations of small organic molecules or of minerals, this ratio is usually generous: ~10. But for a typical protein crystal:

For small molecules, forming well-ordered crystals, B-factors reflect thermal vibrational amplitudes.

	Low resolution ... High					
Resolution in Å	4.0	3.5	3.0	2.5	2.0	1.5
Ratio of observations to parameters	0.3	0.4	0.6	1.1	2.2	3.8

(Resolution measures the fineness of the details that can be distinguished; hence, the lower the number, the higher the resolution.)

In addition to disorder, errors in crystal structures reflect both errors in data measurement and errors in solving the structure. A comparison of four independently-solved structures of interleukin-1β showed an average variation in atomic position of 0.84 Å, higher than the expected experimental error.

Many crystallographers deposit their experimental data along with the solved structures. This permits detailed checks on the results. But, in many cases, the experimental data are not available. How can one then assess the quality of a structure? B-factors provide important clues; high B-factors in an entire region suggest that the region is not well determined. This usually reflects imperfect order in the crystal. Programs can flag stereochemical outliers — exceptions to regularities common to well-determined protein structures. The entries corresponding to the PDB entries in `www.cmbi.kun.nl/gv/pdbreport` describe diagnostic analysis and identification of problems and outliers.

But although outliers are relatively easy to *detect*, it is difficult to decide whether they are correct but unusual features of the structure, or the result of errors in building the model, or the inevitable result of crystal disorder. Proper assessment requires access to the experimental data; and fixing real errors may well require the attention of an experienced crystallographer. The conclusion seems inescapable that structure factors should be archived and available.

Nuclear magnetic resonance (NMR)

NMR is the second major technique for determining macromolecular structure. It produces structures that are correct in topology but generally not as precise as a good X-ray structure determination. Crystallographers report a single structure, or only a small number. NMR spectroscopists usually produce a family of ~10–20 related structures or even more, calculated from the same experimental data. Comparison across such an ensembl indicates precision; regions in which the local variation in structure is small are well defined by the data. This is a rough equivalent of the crystallographer's B-factor.

There are two sources of structural variation among the models reported by NMR spectroscopists. One is genuine dynamic disorder, arising because the conformation is not locked in by crystal packing forces. The other is an uncomfortably low ratio of measurements to parameters that need to be determined. As a result, several different conformations may fit the experimental data comparably well.

Analysis of NMR measurements can distinguish these effects, but is carried out in only a minority of NMR protein structure determinations.

Web resources Protein and nucleic acid structures

The world-wide Protein Data Bank:
http://www.wwPDB.org

Nucleic acid database:
http://ndbserver.rutgers.edu/

online
resource
centre

→

Web resources (*continued*)
BioMagResBank:
http://www.bmrb.wisc.edu/

Searching the Protein Data Bank:
SCOP (Structural Classification of Proteins):
http://scop.mrc-lmb.cam.ac.uk/scop/

OCA:
http://oca.ebi.ac.uk/oca-bin/ocamain

Database of Protein Quaternary Structure:
http://pqs.ebi.ac.uk/

Reports of structure quality:
http://www.cmbi.kun.nl/gv/pdbreport

Classification and assignment of protein function

The Enzyme Commission

The first detailed classification of protein functions was that of the Enzyme Commission (EC). In 1955, the General Assembly of the International Union of Biochemistry (IUB), in consultation with the International Union of Pure and Applied Chemistry (IUPAC), established an International Commission on Enzymes, to systematize nomenclature. The Enzyme Commission published its classification scheme, first on paper and now on the Web: http://www.chem.qmul.ac.uk/iubmb/enzyme/

EC numbers (looking suspiciously like IP numbers) contain four fields, corresponding to a four-level hierarchy. For example, EC 1.1.1.1 corresponds to alcohol dehydrogenase, catalysing the general reaction:

$$\text{an alcohol} + \text{NAD} = \text{the corresponding aldehyde or ketone} + \text{NADH}_2$$

Note that several reactions, involving different alcohols, would share this number; but that the same dehydrogenation of one of these alcohols by an enzyme using the alternative cofactor NADP would be assigned EC 1.1.1.2.

The first field in an EC number indicates to which of the six main divisions (classes) the enzyme belongs:

Class 1.	Oxidoreductases
Class 2.	Transferases
Class 3.	Hydrolases
Class 4.	Lyases
Class 5.	Isomerases
Class 6.	Ligases

The significance of the second and third numbers depends on the class. For Oxidoreductases the second number describes the substrate and the third number the acceptor. For Transferases, the second number describes the class of item transferred, and the third number describes either more specifically what they transfer or in some cases the acceptor. For Hydrolases, the second number signifies the kind of bond cleaved (e.g. an ester bond) and the third number the molecular context (e.g. a carboxylic ester or a thiol ester). (Proteinases are treated slightly differently, with the third number including the mechanism: serine proteinases, thiol proteinases and acid proteinases are classified separately.) For Lyases, the second number signifies the kind of bond formed (e.g. C–C or C–O), and the third number the specific molecular context. For Isomerases, the second number indicates the type of reaction and the third number the specific class of reaction. For Ligases, the second number indicates the type of bond formed and the third number the type of molecule in which it appears—for example, EC 6.1 for C–O bonds (enzymes acylating tRNA), EC 6.2 for C–S bonds (acyl–CoA derivatives), etc. The fourth number gives the specific enzymatic activity.

Specialized classifications are available for some families of enzymes; for instance, the MEROPS database by N.D. Rawlings and A.J. Barrett provides a structure-based classification of peptidases and proteinases. http://www.merops.sanger.ac.uk/

The Gene Ontology™ Consortium protein function classification

In 1999, Michael Ashburner and many co-workers faced the problem of annotating the soon-to-be-completed *Drosophila melanogaster* genome sequence. As a classification of function, the EC classification was unsatisfactory, if only because it was limited to enzymes. Ashburner organized the Gene Ontology Consortium to produce a standardized scheme for describing function.[2]

The Gene Ontology™ Consortium (http://www.geneontology.org) has produced a systematic classification of gene function, in the form of a dictionary of terms, and their relationships.

Organizing concepts of the Gene Ontology project include three categories:

- **Molecular function** A function associated with what an individual protein or RNA molecule does in itself; either a general description such as *enzyme,* or a specific one such as *alcohol dehydrogenase.* This is function from the biochemists' point of view.

- **Biological process** A component of the activities of a living system, mediated by a protein or RNA, possibly in concert with other proteins or RNA molecules; either a general term such as *signal transduction,* or a particular one such as *cyclic AMP synthesis.* This is function from the cell's point of view.

Because many processes are dependent on location. Gene Ontology also tracks:

- **Cellular component** The assignment of site of activity or partners; this can be a general term such as *nucleus* or a specific one such as *ribosome.*

Recall that an ontology is a formal set of well-defined terms with well-defined inter-relationships, i.e. a dictionary and rules of syntax.

2 Described in his memoir: Ashburner, M. (2006). *Won for All: How the* Drosophila *Genome was Sequenced.* Cold Spring Harbor Laboratory Press, Cold Spring Harbor, New York.

An example of the GO classification is shown in Fig. 4.3.

Neither the EC nor the GO classification is an assignment of function to individual proteins. The EC emphasized that: 'It is perhaps worth noting, as it has been a matter of long-standing confusion, that enzyme nomenclature is primarily a matter of naming reactions catalysed, not the structures of the proteins that catalyse them'. Assigning EC or GO numbers to proteins is a separate task. Such assignments appear in protein databases such as PIR or SWISS-PROT.

Comparison of EC and GO classifications

EC identifiers form a strict four-level hierarchy, or tree. For example, isopentenyl-diphosphate Δ-isomerase is assigned EC number 5.3.3.2. The initial 5 specifies the most general category, 5 = isomerases, 5.3 comprises intramolecular isomerases, 5.3.3 those enzymes that transpose C=C bonds, and the full identifier 5.3.3.2 specifies the particular reaction. In the molecular function ontology, GO assigns the identifier

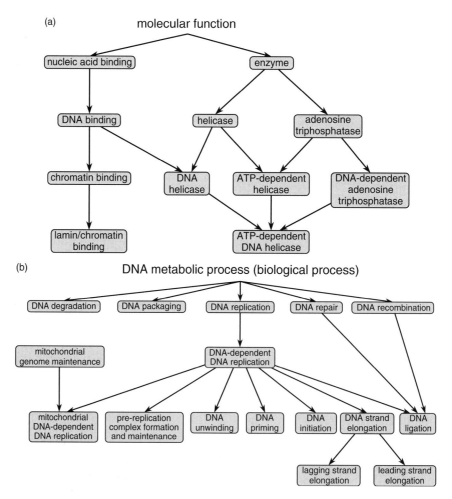

Fig. 4.3

(c)

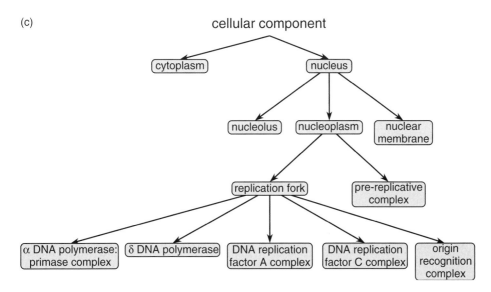

Fig. 4.3 Selected portions of the three categories of Gene Ontology, showing classifications of functions of proteins that interact with DNA. (a) *Biological process:* DNA metabolism. (b) *Molecular function:* including general DNA binding by proteins, and enzymatic manipulations of DNA. (c) *Cellular component:* different places within the cell.

These pictures illustrate the general structure of the Gene Ontology classification. Each term describing a function is a *node* in a graph. Each node has one or more parents and one or more descendants: arrows indicate direct ancestor-descendant relationships. A path in the graph is a succession of nodes, each node the parent of the next. Nodes can have 'grandparents', and more remote ancestors.

Unlike the EC hierarchy, the Gene Ontology graphs are *not* trees in the technical sense, because there can be more than one path from an ancestor to a descendant. For example, there are two paths in (a) from enzyme to ATP-dependent helicase. Along one path helicase is the intermediate node. Along the other path adenosine triphosphatase is the intermediate node.

Although the nodes are shown on discrete levels to clarify the structure of the graph, all the nodes on any given level do not necessarily have a common degree of significance; unlike family, genus and species levels in the Linnaean taxonomic tree, or the ranks in military, industrial, academic, etc., organizations. GO terms could not have such a common degree of significance, given that there can be multiple paths, of different lengths, between different nodes.

0004452 to isopentenyl-diphosphate Δ-isomerase. (The numbers themselves have no specific significance.)

Figure 4.4 compares the EC and GO classifications of isopentenyl-diphosphate Δ-isomerase. The figure shows a path from GO:0004452 to the root node of the molecular function graph, GO:0003674. In this case there are four intervening nodes, progressively more general categories as we move up the figure. Note that the GO description of this enzyme as an oxidoreductase is inconsistent with the EC classification, in

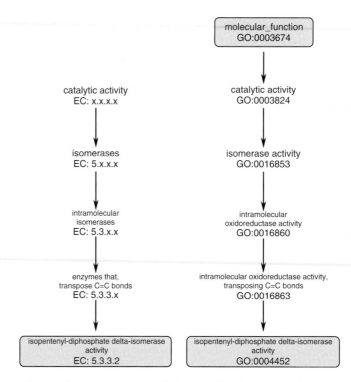

Fig. 4.4 Comparison of Enzyme Commission and Gene Ontology classifications of isopentenyl-diphosphate Δ-isomerase.

which a committed choice between oxidoreductase and isomerase must be made at the highest level of the EC hierarchy.

Specialized, or 'boutique' databases

Many individuals or groups select, annotate and recombine data focused on particular topics, and include links affording streamlined access to information about subjects of interest.

For instance, the protein kinase resource is a specialized compilation that includes sequences, structures, functional information, laboratory procedures, lists of interested scientists, tools for analysis, a bulletin board and links.

The HIV protease database archives structures of human immunodeficiency virus 1 proteinases, human immunodeficiency virus 2 proteinases and simian immunodeficiency virus proteinases, and their complexes; and provides tools for their analysis and links to other sites with AIDS-related information. This database contains some crystal structures not deposited in the PDB.

In the field of immunology:

• IMGT, the international ImMunoGeneTics database, is a high-quality integrated database specializing in immunoglobulins (Ig), T-cell receptors (TcRs) and major histocompatibility complex (MHC) molecules of all vertebrate species. The IMGT server provides a common access to all immunogenetics data. At present, it includes two databases: IMGT/LIGM-DB, a comprehensive database of Ig and TCR gene sequences from human and other vertebrates, with translation for fully annotated sequences, and IMGT/HLA-DB, a database of the human MHC referred to as HLAS (human leucocyte antigens) (http://imgt.cines.fr).

• IEDB, the Immune Epitope Database and Analysis Resource, curated at the La Jolla Institute for Allergy and Immunology, containing data related to antibody and T-cell epitopes (http://beta.immuneepitope.org/home.do).

Expression and proteomics databases

Recall the central dogma: DNA makes RNA makes protein. Genomic databases contain *DNA sequences*. Expression databases record measurements of *mRNA* levels, usually via ESTs (short terminal sequences of cDNA synthesized from mRNA) describing patterns of gene transcription. Proteomics databases record measurements on *proteins,* describing patterns of gene translation.

Comparisons of expression patterns give clues to: (1) the function and mechanism of action of gene products, (2) how organisms coordinate their control over metabolic processes in different conditions—for instance yeast under aerobic or anaerobic conditions, (3) the variations in mobilization of genes at different stages of the cell cycle, or of the development of an organism, (4) mechanisms of antibiotic resistance in bacteria, and consequent suggestion of targets for drug development, (5) the response to challenge by a parasite, and (6) the response to medications of different types and dosages, to guide effective therapy.

There are many databases of ESTs. In most, the entries contain fields indicating tissue of origin and/or subcellular location, state of development, conditions of growth and quantitation of expression level. Within GenBank, the dbEST collection currently contains 43 million entries, from 678 species, led by:

Species with the largest number of entries in dbEST

Species	Number of entries
Homo sapiens (human)	8 119 099
Mus musculus + domesticus (mouse)	4 850 243
Danio rerio (zebrafish)	1 350 105
Bos taurus (cattle)	1 318 208
Arabidopsis thaliana (thale cress)	1 276 692
Xenopus tropicalis (western clawed frog)	1 256 157
Oryza sativa (rice)	1 211 418

Species with the largest number of entries in dbEST

Species	Number of entries
Zea mays (maize)	1 161 241
Triticum aestivum (wheat)	1 050 267
Rattus norvegicus + sp. (rat)	871 163
Ciona intestinalis (sea squint)	686 396
Xenopus laevis (African clawed frog)	677 784
Sus scrofa (pig)	646 434
Gallus gallus (chicken)	599 330
Drosophila melanogaster (fruit fly)	532 557

Some EST collections are specialized to particular tissues (e.g. muscle, tooth) or to species. In many cases there is an effort to link expression patterns to other knowledge of the organism. For instance, the Jackson Lab Gene Expression Information Resource Project for Mouse Development coordinates data on gene expression and developmental anatomy.

Many databases provide connections between ESTs in different species, for instance linking human and mouse homologues, or relationships between human disease genes and yeast proteins. Other EST collections are specialized to a type of protein, for instance cytokines. A large effort is focused on cancer: integrating information on mutations, chromosomal rearrangements and changes in expression patterns, to identify changes during tumour formation and progression.

Although of course there is a close relationship between patterns of transcription and patterns of translation, direct measurements of protein contents of cells and tissues—proteomics—provides additional valuable information. Because of differential rates of translation and turnover of different mRNAs, measurements of proteins directly give a more accurate description of patterns of gene expression than measurements of transcription. Post-translational modifications can be detected *only* by examining the proteins.

Proteome analysis involves separation, identification and quantitative determination of amounts of proteins present in the sample (see Chapter 7). Proteome databases store images of gels, and their interpretation in terms of protein patterns. For each protein, an entry typically records (see Weblem 4.21):

- identification of protein
- relative amount
- function
- mechanism of action
- expression pattern

- subcellular localization
- related proteins
- post-translational modifications
- interactions with other proteins
- links to other databases

Proteomics has become an important field in bioinformatics, which contributes to the development of these databases, and also to the development of algorithms for comparing and analysing the patterns they contain.

Databases of metabolic pathways

Metabolism is the flow of molecules and energy through pathways of chemical reactions. Substrates of metabolic reactions can be macromolecules—proteins and nucleic acids—as well as small compounds such as amino acids and sugars.

The full panoply of metabolic reactions forms a complex network. The *structure* of the network corresponds to a graph in which metabolites are the nodes, and the substrate and product of each reaction form an edge. The *dynamics* of the network depend on the flow capacities of all the individual links, analogous to traffic patterns on the streets of a city.

Some patterns within the metabolic network are linear pathways. Others form closed loops, such as the tricarboxylic acid (Krebs) cycle. Many pathways are highly branched and interlock densely. However, metabolic networks also contain recognizable clusters or blocks; for instance, catabolic and anabolic reactions form clustered subnetworks. There is a relatively high density of internal connections within clusters and relatively few connections between them.

Biochemists have learned a lot about different enzymes in different species. Approximately 40% of the sequences in the SWISS-PROT/TrEMBL databases are enzymes (~10 000 sequences in all). They come from 1925 species. They represent 2569 different enzymatic activities.

Databases organize this information, collecting it within a coherent and logical structure, with links to other databases that provide different data selections and different modes of organization. EcoCyc treats *E. coli*. It is the model for—and linked with—numerous parallel databases, with uniform web interfaces, treating other organisms. BioCyc is the 'umbrella' collection. KEGG, The Kyoto Encyclopedia of Genes and Genomes, contains information from multiple organisms.

> Several databases organize information on metabolic pathways in different organisms.

Database	Home page
EcoCYC	http://ecocyc.org
BioCyc	http://www.biocyc.org
KEGG	http://www.genome.jp/kegg/

EcoCYC

EcoCYC is a database representing what we know about the biology of *Escherichia coli*, strain K-12 MG1655. It contains:

- **The genome** The complete sequence, and for each gene its position, and function if known.
- **Transcription regulation** Operons, promoters, and transcription factors and their binding sites.
- **Metabolism** The pathways, including details of the enzymology of individual steps. For each enzyme, the reaction, activators, inhibitors and subunit structure.
- **Membrane transporters** Transport proteins and their cargo.
- **Links to other databases** of protein and nucleic acid sequence data, literature references, and comparisons of different *E. coli* strains.

A tiny subset of the *E. coli* metabolic network is the pathway for synthesis of methionine from aspartate (see Box).

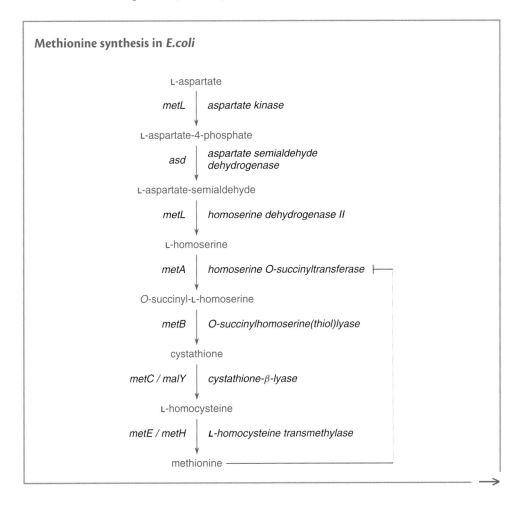

Methionine synthesis in *E.coli*

L-aspartate

metL — aspartate kinase

L-aspartate-4-phosphate

asd — aspartate semialdehyde dehydrogenase

L-aspartate-semialdehyde

metL — homoserine dehydrogenase II

L-homoserine

metA — homoserine O-succinyltransferase

O-succinyl-L-homoserine

metB — O-succinylhomoserine(thiol)lyase

cystathione

metC / malY — cystathione-β-lyase

L-homocysteine

metE / metH — L-homocysteine transmethylase

methionine

→

Methionine synthesis in *E.coli (continued)*

The diagram shows the seven-step synthesis of methionine from L-aspartate. To appreciate the logic of the system, keep in mind that both the reaction sequence (and the associated control networks, not shown here) are embedded in much larger networks.

• The first step, phosphorylation of L-aspartate, is common to the biosynthesis of methionine, lysine, and threonine. *Escherichia coli* contains three aspartate kinases, encoded by three separate genes, each specific for one of the end-product amino acids. They catalyse the same reaction, but are subject to separate regulation.

• The third step, conversion of L-aspartate-semialdehyde to L-homoserine, is common to the methionine and threonine synthesis pathways. Two homoserine dehydrogenases are encoded separately. Regulation of expression of the aspartate kinases and homoserine dehydrogenases suffices to control all three pathways.

• After synthesis, methionine is converted to *S*-adenosylmethionine, a common participant in methyl group transfers. *S*-adenosylmethionine activates the met repressor. In classic feedback inhibition, a product interacts directly with an enzyme that produces one of its precursors. This is a more complicated form of feedback: the product interacts with a repressor, that reduces the expression—not the activity—of enzymes that produce its precursors.

In the EcoCyc web page that contains the information corresponding to this diagram, the items are active. Links to other internal pages expand information about metabolites, cofactors, enzymes, genes and regulators. It is possible to 'zoom' in or out by controlling the level of detail. For instance, asking for less detail than the contents of the preceding diagram would first eliminate the information about the genes and enzymes, then reduce the pathway to an outline showing only critical intermediates:

L-aspartate →→→ homoserine →→→ L-homocysteine → L-methionine

It is also possible to explore in other dimensions. The methionine synthesis pathway is embedded in larger networks. One of these involves synthesis of the amino acids lysine and threonine in addition to methionine, all starting with aspartate:

→

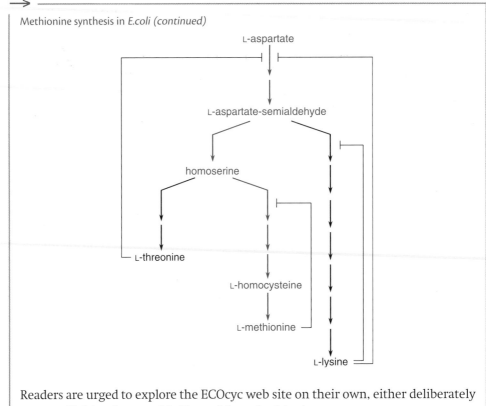

Methionine synthesis in *E.coli (continued)*

Readers are urged to explore the ECOcyc web site on their own, either deliberately or serendipitously, or guided by weblems in this chapter.

Of particular interest for comparative genomics are facilities to compare pathways among different organisms. Alignment and comparison of pathways can expose how pathways have diverged between species. Even if the pathways are the same, in some cases the enzymes are non-homologous.

Pathway comparison can be useful for annotation of genomes. It is often possible to assign function to proteins on the basis of similarity to sequences of proteins of known function in other organisms. However, sometimes there are several weak similarities to other proteins and it is unclear which is the true homologue. Conversely, sometimes an organism has a metabolic pathway but no annotated enzyme for an essential step. Confronting the unannotated proteins with the unassigned functions can sometimes identify the protein that fills the gap in the pathway.

If an enzyme needed for a pathway cannot be identified even by weak sequence similarity, it may be that the organism has evolved a non-homologous enzyme for the task. For example, the archaeon *Methanococcus jannaschii* has a pathway for biosynthesis of chorismate from 4-dehydroquinate. Enzymes for most of the steps have homologues in bacteria and/or eukaryotes. However, shikimate kinase was not identifiable from sequence similarity. Because the metabolic pathway is not interrupted, *M. jannaschii* must have *some* protein with this function. How can it be found?

Although in bacteria, genes consecutive in pathways are often consecutive in operons in the genome, this is not true of *M. jannaschii*. However, the genes for successive steps of the chorismate biosynthesis pathway *are* clustered and consecutive in another archaeon, *Aeropyrum pernix*. It was possible to propose a gene for a shikimate kinase in *A. pernix*, and to identify a homologue of that gene in *M. jannaschii*.

Experiment confirmed the prediction that the *M. jannaschii* gene so identified (*MJ1440*) encodes a shikimate kinase. It has no sequence similarity to bacterial or eukaryotic shikimate kinases. A protein from a different family has been recruited for the archaeal pathway. (For more details, see *Introduction to Genomics,* pp. 378–379.)

The Kyoto Encyclopedia of Genes and Genomes (KEGG)

The Kyoto Encyclopedia of Genes and Genomes (KEGG) collects individual genomes, gene products and their functions, but its special strengths lie in its integration of biochemical and genetic information. KEGG focuses on interactions: molecular assemblies, and metabolic and regulatory networks. It has been developed under the direction of M. Kanehisa.

KEGG organizes five types of data into a comprehensive system:

1. Catalogues of chemical compounds in living cells

2. Gene catalogues

3. Genome maps

4. Pathway maps

5. Orthologue tables

The catalogues of chemical compounds and genes—items 1 and 2—contain information about particular molecules or sequences. Item 3, genome maps, integrates the genes themselves according to their appearance on chromosomes. In some cases, knowing that a gene appears in an operon can provide clues to its function.

Item 4, the pathway maps, describe potential networks of molecular activities, both metabolic and regulatory. A metabolic pathway in KEGG is an idealization corresponding to a large number of possible metabolic cascades. It can generate a real metabolic pathway of a particular organism, by matching the proteins of that organism to enzymes within the reference pathways.

One enzyme in one organism would be referred to in KEGG in its orthologue tables, item 5, which link the enzyme to related enzymes in other organisms. This permits analysis of relationships between the metabolic pathways of different organisms.

KEGG derives its power from the very dense network of links among these categories of information, and additional links to many other databases to which the system maintains access. Two examples of the kinds of questions that can be treated by KEGG are:

◆ It has been suggested that simple metabolic pathways evolve into more complex ones by gene duplication and subsequent divergence. Searching the pathway catalogue for sets of enzymes that share a folding pattern will reveal clusters of linked paralogues.

◆ KEGG can take the set of enzymes from some organism and check whether they can be integrated into known metabolic pathways. A gap in a pathway suggests a missing enzyme or an unexpected alternative pathway. The archaeal shikimate kinase, not homologous to its bacterial counterparts, is an example.

Bibliographic databases

MEDLINE (based at the US National Library of Medicine) integrates the medical literature, including very many papers dealing with subjects in molecular biology that are not overtly clinical in content. It is included in PubMed, a bibliographical database offering abstracts of scientific articles, integrated with other information-retrieval tools of the National Center for Biotechnology Information within the National Library of Medicine (http://www.ncbi.nlm.nih.gov/PubMed/).

One very effective feature of PubMed is the option to retrieve *related articles*. This is a very quick way to 'get into' the literature of a topic. Combined with the use of a general search engine for web sites that do not correspond to articles published in journals, fairly comprehensive information is readily available about most subjects. Here's a tip: if you are trying to start to learn about an unfamiliar subject, try adding the keyword *tutorial* to your search in a general search engine, or the keyword *review* to your search in PubMed.

Another general search engine for the scientific literature is Google Scholar.

Almost all scientific journals now place their tables of contents, and in many cases their entire issues, on web sites. The US National Institutes of Health have established a centralized Web-based library of scientific articles, called PubMed Central (http://www.pubmedcentral.nih.gov/). In collaboration with scientific journals, the NCBI is organizing the electronic distribution of the full texts of published articles.

Surveys of molecular biology databases and servers

It is difficult to explore any topic in molecular biology on the Web without quickly bumping into a list of this nature. Lists of web resources in molecular biology are very common. They contain, to a large extent, the same information, but vary widely in their 'look and feel'. The real problem is that unless they are curated they tend to degenerate into lists of dead links. (A draft of this section contained a reference to a web site that contained a reasonable survey. Returning to it two months later, the name of the site had changed, and over half of the links had disappeared.)

This book does not contain a long annotated list of relevant and recommended sites, for the following reasons: (1) You don't want a long list, you need a short one. (2) The Web is too volatile for such a list to stay useful for very long. *It is much more effective to use a general search engine to find what you want at the moment you want it.*

My advice is: spend some time browsing; it won't take you long to find a site that appears reasonably stable and has a style compatible with your methods of work. Alternatively, here's a site that is comprehensive and shows signs of a commitment to keeping up to date: http://www.expasy.org/alinks.html It is a suitable site for starting a browsing session.

Gateways to archives

Databases in molecular biology maintain facilities for a very wide variety of information retrieval and analysis operations. Categories of these operations include:

1. **Retrieval of sequences from a database** Sequences can be 'called up' either on the basis of features of the annotations, or by patterns found within the sequences themselves.

2. **Sequence comparison** This is not a facility, this is a heavy industry! It was introduced in Chapter 1 and will be discussed in detail in Chapter 5. It includes the very important searches for relatives.

3. **Identification of genes in genome sequences, and translation of protein-coding gene sequences to amino acid sequences**

4. **Simple types of structure analysis and prediction** For example, statistical methods for predicting the secondary structure of proteins from sequences alone, including hydrophobicity profiles—from which the transmembrane proteins can generally be identified. Other sites offer full three-dimensional sequence-to-structure prediction.

5. **Pattern recognition** It is possible to search for all sequences containing a pattern or combination of patterns, expressed as probabilities for finding certain sets of residues at consecutive positions. These patterns may extend over large regions of the sequence. Such patterns reflect the global folding pattern of a protein. Other patterns are short. In DNA sequences, these patterns may reflect recognition sites for enzymes such as those responsible for splicing together interrupted genes. In proteins, short and localized patterns generally identify molecules that share a common function.

6. **Molecular graphics** is necessary to provide intelligible depictions of very complicated systems. Typical applications of molecular graphics include:

   - giving a useful overall impression of a protein folding pattern

   - mapping residues believed to be involved in function, onto the three-dimensional framework of a protein; often this will isolate an active site

   - classifying and comparing the folding patterns of proteins

   - analysing changes between closely related structures, or between two conformational states of a single molecule, and

   - studying the interaction of a small molecule with a protein, in order to attempt to assign function, or for drug development

   - interactive fitting of a model to the noisy and fuzzy image of the molecule that arises initially from the measurements in solving protein structures by X-ray crystallography

   - design and modelling of new structures

Access to databases in molecular biology

How to learn web skills

It would be difficult to learn to ride a bicycle by reading a book describing the sets of movements required, much less a treatise on the theory of the gyroscope. Similarly, the place to learn web skills is at a terminal, running a browser. True enough, but there is always a certain initial period of difficulty and imbalance. Here the goal is only to provide some temporary assistance to get you started. Then, off you go!

This section contains introductions to some of the major databanks and information-retrieval systems in molecular biology. In each case, the illustrations show relatively simple searches and applications. When appropriate, unique features of each system will be emphasized.

ENTREZ

The National Center for Biotechnology Information, a component of the United States National Library of Medicine, maintains databases and avenues of access to them. ENTREZ offers access via many database divisions (see Box).

The ENTREZ database system of the National Center for Biotechnology Information

Name	Contents
Nucleotide	Nucleotide sequences (includes GenBank)
Protein	Amino acid sequences
Genome	Whole genome sequences
Structure	Three-dimensional macromolecular structures
Taxonomy	Organisms in GenBank
SNP	Single-nucleotide polymorphism
Gene	Gene-centred information
HomoloGene	Eukaryotic homology groups
PubChem Compound	Unique small molecule chemical structures
PubChem Substance	Deposited chemical substance records
Genome Project	Genome project information
dbGaP	Genotype and phenotype
UniGene	Gene-oriented clusters of transcript sequences
CDD	Conserved protein domain database
3D Domains	Domains from ENTREZ Structure
UniSTS	Markers and mapping data
PopSet	Population study data sets
GEO Profiles	Expression and molecular abundance profiles
GEO DataSets	Experimental sets of GEO data (GEO = Gene Expression Omnibus)
Cancer Chromosomes	Cytogenetic databases

$\longrightarrow$

The ENTREZ database system of the National Center for Biotechnology Information *(continued)*

Name	Contents
PubChem BioAssay	Bioactivity screens of chemical substances
GENSAT	Gene expression atlas of mouse central nervous system
Probe	Sequence-specific reagents
Protein Clusters	Related protein sequences

For a diagram showing all component ENTREZ databases, and the connections among them, see http://www.ncbi.nlm.nih.gov/Database/datamodel/index.html Links between various databases are a strong point of NCBI's system.

The starting point for retrieval of sequences and structures is called ENTREZ: http://www.ncbi.nlm.nih.gov/Entrez/ Let us pick a molecule—human neutrophil elastase—and search for relevant entries in the different sections of ENTREZ.

Search in the ENTREZ Protein database

Go to http://www.ncbi.nlm.nih.gov/entrez/ Select Protein, enter the search terms HUMAN ELASTASE and click on Go.

The program returns 668 answers. The Box shows 14 of them: the first three, plus selected interesting results from farther down the list. The top hit is LEUKOCYTE ELASTASE PRECURSOR. Other responses include elastases from other species, inhibitors, a leech protein and a transcriptional regulator. (Why should a leech protein and a transcriptional regulator—which presumably interacts with DNA, not protein—show up in a search for human elastase? See Weblem 4.9.) We shall see how to tune the query to eliminate these extraneous responses.

Selected ENTREZ responses to human elastase in the Protein database

1: P08246
Leukocyte elastase precursor (Elastase-2) (Neutrophil elastase) (PMN elastase) (Bone marrow serine protease) (Medullasin) (Human leukocyte elastase) (HLE) gi—119292—sp—P08246—ELNE_HUMAN[119292]

2: 1HNEE
Chain E, Human Neutrophil Elastase (HNE) (E.C.3.4.21.37) (Also Referred To As Human Leucocyte Elastase (HLE)) Complex With Methoxysuccinyl-Ala-Ala-Pro-Ala Chloromethyl Ketone (MSACK) gi—230004—pdb—1HNE—E[230004]

→

Selected ENTREZ responses to human elastase in the Protein database *(continued)*

3: 1PPFE
Chain E, Human Leukocyte Elastase (Hle) (Neutrophil Elastase (Hne)) (E.C.3.4.21. 37) Complex With The Third Domain Of Turkey Ovomucoid Inhibitor (Omtky3) gi—809343—pdb—1PPF—E[809343]

. . .

14: P30740
Leukocyte elastase inhibitor (LEI) (Serpin B1) (Monocyte/neutrophil elastase inhibitor) (M/NEI) (EI) gi—266344—sp—P30740—ILEU_HUMAN[266344]

15: AAB20263
Alzheimer's beta-amyloid precursor protein, Kunitz-type protease inhibitor, neutrophil elastase inhibitor, P1-Val-APP-KD [human, Peptide Partial Mutagenesis, 17 aa] gi—238492—gb—AAB20263.1——bbm—163757—bbs—65057 [238492

. . .

166: NP_835455
pancreas specific transcription factor, 1a [Homo sapiens] gi—30039710—ref—NP_835455.1—[30039710]

167: P23352
Anosmin-1 precursor (Kallmann syndrome protein) (Adhesion molecule-like X-linked) gi—134048661—sp—P23352—KALM_HUMAN[134048661]

168: NP_982283
Notch homolog 2 N-terminal like protein [Homo sapiens] gi—46397353—ref—NP_982283.2—[46397353]

. . .

256: AAH76933
Elastase 2, neutrophil [Xenopus tropicalis] gi—49899920—gb—AAH76933.1 —[49899920]

257: 1FZZA
Chain A, The Crystal Structure Of The Complex Of Non-Peptidic Inhibitor Ono-6818 And Porcine Pancreatic Elastase. gi—16975403—pdb—1FZZ—A [16975403]

258: BAA00166
pancreatic elastase 2 precursor [Sus scrofa] gi—217686—dbj—BAA00166.1— [217686]

. . .

262: NP_493468
human KALlmann syndrome homolog family member (kal-1) [Caenorhabditis elegans] gi—25149859—ref—NP_493468.2—[25149859]

→

→

Selected ENTREZ responses to human elastase in the Protein database *(continued)*

263: AAH95070

Elastase 3 like [Danio rerio] gi—63101424—gb—AAH95070.1—[63101424]

. . .

346: AAD09442

guamerin [Hirudo nipponia] gi—4096732—gb—AAD09442.1—[4096732]

The format of the responses is as follows: in each case, the first line contains an identifier, its form reflecting the source database. For example, in the first response, P08246 is a Swiss-Prot accession number; in the second, 1HNEE signifies chain E of world-wide Protein Data Bank entry 1HNE. The next line gives the name and synonyms of the molecule, and the species of origin. Note that Greek letters are spelt out. The last line gives references to the source databanks: gi = GenInfo Identifier, (see page 25), gb = GenBank accession number, sp = Swiss-Prot, pdb = Protein Data Bank, pir = Protein Identification Resource, dbj = DNA Databank of Japan, ref = the Reference Sequence project of NCBI. The entries retrieved include elastases from human and other species, and also inhibitors of elastase.

Opening the entry that corresponds to the first hit retrieves a file containing the material shown in the next Box. (The entire file is 469 lines long.) The first lines are mostly database housekeeping—accession numbers, molecule name, date of deposition, etc. Then descriptive material such as the source, this case human, with the full taxonomic classification; credit to the scientists who deposited the entry; and literature references. There are extensive cross-references to other databanks. Finally, the particular scientific information: the location of the gene, and its product (CDS = coding sequence), and the sequence (see Exercise 4.2). Again, note that the sequence itself occupies quite a small portion of the entry.

US National Center for Biotechnology ENTREZ Protein database entry for human leukocyte elastase precursor

```
LOCUS       P08246                     267 aa            linear   PRI 01-MAY-2007
DEFINITION  Leukocyte elastase precursor (Elastase-2) (Neutrophil elastase)
            (PMN elastase) (Bone marrow serine protease) (Medullasin) (Human
            leukocyte elastase) (HLE).
ACCESSION   P08246
VERSION     P08246  GI:119292
DBSOURCE    swissprot: locus ELNE_HUMAN, accession P08246;
            class: standard.
            extra accessions:P09649,Q6B0D9,Q6LDP5
            created: Aug 1, 1988.
            sequence updated: Aug 1, 1988.
            annotation updated: May 1, 2007.
            xrefs: Y00477.1, CAA68537.1, M20203.1, AAA36359.1, M20199.1,
            M20200.1, M20201.1, M34379.1, AAA36173.1, AY596461.1, AAS89303.1,
            BC074816.2, AAH74816.1, BC074817.2, AAH74817.1, D00187.1,
            BAA00128.1, X05875.1, CAA29299.1, CAA29300.1, J03545.1, AAA52378.1,
```

→

US National Center for Biotechnology ENTREZ Protein database entry for human leukocyte elastase
precursor *(continued)*

```
                M27783.1, AAA35792.1, ELHUL, 1B0FA, 1H1BA, 1H1BB, 1HNEE, 1PPFE,
                1PPGE
                xrefs (non-sequence databases): UniGene:Hs.99863, MEROPS:S01.131,
                Ensembl:ENSG00000197561, KEGG:hsa:1991, HGNC:3309, MIM: 130130,
                MIM: 162800, DrugBank:BTD00002, LinkHub:P08246,
                ArrayExpress:P08246, GermOnline:ENSG00000197561,
                RZPD-ProtExp:F0319, GO:0009986, GO:0005576, GO:0008367, GO:0019955,
                GO:0042708, GO:0006874, GO:0045079, GO:0050922, GO:0050728,
                GO:0045415, GO:0045416, GO:0043406, GO:0048661, GO:0030163,
                GO:0009411, InterPro:IPR009003, InterPro:IPR001254,
                InterPro:IPR001314, Gene3D:G3DSA:2.40.10.10, PANTHER:PTHR19355,
                Pfam:PF00089, PRINTS:PR00722, SMART:SM00020, PROSITE:PS50240,
                PROSITE:PS00134, PROSITE:PS00135
KEYWORDS        3D-structure; Direct protein sequencing; Disease mutation;
                Glycoprotein; Hydrolase; Polymorphism; Protease; Serine protease;
                Signal.
SOURCE          Homo sapiens (human)
  ORGANISM      Homo sapiens
                Eukaryota; Metazoa; Chordata; Craniata; Vertebrata; Euteleostomi;
                Mammalia; Eutheria; Euarchontoglires; Primates; Haplorrhini;
                Catarrhini; Hominidae; Homo.
REFERENCE       1  (residues 1 to 267)
  AUTHORS       Nakamura,H., Okano,K., Aoki,Y., Shimizu,H. and Naruto,M.
  TITLE         Nucleotide sequence of human bone marrow serine protease
                (medullasin) gene
  JOURNAL       Nucleic Acids Res. 15 (22), 9601-9602 (1987)
   PUBMED       3479752
  REMARK        NUCLEOTIDE SEQUENCE [GENOMIC DNA].

Material omitted ...

COMMENT         On or before Mar 21, 2006 this sequence version replaced
                gi:74757422, gi:74724761, gi:67584.
                [FUNCTION] Modifies the functions of natural killer cells,
                monocytes and granulocytes. Inhibits C5a-dependent neutrophil
                enzyme release and chemotaxis.
                [CATALYTIC ACTIVITY] Hydrolysis of proteins, including elastin.
                Preferential cleavage: Val-|-Xaa > Ala-|-Xaa.
                [TISSUE SPECIFICITY] Bone marrow cells.
                [DISEASE] Defects in ELA2 are a cause of cyclic haematopoiesis (CH)
                [MIM:162800]; also known as cyclic neutropenia. CH is an autosomal
                dominant disease in which blood-cell production from the bone
                marrow oscillates with 21-day periodicity. Circulating neutrophils
                vary between almost normal numbers and zero. During intervals of
                neutropenia, affected individuals are at risk for opportunistic
                infection. Monocytes, platelets, lymphocytes and reticulocytes also
                cycle with the same frequency.
                [SIMILARITY] Belongs to the peptidase S1 family. Elastase
                subfamily.
                [SIMILARITY] Contains 1 peptidase S1 domain.
                [WEB RESOURCE] NAME=GeneReviews;
                URL='http://www.genetests.org/query?gene=ELA2'.
                [WEB RESOURCE] NAME=Wikipedia elastase entry;
                URL='http://en.wikipedia.org/wiki/Elastase'.
FEATURES             Location/Qualifiers
     source          1..267
                     /organism="Homo sapiens"
                     /db_xref="taxon:9606"
     gene            1..267
                     /gene="ELA2"
     Protein         1..267
                     /gene="ELA2"
                     /product="Leukocyte elastase precursor"
```

```
US National Center for Biotechnology ENTREZ Protein database entry for human leukocyte elastase
precursor (continued)

                         /EC_number="3.4.21.37"

Material omitted ...

     Region              30..267
                         /gene="ELA2"
                         /region_name="Mature chain"
                         /experiment="experimental evidence, no additional details
                         recorded"
                         /note="Leukocyte elastase. /FTId=PRO_0000027704."
     Bond                bond(55,71)
                         /gene="ELA2"
                         /bond_type="disulfide"
                         /experiment="experimental evidence, no additional details
                         recorded"
     Region              64..67
                         /gene="ELA2"
                         /region_name="Beta-strand region"
                         /experiment="experimental evidence, no additional details
                         recorded"
     Site                70
                         /gene="ELA2"
                         /site_type="active"
                         /experiment="experimental evidence, no additional details
                         recorded"
                         /note="Charge relay system."
ORIGIN
         1 mtlgrrlacl flacvlpall lggtalasei vggrrarpha wpfmvslqlr gghfcgatli
        61 apnfvmsaah cvanvnvrav rvvlgahnls rreptrqvfa vqrifengyd pvnllndivi
       121 lqlngsatin anvqvaqlpa qgrrlgngvq clamgwgllg rnrgiasvlq elnvtvvtsl
       181 crrsnvctlv rgrqagvcfg dsgsplvcng lihgiasfvr ggcasglypd afapvaqfvn
       241 widsiiqrse dnpcphprdp dpasrth
//
```

Many literature references, and many feature table entries have been omitted. Keywords (site types or region names) associated with feature table entries include: Helical region, Beta-strand region, Domain, Hydrogen bonded turn, Disulphide bridge, Mature chain, Propeptide, Signal, Tryp_SPc (signifying membership in the Trypsin-like Serine Protease family), Variant (e.g. an observed SNP), Substrate-binding site, Charge relay system, Glycosylation site.

Searches in the ENTREZ Gene database

Next we look again for HUMAN ELASTASE, this time in the Gene database. On the ENTREZ page, select CoreNucleotide from the pulldown menu at the left, type the following into the box following the word *for,* and then execute the search:

HOMO SAPIENS[ORGANISM] AND LEUKOCYTE[TITLE] AND ELASTASE[TITLE] NOT INHIBITOR[TEXT WORD]

The search returns two hits, including DNA (see Box) and mRNA.

The full ENTREZ Nucleotide database contains core nucleotide sequences + expressed sequence tags + genome survey sequences (similar to ESTs but genomic in origin).

The gene for human neutrophil elastase in the ENTREZ CoreNucleotide database

```
LOCUS       Y00477                    5292 bp    DNA     linear   PRI 14-NOV-2006
DEFINITION  Human bone marrow serine protease gene (medullasin) (leukocyte
            neutrophil elastase gene).
ACCESSION   Y00477
VERSION     Y00477.1  GI:34529
KEYWORDS    elastase; medullasin; serine protease.
SOURCE      Homo sapiens (human)
  ORGANISM  Homo sapiens
            Eukaryota; Metazoa; Chordata; Craniata; Vertebrata; Euteleostomi;
            Mammalia; Eutheria; Euarchontoglires; Primates; Haplorrhini;
            Catarrhini; Hominidae; Homo.
REFERENCE   1  (bases 1 to 5292)
  AUTHORS   Nakamura,H., Okano,K., Aoki,Y., Shimizu,H. and Naruto,M.
  TITLE     Nucleotide sequence of human bone marrow serine protease
            (medullasin) gene
  JOURNAL   Nucleic Acids Res. 15 (22), 9601-9602 (1987)
  PUBMED    3479752
REFERENCE   2  (bases 1 to 5292)
  AUTHORS   Naruto,M.
  TITLE     Direct Submission
  JOURNAL   Submitted (09-NOV-1987) Naruto M., Basic Research Laboratories,
            Toray Insustries, Inc., 1111 Tebiro, Kamakura 248, Japan
COMMENT     This cDNA encodes the full protein sequence of human leukocyte
            (neutrophil) elastase (HLE), which was reported by Sinha et al. in
            PNAS USA 84:2228-2232(1987).
FEATURES             Location/Qualifiers
     source          1..5292
                     /organism="Homo sapiens"
                     /mol_type="genomic DNA"
                     /db_xref="taxon:9606"
                     /clone_lib="tonsil genomic library in lambda gt WES lambda
                     B"
     repeat_region   287..551
                     /note="tandemly arranged direct repeats"
     CAAT_signal     1114..1118
     TATA_signal     1230..1234
     CDS             join(1287..1353,1786..1942,2173..2314,4485..4715,
                     4882..5088)
                     /codon_start=1
                     /product="serine protease"
                     /protein_id="CAA68537.1"
                     /db_xref="GI:296665"
                     /db_xref="GDB:118792"
                     /db_xref="GOA:P08246"
                     /db_xref="HGNC:3309"
                     /db_xref="InterPro:IPR001254"
                     /db_xref="InterPro:IPR001314"
                     /db_xref="InterPro:IPR009003"
                     /db_xref="PDB:1B0F"
                     /db_xref="PDB:1H1B"
                     /db_xref="PDB:1HNE"
                     /db_xref="PDB:1PPF"
                     /db_xref="PDB:1PPG"
                     /db_xref="UniProtKB/Swiss-Prot:P08246"
                     /translation="MTLGRRLACLFLACVLPALLLGGTALASEIVGGRRARPHAWPFM
                     VSLQLRGGHFCGATLIAPNFVMSAAHCVANVNVRAVRVVLGAHNLSRREPTRQVFAVQ
                     RIFENGYDPVNLLNDIVILQLNGSATINANVQVAQLPAQGRRLGNGVQCLAMGWGLLG
                     RNRGIASVLQELNVTVVTSLCRRSNVCTLVRGRQAGVCFGDSGSPLVCNGLIHGIASF
                     VRGGCASGLYPDAFAPVAQFVNWIDSIIQRSEDNPCPHPRDPDPASRTH"
     sig_peptide     join(1287..1353,1786..1805)
     mat_peptide     join(1806..1942,2173..2314,4485..4715,4882..5085)
                     /product="unnamed"
     exon            <1287..1353
                     /number=1
     intron          1354..1785
```

→

The gene for human neutrophil elastase in the ENTREZ CoreNucleotide database *(continued)*

```
                        /number=1
       exon             1786..1942
                        /number=2
       intron           1943..2172
                        /number=2
       exon             2173..2314
                        /number=3
       intron           2315..4484
                        /number=3
       repeat_region    2538..2957
                        /note="tandemly arranged direct repeats"
       exon             4485..4715
                        /number=4
       intron           4716..4881
                        /number=4
       exon             4882..>5088
                        /number=5
       polyA_signal     5146..5151
ORIGIN
         1 ttgtcagagc cccagctggt gtccagggac tgaccgtgag cctgggtgaa agtgagttcc
        61 ccgttggagg caccagacga ggagaggatg gaaggcctgg cccccaagaa tgagccctga
       121 ggttcaggag cggctggagt gagccgcccc cagatctccg tccagctgcg ggtcccagag
       181 gcctgggtta cactcggagc tcctggggga ggcccttgac gtgctcagtt cccaaacagg
       241 aaccctggga aggaccagag aagtgcctat tgcgcagtga gtgcccgaca cagctgcatg
       301 tggccggtat cacagggccc tgggtaaact gaggcaggcg acacagctgc atgtggccgg
       361 tatcacaggg ccctgggtaa actgaggcag gcgacacagc tgcatgtggc cggtatcaca
       421 gggccctggg taaactgagg caggcgacac agctgcatgt ggccggtatc acagggccct
       481 gggtaaactg aggcaggcga cacagctgca tgtggccggt atcacggggc cctggataaa
       541 cagaggcagg cgaggccacc cccatcaagt ccctcaggtc taggtttggc caggtttgga
       601 aaaacacagc aacgctcggt aaatctgaat ttcgggtaag tatatcctgg gcctcatttg
       661 gaagagactt agattaaaaa aaaaacgtcg agaccagccc ggccaacacg tgaaaccccg
       721 tctctactaa aaatacaaaa aattagccag gcgcagtgct cacgcctgtg atcccagcac
       781 tctgggaggt gaggcaggcg gatcacccga ggtcagctgt tcaagaccag cctggccgag
       841 tgggcgaaac actgtctcta ctacaaatac aaaaattagc cgggagtgga ggcaggtgcc
       901 tgtaatctca gctattcagg aggctgaggc aggagaatca cttgaacctg ggaggcggag
       961 gttgccgtga gccgggatca cgccaccgca ctccagcctg ggcgatagag caagactctg
      1021 tctccaaaaa aataaattaa aaaacccaca ttgattatct gacatttgaa tgcgattgtg
      1081 catcctgaat tttgtctgga ggccccaccc gagccaatcc agcgtcttgt cccccttctc
      1141 ccccttttca tcaacgcctg tgccagggga gaggaagtgg agggcgctgg ccggccgtgg
      1201 ggcaatgcaa cggcctccca gcacagggct ataagaggag ccgggcgggc acggaggggc
      1261 agagaccccg gagccccagc cccaccatga ccctcgggcg ccgactcgcg tgtcttttcc
      1321 tcgcctgtgt cctgccggcc ttgctgctgg ggggtgagtt tttgagtcca acctcccgct
      1381 gctccctctg tcccgggttc tgttcccacc tctccataga gggcccccacc agtgtgggtc
      1441 cctcatcctc acaggggagg tgccagctgg gacaaggaga ccagaagaga ctgaggttct
      1501 gagcggtgaa gccaccacca ggagcccaga gttggggttt gaaaaccggg gagggggggg
      1561 gtggcaggtc gccctctggg ttcaagtcca ggtctgtctg tgccttggag gggcaccgtg
      1621 gggaggtccc tttgcctctc cgtgcctcag tttcctcatc tgaacaacag gggtgcgaac
      1681 ggccccgatc ccgtgggttc ccggtggggg atccagaggc cccgtggccg ggaggggaca
      1741 ggctccttgg caggcactca gcacccgcac ccggtgtgtc cccaggcacc gcgctggcct
      1801 cggagattgt ggggggccgg cgagcgcggc cccacgcgtg gcccttcatg gtgtccctgc
      1861 agctgcgcgg aggccacttc tgcggcgcca ccctgattgc gcccaacttc gtcatgtcgg
      1921 ccgcgcactg cgtggcgaat gtgtgagtag ccgggagtgt gcgcgcccgg ctcggacccc
      1981 gcgtcccggt ctgtgaggtg ggtgggggga ggccggggcc ggggctgctg gcgggggggg
      2041 gtccgtccag ggcccgcggg gcccctcgag caccttcgcc ctcaggcccg tcgccggatg
      2101 gggacgacaa ggcgcggctg agcccccacc cccggggccg ccccctgagcc ccgcctctcc
      2161 ctcttttggc agaaacgtcc gcgcggtgcg ggtggtcctg ggagcccata acctctcgcg
      2221 gcgggagccc acccggcagg tgttcgccgt gcagcgcatc ttcgaaaacg gctacgaccc
      2281 cgtaaacttg ctcaacgaca tcgtgattct ccaggtgccg ccgggcgggc ggggcgagg
      2341 ggcggaggcc agaggcctgg ggaggtgga ggcctgggga gggtggaggc tgcgacggag
      2401 gggcgcgtcg gggccgctcg tggggacctg gggtggcatc gtgggctggg tggtcccctc
      2461 tccgcgcctc ggtctgcacc tctgtgaaac gggaaaatac ccgccatggg ccgttgaggg
      2521 gttaaatgag atcctgcagg gaggccccga tctgctgtca atcaacaaac ttactgagaa
      2581 gggaggcccc gatctgttgt caatcaacaa acttactgag aaggaggcc ccgatctgtt
      2641 gtcaatcaac aaacttactg agaagggagg ccccgatctg ctgtcaatca acaaacttac
      2701 tgagaaggga ggccccgatg ttgtcaatca acaaacttac tgagaaggga ggccccgatc
```

→

The gene for human neutrophil elastase in the ENTREZ CoreNucleotide database *(continued)*

```
2761 tgctgtcaat caaccaaact tactgagaag ggaggccccg atctgctgtc aatcatcaaa
2821 cttactgaga agggaggccc cgatctgctg tcaatcaaca aacttactga gaagggaggc
2881 ccccgatctg ttgtcaatca acaaacttac tgagaaggga ggccccgatc tgctgtcaat
2941 caacaaactt actgagattc tgtgtgtctc tccattcacc agtcctgtgg cccagggcag
3001 gggccgcctc tgtctttggg aaaaggggca aaagtcccca cctttccacc cctgtccgcg
3061 gcttgcagtt ctggttattt cctgggcgcc gggccccgtg gctcaggcct gtcatcccag
3121 cactttggga ggctgaggcg ggtggatcac gaggtcaggt gttcgagacc agcctgagca
3181 acatagtgaa accccgtctc tactaaaata cacaaaaaaa aaattagccg agtgtggttg
3241 tgggtgcctg taatgccaac tactcaggag gctgaggaag gagaatcgct tgaaccccgg
3301 aggcggagat tgcagtgagc tgagatcaca ccactgcact ccagcctggg tctcaaaaaa
3361 aaaaaaaaag attcctccct gggaagggtt agagggagag tttccttgtc actaagtttt
3421 ctcatagctc tcacccagtg cagtggcgcg atcgcagctc actacacctc catctcctgg
3481 gctcaagcca ccctctcagc ttggaatggg gggtagctgg aaccacaggt gccaccacgt
3541 ggtccaccac gtctggctaa tatatatata tacacacaca catacatata ttataaataa
3601 taaatatata ttttatttaa ataaaatata taatatttat aattattta taattataat
3661 aatatttata taattataaa tatcatttat aattataata tttattattt tataaaataa
3721 taaatataaa atatataaaa atattttat aaataataaa atatatatat acacacatat
3781 atatatattt tttgagacaa gtctcgctct gtcgcccagg ctggagcgca gtgcacaatc
3841 tcactcactg cacctccgcc tcccaggttc aagcgattct cctgcctcag cctcccaggt
3901 agctgggact acaggcgccc gccaccacgc ctggctaatt tttggtattg ttagtagaga
3961 cggggtttaa ccatgttagc caggatggtc ttgatctcct gacctttga ttggcccacc
4021 tcagcctccc aaaatgctgg gattataggc gtgagcaccg cacctggcaa ttttttttta
4081 ttattttgt agacatgggg ctttgccaca ttgcccaggc tggtcttgaa tgcctggcct
4141 ggcctaagtg atcctcctgc ctcgccctcc caaagtgctg ggcttacaag catgagccac
4201 cgcgcccggc tgtagttttt ttgttaactg agcacctact gcttcctgca ctcaagccac
4261 atccagggac aacctccaac gccctgagcc ttggtgacgg ctcccactct acagatgggg
4321 aaaccgaggc ttgccttggg gagcagagtg tggggtgggt atcctgccct gcaggatccc
4381 agaaccacag tggaacctga gatgggggaaa ctgaggcccg gagaggggag ggtcatcatc
4441 actgccccgt gtgacgcgct gacgatctgt ccccaccgcc acagctcaac gggtcggcca
4501 ccatcaacgc caacgtgcag gtggcccagc tgccggctca gggacgccgc ctgggcaacg
4561 gggtgcagtg cctggccatg ggctggggcc ttctgggcag gaaccgtggg atcgccacgcg
4621 tcctgcagga gctcaacgtg acggtggtga cgtccctctg ccgtcgcagc aacgtctgca
4681 ctctcgtgag gggccggcag gccggcgtct gtttcgtacg tgccctgggt gtccctctgc
4741 tccccacccg ctcccagccc ggtactgcag caacaggcac cgtggctaga ccctaggatg
4801 ggacttccca accctgacac gtcggcgggc aggtgggcag ggcctcgcag tccagcttcc
4861 ccaccttgtc tgcctccaca ggggactcc ggcagccct tggtctgcaa cgggctaatc
4921 cacggaattg cctccttcgt ccggggaggc tgcgcctcag ggctctaccc cgatgccttt
4981 gccccggtgg cacagtttgt aaactggatc gactctatca tccaacgctc cgaggacaac
5041 ccctgtcccc accccgggga cccggacccg gccagcagga cccactgaga agggctgccc
5101 gggtcacctc agctgcccac acccacactc tccagcatct ggcacaataa acattctctg
5161 ttttgtagaa tgtgtttgat gctccttggc tgtgtgattg ggtgttgaaa atggtcagta
5221 ggtcgggcgt ggtggctcac acctgtaatc ccagcacttt gggaggttga ggcaggcgga
5281 tcacttgagc tc
```

//

Compare this file with the result of searching in the Protein database (see Exercise 4.5).

Searches in the ENTREZ structure database

Is the three-dimensional structure of human elastase known? Select the Structure database and rerun the search. The program returns 11 answers:

2OXZ	HUMAN MMP-12 IN COMPLEX WITH TWO PEPTIDES PQG AND IAG
2OXW	HUMAN MMP-12 COMPLEXED WITH THE PEPTIDE IAG
2OXU	UNINHIBITED FORM OF HUMAN MMP-12
1HAX	SNAPSHOTS OF SERINE PROTEASE CATALYSIS: (A) ACYL-ENZYME INTERMEDIATE BETWEEN PORCINE PANCREATIC ELASTASE AND HUMAN BETA-CASOMORPHIN-7 AT PH 5
1Z3J	SOLUTION STRUCTURE OF MMP12 IN THE PRESENCE OF N-ISOBUTYL-N-4-METHOXYPHENYLSULFONYL]GLYCYL HYDROXAMIC ACID (NNGH)
1YCM	SOLUTION STRUCTURE OF MATRIX METALLOPROTEINASE 12 (MMP12) IN THE PRESENCE OF N-ISOBUTYL-N-[4-METHOXYPHENYLSULFONYL]GLYCYL HYDROXAMIC ACID (NNGH)
1W98	THE STRUCTURAL BASIS OF CDK2 ACTIVATION BY CYCLIN E
1JK3	CRYSTAL STRUCTURE OF HUMAN MMP-12 (MACROPHAGE ELASTASE) AT TRUE ATOMIC RESOLUTION
1HAZ	SNAPSHOTS OF SERINE PROTEASE CATALYSIS: (C) ACYL-ENZYME INTERMEDIATE BETWEEN PORCINE PANCREATIC ELASTASE AND HUMAN BETA-CASOMORPHIN-7 JUMPED TO PH 9 FOR 1 MINUTE
1B0F	CRYSTAL STRUCTURE OF HUMAN NEUTROPHIL ELASTASE WITH MDL 101, 146
1QIX	PORCINE PANCREATIC ELASTASE COMPLEXED WITH HUMAN BETA-CASOMORPHIN-7

The designations 2OXZ, etc. are entry codes from the Protein Data Bank.

OOPS!—we may not realize it, but we have missed many useful entries. There are many elastase structures solved in complex with inhibitors, which we have asked the system to reject. Deleting NOT INHIBITOR and rerunning the query returns 29 structures:

2OXZ	HUMAN MMP-12 IN COMPLEX WITH TWO PEPTIDES PQG AND IAG
2OXW	HUMAN MMP-12 COMPLEXED WITH THE PEPTIDE IAG
2OXU	UNINHIBITED FORM OF HUMAN MMP-12
2HU6	CRYSTAL STRUCTURE OF HUMAN MMP-12 IN COMPLEX WITH ACETO-HYDROXAMIC ACID AND A BICYCLIC INHIBITOR
1HAX	SNAPSHOTS OF SERINE PROTEASE CATALYSIS: (A) ACYL-ENZYME IN-TERMEDIATE BETWEEN PORCINE PANCREATIC ELASTASE AND HUMAN BETA-CASOMORPHIN-7 AT PH 5
2D26	ACTIVE SITE DISTORTION IS SUFFICIENT FOR PROTEINASE INHIBIT SECOND CRYSTAL STRUCTURE OF COVALENT SERPIN-PROTEINASE COM-PLEX

→

1Z3J	SOLUTION STRUCTURE OF MMP12 IN THE PRESENCE OF N-ISOBUTYL-N-4-METHOXYPHENYLSULFONYL]GLYCYL HYDROXAMIC ACID (NNGH)
1YCM	SOLUTION STRUCTURE OF MATRIX METALLOPROTEINASE 12 (MMP12) IN THE PRESENCE OF N-ISOBUTYL-N-[4-METHOXYPHENYLSULFONYL]GLYCYL HYDROXAMIC ACID (NNGH)
1Y93	CRYSTAL STRUCTURE OF THE CATALYTIC DOMAIN OF HUMAN MMP12 COMPLEXED WITH ACETOHYDROXAMIC ACID AT ATOMIC RESOLUTION
1W98	THE STRUCTURAL BASIS OF CDK2 ACTIVATION BY CYCLIN E
1UTZ	CRYSTAL STRUCTURE OF MMP-12 COMPLEXED TO (2R)-3-([4-[(PYRIDIN-4-YL)PHENYL]-THIEN-2-YLCARBOXAMIDO)(PHENYL)PROPANOIC ACID
1UTT	CRYSTAL STRUCTURE OF MMP-12 COMPLEXED TO 2-(1,3-DIOXO-1,3-DIHYDRO-2H-ISOINDOL-2-YL)ETHYL-4-(4-ETHOXY[1,1-BIPHENYL]-4-YL)-4-OXOBUTANOIC ACID
1ROS	CRYSTAL STRUCTURE OF MMP-12 COMPLEXED TO 2-(1,3-DIOXO-1,3-DIHYDRO-2H-ISOINDOL-2-YL)ETHYL-4-(4-ETHOXY[1,1-BIPHENYL]-4-YL)-4-OXOBUTANOIC ACID
1RMZ	CRYSTAL STRUCTURE OF THE CATALYTIC DOMAIN OF HUMAN MMP12 COMPLEXED WITH THE INHIBITOR NNGH AT 1.3 A RESOLUTION
1OS9	BINARY ENZYME-PRODUCT COMPLEXES OF HUMAN MMP12
1OS2	TERNARY ENZYME-PRODUCT-INHIBITOR COMPLEXES OF HUMAN MMP12
1LQ8	CRYSTAL STRUCTURE OF CLEAVED PROTEIN C INHIBITOR
1H1B	CRYSTAL STRUCTURE OF HUMAN NEUTROPHIL ELASTASE COMPLEXED WITH AN INHIBITOR (GW475151)
1JIZ	CRYSTAL STRUCTURE ANALYSIS OF HUMAN MACROPHAGE ELASTASE MMP-12
1JK3	CRYSTAL STRUCTURE OF HUMAN MMP-12 (MACROPHAGE ELASTASE) AT TRUE ATOMIC RESOLUTION
1HAZ	SNAPSHOTS OF SERINE PROTEASE CATALYSIS: (C) ACYL-ENZYME INTERMEDIATE BETWEEN PORCINE PANCREATIC ELASTASE AND HUMAN BETA-CASOMORPHIN-7 JUMPED TO PH 9 FOR 1 MINUTE
1B0F	CRYSTAL STRUCTURE OF HUMAN NEUTROPHIL ELASTASE WITH MDL 101, 146
1QIX	PORCINE PANCREATIC ELASTASE COMPLEXED WITH HUMAN BETA-CASOMORPHIN-7
2REL	SOLUTION STRUCTURE OF R-ELAFIN, A SPECIFIC INHIBITOR OF ELASTASE, NMR, 11 STRUCTURES
1FLE	CRYSTAL STRUCTURE OF ELAFIN COMPLEXED WITH PORCINE PANCREATIC ELASTASE
1FUJ	PR3 (MYELOBLASTIN)

> 1PPG HUMAN LEUKOCYTE ELASTASE (HLE) (E.C.3.4.21.37) COMPLEX WITH MEO-SUCCINYL-ALA-ALA-PRO-VAL CHLOROMETHYLACETONE
> 1PPF HUMAN LEUKOCYTE ELASTASE (HLE) (NEUTROPHIL ELASTASE (HNE)) (E.C.3.4.21.37) COMPLEX WITH THE THIRD DOMAIN OF TURKEY OVOMUCOID INHIBITOR (OMTKY3)
> 1HNE HUMAN NEUTROPHIL ELASTASE (HNE) (E.C.3.4.21.37) (ALSO REFERRED TO AS HUMAN LEUCOCYTE ELASTASE (HLE)) COMPLEX WITH METHOXYSUCCINYL-ALA-ALA-PRO-ALA CHLOROMETHYL KETONE (MSACK)

On the other hand, if we search the PDB for elastase (see below), we find 132 structures.

Searches in the bibliographic database PubMed

Perhaps it is time to look at what people have had to say about our molecule. Of course the literature on elastase is huge. A search in PubMed for HUMAN ELASTASE returns 8709 entries. To prune the results, let us try to find citations to articles describing the role of elastase in disease. A search for HUMAN ELASTASE DISEASE returns 1892 entries. What about specific elastase *mutants* related to human disease? A search for HUMAN ELASTASE DISEASE MUTATION returns 71 articles, in reverse chronological order. Here are the first 10:

> 1. Karlsson J, Carlsson G, Ramme KG, Hagglund H, Fadeel B, Nordenskjold M, Henter JI, Palmblad J, Putsep K, Andersson M. Low plasma levels of the protein pro-LL-37 as an early indication of severe disease in patients with chronic neutropenia. Br J Haematol. 2007 Apr;137(2):166–9.
>
> 2. Haug U, Hillebrand T, Bendzko P, Low M, Rothenbacher D, Stegmaier C, Brenner H. Mutant-enriched PCR and allele-specific hybridization reaction to detect K-ras mutations in stool DNA: high prevalence in a large sample of older adults. Clin Chem. 2007 Apr;53(4):787-90. Epub 2007 Feb 22.
>
> 3. Thomas M, Jayandharan G, Chandy M. Molecular screening of the neutrophil elastase gene in congenital neutropenia. Indian Pediatr. 2006 Dec;43(12):1081-4.
>
> 4. Schmid M, Fellermann K, Fritz P, Wiedow O, Stange EF, Wehkamp J. Attenuated induction of epithelial and leukocyte serine antiproteases elafin and secretory leukocyte protease inhibitor in Crohn's disease. J Leukoc Biol. 2007 Apr;81(4):907-15. Epub 2007 Jan 2.

→

5. Archer H, Jura N, Keller J, Jacobson M, Bar-Sagi D. A mouse model of hereditary pancreatitis generated by transgenic expression of R122H trypsinogen. Gastroenterology. 2006 Dec;131(6):1844-55. Epub 2006 Oct 1.

6. Selig L, Sack U, Gaiser S, Kloppel G, Savkovic V, Mossner J, Keim V, Bodeker H. Characterisation of a transgenic mouse expressing R122H human cationic trypsinogen. BMC Gastroenterol. 2006 Oct 27;6:30.

7. Thusberg J, Vihinen M. Bioinformatic analysis of protein structure-function relationships: case study of leukocyte elastase (ELA2) missense mutations. Hum Mutat. 2006 Dec;27(12):1230-43.

8. Welte K, Zeidler C, Dale DC. Severe congenital neutropenia. Semin Hematol. 2006 Jul;43(3):189-95. Review.

9. Kameda K, Matsunaga T, Abe N, Fujiwara T, Hanada H, Fukui K, Fukuda I, Osanai T, Okumura K. Increased pericardial fluid level of matrix metalloproteinase-9 activity in patients with acute myocardial infarction: possible role in the development of cardiac rupture. Circ J. 2006 Jun;70(6):673-8.

10. Carlsson G, Aprikyan AA, Ericson KG, Stein S, Makaryan V, Dale DC, Nordenskjold M, Fadeel B, Palmblad J, Hentera JI. Neutrophil elastase and granulocyte colony-stimulating factor receptor mutation analyses and leukaemia evolution in severe congenital neutropenia patients belonging to the original Kostmann family in northern Sweden. Haematologica. 2006 May;91(5):589-95.

There are references to a relationship between mutations in neutrophil elastase and neutropenia—a low level of a type of white blood cells called neutrophils. To pursue this idea, we can look for elastase in the database of human genetic disease.

Online Mendelian Inheritance in Man (OMIM^TM)

OMIM is a database of human genes and genetic disorders. It was originally compiled by V.A. McKusick, M. Smith and colleagues and published on paper. The National Center for Biotechnology Information (NCBI) of the US National Library of Medicine has developed it into a database accessible from the Web, and introduced links to other archives of related information, including sequence databanks and the medical literature. OMIM is now well integrated with the NCBI information-retrieval system ENTREZ. A related database, the OMIM Morbid Map, treats genetic diseases and their chromosomal locations.

The response to ELASTASE in a search of OMIM describes the results linking mutations in the gene to both cyclic and congenital (non-cyclic) neutropenia. OMIM lists nine allelic variants (many more are known). Five are associated with cyclic neutropenia, of which three cause amino acid substitutions, one is in a splice site and one is in an intron. Four variants, all substitutions, are associated with severe congenital neutropenia.

The collection of results on elastase that we have assembled would support research on the system; for instance, we could map elastase mutants onto the structure of the molecule to see whether we could derive clues to the causes of cyclic and non-cyclic neutropenia (see Weblem 6.12).

The Sequence Retrieval System (SRS)

SRS, written originally by T. Etzold, and developed by Lion Bioscience, is an integrated system for information retrieval from many sequence databases, and for passing the sequences retrieved to analytic tools such as sequence comparison and alignment programs.

SRS can search 200 databases of protein and nucleotide sequences, metabolic pathways, three-dimensional structures and functions, genomes, and disease and phenotype information (see http://srs.ebi.ac.uk/srsbin/cgi-bin/wgetz?-page+databanks). These include many small databases such as the Prosite and Blocks databases of protein structural motifs, transcription factor databases, and databases specialized to certain pathogens. There are many publicly available SRS servers, which vary to some extent in the complement of databases which they can search. See http://downloads.biowisdomsrs.com/publicsrs.html

In addition to the number and variety of databases to which it offers access, SRS offers tight links among the databases, and fluency in launching applications. A search in a single database component can be extended to a search in the complete network; that is, entries in all databases that pertain to a given protein, can be found easily. Similarity searches and alignments can be launched directly, without saving the responses in an intermediate file.

Other options on the search results page allow you to create and download reports on the selected matches. This might be simply a listing of the sequences, or the result of a more complex analysis of the results. For example, the protein databases offer a hydrophobicity plot.

The parent URL of SRS is: http://srs.ebi.ac.uk/
http://www.lionbioscience.ac.psiweb.com/publicsrs.html contains a list of the many mirror sites.

To search for human elastase, open an SRS session and select SWISSPROT. Enter HUMAN ELASTASE as a simple query, and click QUICK SEARCH. The program returns:

RootLibs	acc	des	sl
SWISSPROT:EL1_HUMAN	P11423	ELASTASE 1 (EC 3.4.21.36) (FRAGMENT).	68
SWISSPROT:EL2A_HUMAN	P08217	ELASTASE 2A PRECURSOR (EC 3.4.21.71).	269
SWISSPROT:EL2B_HUMAN	P08218	ELASTASE 2B PRECURSOR (EC 3.4.21.71).	269
SWISSPROT:EL3A_HUMAN	P09093	ELASTASE IIIA PRECURSOR (EC 3.4.21.70) (PROTEASE E).	270
SWISSPROT:EL3B_HUMAN	P08861	ELASTASE IIIB PRECURSOR (EC 3.4.21.70) (PROTEASE E).	270
SWISSPROT:ELNE_HUMAN	P08246 P09649	LEUKOCYTE ELASTASE PRECURSOR (EC 3.4.21.37) (NEUTROPHIL ELASTASE) (PMN ELASTASE) (BONE MARROW SERINE PROTEASE) (MEDULLASIN).	267 267
SWISSPROT:ILEU_HUMAN	P30740	LEUKOCYTE ELASTASE INHIBITOR (LEI) (MONOCYTE/NEUTROPHIL ELASTASE INHIBITOR) (M/NEI) (EI).	379
SWISSPROT:ELAF_HUMAN	P19957	ELAFIN PRECURSOR (ELASTASE-SPECIFIC INHIBITOR) (ESI) (SKIN-DERIVED ANTILEUKOPROTEINASE) (SKALP).	117 117

To demonstrate the launching of applications, let us retrieve the sequences of mammalian elastases, and perform multiple sequence alignment using CLUSTAL-W. In the SRS session, select the library UniProtKB and, in the Standard Query form, search for ALL TEXT: ELASTASE!INHIBITOR!FRAGMENT and TAXONOMY: MAMMALIA. (The ! signifies NOT. The goal is to retrieve entries in which the description points to mammalian elastases, but does not contain INHIBITOR and does not contain FRAGMENT. We want only complete elastase molecules, no inhibitors or fragments need apply.)

The program will return approximately 50 responses. (The screen shot shows only the first dozen.)

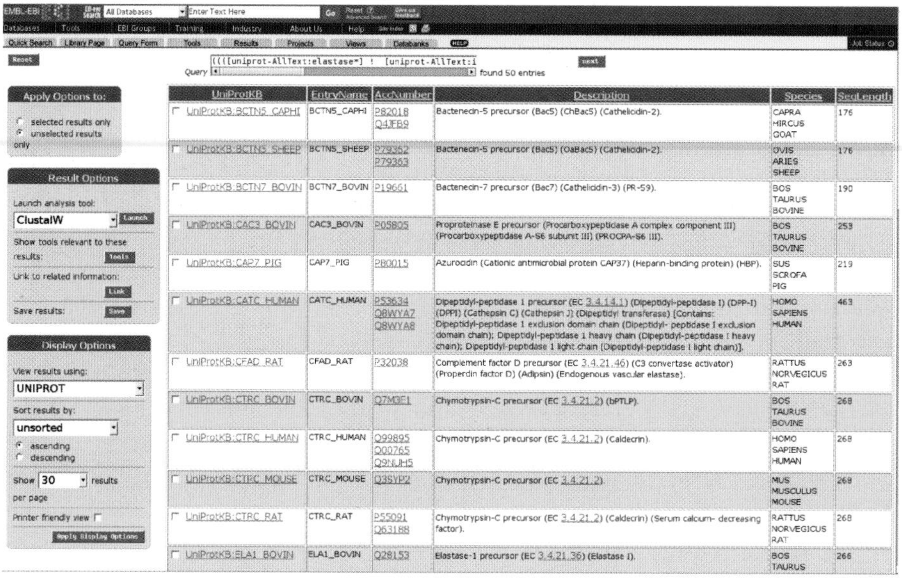

Select some or all of them by clicking in the boxes at the left. Look for a pulldown menu under the heading LAUNCH ANALYSIS TOOL, set it to CLUSTALW, and click on launch. The program will show you the scaffolding of the input to CLUSTAL-W that it created; this gives you the opportunity to alter the values of parameters from their defaults, or to discontinue the run if for any reason you are dissatisfied. Initiate the calculation by clicking on LAUNCH again, and wait.

The results of one such alignment are shown in Plate IV.

The Protein Identification Resource (PIR)

The Protein Identification Resource is an effective combination of a carefully curated database, information-retrieval access software and a workbench for investigations of sequences. The PIR describes itself as an integrated protein informatics resource for genomic and proteomic research. Think of this as an analysis package sitting on top of a retrieval system. Its functionality includes browsing, searching and similarity analysis, and links to other databases. Users may:

◆ Browse by annotations

- Search selected text fields for different annotations, such as Superfamily, Family, Title, Species, Taxonomy group, Keywords and Domains

- Analyse sequences using BLAST or FASTA Searches, Pattern Match, Multiple alignment

- Global and Domain Search, and Annotation-sorted Search

- View Statistics for Superfamily, Family, Title, Species, Taxonomy group, Keywords, Domains, Features

- View Links to other databases, including PDB, COG, KEGG, WIT and BRENDA

- Select Specialized Sequence Groups such as Human, Mouse, Yeast and *E. coli* genomes

A URL for a search of PIR by Text terms is: http://pir.georgetown.edu/pirwww

One feature of the PIR International system is the search for a specific peptide. Looking at the alignment of mammalian elastases in Plate IV, we note at positions 220–228 a conserved motif: most of the sequences contain CNGDSGGPLN. In the PIR, we can select PEPTIDE SEARCH in *i*ProClass, and retrieve exact matches for the subsequence CNGDSGGPLNgiving:

Recognizing proteins from sequences of fragments also has applications in proteomics; see Chapter 7.

```
   1       ELRT2  pancreatic elastase II (EC 3.4.21.71)
           214 - 223 GVTSSCNGDSGGPLNCQASN
   2       CPBOA3  procarboxypeptidase A complex compon
           183 - 192 DTRSGCNGDSGGPLNCPAAD
   3       S68826  pancreatic elastase (EC 3.4.21.36) i
           212 - 221 GVISACNGDSGGPLNCQLEN
   4       S68825  pancreatic elastase (EC 3.4.21.36) i
           212 - 221 GVISACNGDSGGPLNCQLEN
   5       A29934  pancreatic elastase (EC 3.4.21.36) I
           213 - 222 YIRSGCNGDSGGPLNCPTED
   6       B26823  pancreatic elastase II (EC 3.4.21.71
           212 - 221 GVISSCNGDSGGPLNCQASD
   7       C26823  pancreatic elastase II (EC 3.4.21.71
           212 - 221 GVICTCNGDSGGPLNCQASD
   8       A26823  pancreatic elastase II (EC 3.4.21.71
           212 - 221 GIISSCNGDSGGPLNCQGAN
   9       A25528  pancreatic elastase II (EC 3.4.21.71
           214 - 223 GVTSSCNGDSGGPLNCRASN
  10       JQ1473  pancreatic elastase (EC 3.4.21.36) I
           212 - 221 GVISACNGDSGGPLNCQAED
  11       B29934  pancreatic elastase (EC 3.4.21.36) I
           213 - 222 DIRSGCNGDSGGPLNCPTED
  12       S29239  chymotrypsin (EC 3.4.21.1) 1 precurs
           219 - 228 GGKSTCNGDSGGPLNLNGMT
  13       T10495  chymotrypsin (EC 3.4.21.1) BII - pen
           214 - 223 GGKGTCNGDSGGPLNLNGMT
```

Note that the molecule names are truncated, which can sometimes create misleading sitations, especially if one tries to analyse the output with a computer program, with which it is often harder to see the obvious. For instance, it might appear that an identical 10-residue subsequence appears in carboxypeptidase, a molecule entirely unrelated to elastase. But entry CPBOA3, the second response, is actually the molecule bovine procarboxypeptidase A complex component III, an elastase homologue. Chymotrypsin is of course a close homologue of elastase.

Returning to the alignment table (Plate IV), variations in the pattern appear in some molecules. The more general search for C[RNQF]GDSG[GS]PL[HNV], in which [XYZ] means a position containing X or Y or Z, would pull out all the mammalian elastases in the alignment, plus a total of 82 sequences in all. Even these are not all the sequences related to elastase in the databank, as one could find by running a PSI-BLAST search for any of the sequences, or, remaining strictly within PIR, by looking up elastase in the PROT-FAM database. The pattern matches 20 families, all serine proteinases.

We are well on the way to generating a complete list of homologues.

ExPASy — Expert Protein Analysis System

ExPASy is the information-retrieval and analysis system of the Swiss Institute of Bioinformatics, which (in collaboration with the European Bioinformatics Institute) also produces the protein sequence databases SWISS-PROT and TrEMBL. TrEMBL contains translations of nucleotide sequences from the EMBL Nucleotide Database not yet fully integrated into SWISS-PROT.

Opening the main web page of ExPASy (http://www.expasy.org) and selecting SWISS-PROT and TrEMBL gives access to a set of information-retrieval tools. There is also the option of searching SWISS-PROT directly. If we select FULL TEXT SEARCH and probe SWISS-PROT with the single term ELASTASE, we find ELNE_HUMAN, the real goal of our search, and 180 other hits — 53 from SWISS-PROT and 127 from TrEMBL. These include many inhibitors. One elastase homologue found is from the blood fluke: CERC_SCHMA. Both sequences are precursors; in the following alignment of these two sequences, upper case letters indicate the mature enzyme:

```
CERC_SCHMA  --msnrwrfvvvvtlftycltfervstwlIRSGEPVQHPAEFPFIAFLTTER-TMCTGSL  57
ELNE_HUMAN  mtlgrrlaclflacvlpalllggtalaseIVGGR-RARPHAWPFMVSLQLRGGHFCGATL  59
            :..*  :.:. ::.  *  . : *.*.  :* :**:. *   :* .:*
CERC_SCHMA  VSTRAVLTAGHCVCSPLPVIRVSFLTLRNGDQQGIHHQPSGVKVAPGYMPSCMSARQRRP  117
ELNE_HUMAN  IAPNFVMSAAHCVAN----VNVRAVRVVLGAHNLSRREP----TRQVFAVQRIFENGYDP  111
            ::.. *::*.***..   :.*  :  : ::*  . .  : .   *
CERC_SCHMA  IAQTLSGFDIAIVMLAQMVNLQSGIRVISLPQPSDIPPPGTGVFIVGYGRDDNDRDPSRK  177
ELNE_HUMAN  VNLLN---DIVILQLNGSATINANVQVAQLPAQGRRLGNGVQCLAMGWGLLGRNRG----  164
            :      **.*: *    .:::.::* .**   .     *. : :*:* ..:*.
CERC_SCHMA  NGGILKKGRATIMECRHATNGNPICVKAGQNFGQLPAPGDSGGPLLPS-LQGPVLGVVSH  236
ELNE_HUMAN  IASVLQELNVTVVTS-LCRRSNVCTLVRGRQAG--VCFGDSGSPLVCNGLIHGIASFVRG  221
            ..:*:: ..*:: ..*   :   *:: *   **** .**:.  *   : ..*
CERC_SCHMA  GVTLPNLPDIIVEYASVARMLDFVRSNI------------------  264
ELNE_HUMAN  GCASGLYPDAFAPVAQFVNWIDSIIQRSEDNPCPHPRDPDPASRTH  267
            *  :   ** :.  *.... :*  : ..
```

The structure of human neutrophil elastase is known from X-ray crystallography, but that of the blood fluke elastase is not.

One of the facilities of the ExPASy server is the link to SWISS-MODEL, an automatic web server for building homology models. Opening SWISS-MODEL and choosing FIRST APPROACH MODE (the simplest), we can simply enter the SWISS-PROT code

CERC_SCHMA, and launch the application. Model building is not a trivial operation, so the job is done offline and the results sent by e-mail.

We shall discuss SWISS-MODEL further in Chapter 6.

Where do we go from here?

We have visited only a few of the many databanks in molecular biology accessible on the Web. In the short term, readers will explore these sites and others, and become familiar with not only the contents of the Web but its dynamics—the appearance and disappearance of sites and links. There are various biological metaphors for the Web—as an ecosystem, that is evolving, or that is growing polluted by dead sites and links to dead sites.

Databanks are developing more effective avenues of intercommunication, to the point where ever more intimate links shade into apparent coalescence. The time is not far off when there will be one molecular biology databank, with many avenues of access. Scientists will be able to configure their own access to selected slices and views of the information, creating personal 'virtual databases'.

Recommended reading

Each year the January issue of the journal *Nucleic Acids Research* contain a set of articles on databases in molecular biology. This should be kept at hand for ready reference.

Bishop, M.J. (1999). *Genetics Databases.* Academic Press, London. [A compendium of databases, access and analysis.]

Doolittle, R.F. (1981). Similar amino acid sequences: chance or common ancestry? *Science* **214**, 149–159. [Some basic ideas about the relationship between sequence similarity and homology.]

Hubbard T.J. *et al.* (2007). Ensembl 2007. *Nucleic Acids Research*, **35**, D610–D617. [Recent description of Ensembl.]

http://www.ornl.gov/sci/techresources/Human_Genome/posters/chromosome/sequence.shtml [Tutorial covering accessing records in NCBI's sequence databases, with links to tutorials about other ENTREZ databases.]

http://www.nlm.nih.gov/bsd/pubmed_tutorial/m1001.html [NCBI tutorial on use of PubMed]

Likić, V.A. (2006). Databases of metabolic pathways. Biochemistry and Molecular Biology Education, **6**, 408–412. [Expository comparison of BioCyc and KEGG.]

Exercises, Problems and Weblems

Exercises

Exercise 4.1 A database of vehicles has entries for the following: bicycle, tricycle, motorcycle, car. It stores only the following information about each entry: (1) how many wheels (a number), and (2) source of propulsion = human or engine. For every possible pair of vehicles, devise a logical combination of query terms referring to either the exact value or the range in the number of wheels, and to the source of propulsion, that will return the two selected vehicles and no others.

Exercise 4.2 A box on pages 223–5 showed the NCBI protein entry for human elastase 1 precursor. On a photocopy of this page, indicate which items are (a) purely database housekeeping, (b) peripheral data such as literature references, (c) the results of experimental measurements, (d) information inferred from experimental measurements, or (e) links to other databases exclusive of literature references.

Exercise 4.3 Why did the search in the ENTREZ structure database return 29 structures, but the search in the PDB returned many more? (See page 229.)

Exercise 4.4 Write a PERL script to extract the amino acid sequence or the encoded protein from an entry in the EMBL nucleotide sequence database, as shown in the Box, pages 191–2, and convert it to FASTA format.

Exercise 4.5 Compare the file retrieved by a search in NCBI for human elastase under Protein (pages 223–5) and Nucleotide (pages 226–8). On photocopies of these two pages, mark with a highlighter all information that the two files have in common.

Exercise 4.6 What taxon contains the latest common ancestor of the human and the aardvark? (Compare information in boxes on pages 191 and 224.)

Exercise 4.7 The Box on pages 223–5 contains the amino acid sequence of human elastase 1 precursor. What sequence differences are there between this and the mature protein?

Problems

Problem 4.1 Based on Fig. 4.3a: (a) Are all ATP-dependent DNA helicases DNA-binding proteins? (b) Are any ATP-dependent DNA helicases chromatin-binding proteins? (c) Are any ATP-dependent helicases DNA-binding proteins?

Problem 4.2 The multiple sequence alignment of mammalian elastases in Plate IV contains 34 conserved residues.

(a) How many residues are conserved, in the alignment shown in Plate IV, between EL2_PIG and EL2_RAT?

(b) How many residues are conserved, in the alignment shown in Plate IV, between EL2_BOVINE and EL2_MOUSE?

(c) How many of the positions found in parts (a) and (b) are common?

(d) How many positions found in (a) are *not* conserved in the full alignment in Plate IV?

(e) How many positions found in (b) are *not* conserved in the full alignment?

(f) How many positions found in (c) are *not* conserved in the full alignment?

The point of this problem is to compare the efficacy of detection of conservation patterns between pairwise and multiple sequence alignments. In principle, the reader should have been required to perform pairwise realignments of each pair of sequences treated separately. However, for sequences this closely related, that would not make a very great difference. For distantly related sequences, it would have been essential.

Weblems

Weblem 4.1 Retrieve the complete SWISS-PROT entry for bovine pancreatic trypsin inhibitor (*not* pancreatic secretory trypsin inhibitor) and the complete PIR entry for this protein. What information does each have that the other does not? (Note: the Box on pages 195–7 contains only *part* of the entry.)

Weblem 4.2 Find a list of official and unofficial mirror sites of the Protein Data Bank. Which is closest to you?

Weblem 4.3 Find all structures of sperm whale myoglobin in the Protein Data Bank and draw a histogram of their dates of deposition. Consider only the year of deposition.

Weblem 4.4 Find protein structures determined by Peter Hudson, alone or with colleagues.

Weblem 4.5 Design a search string for use with the RCSB home page site search facility, that would return *E. coli* thioredoxin structures but *not* Staphylococcal nuclease structures.

Weblem 4.6 For what fraction of structures determined by X-ray crystallography deposited in the Protein Data Bank have structure factor files also been deposited?

Weblem 4.7 Protein Data Bank entry 8XIA contains the structure of one monomer of D-xylose isomerase from *Streptomyces rubiginosus*. What is the probable quaternary structure? How was the geometry of the assembly corresponding to the probable quaternary structure derived from the coordinates in the entry?

Weblem 4.8 Find structural neighbours of Protein Data Bank entry 2TRX (*E. coli* thioredoxin), according to SCOP, CATH, FSSP and CE. Which if any structures do *all* these classifications consider structural neighbours of 2TRX? Which structures are considered structural neighbours in some but not all classifications?

Weblem 4.9 Why did an ENTREZ search in the protein category for HUMAN ELASTASE return a transcriptional regulator?

Weblem 4.10 What is the relationship between the elastase sequences recovered from searching the NCBI and the PIR?

Weblem 4.11 Using SWISS-PROT directly, or SRS, recover the SWISS-PROT entry for human elastase. What information does this file contain that does not appear in (a) the corresponding entry in ENTREZ (Protein) and (b) the corresponding entry in PIR?

Weblem 4.12 What homologues of human neutrophil elastase can be identified by one round of PSI-BLAST in the non-redundant database (nr)? (a) With E-value $< 10^{-100}$. (b) With E-value between 0.0001 and 0.00001.

Weblem 4.13 List five structures returned from the RCSB after a search for elastase, that are not listed in the box on page 229.

Weblem 4.14 Search for structures of elastases using the Data Bank facilities of OCA. Compare the results with those from ENTREZ/structure, described in the text.

Weblem 4.15 Which gene in *C. elegans* encodes a protein similar in sequence to human elastase?

Weblem 4.16 What is the chromosomal location of the human gene for glucose-6-phosphate dehydrogenase?

Weblem 4.17 Pseudogenes in eukaryotes can be classified into those that arose by gene duplication and divergence, and those reinserted into the genome from mRNA by a retrovirus, called *processed* pseudogenes. Processed pseudogenes can be identified by the absence of introns. Which if any of the pseudogenes in the human globin gene clusters are processed pseudogenes?

Weblem 4.18 Preliminary genetic analysis on the way to isolating the gene associated with cystic fibrosis bracketed it between the MET oncogene and RFLP D7S8. It was then estimated that this region contained 1–2 million bp, and might contain 100–200 genes. (a) How many base pairs long did this region actually turn out to be? (b) How many expressed genes is this region now believed to contain?

Weblem 4.19 The gene for Berardinelli–Seip syndrome was initially localized between two markers on chromosome band 11q13—D11S4191 and D11S987. How many base pairs are there in the interval between these two markers?

Weblem 4.20 Is there a database available on the Web that collects structural and thermodynamic information on protein–nucleic acid interactions?

Weblem 4.21 The yeast proteome database contains an entry for cdc6, the protein that regulates initiation of DNA replication.

(a) On what chromosome is the gene for yeast cdc6?

(b) What post-translational modification does this protein undergo to reach its mature active state?

(c) What are the closest known relatives of this protein in other species?

(d) With what other proteins is yeast cdc6 known to interact?

(e) What is the effect of distamycin A on the activity of yeast cdc6?

(f) What is the effect of actinomycin D on the the activity of yeast cdc6?

Weblem 4.22 Find a Web site that purports to contain a useful list of molecular biology databases and computation servers. Report the URL of the site. Choose 10 of the sites listed, at random. How many dead links did you encounter? (Count as a live link a window that opens with a message indicating that you will be transferred to a newer site, provided that this transfer successfully takes you to the listed project.)

Weblem 4.23 Find the entry for peptidylglycine monooxygenase in BRENDA http://www.brenda-enzymes.org/.

(a) What is the Michaelis constant (K_M) for the enzyme in the pig?

(b) What is the pH optimum for the enzyme in the pig?

Weblem 4.24 (a) What is the Gene Ontology classification of peptidylglycine monooxygenase, under molecular function? (b) Show one path from this enzyme back to the root of the molecular function graph.

Weblem 4.25 An enzyme with a function related to peptidylglycine monooxygenase (EC 1.14.17.3) linked to it in the ENZYME DB, is 1-aminocyclopropane-1-carboxylate oxidase (EC 1.14.17.4).

(a) What is the lowest common ancestor of the two reactions in the EC classification?

(b) What is the lowest common ancestor of the two reactions in the Gene Ontology classification?

(c) Are these two enzymes closely related in the Gene Ontology classification?

Weblem 4.26 The node corresponding to DNA metabolic process is a descendant of the root node in the Gene Ontology biological process scheme. Draw a graph showing the root node, Biological Process, the DNA metabolic process node, and all the pathways connecting them directly through a series of parent–child relationships.

Weblem 4.27 If you search the ENTREZ protein database with: HOMO SAPIENS[ORGA NISM] AND ELASTASE[TEXT WORD] NOT INHIBITOR[TEXT WORD] how many hits are returned? Comparing the responses to this query in the ENTREZ CoreNucleotide and Protein databases, do all the responses from the CoreNucleotide database correspond to (in the sense of encode) a response from the Protein database?

Alignments and phylogenetic trees

Learning goals

1. Understanding the concept of sequence alignment: the assignment of residue-residue correspondences.

2. Knowing how to construct and interpret dotplots, and the relationship between dotplots and alignments.

3. Being able to define and distinguish the Hamming distance and Levenshtein distance as measures of dissimilarity of character strings.

4. Understanding the basis of scoring schemes for string alignment, including substitution matrices and gap penalties.

5. Appreciating the difference between global alignments and local alignments, and understanding the use of approximate methods for quick screening of databases.

6. Understanding the significance of Z-scores, and knowing how to interpret P-values and E-values returned by database searches.

7. Being able to interpret multiple alignments of amino acid sequences, and to make inferences from multiple sequence alignments about protein structures.

8. Being able to define and distinguish the concepts of homology, similarity, clustering and phylogeny.

9. Becoming expert in the use of PSI-BLAST and related programs.

10. Appreciating the use of profile methods and Hidden Markov Models in database searching.

11. Understanding the contents and significance of phylogenetic trees, and the methods available for deriving them, including maximum parsimony and maximum likelihood; knowing the role and use of an outgroup in derivation of a phylogenetic tree.

Introduction to sequence alignment

Given two or more sequences, we wish to:

♦ measure their similarity

♦ determine the residue–residue correspondences

♦ observe patterns of conservation and variability

♦ infer evolutionary relationships

If we can do these, we will be in a good position to go fishing in databanks for related sequences. A major application is to the annotation of genomes, involving assignment of structure and function to as many genes as possible.

How can we define a quantitative measure of sequence similarity? Before comparing the nucleotides or amino acids that appear at corresponding positions in two or more sequences, we must first assign those correspondences. *Sequence alignment is the identification of residue–residue correspondences.* It is *the* basic tool of bioinformatics.

Any assignment of correspondences that preserves the *order* of the residues within the sequences is an alignment. Gaps may be introduced.

Given two text strings: first string = a b c d e
second string = a c d e f

a reasonable alignment would be: a b c d e -
a - c d e f

We must define criteria so that an algorithm can choose the *best* alignment. For the sequences gctgaacg and ctataatc:

An uninformative alignment: - - - - - - - g c t g a a c g
c t a t a a t c - - - - - - -

An alignment without gaps: g c t g a a c g
c t a t a a t c

An alignment with gaps: g c t g a - a - - c g
- - c t - a t a a t c

And another: g c t g - a a - c g
- c t a t a a t c -

Most readers would consider the last of these alignments the best of the four. To confirm this, and to decide whether it is the best of *all* possibilities, we need a way to examine all possible alignments systematically. Then we need to compute a score reflecting the quality of each possible alignment. Then we can identify the alignment with the optimal score. In many cases, the optimal alignment is not unique: several different alignments may give the same best score. Moreover, even minor variations in the scoring scheme may change the ranking of alignments, causing a different one to emerge as the best.

These examples illustrate **pairwise sequence alignments.** However, usually we can find large families of similar sequences, by identifying homologues in different species. A mutual alignment of more than two sequences is called a **multiple sequence alignment.** Multiple sequence alignments are much more informative than pairwise sequence alignments, in terms of revealing patterns of conservation.

The dotplot

The **dotplot** is a simple picture that gives an overview of the similarities between two sequences. Less obvious is its close relationship to alignments.

The dotplot is a table or matrix. The rows correspond to the residues of one sequence and the columns to the residues of the other sequence. In its simplest form, the positions in the dotplot are left blank if the residues are different, and filled if they match. Stretches of similar residues show up as diagonals in the upper left–lower right (Northwest–Southeast) direction.

Example 5.1

Dotplot showing identities between short name (DOROTHYHODGKIN) and full name (DOROTHYCROWFOOTHODGKIN) of a famous protein crystallographer.

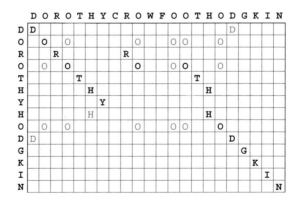

	D	O	R	O	T	H	Y	C	R	O	W	F	O	O	T	H	O	D	G	K	I	N
D	D																	D				
O		O		O						O			O	O			O					
R			R						R													
O		O		O						O			O	O			O					
T					T										T							
H						H										H						
Y							Y															
H						H										H						
O		O		O						O			O	O			O					
D	D																	D				
G																			G			
K																				K		
I																					I	
N																						N

Letters corresponding to *isolated* matches are shown in non-bold type. The longest matching regions, shown in boldface, are the first and last names DOROTHY and HODGKIN. Shorter matching regions, such as the OTH of dorOTHy and crowfoOTHodgkin, or the RO of doROthy and cROwfoot, are noise.

Example 5.2

Dotplot showing identities between a repetitive sequence (ABRACADAB-RACADABRA) and itself. The repeats appear on several subsidiary diagonals parallel to the main diagonal.

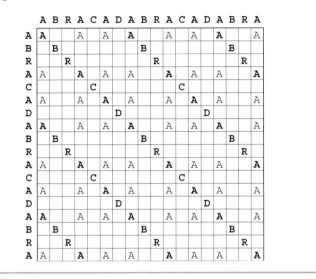

Example 5.3

Dotplot showing identities between the palindromic sequence MAX I STAY AWAY AT SIX AM and itself. The palindrome reveals itself as a stretch of matches *perpendicular* to the main diagonal.

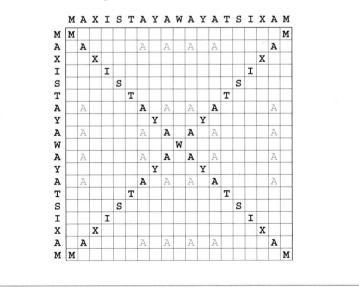

→

→

This is not just word play—regions in DNA recognized by transcriptional regulators or restriction enzymes have sequences related to palindromes, crossing from one strand to the other:

*Eco*RI recognition site: GAATTC

CTTAAG

Within each strand a region is followed by its reverse complement (see Exercise 5.9 and Problem 5.9). Longer regions of DNA or RNA containing inverted repeats of this form can form stem–loop structures. In addition, some transposable elements in plants contain true (approximate) palindromic sequences—inverted repeats of non-complemented sequences, on the *same* strand; the following example appears in the wheat dwarf virus genome: ttttcgtgagtgcgcggaggctttt.

The dotplot gives a quick pictorial statement of the relationship between two sequences. Obvious features of similarity stand out. For example, a dotplot relating the mitochondrial ATPase-6 genes from a lamprey (*Petromyzon marinus*) and dogfish shark (*Scyliorhinus canicula*) shows that the similarity of the sequences is weakest near the beginning. This gene codes for a subunit of the ATPase complex. In the human, mutations in this gene cause Leigh syndrome, a neurological disorder of infants produced by the effects of impaired oxidative metabolism on the brain during development.

A disadvantage of the dotplot is that its 'reach' into the realm of distantly related sequences is poor. In analysing sequences, one should always look at a dotplot to be sure of not missing anything obvious, but be prepared to apply more subtle tools.

ATPases lamprey / dogfish shark

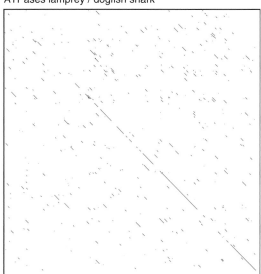

Often regions of similarity may be displaced, to appear on parallel but not collinear diagonals. This indicates that insertions or deletions have occurred in the segments between the similar regions. A dotplot relating the PAX-6 protein of mouse and the eyeless protein of *Drosophila melanogaster* shows three extended regions of similarity with different lengths of sequence between them, two near the beginning of the sequences and one near the middle. Between the second and third of them, there is a longer intervening region in the mouse than in the *Drosophila* sequence.

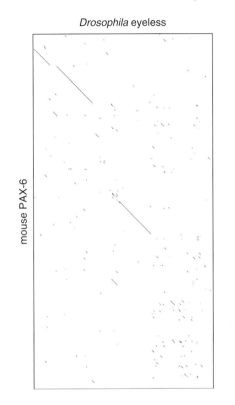

Drosophila eyeless

mouse PAX-6

Filtering the results can reduce the noise in a dotplot. In the comparison of the ATPase sequences, dots were not shown unless they were at the centre of a consecutive region of 15 residues containing at least six matches. The PERL program for dotplots (see Box) allows the user to set values for a **window** (length of region of consecutive residues) and a **threshold** (number of matches required within the window).

Web resources   Dotplots

E.L. Sonnhammer's program Dotter computes and displays dotplots. It allows the user to control the calculation and alter the appearance of the display by adjusting parameters interactively.

http://www.cgr.ki.se/cgr/groups/sonnhammer/Dotter.html

To use the full set of features of Dotter it is necessary to install it locally.

online resource centre

A web site that offers interactive dotplotting is:

http://myhits.isb-sib.ch/cgi-bin/dotlet

A PERL program to draw dotplots

The program shown reads:

1. A general title for the job, printed at the top of the output drawing. (First line of input.)

2. Parameters specifying the filtering parameters *window* and *threshold* (second line of input). A dot will appear in the dotplot if it is in the centre of a stretch of residues of length *window* in which the number of matches is ≥ *threshold*.

3. The two sequences, each beginning with a title line and ending with an*.

The program draws a dotplot similar to those shown in the text. The output is in a graphical language called PostScript™, which can be displayed or printed on many devices, or converted to the common pdf format.

```perl
#!/usr/bin/perl
#dotplot.pl -- reads two sequences and prints dotplot

# read input

$/ = "";
$_ = <DATA>; $_ =~ s/#(.*)\n/\n/g;
$_ =~ /^(.*)\n\s*(\d+)\s+(\d+)\s*\n(.*)\n([A-Z\n]*)\*\s*\n(.*)\n([A-Z\n]*)\*/;
$title = $1; $nwind = $2; $thresh = $3;
$seqt1 = $4; $seq1 = $5; $seqt2 = $6; $seq2 = $7;
$seq1 =~ s/\n//g; $seq2 =~ s/\n//g; $n = length($seq1); $m = length($seq2);

# postscript header

print <<EOF;
%!PS-Adobe-
/s /stroke load def /l /lineto load def /m /moveto load def /r /rlineto load def
/n /newpath load def /c /closepath load def /f /fill load def
1.75 setlinewidth 30 30 translate /Helvetica findfont 20 scalefont setfont
EOF

#print matrix

$dx = 500.0/$n;  $mdx = -$dx; $dy = 500.0/$m;
if ($dy < $dx) {$dx = $dy;} $dy = $dx; $xmx = $n*$dx; $ymx = $m*$dx;
print "0 510 m ($title NWIND = $nwind) show\n";
printf "0 0 m 0 %9.2f 1 %9.2f %9.2f 1 %9.2f 0 1 c s\n", $ymx,$xmx,$ymx,$xmx;

for ($k = $nwind - $m + 1; $k < $n - $nwind; $k++) {
    $i = $k; $j = 1; if ($k < 1) {$i = 1; $j = 2 - $k;}
    while ($i <= $n - $nwind && $j <= $m - $nwind) {
        $_ = (substr($seq1,$i -1,$nwind) ^ substr($seq2,$j -1,$nwind));
        $mismatch = ($_ =~ s/[^\x0]//g);
```

→

A PERL program to draw dotplots *(continued)*

```
        if ($mismatch < $thresh) {
            $xl = ($i - 1)*$dx; $yb = ($m - $j)*$dy;
            printf "n %9.2f %9.2f m %9.2f 0 r 0 %9.2f r %9.2f 0 r c f\n",
                $xl,$yb,$dx,$dy,$mdx;
        }
        $i++; $j++;
    }
}
print "showpage\n";


__END__
ATPases lamprey / dogfish                   #TITLE
15 6                                        #WINDOW, THRESHOLD
Petromyzon marinus mitochondrion            #SEQUENCE 1
ATGACACTAGATATCTTTGACCAATTTACCTCCCCAACA
ATATTTGGGCTTCCACTAGCCTGATTAGCTATACTAGCCCCCTAGCTTA
ATATTAGTTTCACAAACACCAAAATTTATCAAATCTCGTTATCACACACTA
CTTACACCCATCTTAACATCTATTGCCAAACAACTCTTTCTTCCAATAAAC
CAACAAGGGCATAAATGAGCCTTAATTTGTATAGCCTCTATAATATTTATC
TTAATAATTAATCTTTTAGGATTATTACCATATACTTATACACCAACTACC
CAATTATCAATAAACATAGGATTAGCAGTGCCACTATGACTAGCTACTGTC
CTCATTGGGTTACAAAAAAAACCAACAGAAGCCCTAGCCCACTTATTACCA
GAAGGTACCCCAGCAGCACTCATTCCCATATTAATTATCATTGAAACTATT
AGTCTTTTTATCCGACCTATCGCCCTAGGAGTCCGACTAACCGCTAATTTA
ACAGCTGGTCACTTACTTATACAACTAGTTTCTATAACAACCTTTGTAATA
ATTCCTGTCATTTCAATTTCAATTATTACCTCACTACTTCTTCTATTA
CTAACAATTCTGGAGTTAGCTGTTGCTGTAATCCAGGCATATGTATTTATT
CTACTTTTAACTCTTTATCTGCAAGAAAACGTTT*
Scyliorhinus canicula mitochondrion        #SEQUENCE 2
ATGATTATAAGCTTTTTTGATCAATTCCTAAGTCCCTCCTTTCTAGGA
ATCCCACTAATTGCCCTAGCTATTTCAATTCCATGATTAATATTTCCAACACCAACC
AATCGTTGACTTAATAATCGATTATTAACTCTTCAAGCATGATTTATTAACCGATTTATT
TATCAACTAATACAACCCATAAATTTAGGAGGACATAAATGAGCTATCTTATTTACAGCC
CTAATATTATTTTTAATTACCATCAATCTTCTAGGTCTCCTTCCATATACTTTTACGCCT
ACAACTCAACTTTCTCTTAATATAGCCTTTGCCCTGCCCTTATGGCTTACAACTGTATTA
ATTGGTATATTTAATCAACCAACCATTGCCCTAGGGCACTTATTACCTGAAGGTACCCCA
ACCCCTTTAGTACCAGTACTAATCATTATCGAAACCATCAGTTTATTTATTCGACCATTA
GCCTTAGGAGTCCGATTAACAGCCAACTTAACAGCTGGACATCTCCTTATACAATTAATC
GCAACTGCGGCCTTTGTCCTTTTAACTATAATACCAACCGTGGCCTTACTAACCTCCCTA
GTCCTGTTCCTATTGACTATTTTAGAAGTGGCTGTAGCTATAATTCAAGCATACGTATTT
GTCCTTCTTTTAAGCTTATATCTACAAGAAAACGTATAA*
```

Dotplots and sequence alignments

The dotplot captures in a single picture not only the overall similarity of two sequences, but also the complete set and relative quality of different possible alignments. Any path through the dotplot from upper left to lower right, moving from each point only in East, South or Southeast directions, corresponds to a possible alignment. If two sequences are closely related, the alignment can be read directly off the dotplot.

Figure 5.1 shows an example based on the Dorothy Hodgkin dotplot.

If the direction of the 'move' between successive cells is diagonal, two pairs of successive residues appear in the alignment without an insertion between them. If the direction of the move is horizontal, a gap is introduced in the sequence indexing the rows. If the direction of the move is vertical, a gap would be introduced in the sequence indexing the columns. Note that no moves can be directed up or to the left,

For pictures of
the superposed
structures, see
*Introduction to
Protein Science.*

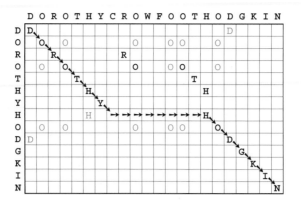

Fig. 5.1 Any path through the dotplot from upper left to lower right passes through a succession of cells, *each of which picks out a pair of positions, one from the row and one from the column, that are matched in the alignment that corresponds to that path; or that indicates a gap in one of the sequences.* The path need not pass through filled-in points only. However, the more filled-in points on the path, the more matching residues in the alignment.

as this would correspond to aligning several residues of one sequence with only one residue of the other, or to introduce gaps in both sequences. The path indicated by the arrows corresponds to the obvious alignment:

```
DOROTHY-------HODGKIN
DOROTHYCROWFOOTHODGKIN
```

Another way to think of a path through the dotplot is as an *edit script*; that is, a prescription of a series of operations that transforms the sequence that indexes the columns—the 'horizontal' sequence—into the sequence that indexes the rows—the 'vertical' sequence. Each move tells us to perform an operation—a substitution, an insertion or a deletion. When the end of the path is reached, the effect will be to change one sequence into the other. In many cases, several different sequences of edit operations may convert one string into the other in the same number of steps, but they induce different alignments.

It should be emphasized that although a sequence of edit operations derived from an optimal alignment *may* correspond to an actual set of evolutionary events, it is impossible to *prove* that it does. The larger the edit distance, the larger the number of reasonable evolutionary pathways between two sequences. An alignment does not contain any information about the *order* of occurence of the sequence changes during evolution.

Case Study 5.1 **Dotplots and alignments**

Let us compare the appearance of dotplots between pairs of proteins with increasingly more distant relationships. Figure 5.2 shows the dotplot comparisons

Case Study 5.1 *(continued)*

of the sulphydryl proteinase papain from papaya, with four homologues—the close relative, kiwi fruit actinidin, and successively more distant relatives, human procathepsin L, human cathepsin B and *Staphylococcus aureus* staphopain. The sequence alignments are also shown. As the sequences progressively diverge, it becomes more and more difficult to spot the correct alignment in the dotplot. The alignments shown were derived from comparisons of the structures.

```
NPOS = 219 NIDENT = 102  %IDENT = 46.58

IPEYVDWRQKGAVTPVKNQGSCGSCWAFSAVVTIEGIIKIRTGNLNQYSEQELLDCDR--
 | |||||  |||   | || ||  ||||||  | ||| ||  | |   ||||| || |
LPSYVDWRSAGAVVDIKSQGECGGCWAFSAIATVEGINKITSGSLISLSEQELIDCGRTQ

RSYGCNGGYPWSALQ-LVAQYGIHYRNTYPYEGVQRYCRSREKGPYAAKTDGVRQVQPYN
  || |||   || |||        ||      |||         |        | |
NTRGCDGGYITDGFQFIINDGGINTEENYPYTAQDGDCDVALQDQKYVTIDTYENVPYNN

QGALLYSIANQPVSVVLQAAGKDFQLYRGGIFVGPCGNKVDHAVAAVGYGP----NYILI
 ||       |||||| | |||  |  |  ||| |||| |||| |||  ||||    |
EWALQTAVTYQPVSVALDAAGDAFKQYASGIFTGPCGTAVDHAIVIVGYGTEGGVDYWIV

KNSWGTGWGENGYIRIKRGTGNSYGVCGLYTSSFYPVKN
|||| |  ||| || || |   | || |    ||||
KNSWDTTWGEEGYMRILRNVGGA-GTCGIATMPSYPVKY
```

PAPA_CARPA / ACTN_ACTCH

Fig 5.2a Alignment of papaya papain and kiwi fruit actinidin, with the corresponding dotplot.

Case Study 5.1 *(continued)*

```
NPOS = 220 NIDENT  = 81  %IDENT = 36.82

IPEYVDWRQKGAVTPVKNQGSCGSCWAFSAVVTIEGIIKIRTGNLNQYSEQELLDCD--R
     |||  ||  ||||||||  |||  |||||      ||     || |   ||| ||
V----DWREKGYVTPVKNQGQCGSSWAFSATGALEGQMFRKTGRLISLSEQNLVDCSGPE

RSYGCNGGYPWSALQLVAQY-GIHYRNTYPYEGVQRYCRSREKGPYAAKTDGVRQVQPYN
  |||||     |  |  |       |    |        |     |         |
GNEGCNGGLMDYAFQYVQDNGGLDSEESYPYEATEESCKYNPKYS-VANDAGFVDIPKQE

QGALLYSIANQPVSVVLQAAGKDFQLYRGGIFVGP--CGNKVDHAVAAVGYG---PNYIL
       |  ||    |     |   |   ||   |         || |   ||||    | |
KALMKAVATVGPISVAIDAGHESFLFYKEGIYFEPDCSSEDMDHGVLVVGYGFESNKYWL

IKNSWGTGWGENGYIRIKRGTGNSYGVCGLYTSSFYPVKN
 |||||    ||    ||              |     ||           ||
VKNSWGEEWGMGGYVKMAKDRRN-H--CGIASAASYPTV-
```

PAPA_CARPA / CATL_HUMAN

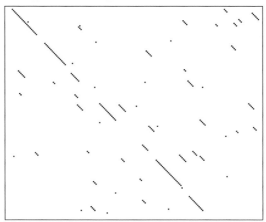

Fig 5.2b Alignment of papaya papain and human procathepsin L, with the corresponding dotplot. This dotplot shows that there are several similar regions, but it would be difficult to generate a complete sequence alignment from the dotplot.

```
NPOS = 251  NIDENT  = 66  %IDENT = 26.29

IPEYVD-WRQKGAVTPVKNQGSCGSCWAFSAVVTIEGIIKIRTGNLNQYSEQELLD-C-D
   | |           |||||||||||  |     |    |    |   |   |
--DAREQWPQCPTIKEIRDQGSCGSCWAFGAVEAISDRICIHTNVSVEVSAEDLLTCCGS
+
RRSYGCNGGYP------WSALQLVAQYGI--HYRN-TY-----P--YEGVQRYCRSREKG
  |||||||        |     ||       |          |           |
MCGDGCNGGYPAEAWNFWTRKGLVSGGLYESHVGCRPYSIPPCEHHVNGSRPPCTGEGDT

PYAAK------TDGVRQVQPYNQGALLYSIANQPVSV-V-----LQ---AAGKDFQLYRG
 |   |         |    |        |                     || ||
PKCSKICEPGYSPTYKQDKHYGYNSYSVSNSEKDIMAEIYKNGPVEGAFSVYSDFLLYKS
```

Case Study 5.1 *(continued)*

```
GIFVGPCGNKV-DHAVAAV--GY--GPNYILIKNSWGTGWGENGYIRIKRGTGNSYGVCG
|       |     ||         |      |  |     |||  |  ||  ||    |  ||
GVYQHVTGEMMGGHAIRILGWGVENGTPYWLVANSWNTDWGDNGFFKILRGQ-DHCGIES

LYTSSFYPVKN
          |
EVVAGI-PRTD
```

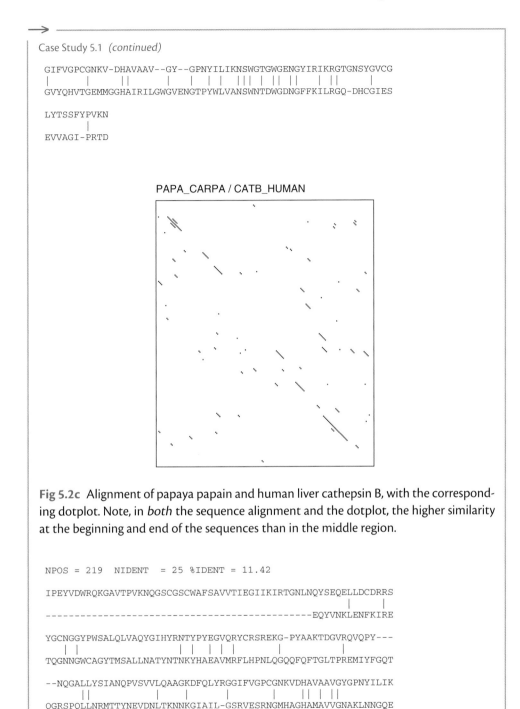

PAPA_CARPA / CATB_HUMAN

Fig 5.2c Alignment of papaya papain and human liver cathepsin B, with the corresponding dotplot. Note, in *both* the sequence alignment and the dotplot, the higher similarity at the beginning and end of the sequences than in the middle region.

```
NPOS = 219   NIDENT  = 25  %IDENT = 11.42

IPEYVDWRQKGAVTPVKNQGSCGSCWAFSAVVTIEGIIKIRTGNLNQYSEQELLDCDRRS
                                                          |   |
--------------------------------------------EQYVNKLENFKIRE

YGCNGGYPWSALQLVAQYGIHYRNTYPYEGVQRYCRSREKG-PYAAKTDGVRQVQPY---
    |  |       |      |   |  |   |         |        |
TQGNNGWCAGYTMSALLNATYNTNKYHAEAVMRFLHPNLQGQQFQFTGLTPREMIYFGQT

--NQGALLYSIANQPVSVVLQAAGKDFQLYRGGIFVGPCGNKVDHAVAAVGYGPNYILIK
    ||           |        |       |        ||  | |
QGRSPQLLNRMTTYNEVDNLTKNNKGIAIL-GSRVESRNGMHAGHAMAVVGNAKLNNGQE

NSWGTGWGENGYIRIKRGTGNSYGVCGLYTSSFYPVKN-
        ||                    |
VIIIWNPWDNGFMTQDAKNNVIPVSNGDHYQWYSSIYGY
```

Case Study 5.1 *(continued)*

PAPA_CARPA / STPA_STAAU

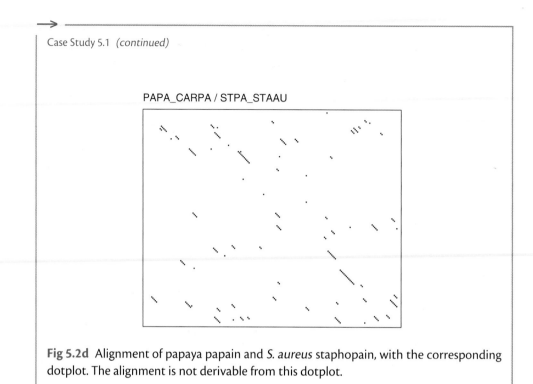

Fig 5.2d Alignment of papaya papain and *S. aureus* staphopain, with the corresponding dotplot. The alignment is not derivable from this dotplot.

Measures of sequence similarity

To go beyond 'alignment by eyeball' via dotplots, we must define quantitative measures of sequence similarity and difference.

Given two character strings, two measures of the distance between them are:

1. The **Hamming distance**, defined between two strings of equal length, is the number of positions with mismatching characters.

2. The **Levenshtein**, or **edit distance**, defined between two strings of not necessarily equal length, is the minimal number of 'edit operations' required to change one string into the other. An edit operation is a deletion, insertion or alteration of a single character in either sequence. A given sequence of edit operations induces a unique alignment, but not vice versa.

For example:

```
agtc                      Hamming distance = 2
cgta
```

```
ag-tcc                    Levenshtein distance = 3
cgctca
```

For applications to molecular biology, recognize that certain changes are more likely to occur than others. For example, amino acid substitutions tend to be conservative: the replacement of one amino acid by another with similar size or physicochemical properties is more likely to have occurred than its replacement by another amino acid with greater differences. Or, the deletion of a succession of contiguous bases or amino acids is a more probable event than the independent deletion of the same number of bases or amino acids at non-contiguous positions in the sequences. Therefore, we may wish to assign variable weights to different edit operations. A computer program can then determine not just minimal edit distances but optimal alignments. It can score each path by adding up the scores of the individual steps. For substitutions, it adds the score of the mutation, depending on the pair of residues involved. For horizontal and vertical moves, it adds a suitable gap penalty.

Scoring schemes

A scoring system must account for residue substitutions, and insertions or deletions. Deletions, or gaps in a sequence, will have scores that depend on their lengths.

Hamming and Levenshtein distances measure the *dissimilarity* of two sequences: *similar* sequences give *small* distances and *dissimilar* sequences give *large* distances. It is common in molecular biology to define scores as measures of sequence *similarity*. Then *similar* sequences give *high* scores and *dissimilar* sequences give *low* scores. These are equivalent formulations. Algorithms for optimal alignment can seek either to minimize a dissimilarity measure or to maximize a scoring function.

For nucleic acid sequences, it is common to use a simple scheme for substitutions: +1 for a match, −1 for a mismatch, or a more complicated scheme based on the higher frequency of transition mutations than transversion mutations.

Example 5.4

Transition mutations (*purine* ↔ *purine* and *pyrimidine* ↔ *pyrimidine*, i.e. a ↔ g and t ↔ c) are more common than *transversions* [*purine* ↔ *pyrimidine*, i.e. (a,g) ↔ (t,c)]. Suggest a substitution matrix that reflects this. (The higher the value in the matrix, the more favourable the contribution to the alignment score.)

One possibility is:

	a	g	t	c
a	20	10	5	5
g	10	20	5	5
t	5	5	20	10
c	5	5	10	20

For proteins, a variety of scoring schemes have been proposed. We might group the amino acids into classes of similar physicochemical type, and score +1 for a match within residue class, and −1 for residues in different classes. We might try to devise a more precise substitution score from a combination of properties of the amino acids. Alternatively, we might try to let the proteins teach us an appropriate scoring scheme. M.O. Dayhoff did this first, by collecting statistics on substitution frequencies in the protein sequences then known. Her results were used for many years to score alignments. They have been superseded by newer matrices based on the very much larger set of sequences that has subsequently become available.

Derivation of substitution matrices: PAM matrices

As sequences diverge, mutations accumulate. To measure the relative probability of any particular substitution, for instance Serine → Threonine, we can count the number of Serine → Threonine changes in pairs of aligned homologous sequences. We could use the relative frequencies of such changes to form a scoring matrix for substitutions. A common change should score higher than a rare one. But, what if there have been multiple substitutions at certain sites? This will bias the statistics. We can avoid this problem by restricting our samples to sequences that are sufficiently similar that we can assume that no position has changed more than once.

A measure of sequence divergence is the PAM: 1 PAM = 1 Percent Accepted Mutation. Thus, two sequences 1 PAM apart have 99% identical residues. For pairs of sequences within the 1 PAM level of divergence, it is likely that there has been no more than one change at any position. Collecting statistics from pairs of sequences as closely related as this, and correcting for different amino acid abundances, produces the **1 PAM substitution matrix.**

To produce a matrix appropriate for more widely divergent sequences, we can take powers of this matrix. The PAM250 level, corresponding to ~20% overall sequence identity, is the lowest sequence similarity for which we can hope to produce a correct alignment by simple pairwise sequence comparison alone. It is therefore the appropriate level to choose for practical work. (Several authors have derived substitution matrices appropriate in different ranges of overall sequence similarity.)

The occurrence of reversions, either directly or via one or more other changes, produces an apparent slowdown in mutation rates as sequences progressively diverge. The relationship between PAM score and percentage sequence identity is:

PAM	0	30	80	110	200	250
% identity	100	75	50	60	25	20

The PAM250 matrix of M.O. Dayhoff is shown in the box. It expresses scores as *log-odds* values:

$$\text{Score of mutation } i \leftrightarrow j = \log_{10} \frac{\text{observed } i \leftrightarrow j \text{ mutation rate}}{\text{mutation rate expected from amino acid frequencies}}$$

The numbers are multiplied by 10, simply to avoid decimal points. The matrix entries reflect the probabilities of mutational events. A value of +2—for instance, C ↔ S—implies that in related sequences the mutation would be expected to occur 1.6 times more frequently than random. The calculation is as follows: the matrix entry 2 corresponds to the actual value 0.2 because of the scaling. The value 0.2 is $\log_{10}$ of the relative expectation value of the mutation. Therefore, this expectation value is $10^{0.2} = 1.6$.

The probability of two independent mutational events is the product of their probabilities. By using logs, we have scores that we can add up rather than multiply, a computational convenience.

The BLOSUM matrices

S. Henikoff and J.G. Henikoff developed the family of BLOSUM matrices for scoring substitutions in amino acid sequence comparisons. Their goal was to replace the Dayhoff matrix with one that would perform best in identifying distant relationships, making use of the much larger amount of sequence data that had become available since Dayhoff's work.

The BLOSUM matrices are based on the BLOCKS database of aligned protein sequences; hence the name BLOcks SUbstitution Matrix. From regions of closely related proteins alignable without gaps, Henikoff and Henikoff calculated the ratio of the number of observed pairs of amino acids at any position to the number of pairs expected from the overall amino acid frequencies. Like the Dayhoff matrix, the results are expressed as log-odds. In order to avoid overweighting closely related sequences, the Henikoffs replaced groups of proteins that have sequence identities higher than a threshold by either a single representative or a weighted average. The threshold 62% produces the commonly used BLOSUM62 substitution matrix (see Box). This is offered by all programs as an option, and is the default in most. BLOSUM matrices have replaced the Dayhoff matrix for most applications.

Substitution matrices used for scoring amino acid sequence similarity

The entries are in alphabetical order of the THREE-letter amino acid names. Only the lower triangles of the matrices are shown, as the substitution probabilities are taken as symmetric. (This is not because we are sure that the rate of any substitution is the same as the rate of its reverse, but because we cannot determine the differences between the two rates.)

The Dayhoff PAM250 matrix (MDM78):

	A	R	N	D	C	Q	E	G	H	I	L	K	M	F	P	S	T	W	Y	V
Ala (A)	2																			
Arg (R)	−2	6																		
Asn (N)	0	0	2																	
Asp (D)	0	−1	2	4																
Cys (C)	−2	−4	−4	−5	12															
Gln (Q)	0	1	1	2	−5	4														
Glu (E)	0	−1	1	3	−5	2	4													
Gly (G)	1	−3	0	1	−3	−1	0	5												
His (H)	−1	2	2	1	−3	3	1	−2	6											
Ile (I)	−1	−2	−2	−2	−2	−2	−2	−3	−2	−5										
Leu (L)	−2	−3	−3	−4	−6	−2	−3	−4	−2	2	6									
Lys (K)	−1	3	1	0	−5	1	0	−2	0	−2	−3	5								
Met (M)	−1	0	−2	−3	−5	−1	−2	−3	−2	2	4	0	6							
Phe (F)	−3	−4	−3	−6	−4	−5	−5	−5	−2	1	2	−5	0	9						
Pro (P)	1	0	0	−1	−3	0	−1	0	0	−2	−3	−1	−2	−5	6					
Ser (S)	1	0	1	0	0	−1	0	1	−1	−1	−3	0	−2	−3	1	2				
Thr (T)	1	−2	0	0	−2	−1	0	0	−1	0	−2	0	−1	−3	0	1	3			
Trp (W)	−6	2	−4	−7	−8	−5	−7	−7	−3	−5	−2	−3	−4	0	−6	−2	−5	17		
Tyr (Y)	−3	−4	−2	−4	0	−4	−4	−5	0	−1	−1	−4	−2	7	−5	−3	−3	0	10	
Val (V)	0	−2	−2	−2	−2	−2	−2	−1	−2	4	2	−2	2	−1	−1	−1	0	−6	−2	4

→

Substitution matrices used for scoring amino acid sequence similarity *(continued)*

The BLOSUM62 matrix:

	A	R	N	D	C	Q	E	G	H	I	L	K	M	F	P	S	T	W	Y	V
Ala (A)	4																			
Arg (R)	−1	5																		
Asn (N)	−2	0	6																	
Asp (D)	−2	−2	1	6																
Cys (C)	0	−3	−3	−3	9															
Gln (Q)	−1	1	0	0	−3	5														
Glu (E)	−1	0	0	2	−4	2	5													
Gly (G)	0	−2	0	−1	−3	−2	−2	6												
His (H)	−2	0	1	−1	−3	0	0	−2	8											
Ile (I)	−1	−3	−3	−3	−1	−3	−3	−4	−3	4										
Leu (L)	−1	−2	−3	−4	−1	−2	−3	−4	−3	2	4									
Lys (K)	−1	2	0	−1	−3	1	1	−2	−1	−3	−2	5								
Met (M)	−1	−1	−2	−3	−1	0	−2	−3	−2	1	2	−1	5							
Phe (F)	−2	−3	−3	−3	−2	−3	−3	−3	−1	0	0	−3	0	6						
Pro (P)	−1	−2	−2	−1	−3	−1	−1	−2	−2	−3	−3	−1	−2	−4	7					
Ser (S)	1	−1	1	0	−1	0	0	0	−1	−2	−2	0	−1	−2	−1	4				
Thr (T)	0	−1	0	−1	−1	−1	−1	−2	−2	−1	−1	−1	−1	−2	−1	1	5			
Trp (W)	−3	−3	−4	−4	−2	−2	−3	−2	−2	−3	−2	−3	−1	1	−4	−3	−2	11		
Tyr (Y)	−2	−2	−2	−3	−2	−1	−2	−3	2	−1	−1	−2	−1	3	−3	−2	−2	2	7	
Val (V)	0	−3	−3	−3	−1	−2	−2	−3	−3	3	1	−2	1	−1	−2	−2	0	−3	−1	4

Scoring insertions/deletions, or 'gap weighting'

To form a complete scoring scheme for alignments, we need, in addition to the substitution matrix, a way of scoring gaps. How important are insertions and deletions, relative to substitutions? Distinguish gap initiation:

```
aaagaaa
aaa-aaa
```

from gap extension:

```
aaagggaaa
aaa----aaa
```

For aligning DNA sequences, CLUSTAL-W recommends use of the identity matrix for substitution (+1 for a match, 0 for a mismatch) and gap penalties 10 for gap initiation and 0.1 for gap extension by one residue. For aligning protein sequences, the recommendations are to use the BLOSUM62 matrix for substitutions, and gap penalties 11 for gap initiation and 1 for gap extension by one residue.

Computing the alignment of two sequences

Now that we have a scoring scheme, we can apply it to finding optimal alignments—we seek the alignment that maximizes the score. A famous algorithm to determine the global optimal alignments of two sequences is based on a mathematical technique called **dynamic programming**. (Details appear in the next section.) This algorithm has been extremely important in molecular biology. Some noteworthy features are:

* The good news is that the method is guaranteed to give an optimal *global* alignment. It will find the *best* alignment score, given the choice of parameters—substitution matrix and gap penalty—with no approximation.

- The bad news is that many alignments may give the same optimal score. And none of these need correspond to the biologically correct alignment. For instance, in comparing the α- and β-chains of chicken haemoglobin, W. Fitch and T. Smith found 17 alignments, all of which give the same optimal score, one of which is correct (on the basis of the structures, the court of last resort). There are 1317 alignments with scores within 5% of the optimum.

- Another item of bad news is technical: the time required to align two sequences of lengths n and m is proportional to $n \times m$, because this is the size of the edit matrix that must be filled in. This means that the dynamic programming method is not convenient to use for searching in an entire sequence database for a match to a probe sequence, and even less convenient for 'all-against-all' alignments. The database search problem is in effect the problem of matching a probe sequence to a region of a very long sequence, the length of the entire database.

Variations and generalizations

Variations of the dynamic programming method apply to three related alignment questions: entire sequence against entire sequence, region of one sequence against entire other sequence, and region of one sequence against region of other sequence (see page 26). The global alignment algorithm was first applied to biological sequence alignment by S.B. Needleman and C.D. Wunsch. T. Smith and M. Waterman modified it to identify local matches.

Approximate methods for quick screening of databases

It is routine to screen genes from a new genome against the databases, for similarity to other sequences. Approximate methods can detect close relationships well and quickly but are inferior to the exact ones in picking up very distant relationships. In practice, they give satisfactory performance in the many cases in which the probe sequence is fairly similar to one or more sequences in the databank. They are therefore worth trying first.

We have already seen an example of PSI-BLAST in Chapter 1.

A typical approximation approach would take a small integer k, and determine all instances of each k-tuple of residues in the probe sequence that occur in any sequence in the database. A candidate sequence is a sequence in the databank containing a large number of matching k-tuples, with equivalent spacing in probe and candidate sequences.

There are several variations on this theme, including the original BLAST (Basic Local Alignment Search Tool) program and its variants (see Box, page 261). For a selected set of candidate sequences, *approximate* optimal alignment calculations are then carried out, with the time- and space-saving restriction that the paths through the matrix that can be considered are restricted to bands around the diagonals containing the many matching k-tuples. It is clearest to show the procedure in terms of a dotplot (see Fig. 5.3).

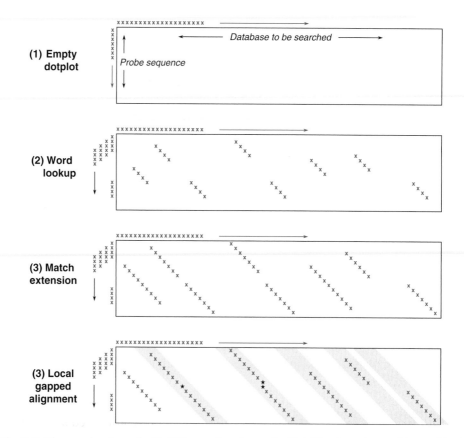

Fig. 5.3 The mechanism of a BLAST search, schematic. BLAST solves the problem of finding matches of a probe sequence in a full genome or a full database, much longer than the probe sequence.

(1) The 'playing field' of the algorithm is the outline of a dotplot, just as if the problem were going to be solved by application of an exact alignment method.

(2) BLAST first divides the probe sequence into fixed-length words of length k, here $k = 4$. It then identifies all exact occurrences of these words in the full database — no mismatches, no gaps. Note that the same four-letter word may occur several times in the probe sequence (shown here in blue), and of course each four-letter word may match many times within the database. It is possible to do this step quickly after pre-processing the database to record the sites of appearance of all four-letter words.

(3) Starting with each match, BLAST tries to extend the match in both directions. Still no mismatches, no gaps allowed.

(4) Given the extended matches, BLAST tries to put them together by doing alignments *allowing* mismatches and gaps, but only within limited regions containing the preliminary matches (grey areas). The result of this step is to add to the matches the positions shown as ★. This produces longer matching regions.

It is the restriction of the more complex matching procedure to relatively small regions, rather than applying it to the entire matrix, that gives the method its speed. The price to pay is that if a combined match lies outside the grey area, the method will miss it. In the example illustrated, the matching regions at the right of the matrix, will not be combined, but reported as separate hits..

BLAST programs come in several flavours

Program	Type of query sequence	Search in database of
BLASTP	amino acid sequence	protein sequences
BLASTX	translated nucleotide sequence	protein sequences
TBLASTN	amino acid sequence	translated nucleotide sequences
TBLASTX	translated nucleotide sequence	translated nucleotide sequences
PSI-BLAST	amino acid sequence	protein sequence database

All these programs compare amino acid sequences with amino acid sequences, using by default the BLOSUM62 matrix. Searches involving nucleotide sequences, either as query sequence or in the database searched, are carried out by translating nucleotide sequences to amino acid sequences in all six possible reading frames. BLASTN compares nucleic acid query sequences with nucleic acid databanks directly.

The dynamic-programming algorithm for optimal pairwise sequence alignment

A chart implicitly containing all possible alignments can be constructed as a matrix similar to that used in drawing the dotplot. The residues of one sequence index the rows, the residues from the other sequence index the columns. Any path through the matrix from upper left to lower right corresponds to an alignment. The task is to find the path that has the lowest cost, and the difficulty is that there are a very large number of paths to consider.

As an illustration, suppose you wanted to drive from Malmö in southern Sweden to Tromsø in Northern Norway. Your route will consist of a number of segments, taking you through a succession of intermediate cities (see Fig. 5.4). There are many choices of different combinations of segments to produce a complete, continuous path.

The computational approach to finding the optimal path begins by assigning a numerical measure of the 'cost' to each of the possible individual segments of the journey. This 'cost' is not simply the financial outlay, but a more general estimate of your relative preferences for different portions of the route. The distance travelled will clearly be an important component of the cost, but other factors such as the quality of the roads and the opportunities for sightseeing may also contribute. For any route selected, the overall cost of the trip is the sum of the costs of the individual segments. Clearly it is inefficient to repeat any leg of the journey, or to visit any city twice, so we will agree that every intermediate stop will be North of the previous one. This formalism is expressed in terms of minimizing a cost rather than maximizing a score; for our purposes the two approaches are equivalent. An algorithm can explore the possible combinations to determine an optimal overall route.

Optional section. Readers in doubt may consider the remarks in Lesk, A.M. (1988). TATA for now. . . *Trends in Biochemical Sciences*, **13**, 410.

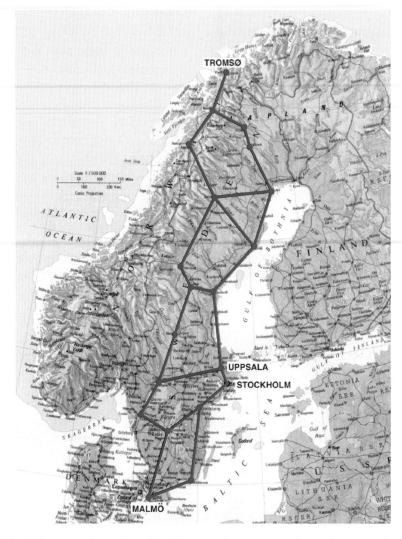

Fig. 5.4 Possible routes from Malmö to Tromsø. How can you determine an optimal route? (© Collins Bartholemew Ltd. 1980.)

Here is an abstract version of the problem, which illuminates the essential idea of dynamic programming.

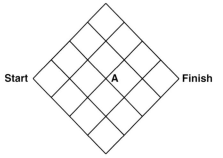

Consider first: how many paths from Start to Finish pass through A? There are six paths from Start to A. (Write them all down.) Therefore, by symmetry, there are six paths from A to finish, and a total of 36 paths from Start to Finish passing through A. (Why?) Assuming that we have assigned costs to the individual steps, do we have to check all 36 paths to find the path of minimum cost that goes from Start to Finish, passing through A? No—here is the crucial observation: *the choice of the best path from A to Finish is independent of the choice of path from the Start to A.* If we determine the best of the six paths from Start to A, and we determine the best of the six paths from A to Finish, the best path from Start to Finish passing through A is: the best path from Start to A *followed by* the best path from A to finish. No more than 12 of the paths through A need be considered.

Even greater simplification is possible by systematically resubdividing the problem. The dynamic programming method for finding the optimal path through the matrix is based on this idea.

A statement of the optimal alignment problem and the dynamic programming solution are as follows: given two character strings, possibly of unequal length: $A = a_1 a_2 \cdots a_n$ and $B = b_1 b_2 \cdots b_m$, where each a_i and b_j is a member of an alphabet set $\mathcal{A}$, consider sequences of edit operations that convert A and B to a common sequence. Individual edit operations include:

Substitution of b_j for a_i—represented (a_i, b_j).

Deletion of a_i from sequence A—represented (a_i, ϕ).

Deletion of b_j from sequence B—represented as (ϕ, b_j).

If we extend the alphabet set to include the null character ϕ: $\mathcal{A}^+ = \mathcal{A} \cup \{\phi\}$, a sequence of edit operations is a set of ordered pairs (x, y), with $x, y \in \mathcal{A}^+$.

A *cost function, d*, is defined on edit operations:

$d(a_i, b_j) = $ cost of a mutation in an alignment in which position i of sequence A corresponds to position j of sequence B, and the mutation substitutes $a_i \leftrightarrow b_j$.

$d(a_i, \phi)$ or $d(\phi, b_j) = $ cost of a deletion or insertion.

Define the minimum weighted distance between sequences A and B as

$$D(A, B) = \min_{A \to B} \Sigma d(x, y)$$

where $x, y \in \mathcal{A}^+$ and the minimum is taken over all sequences of edit operations that convert A and B into a common sequence.

The problem is to find $D(A, B)$ and one or more of the alignments that correspond to it.

An algorithm that solves this problem, requiring execution time proportional to the product of the lengths of the two sequences, creates a matrix $\mathcal{D}(i, j), i = 0, \cdots n; j = 0, \cdots m$, such that $\mathcal{D}(i, j)$ is the minimal distance between the strings that consist of the first i characters of A and the first j characters of B. Then $\mathcal{D}(n, m)$ will be the required minimal distance $D(A, B)$.

The algorithm computes $\mathcal{D}(i,j)$ by recursion. The value of $\mathcal{D}(i,j)$ corresponds to the conversion of the initial subsequences $A_i = a_1a_2\cdots a_i$ and $B_j = b_1b_2\cdots b_j$ into a common sequence by L edit operations $S_k, k = 1,\cdots L$, which can be considered to be applied in increasing order of position in the strings. Consider *undoing* the last of these edit operations. The resulting truncated sequence of edit operations, $S_k, k = 1,\cdots L-1$, is a sequence of edit operations for converting a substring of A_i and a substring of B_j into a common result. What is more, it must be an *optimal* sequence of edit operations for these substrings, for if some other sequence S'_k were a lower-cost sequence of operations for these substrings, then S'_k followed by S_L would be a lower-cost sequence of operations than S_k for converting A_i to B_j. Therefore, there should be a recursive method for calculating the $\mathcal{D}(i,j)$.

Recognize the correspondence of steps between adjacent squares in the matrix, and individual edit operations (see Fig. 5.1):

$(i-1,j-1) \rightarrow (i,j)$ corresponds to the substitution $a_i \rightarrow b_j$.
$(i-1,j) \rightarrow (i,j)$ corresponds to the deletion of a_i from A.
$(i,j-1) \rightarrow (i,j)$ corresponds to the insertion of b_j into A at position i.

Sequences of edit operations correspond to stepwise paths through the matrix

$$(i_0,j_0) = (0,0) \rightarrow (i_1,j_1) \rightarrow \cdots (n,m)$$

where $0 \leq i_{k+1} - i_k \leq 1$ (for $0 \leq k \leq n-1$), $0 \leq j_{k+1} - j_k \leq 1$ (for $0 \leq k \leq m-1$). Considering the possible sequences of edit operations and the corresponding paths through the matrix, the predecessor of an optimal string of edit operations leading from $(0,0)$ to (i,j), where $i,j > 0$, must be an optimal sequence of edit operations leading to one of the cells $(i-1,j), (i-1,j-1)$ or $(i,j-1)$; and, correspondingly, $\mathcal{D}(i,j)$ must depend only on the values of $\mathcal{D}(i-1,j), \mathcal{D}(i-1,j-1)$ and $\mathcal{D}(i,j-1)$, (together of course with the parameterization specified by the cost function d).

The algorithm is then as follows:
Compute the $(m+1) \times (n+1)$ matrix $\mathcal{D}$ by applying:

1. the initialization conditions on the top row and left column:

$$\mathcal{D}(i,0) = \sum_{k=0}^{i} d(a_k, \phi)$$

$$\mathcal{D}(0,j) = \sum_{k=0}^{j} d(\phi, b_k)$$

These values impose the gap penalty on unmatched residues at the beginning of either sequence

and then

2. the recurrence relationships:

$$\mathcal{D}(i,j) = \min\{\mathcal{D}(i-1,j) + d(a_i, \phi), \mathcal{D}(i-1,j-1) + d(a_i, b_j), \mathcal{D}(i,j-1) + d(\phi, b_j)\}$$

for $i = 1, \cdots n; j = 1, \cdots m$. This means: consider all three possible steps to $\mathcal{D}(i, j)$:

Operation	Cumulative cost
Insert a gap in sequence A	$\mathcal{D}(i - 1, j) + d(a_i, \phi)$
Substitute $a_i \leftrightarrow b_j$	$\mathcal{D}(i - 1, j - 1) + d(a_i, b_j)$
Insert a gap in sequence B	$\mathcal{D}(i, j - 1) + d(\phi, b_j)$

From these, choose the minimal value of the cumulative cost. For each cell, record not only the value $\mathcal{D}(i, j)$ but a pointer back to (one or more of) the cell(s) $(i - 1, j), (i - 1, j - 1)$ or $(i, j - 1)$ selected by the minimization operation. Note that more than one predecessor may give the same value.

When the calculations are complete, $\mathcal{D}(n, m)$ is the optimal distance $D(A, B)$. An alignment corresponding to the sequence of edit operations recorded by the pointers can be recovered by tracing a path back through the matrix from (n, m) to $(0, 0)$. This alignment corresponding to the minimal distance $D(A, B) = \mathcal{D}(n, m)$ may well not be unique.

Example 5.5

Align the strings $A = $ ggaatgg and $B = $ atg, according to the simple scoring scheme: match $= 0$, mismatch $= 20$, insertion or deletion $= 25$.

Here is the state of play after the top row and leftmost column have been initialized (italic), and the element in the second row and second column has been entered as **20** (boldface):

	ϕ	a	t	g
ϕ	*0*	*25*	*50*	*75*
g	*25*	**20**		
g	*50*			
a	*75*			
a	*100*			
t	*125*			
g	*150*			
g	*175*			

The value of **20** was chosen as the minimum of $25 + 25$ (horizontal move, or insert gap into string atg), $0 + 20$ (substitution a $\leftrightarrow$ g) and $25 + 25$ (vertical move, or insert gap into string ggaatgg). Because the substitution (the diagonal move) provided the minimal value, the cell containing 0 in the upper left-hand corner of the matrix is the predecessor of the cell in which we have just entered the 20. For traceback purposes, we would also draw an arrow from the value of 20 just entered, back to the 0 at the upper left. (If two or even three of the possible moves produce the same value, the resulting cell has multiple predecessors.)

$\longrightarrow$

Here is matrix after completion of the calculation:

```
      φ       a       t       g
  φ   0   ←25   ←50   ←75
      ↑    ↖     ↖      ↖
  g  25    20   ←45    50
      ↑   ↖↑     ↖      ↖
  g  50    45    40    45
      ↑    ↖    ↖↑      ↖
  a  75    50    65    60
      ↑   ↖↑     ↖     ↖↑
  a 100    75    70    85
      ↑     ↑    ↖      ↖
  t 125   100    75    90
      ↑     ↑     ↑   ↖
  g 150   125   100    75
      ↑     ↑     ↑   ↖↑
  g 175   150   125   100
```

It includes the traceback information in the form of arrows pointing from each cell to its predecessor(s). For some applications we may need only the value of $D(A, B)$ but not an alignment; if so, it is unnecessary to save the pointers. Boldface arrows delineate the paths of optimal alignment retracing a trail of predecessors from lower right, back to upper left. In some cases, one cell may show two predecessors. These correspond to alternative alignments with the same score.

There are two cells at which the traceback path branches. This gives a total of four optimal alignments with equal score:

```
    ggaatgg         ggaatgg         ggaatgg         ggaatgg
    ---atg-         ---at-g         --a-tg-         --a-t-g
```

With a gap-weighting scheme that assigned a smaller penalty to gap extension than to gap initiation the first two of these would score better than the others, because they contain the smallest number of gaps, irrespective of the length of the individual gaps. However, more sophisticated gap-weighting schemes require more complicated recurrence formulas for filling the matrix.

This algorithm determines the optimal *global* alignment of two sequences. It is inappropriate for detection of local regions of high similarity within two sequences, or for probing a long sequence with a short fragment, because it imposes gap penalties *outside* the similar regions. The method of T. Smith and M. Waterman solves this problem. Their modifications of the basic dynamic programming algorithm find optimal local alignments; that is, they select the substrings from both sequences that are most similar to each other. Their changes affect:

→

1. **Initialization of the matrix**—setting the values of the top row and left column. In the Smith–Waterman method, the top row and left column are set to 0. As a result, either sequence can slide along the other before alignment starts, without incurring any gap penalty against the residues it passes by.

2. **Filling in the matrix** In global alignment, at each step a choice is forced among match, insertion or deletion, even if none of these choices is attractive and even if a succession of unattractive choices degrades the score along a path containing a well-fitting local region. The Smith–Waterman method adds the fourth option: end the region being aligned.

3. **Scoring and traceback** The score of a global alignment is the number in the matrix element at the lower right. In the Smith–Waterman method it is the optimal value encountered, wherever in the matrix it appears. For global alignment, traceback to determine the actual alignment starts at the lower-right cell. In the Smith–Waterman method it starts at the cell containing the optimal value and continues back only as far as the region of local similarity continues.

The Smith–Waterman method would report a unique global optimum for our example:

```
ggaatgg
  atg
```

Note that no gaps appear outside the region matched.

(Example adapted from: Tyler, E.C., Horton, M.R. and Krause, P.R. (1991). A review of algorithms for molecular sequence comparison. *Computers and Biomedical Research*, **24**, 72–96.)

Significance of alignments

Suppose alignment reveals an intriguing similarity between two sequences. Is the similarity significant or could it have arisen by chance? (We raised this question in Chapter 1). For some simple phenomena—tossing a coin or rolling dice—it is possible to calculate exactly the expected distribution of results, and the likelihood of any particular result. For sequences, it is not trivial to define the population from which the alignment is selected. For instance, to take random strings of nucleotides or amino acids as controls ignores the bias arising from non-random composition.

A practical approach to the problem is as follows: if the score of the alignment observed is no better than might be expected from the corresponding alignment of a *random permutation* of the sequence, then it is likely to have arisen by chance. We may randomize one of the sequences, many times, realign each result to the second sequence (held fixed) and collect the distribution of resulting scores. Figure 5.5 shows a typical result. For database searches, we would use the population of results returned from the entire database as the population with which to measure the statistics.

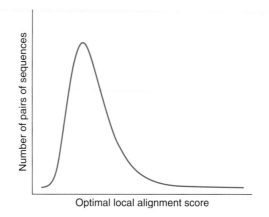

Fig. 5.5 Optimal local alignment scores for pairs of random amino acid sequences of the same length follow an extreme value distribution. For any score x, the probability of observing a score $\geq x$ is:

$$P(\text{Score} \geq x) = 1 - \exp(-Ke^{-\lambda x}),$$

where K and λ are parameters related to the position of the maximum and the width of the distribution. Note the long tail at the right. This means that a score several standard deviations above the mean has a higher probability of arising by chance (i.e. it is *less* significant) than if the scores followed a normal distribution.

Clearly if the randomized sequences score as well as the original one, the alignment is unlikely to be significant. We can measure the mean and standard deviation of the scores of the alignments of randomized sequences, and ask whether the score of original sequence is unusually high. The Z-score reflects the extent to which the original result is an outlier from the population:

$$\text{Z-score} = \frac{\text{score} - \text{mean}}{\text{standard deviation}}$$

A Z-score of 0 means that the observed similarity is no better than the average of random permutations of the sequence, and might well have arisen by chance. Other values used as measures of significance are $P = $ the probability that the observed match could have happened by chance, and, for database searching, $E = $ the number of matches as good as the observed one that would be expected to appear by chance in a database of the size probed (see Box).

How to play with matches but not get burned

Pairwise alignments and database searches often show tenuous but tantalizing sequence similarities. How can we decide whether we are seeing a true relationship?

$\longrightarrow$

→

How to play with matches but not get burned *(continued)*

Statistics cannot answer biological questions directly, but can tell us the likelihood that a similarity as good as the one observed would appear, just by chance, among unrelated sequences. To do this, we want to compare our result with alignments of the same sequences to a large population. This 'control' population should be similar in general features to our aligned sequences, but should contain few sequences related to them. Only if the observed match stands out from the population can we regard it as significant.

To what population of sequences should we compare our alignment? For pairwise alignments, we can pick one of the two sequences, make many scrambled copies of it using a random-number generator, and align each permuted copy to the second sequence. For probing a database, the entire database provides a comparison population.

Alignments of our sequence to each member of the control population generates a large set of scores. How does the score of our original alignment rate? Several statistical parameters have been used to evaluate the significance of alignments:

♦ The Z-score is a measure of how unusual our original match is, in terms of the mean and standard deviation of the population scores. If the original alignment has score S,

$$\text{Z-score of S} = \frac{S - \text{mean}}{\text{standard deviation}}$$

A Z-score of 0 means that the observed similarity is no better than the average of the control population, and might well have arisen by chance. The higher the Z-score, the greater the probability that the observed alignment has not arisen simply by chance. Experience suggests that Z-scores ≥ 5 are significant.

♦ Many programs report $P =$ the probability that the alignment is better than random. The relationship between Z and P depends on the distribution of the scores from the control population, which do *not* follow the normal distribution.

A rough guide to interpreting P-values:

$P \leq 10^{-100}$	exact match
P in range $10^{-100} - 10^{-50}$	sequences very nearly identical, e.g. alleles or SNPs
P in range $10^{-50} - 10^{-10}$	closely related sequences, homology certain
P in range $10^{-5} - 10^{-1}$	usually, distant relatives
$P > 10^{-1}$	match probably insignificant

♦ For database searches, some programs (including PSI-BLAST) report E-values. The E-value of an alignment is the expected number of sequences that give the same Z-score or better if the database is probed with a random sequence. E is found by multiplying the value of P by the size of the database probed. Note that E but not P depends on the size of the database. Values of P are between 0 and 1.0. Values of E are between 0 and the number of sequences in the database searched.

→

→

How to play with matches but not get burned *(continued)*

A rough guide to interpreting *E*-values:

$E \leq 0.02$	sequences probably homologous
E between 0.02 and 1	homology unproven but can't be ruled out
$E > 1$	you'd have to expect this good a match just by chance

Statistics are a useful guide, but not a substitute for thinking carefully about the results, and further analysis of ones that look promising!

Many 'rules of thumb' are expressed in terms of percentage identical residues in the optimal alignment. If two proteins have >45% identical residues in their optimal alignment, they will have very similar structures, and are very likely to have a common or at least a similar function. If two proteins have >25% identical residues, they are likely to have a similar general folding pattern. On the other hand, observations of a lower degree of sequence similarity cannot rule out homology. Recall R.F. Doolittle's definition of the region of 18–25% sequence identity as the 'twilight zone' in which the suggestion of homology is tantalizing but dangerous. Below the twilight zone is a region where pairwise sequence alignments tell very little. Lack of significant sequence similarity does not preclude genuine homology.

Although the twilight zone is a treacherous region, we are not entirely helpless. In deciding whether there is a genuine relationship, the 'texture' of the alignment is important—are the similar residues isolated and scattered throughout the sequence; or are there 'icebergs'—local regions of high similarity (another term of Doolittle's), which may correspond to a shared active site? We may need to rely on other information, about shared ligands or function. Of course if the structures are known, we could examine them directly.

Some illustrative examples:

♦ Sperm whale myoglobin and lupin leghaemoglobin have 15% identical residues in optimal alignment. This is even below Doolittle's definition of the twilight zone. But we also know that both molecules have similar three-dimensional structures, both contain a haem group and both bind oxygen. They are indeed distantly related homologues.

♦ The sequences of the N- and C- terminal halves of rhodanese have 11% identical residues in optimal alignment. If these appeared in independent proteins, one could not conclude from the sequences alone that they were related. However, their appearance in the same protein suggests that they arose via gene duplication and divergence. The striking similarity of their structures confirms their relationship.

♦ As a cautionary note, consider the proteinases chymotrypsin and subtilisin. They have 12% identical residues in optimal alignment. These enzymes have a common function, and a common catalytic triad. However, they have dissimilar folding patterns, and are not related. Their common function and mechanism is an example of convergent evolution. This case serves as a warning against special pleading for relationships between proteins with dissimilar sequences on the basis of similarities of function and mechanism!

Multiple sequence alignment

'One amino acid sequence plays coy; a pair of homologous sequences whisper; many aligned sequences shout out loud'. In nature, even a single sequence contains all the information necessary to dictate the fold of the protein. How does a multiple sequence alignment make that information more intelligible and useful? Alignment tables expose patterns of amino acid conservation, from which distant relationships may be more reliably detected. Structure prediction tools also give more reliable results when based on multiple sequence alignments than on single sequences.

Visual examination of multiple sequence alignment tables is one of the most profitable activities that a molecular biologist can undertake away from the lab bench. Don't even THINK about not displaying them with different colours for amino acids of different physiochemical type. A reasonable colour scheme (not the only possible one) is:

Colour	Residue type	Amino acids
Yellow	Small non-polar	Gly, Ala, Ser, Thr
Green	Hydrophobic	Cys, Val, Ile, Leu, Pro, Phe, Tyr, Met, Trp
Magenta	Polar	Asn, Gln, His
Red	Negatively charged	Asp, Glu
Blue	Postively charged	Lys, Arg

To be informative, a multiple alignment should contain a distribution of closely and distantly related sequences. If all the sequences are very closely related, the information they contain is largely redundant, and few inferences can be drawn. If all the sequences are very distantly related, it will be difficult to construct an accurate alignment (unless all the structures are available), and in such cases the quality of the results, and the inferences they might suggest, are questionable. Ideally, one has a complete range of similarities, including distantly related examples linked through chains of close relationships.

> **Case Study 5.2 Structural inferences from multiple sequence alignments of thioredoxins**
>
> Thioredoxins are enzymes found in all cells. They participate in a broad range of biological processes, including cell proliferation, blood clotting, seed germination, insulin degradation, repair of oxidative damage and enzyme regulation. The common mechanism of these activities is the reduction of protein disulphide bonds.
>
> $\longrightarrow$

Case Study 5.2 *(continued)*

Plate V shows a multiple sequence alignment of 16 thioredoxins. The structure of *E. coli* thioredoxin contains a central five-stranded β-sheet flanked on either side by α-helices; these helices and strands are indicated by the symbols α and β. Other thioredoxins are expected to share most but not all of the secondary structure of the *E. coli* enzyme. The plate also shows a summary of the alignment as a *sequence logo,* in which letters of different sizes indicate different proportions of amino acids. (T. Schneider and M. Stephens designed sequence logos; this example was produced using the webserver at http://weblogo.berkeley.edu/logo.cgi)

Structural and functional features of thioredoxins that we might hope to identify from the multiple sequence alignment include (see Fig. 5.6 and Plate V):

- **The most highly conserved regions probably correspond to the active site.** The disulphide bridge between residues 32 and 35 in *E. coli* thioredoxin is part of a WCGPC[K or R] motif conserved in the family. Other regions conserved in the sequences, including the PT at residues 75–77 and the GA at residues 92–93, are involved in substrate binding.

- **Regions rich in insertions and deletions probably correspond to surface loops. A position containing a conserved Gly or Pro probably corresponds to a turn.** Turns correlated with insertions and deletions occur at positions 9, 20, 60 and 95. The conserved glycine at position 92 in *E. coli* thioredoxin is indeed part of a turn. It is in an unusual mainchain conformation, one that is easily accessible only to glycine (see Chapter 6). The conserved proline at position 76 in *E. coli* thioredoxin is also associated with a turn. It is in another unusual mainchain conformation, this one easily accessible only to proline.

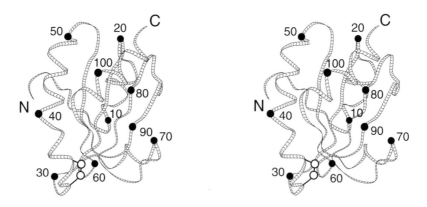

Fig 5.6 The structure of *E. coli* thioredoxin [2TRX] (see also Plate V). Residue numbers correspond to those in the multiple sequence alignment table. The N- and C-termini are also marked. Spheres indicate the positions of the Cα atoms of every tenth residue. The reactive disulphide bridge between Cys32 and Cys35 appears between the numbers 30 and 60.

→

Case Study 5.2 *(continued)*

◆ **A conserved pattern of hydrophobicity with spacing 2 (i.e. every other residue)—with the intervening residues more variable and including hydrophilic residues—suggests a β-strand on the surface**. This pattern is observable in the β-strand between residues 50 and 60.

◆ **A conserved pattern of hydrophobicity with spacing ~4 suggests a helix**. This pattern is observable in the region of helix between residues 40 and 49.

Thioredoxins are members of a superfamily including many more distantly related homologues. These include glutaredoxin (hydrogen donor for ribonucleotide reduction in DNA synthesis), protein disulphide isomerase (which catalyses exchange of mismatched disulphide bridges in protein folding), phosducin (a regulator of G-protein signalling pathways) and glutathione s-transferases (chemical defence proteins). Implicit in the multiple sequence alignment table of the thioredoxins themselves are patterns that should be applicable to identifying these more distant relatives.

Applications of multiple sequence alignments and database searching

Searching in databases for homologues of known proteins is a central theme of bioinformatics. Indeed it brooked no delay; we introduced it in Chapter 1 with the application of PSI-BLAST. We reconsider database searching here, with the goal of trying to understand how we can best use available information to build effective procedures. The goals are high **sensitivity**—picking up even very distant relationships—and high **selectivity**—minimizing the number of sequences reported that are not true homologues. Here we discuss how to apply multiple sequence alignments to this problem. In Chapter 6 we shall discuss how to apply structural information in addition.

During the last decade, great progress has been made in devising methods for applying multiple sequence alignments of known proteins to identify related sequences in database searches. The results are central to contemporary applications of bioinformatics, including the interpretion of genomes. Three important methods are: Profiles, PSI-BLAST and Hidden Markov Models (HMMs).

Profiles

Profiles express the patterns inherent in a multiple sequence alignment of a set of homologous sequences. They have several applications:

◆ They permit greater accuracy in alignments of distantly related sequences.

◆ Sets of residues that are highly conserved are likely to be part of the active site, and give clues to function.

◆ The conservation patterns facilitate identification of other homologous sequences.

◆ Patterns in the sequences are useful in classifying subfamilies within a set of homologues.

◆ Sets of residues that show little conservation, and are subject to insertion and deletion, are likely to be in surface loops. This information has been applied to vaccine design, because such regions are likely to elicit antibodies that will cross-react well with the native structure.

◆ Most structure-prediction methods are more reliable if based on a multiple sequence alignment than on a single sequence. Homology modelling, for instance, depends crucially on correct sequence alignments, and can make effective use of the conformational variation seen in multiple parent structures.

To use profile patterns to identify homologues, the basic idea is to match the query sequences from the database against the sequences in the alignment table, giving higher weight to positions that are conserved than to those that are variable. If a region is absolutely conserved, such as the WGCPC motif in thioredoxins, the procedure should all but insist on finding it. But the risk of being too compulsive is to miss interesting distant relatives; some leeway should be allowed.

What is needed is a quantitative measure of conservation. For each position in the table of aligned sequences, take an inventory of the distribution of amino acids. For instance, for positions 25–30 of the thioredoxin alignment:

Residue number	Number of occurrences																			
---	A	C	D	E	F	G	H	I	K	L	M	N	P	Q	R	S	T	V	W	Y
25	1								2									13		
26			16																	
27					16															
28																7	1		5	3
29	16																			
30			1	4									2		1	7	1			

Given a query sequence representing a potential thioredoxin homologue, we want to evaluate its similarity to the query sequence, in such a way that agreement with the known sequences at the absolutely conserved positions—for instance 26, 27 and 29—contributes a very high score, and disagreement at these positions contributes a very low score. For moderately conserved positions, such as 28, we want a modest positive contribution to the score if the query sequence has an S or a W at this position, and a smaller contribution if it has T or Y. The general idea is to score each residue from the query sequence based on the amino acid distribution at that position in the multiple sequence alignment table.

It is tempting to use the inventories as scores directly. For example, if the residues in a query sequence that correspond to positions 25–30 in thioredoxin contain the sequence VDFSAE, this fragment would score $13 + 16 + 16 + 6 + 16 + 4 = 71$. This is almost the greatest value possible. The alternative query sequence ACGVAP would score $1 + 0 + 0 + 5 + 16 + 2 = 24$, a much lower value. Of course for each query sequence we have to test all possible alignments with the multiple alignment table, and take the largest total score. The highest scoring sequences best fit the patterns implicit in the table.

This simple approach would work if our table contained a large and unbiased sample of thioredoxin sequences. But only in that case would the simple inventory give a correct picture of the *potential* distributions of residues at each position. If our sample were small, the pattern derived would be unlikely to reflect the complete repertoire. Or, if the sample contained a large subset of similar sequences, these would be over-represented in the inventories. For instance, we can see in Plate V that vertebrate thioredoxins form a very closely related set. If we included 20 more vertebrate thioredoxins in the alignment, the profile would recognize only vertebrate thioredoxins effectively.

Substitution matrices suggest how to make the inventory 'fuzzy' and thereby more general.

The observed amino acid distribution at any residue position is a 20-membered array: $(a_1, a_2, a_3, \ldots a_{20})$, where a_i is the number of amino acids of type i observed at that position. (For position 25 of the thioredoxins, $a_1 = 1$ because 1 alanine is observed, and $a_{18} = 13$, representing the valines.) Then in the simple 'inventory scoring' scheme, the score of an alanine at position 25 is just 1; the score of a valine is 13; in general, the score of an amino acid of type i is a_i. In this scheme the rows of the inventory itself provide the arrays a needed for scoring each position.

A better scoring scheme would evaluate any amino acid according to its chance of being substituted for one of the observed amino acids. If $D(i,j)$ is an amino acid substitution matrix — BLOSUM62, for example — then amino acid i could score $a_1 D(i, 1) + a_2 D(i, 2) \ldots a_{20} D(i, 20)$. This scheme distributes the score among observed amino acids, weighted according to the substitution probability. An amino acid in the query sequence could score high *either* if it appears frequently in the inventory at this position, *or* if it has a high probability of arising by mutation from residue types that are common at this position. This approach is more effective in detection of distant relatives from a limited set of known sequences. In this case, the scoring vector for amino acids at any position, is a row in the product of the substitution matrix and the rows of the inventory array. An even better approach is to use as the amino acid distribution a combination of the observed inventory and a general background level of amino acid composition.

The result is a set of probability scores for each amino acid (or gap) at each position of the alignment, called a **position-specific scoring matrix**. An alternative method of deriving a position-specific scoring matrix, based on three-dimensional structures, is described in Chapter 6.

Given a query sequence, and the position-specific scoring matrices derived from a profile, the calculations required to find the optimal score over all alignments of the query sequence with the profile are extensions of the dynamic programming methods for aligning two sequences.

A weakness of simple profiles is that the multiple sequence alignment must be provided in advance, and is taken as fixed. PSI-BLAST and Hidden Markov Models (page 278) gain power by integrating the alignment step with the collection of statistics.

PSI-BLAST

PSI-BLAST is a program that searches a databank for sequences similar to a query sequence. It is a development of the earlier program BLAST. The BLAST program and its variants check each entry in the databank *independently* against a query sequence. PSI-BLAST begins with such a one-at-a-time search. It then derives pattern information from a multiple sequence alignment of the initial hits, and reprobes the database using the pattern. Then it repeats the process, fine-tuning the pattern in successive cycles (see Fig. 5.7).

The problem that BLAST was originally designed to solve is that full-blown dynamic programming methods are rather slow for complete searches in a large databank. Often the databank contains close matches to the query sequence. Less-sensitive but faster programs, such as BLAST, are quite capable of identifying the close matches. If that is what you want, fine. For example, if you want to search for homologues of a mouse protein in the human genome, the similarity is likely to be high and an approximate method likely to find it. But if you want to search for homologues of a human protein in *C. elegans* or yeast, the relationship may be more tenuous. More sophisticated, slower, methods may be required. (It may come as a surprise, but computer time requirements are still a consideration. For although computing is becoming less expensive, the sizes of the databanks and the number of searches desired, on a World-Wide basis, are growing. The net effect is that the pressure on computing resources is increasing.)

The method used by BLAST goes back, in a sense, to the dotplot approach, checking for well-matching local regions. For each entry in the database, it checks for short contiguous regions that match a short contiguous region in the query sequence, using a substitution scoring matrix but allowing no gaps. An approach in which candidate regions of *fixed length* are identified initially can be made very fast by the use of lookup tables.

Once BLAST identifies a well-fitting region, it tries to extend it. In some versions gaps are allowed. The output of BLAST is the set of local segment matches. In an example from Chapter 1:

```
My.care.is.loss.of.care,.by.old.care.done,
   |||||||||     ||||||||||||||     ||||||| ||
Your.care.is.gain.of.care,.by.new.care.won
```

even a very simple algorithm could pick up all matching regions of five contiguous residues and then combine and extend them.

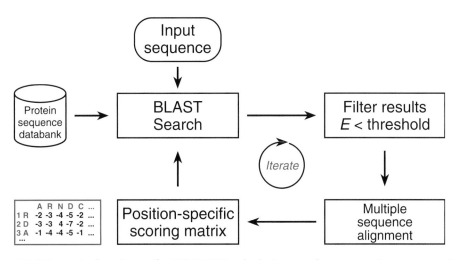

Fig. 5.7 Schematic flowchart of a PSI-BLAST calculation, to detect protein sequences in a database similar to a probe sequence. The user submits an input sequence, and chooses a protein sequence databank to probe.

First, using the input sequence and a standard substitution matrix such as BLOSUM62, a BLAST calculation identifies similar sequences in the database, and assigns a statistical measure of significance, E, to each 'hit'. For each sequence retrieved from the database, E is the number of sequences of equal or higher similarity to the probe sequence that would be expected to be found in the database, just by chance.

The program will select those sequences for which E is no greater than a specified threshold, often chosen as 0.005, and perform a multiple sequence alignment of them.

By counting the relative frequencies of different amino acids in each column of the multiple sequence alignment, the program will derive a position-specific scoring matrix. The box at the lower left shows a part of a position-specific scoring matrix. The columns are labelled by the 20 natural amino acids. The rows are labelled by the sequence to be scored by the matrix. In this case the N-terminal sequence of the sequence to be scored is RDA. . . The entries in the column are the log-odd scores of finding any amino acid at any position in the multiple alignment. For instance, the entry under A in row 3 is -1; therefore, the probability of finding an A at the third position is proportional to 10^{-1}.

To find the score of the sequence, add up the value in the R column of the first row, the D column of the second row, the A column of the third row, etc., to give: $10^{-3} + 10^{-7} + 10^{-1}$. In this example the probabilities are expressed unscaled and as logs to the base 10. Note that the sequence being scored may contain gaps.

This matrix can be used as an alternative to the input sequence and substitution matrix in a BLAST search. Each subsequent BLAST search, based on the matrix derived in the previous step, will return a different set of 'hits'. With a sensible choice of input parameters the procedure will usually converge, to produce a more reliable set of similar sequences than would be returned by the simple BLAST search of the input sequence performed in the first step.

A flowchart for PSI-BLAST

1. Probe each sequence in the chosen database independently for local regions of similarity to the query sequence, using a BLAST-type search but allowing gaps.

2. Collect significant hits. Construct a multiple sequence alignment table between the query sequence and the significant local matches.

3. Form a profile from the multiple sequence alignment.

4. Reprobe the database with the profile, still looking only for local matches.

5. Decide which hits are statistically significant and retain these only.

6. Go back to step 2, until a cycle produces little or no change. This accounts for the 'Iterated' in the program title.

PSI-BLAST, using iterated pattern search (see Box and Fig. 5.7), is much more powerful than simple pairwise BLAST in picking up distant relationships.

PSI-BLAST correctly identifies three times as many homologues as BLAST in the region below 30% sequence identity. It is therefore a very useful method for analysing whole genomes. PSI-BLAST was able to match protein domains of known structure to 39% of the genes in *M. genitalium,* 24% of the genes in yeast and 21% of the genes in *C. elegans.*

The only methods based entirely on sequence analysis that do better than PSI-BLAST are HMMs. These are described in the next section. To achieve significantly better performance it is necessary to make explicit use of structural information. This is discussed in the next chapter.

Hidden Markov Models

An HMM is a computational structure for describing the subtle patterns that define families of homologous sequences. HMMs are powerful tools for detecting distant relatives, and for prediction of protein folding patterns. They are the only method based entirely on sequences—that is, without explicitly using structural information—competitive with PSI-BLAST for identifying distant homologues. They also perform well at identifying the folding pattern of a protein from the amino acid sequence, as assessed in CASP programmes.

Within an HMM is a multiple sequence alignment. However, HMMs are usually presented as *procedures for generating sequences.* A conventional multiple sequence alignment table could also be used to generate sequences, by selecting amino acids at successive positions, each amino acid being chosen from a position-specific probability distribution derived from the profile. But HMMs are more general than profiles:

1. They include the possibility of introducing gaps into the generated sequence, with position-dependent gap penalties.

2. Application of profiles requires that the multiple sequence alignment be specified up front; the pattern statistics are then derived from the alignment. HMMs carry out the alignment and the assignment of probabilities together.

The internal structure of an HMM shows the mechanism for generating sequences (Fig. 5.8). Begin at Start, and follow some chain of arrows until arriving at End. Each arrow takes you to a state of the system. At each state (1) you take some action—emit a residue, perhaps—and (2) choose an arrow to take you to the next state. The action and the choice of successor state are governed by sets of probabilities. Associated with each state that emits a residue are: one probability distribution for the 20 amino acids, and a second probability distribution for the choice of successor state. Both of these probability distributions are calibrated to encode information about a particular sequence family. In this way, the same general computational framework can be specialized to many different sequence families.

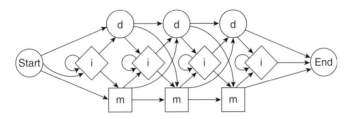

Fig. 5.8 The structure of a Hidden Markov Model (HMM). Corresponding to each residue position in a multiple sequence alignment, the HMM contains a match state (m) and a delete state (d). Insert states (i) appear between residue positions, and at the beginning and end.

◆ *Match states* emit a residue. Here the term match means only that there is *some* amino acid both in the model underlying the HMM and in the sequence emitted, not that these are necessarily the *same* amino acid. The probability of emitting each of the 20 amino acids in each of the match states is a property of the model. As with profiles, the probabilities are position dependent.

◆ *Delete states* skip a column in the multiple sequence alignment. Arriving at a delete state from a match or insert state corresponds to gap opening, and the probabilities of these transitions reflect a position-specific gap-opening penalty. Arriving at a delete state from a previous delete state corresponds to gap extension.

◆ *Insert states* appear between two successive positions in the alignment. If the system enters an insert state, a new residue that does not correspond to a position in the alignment table appears in the emitted sequence. An insert state can be followed by itself, to insert more than one residue. The succession of residues emitted from match and insert states generates the output sequence.

After taking the action appropriate to the type of state (m, d or i), another probability distribution governs the choice of the next state. In every possible succession of states, every column of the embedded alignment must be visited, and either matched or deleted—there is no way to traverse the network without passing through either an m state or a d state at each position.

The dynamics of the system is such that only the current state influences the choice of its successor—the system has no 'memory' of its history. This is characteristic of processes studied by the nineteenth century Russian mathematician A.A. Markov. Distinguish the succession of states from the succession of amino acids emitted to form the output sequence. Several paths through the system can generate the same sequence. Only the succession of characters emitted is visible; the state sequence that generated the characters remains internal to the system, i.e. hidden. By the probability distributions associated with the individual states the system captures—or models—the patterns inherent in a family of sequences. Hence the name, Hidden Markov Model.

Software for applying HMMs to biological sequence analysis can achieve:

1. **Training** Given a set of unaligned homologous sequences, it can align them and adjust the transition and residue output probabilities to define an HMM capturing the patterns inherent in the sequences submitted.

2. **Detection of distant homologues** Given an HMM and a test sequence, calculate the probability that the HMM would generate the test sequence. If an HMM trained on a known family of sequences would generate the test sequence with relatively high probability, the test sequence is likely to belong to the family.

3. **Alignment of additional sequences** The probability of any sequence of states can be computed from the individual state-to-state transition probabilities. Finding the most likely succession of states that the HMM would use to generate one or more test sequences reveals their optimal alignment to the family.

**online
resource
centre**

Web resources   Hidden Markov Models

Two research groups specializing in biological applications of Hidden Markov Models run web servers and distribute their programs:

R. Hughey, K. Karplus and D. Haussler (University of California at Santa Cruz.):
http://cse.ucsc.edu/research/compbio/sam.html
http://cse.ucsc.edu/research/compbio/HMM-apps/HMM-applications.html

R. Hughey, K. Karplus and D. Haussler (University of California at Santa Cruz.):
http://cse.ucsc.edu/research/compbio/sam.html

S. Eddy (Washington University, St. Louis, MO, U.S.A.)
http://hmmer.wustl.edu/

Results of analysis of known sequences and structures are also available on the web:

Pfam is a database of multiple sequence alignments and HMMs for many protein domains, developed by A. Bateman,

---

→

Web resources (*continued*)

E. Birney, R. Durbin, S.R. Eddy, K.L. Howe and E.L. Sonnhammer:
http://www.sanger.ac.uk/Software/Pfam.

Phylogeny

We have now seen several examples of evolution, in proteins and in genomes. These represent the extension to the molecular level of concerns that have occupied biologists since Darwin and even before. The basic principle is that *the origin of similarity is common ancestry.* Although there are exceptions, arising from convergent evolution or horizontal gene transfer, the importance of this principle both for rationalizing contemporary observations and giving a window into the history of life cannot be underestimated.

The field of phylogeny has the goals of working out the relationships among species, populations, individuals or genes. (The general term is 'taxa.') The *observable* taxa — for instance the extant species for which we wish to work out the pattern of ancestry, are called the 'operational taxonomic units', abbreviated to OTUs. Relationship is taken in the literal sense of kinship or genealogy, that is, assignment of a scheme of descendants of a common ancestor (see Box). Evolutionary relationships give us a glimpse at the historical development of life (see Box: *Time scale of Earth history*). Although molecules themselves cannot be dated, the evolutionary events as observed on the molecular level can be calibrated with the fossil record.

Concepts related to biological classification (see also page 22) and phylogeny

Homology means, specifically, descent from a common ancestor.

Similarity is the measurement of resemblance or difference, independent of the source of the resemblance. Similarity is observable in data collectable *now*, and involves no historical hypotheses. In contrast, assertions of homology require inferences about historical events which are usually unobservable.

Clustering is bringing together similar items, distinguishing classes of objects that are more similar to one another than they are to other objects outside the classes. Most people would agree about degrees of similarity, but clustering is more subjective. When classifying objects, some people prefer larger classes, tolerating wider variation; others prefer smaller, tighter, classes. They are called *groupers* or *splitters.*

→

Concepts related to biological classification (see also page 22) and phylogeny *(continued)*

Hierarchical clustering is the formation of clusters of clusters of . . .

Phylogeny is the description of biological relationships, usually expressed as a tree. A statement of phylogeny among objects *assumes* homology and *depends* on classification. Phylogeny states a topology of the relationships based on classification according to similarity of one or more sets of characters, or on a model of evolutionary processes. In many cases, phylogenetic relationships based on different characters are consistent, and support one another. If different characters induce inconsistent phylogenetic relationships, they are all dubious. Note that the same similarity data may be consistent with different possible topologies or trees.

Time scale of Earth history

Geological eras (e.g. Cenozoic), periods (e.g. Jurassic) and cataclysmic events (e.g. asteroid impact: mass extinction) in black. First appearance of, or prevalence of, different life forms in blue. mya = millions of years ago.

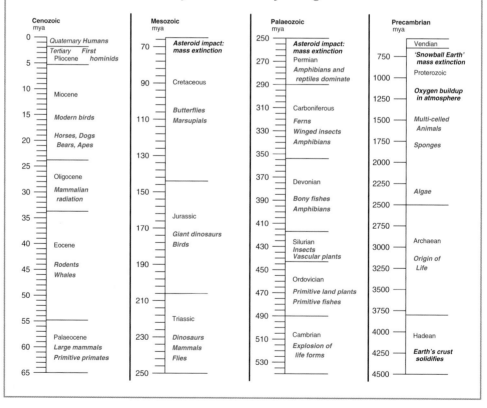

The results of phylogenetic analyses are usually presented in the form of an evolutionary tree. The taxonomy of the ratites—large flightless birds—is a typical example (Fig. 5.9a). The ancestor of the ratites is believed to be a bird that could fly, probably related to the extant tinamous.

Such a tree, showing all descendants of a single original ancestral species, is said to be **rooted**. (The root of the tree typically appears at the top or the side; botanists will have

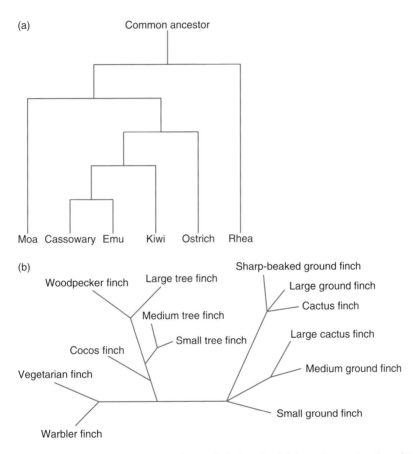

Fig. 5.9 (a) Phylogenetic tree of ratites (large flightless birds) based on mitochondrial DNA sequences. The common ancestor is at the *root* of this tree. A surprising implication of these DNA sequences is that the moa and kiwi are not the closest relatives, and therefore that New Zealand must have been colonized twice by ratites or their ancestors. (b) *Unrooted* tree of relationships among finches from the Galapagos and Cocos Islands. Darwin studied the Galapagos finches in 1835, noting the differences in the shapes of their beaks and the correlation of beak shape with diet. Finches that eat fruits have beaks like those of parrots, and finches that eat insects have narrow, prying beaks. These observations were seminal to the development of Darwin's ideas. As early as 1839 he wrote, in *The Voyage of the Beagle*, 'Seeing this gradation and diversity of structure in one small, intimately related group of birds, one might really fancy that from an original paucity of birds in this archipelago, one species had been taken and modified for different ends'.

to get used to this.) Alternatively, we may be able to specify relationships but not order them according to a history. The relationships among the finches of the Galapagos Islands, studied by Darwin, plus a related species from the nearby Cocos Island, are shown in an **unrooted tree** (Fig. 5.9b). Addition of data from species on the South American mainland ancestral to the island finches might allow us to **root the tree**.

Statement of a tree of relationships may reveal only the connectivity or topology of the tree, in which case the lengths of the branches contain no information. A more ambitious goal is to show the distances between taxa quantitatively, for instance to label the branches with the time since divergence from a common ancestor.

Determination of taxonomic relationships from molecular properties

Given a set of data that characterizes different groups of organisms—for example, DNA or protein sequences, or protein structures, or shapes of teeth from different species of animals—how can we derive information about the relationships among the organisms in which they were observed? It is rare for species relationships and ancestry to be directly observable. Evolutionary trees determined from genetic data are often based on inferences from the patterns of similarity. We generally assume that the more similar the characters the more closely related the species, although this is a dangerous assumption. Nevertheless, from the relationships among the characters we wish to infer patterns of ancestry: the *topology* of the phylogenetic relationships (informally, the 'family tree').

To what extent do the topologies of the relationships depend on the choice of character? In particular, are there *systematic* discrepancies between the implications of molecular and palaeontological analysis?

Molecular approaches to phylogeny developed against a background of traditional taxonomy, based on a variety of morphological characters, embryology and, for fossils, information about the geological context (stratigraphy). The classical methods have some advantages. Traditional taxonomists have much less restricted access to extinct organisms, via the fossil record. They can *date* appearances and extinctions of species by geological methods.

A crucial event in the acceptance of molecular methods occurred in 1967 when V.M. Sarich and A.C. Wilson dated the time of divergence of humans from chimpanzees at 5 mya, based on immunological data. At that time palaeontologists had dated this split at 15 mya, and were reluctant to accept the molecular approach. Reinterpretation of the fossil record led to acceptance of a more recent split, and broke the barrier to general acceptance of molecular methods. It is now generally accepted that human and chimpanzee lineages diverged between ∼6 and 8 mya.

Indeed, many molecular properties have been used for phylogenetic studies, some surprisingly long ago. Serological cross-reactivity was used from the beginning of the last century until superseded by direct use of sequences. In one of the most premature scientific studies I know of, E.T. Reichert and A.P. Brown published, almost a century ago, a phylogenetic analysis of fishes based on haemoglobin crystals. Their work was based on Stenö's law (1669), which states that although different crystals of the same substance have different dimensions—some are big, some small—they have the same interfacial angles, reflecting the similarity in microscopic arrangement and packing of the atomic or molecular units within the crystals. Reichert and Brown demonstrated that the interfacial angles of crystals of haemoglobins isolated from different species showed patterns of similarity and divergence parallel to the species' taxonomic relationships.

Reichert and Brown's results are replete with significant implications. They show that proteins have definite, fixed shapes, an idea by no means recognized at the time. They imply that as species progressively diverge, the structures of their haemoglobins also progressively diverge. In 1909, no one had a clue about nucleic acid or protein sequences. In principle, therefore, the recognition of evolution of protein structures preceded, by several decades, the idea of evolution of sequences.

Today, DNA sequences provide the best measures of similarities among species for phylogenetic analysis. The data are digital. It is even possible to distinguish selective from non-selective genetic change, using the third position in codons, or untranslated regions such as pseudogenes, or the ratio of synonymous to non-synonymous codon substitutions. Many genes are available for comparison. This is fortunate because, given a set of species to be studied, it is necessary to find genes that vary at an appropriate rate. Genes that remain almost constant among the species of interest provide no discrimination of degrees of similarity. Genes that vary too much cannot be aligned. There is an analogous situation in radioactive dating requiring choice of an isotope with a half-life of the same general magnitude as the time interval to be determined.

Fortunately genes vary widely in their rates of change. The mammalian mitochondrial genome, a circular double-stranded DNA molecule approximately 16 000 bp long, provides a useful fast-changing set of sequences for study of evolution among closely related species. In contrast, rRNA sequences were used by C. Woese to identify the three major kingdoms: Archaea, Bacteria and Eukarya (see Fig. 1.2).

Conversely, different rates of change of sequences of different genes can lead to different and even contradictory results in phylogenetic studies. This is especially true if what we want is not just the topology of the relationships but the branch lengths. In addition, horizontal gene transfer, and convergent evolution, are competing phenomena — that is, competing with descent — that interfere with the deduction of phylogenetic relationships.

One limitation of sequence methods is the paucity of access to extinct species. Some subfossil remains of species which became extinct as recently as the last century or two have legible DNA, including specimens of the quagga (a relative of the zebra) and the thylacine (Tasmanian 'wolf', a marsupial), and some New Zealand birds (including moas). We have seen an example of a sequence from the mammoth. Some DNA sequences from Neanderthal man have been recovered from an individual who died approximately 30 000 years ago.

But *Jurassic Park* remains fiction!

It has been possible to sequence DNA amplified from ancient samples, given adequate preservation in continuously frozen environments. Most of the samples that provide useful sequences come from locations near the poles. A few are from mountaintop glaciers even in the tropics.

Holders of the current age record for extraction and analysis of ancient DNA are E. Willerslev and co-workers[1] who have sequenced DNA fragments from ice cores extracted from Greenland. The longest of these cores extended to depths of 2 and 3 km, reaching back 500 000 years or more.

Geologists have reconstructed ancient environments and ecologies from fossil material. Actual sequence data offer potentially a much sharper focus. One of the ice core samples, from the Dye 3 site in South-central Greenland (65° 11′ N, 45° 50′ W), contained sequences from several species of trees, including alder, spruce, pine and yew. DNA from animals included butterflies and moths (definitely) and beetles, flies, spiders and brushfoots (probably). These flora and fauna do *not* characterize the current ecology of Southern Greenland. They suggest that at the time the samples were laid down the temperature did not fall below –17°C in the winter and rose higher than +10°C in the summer. This corresponds to current conditions in Helsinki, Finland, or Murmansk, Russia, warmer than current conditions in Southern Greenland (see http://worldclimate.com).

Phylogenetic trees

We describe phylogenetic relationships as trees. In computer science, a tree is a particular kind of graph. A graph is a structure containing nodes (abstract points) connected by edges (represented as lines between the points) (see Box). In a **directed graph**, each edge has a direction, like a one-way street. A **path** from one node to another is a consecutive set of edges beginning at one point and ending at the other, like our trip from Malmö to Tromsø. A **connected graph** is a graph containing at least one path between any two nodes. From these we can define a **tree**: a connected graph in which there is *exactly* one path between every two points. A particular node may be selected as a **root**; but this is not necessary—abstract trees may be rooted or unrooted (see Fig. 5.9). Unrooted trees show the topology of relationship but not the pattern of descent. A rooted tree in which every node has two descendants is called a **binary tree** (see Box: *A Perl program to draw binary trees*). A **completely connected** graph has an edge between every pair of nodes.

Another special kind of graph is a **directed graph** in which each edge is a one-way street. Examples include the Gene Ontology subnetworks (Fig. 4.3), the HMM diagram shown in Fig. 5.8, and the neural networks illustrated in Chapter 6. Rooted phylogenetic trees are, implicitly, directed graphs, the ancestor–descendant relationship implying the direction of each edge.

It may be possible to assign numbers to the edges of a graph to signify, in some sense, a 'distance' between the nodes connected by the edges. The graph may then be drawn to scale, with the sizes of the edges proportional to the assigned lengths. The length of a path through the graph is the sum of the edge lengths.

1 Willerslev, E. *et al.* (2007). Ancient biomolecules from deep ice cores reveal a forested Southern Greenland. *Science*, **317**, 111–113.

In phylogenetic trees, edge lengths signify either some measure of the dissimilarity between two species, or the length of time since their separation. The assumption that differences between properties of living species reflect their divergence times will be true only if the rates of divergence are the same in all branches of the tree. Many exceptions are known; for instance, among mammals, rodents show, for many proteins, relatively fast evolutionary rates (see Weblem 5.8).

Glossary of terms related to graphs

Graph An abstract structure containing *nodes* (points) and *edges* (lines connecting points).

Path A consecutive set of edges.

Connected graph A graph in which there is at least one path between every two nodes.

Tree A connected graph with exactly one path between every two points.

Edge length A number assigned to each edge signifying in some sense the distance between the nodes connected by the edge.

Path length The sum of the lengths of the edges that comprise the path.

Broadly, there are two approaches to deriving phylogenetic trees. One approach makes no reference to any historical model of the relationships. Proceed by measuring a set of distances between species, and generate the tree by a hierarchical clustering procedure. This is called the **phenetic** approach. The alternative, the **cladistic** approach, is to consider possible pathways of evolution, infer the features of the ancestor at each node, and choose an optimal tree according to some model of evolutionary change. Phenetics is based on similarity; cladistics is based on genealogy.

A PERL program to draw binary trees

The input: (A((BC)D)(EF)) produces the following output, as a PostScript file, which can be printed on most printers and displayed on most terminals.

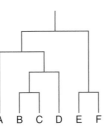

$\longrightarrow$

→

A PERL program to draw binary trees *(continued)*

This representation of a tree is about as simple as possible. Many programs are available on the Web to make fancier drawings, including those with curved lines, such as the one that appears on the cover of this book, or three-dimensional representations for clarity in depiction of very complex trees.

```perl
#!/usr/bin/perl
#drawtree.pl -- draws binary trees (root at top)
#usage:  echo '(A((BC)D)(EF))' | drawtree.pl > output.ps

print <<EOF;
%!PS-Adobe-
%%BoundingBox: atend
/n /newpath load def /m /moveto load def /l /lineto load def
/rm /rmoveto load def /rl /rlineto load def /s /stroke load def
1.0 setlinewidth 50 100 translate 2 2 scale
/Helvetica findfont 10 scalefont setfont
EOF

$tree = <>; chop($tree); $_ = reverse($tree); s/[()]//g;

$x = 0; $y = 0;
while ($nd = chop()) {
    print "$x $y m ($nd) stringwidth pop -0.5 mul 0 rm ($nd) show\n";
    $xx{$nd} = $x; $x+=20; $yy{$nd} = 10;
}

while ($tree =~ s/\(?([A-Z])([A-Z])\)?/$1/) {
    print "n $xx{$1} $yy{$1} m\n";
    ($yy{$1} > $yy{$2}) || {$yy{$1} = $yy{$2}}; $yy{$1} += 20;
    print "$xx{$1} $yy{$1} l $xx{$2} $yy{$1} l $xx{$2} $yy{$2} l s\n";
    $xx{$1} = 0.5*($xx{$1} + $xx{$2});
}
print "n $xx{$tree} $yy{$tree} m 0 20 rl s showpage\n";

$rx = 2*$x + 30; $yt = 2*$yy{$tree} + 146;
print "%%BoundingBox: 40 95 $rx $yt\n";
```

Clustering methods

Phenetic, or clustering, approaches to determination of phylogenetic relationships are explicitly non-historical. Indeed, hierarchical clustering methods are perfectly capable of producing a tree even in the absence of evolutionary relationships. A department store has goods clustered into sections according to the type of product—for instance, clothing or furniture—and subclustered into more closely related subdepartments, such as men's and women's shoes. Men's and women's shoes have a common ancestor, but there is no implication that shoes and furniture do.

A simple clustering procedure works as follows: given a set of species, determine for all pairs a measure of the similarity or difference between them. This could depend on a physical body trait such as the difference between the average adult height of members of two species. Or one could use the number of different bases in alignments of mitochondrial DNA. To create a tree from the set of dissimilarities, first choose the two most closely related species and insert a node to represent their common ancestor. Then replace the two selected species by a set containing both, and replace the distances from the pair to the others by the average of the distances of the two selected species to the others. Now we have a set of pairwise dissimilarities, not between individual species, but between sets of species. (Regard each remaining individual species as a set containing only one element.) Then repeat the process, as in the following example.

Example 5.6

Consider four species characterized by homologous sequences ATCC, ATGC, TTCG and TCGG. Taking the number of differences as the measure of dissimilarity between each pair of species, use a simple clustering procedure to derive a phylogenetic tree.

The distance matrix is:

	ATCC	ATGC	TTCG	TCGG
ATCC	0	1	2	4
ATGC		0	3	3
TTCG			0	2
TCGG				0

Because the matrix is symmetric, we need fill in only the upper half. The smallest distance is 1 (in blue), between ATCC and ATGC. Therefore, our first cluster is {ATCC, ATGC}. The tree will contain the fragment:

ATCC   ATGC

The reduced distance matrix is:

	{ATCC, ATGC}	TTCG	TCGG
{ATCC, ATGC}	0	$\frac{1}{2}(2+3)=2.5$	$\frac{1}{2}(4+3)=3.5$
TTCG		0	2
TCGG			0

The next cluster is {TTCG, TCGG}, distance 2. Finally, linking the clusters {ATCC, ATGC} and {TTCG, TCGG} gives the tree:

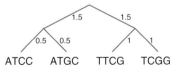

ATCC   ATGC   TTCG   TCGG

Branch lengths have been assigned according to the rule:

branch length of edge between nodes X and Y = $\frac{1}{2}$ distance between X and Y

Whether the branch lengths are truly proportional to the divergence times of the taxa represented by the nodes must be determined from external evidence.

This process of tree building is called the UPGMA method (Unweighted Pair Group Method with Arithmetic mean). A modification of the UPGMA method by N. Saitou and M. Nei, called Neighbour-Joining, is designed to correct for unequal rates of evolution in different branches of the tree.

Cladistic methods

Cladistic methods deal explicitly with the patterns of ancestry implied by the possible trees relating a set of taxa. Their aim is to select the correct tree by utilizing an explicit model of the evolutionary process. The most popular cladistic methods in molecular phylogeny are the **maximum parsimony** and **maximum likelihood** approaches. They are specialized to sequence data, starting from a multiple sequence alignment. Neither maximum parsimony nor maximum likelihood could be applied to anatomic characters such as average adult height.

The *maximum parsimony* method of W. Fitch defines an optimal tree as the one that postulates the fewest mutations. For instance, given species characterized by homologous sequences ATCG, ATGG, TCCA and TTCA, the tree:

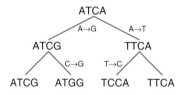

postulates four mutations. An alternative tree:

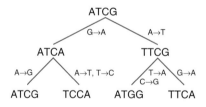

postulates seven mutations. Note that the second tree implies that the G → A mutation in the fourth position occurred twice independently. The former tree is optimal according to the maximum parsimony method, because no other tree involves fewer mutations. In many cases, several trees may postulate the same number of mutations, fewer than any other tree. For such cases the maximum parsimony approach does not give a unique answer.

The *maximum likelihood* method assigns quantitative probabilities to mutational events, rather than merely counting them. Like maximum parsimony, maximum likelihood reconstructs ancestors at all nodes of each tree considered; but it also assigns branch lengths based on the probabilities of the mutational events postulated. For each possible tree topology, the assumed substitution rates are varied to find the parameters that give the highest likelihood of producing the observed sequences. The optimal tree is the one with the maximum likelihood of generating the observed data.

Both maximum parsimony and maximum likelihood methods are superior to clustering techniques. This has been demonstrated with cases where independent evidence—for instance, from classical palaeontology—provides a correct answer, and also with simulated data—computed generation of evolving sequences.

Reconstruction of ancestral sequences

Maximum likelihood methods of phylogenetic tree construction determine, as part of their procedures, the sequences expected in the ancestors of extant species. L. Pauling and E. Zuckerkandl suggested in a 1963 paper that if we could determine the sequences of proteins from extinct organisms, they could be recreated by synthesis.[2] In some cases, such a synthesized sequence shows the expected biological activity. Although this does not prove that it is the correct ancestral sequence, of course the result is gratifying.

Steroid receptors are vertebrate proteins that detect hormones, including androgen, oestrogen, progesterone, glucocorticoid and mineralocorticoid receptors. Upon activation they translocate from the cytoplasm to the nucleus and control gene expression. They belong to a larger superfamily of receptors with affinity for a wider variety of ligands, including thyroid hormones, prostaglandins and 1,25-dihydroxy vitamin D. It is likely that the entire superfamily diverged from a single ancestral protein.

Higher vertebrates contain two closely related receptors:

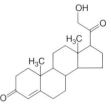

11-deoxycorticosterone

Receptor	Ligand	Physiological role includes
Mineralocorticoid receptor	11-Deoxycorticosterone (teleost fish) aldosterone (tetrapods)	Electrolyte homeostasis
Glucocorticoid receptor	Cortisol	Regulation of metabolism, inflammation and immunity

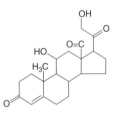

aldosterone

J.W. Thornton and colleagues have studied the species distribution and the evolution of these receptors.[3] The most primitive species in which natural steroid receptors appear are lamprey and hagfish. The receptors in these species show high affinity for 11-deoxycorticosterone, cortisol and aldosterone. It is likely that 11-deoxycorticosterone is the natural ligand of the ancestral homologue. A gene duplication before the emergence of tetrapods gave rise to receptors with differential specificity for cortisol and aldosterone (see Fig. 5.10).

These specificity profiles must be interpreted in light of the fact that aldosterone is not present in primitive vertebrates. The affinity *preceded* the ligand. Aldosterone was first synthesized along the line leading to tetrapods, when a mutation in the

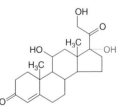

cortisol

2 Pauling, L. and Zuckerkandl, E. (1963). Chemical paleogenetics: molecular restoration studies of extinct forms of life. *Acta Chemica Scandinavica* **17**, Suppl. 1, 9–16.
3 Bridgham, J.T., Carroll, S.M. and Thornton, J.W. (2006). Evolution of hormone-receptor complexity by molecular exploitation. *Science*, 322, 97–101; Ortlund, E.A., Bridgham, J.T., Redinbo, M.R. and Thornton, J.W. (2007). Crystal structure of an ancient protein: evolution by conformational epistasis. *Science*, **317**, 1544–1548.

How could a lock evolve millions of years before the appearance of the key?

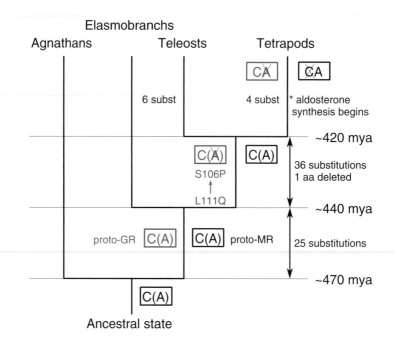

Fig. 5.10 Evolutionary pathway of steroid receptors, inferred from ancestral sequence reconstruction, and synthesis and assay of predicted sequences. The ancestor of the vertebrate steroid receptor probably had affinity for aldosterone (A) as well as other steroids (C), notably 11-deoxycorticosterone and cortisol. This affinity is indicated by C(A), with the (A) in parenthesis because aldosterone was not a natural ligand at that time. After gene duplication to form proto-glucocorticord (proto-GR) and proto-mineralocorticoid (proto-MR) receptors, the protein diverged to attain differential specificity. In tetrapods, subsequent to the origin of aldosterone synthesis, two families of receptor show selectivity for aldosterone (MR) or cortisol. mya = millions of years ago.

cytochrome P-450 11-β hydroxylase gene produced a protein that would hydroxylate 11-deoxycorticosterone. The tetrapod glucocorticoid receptor must have *lost* aldosterone affinity (see Fig. 5.10).

Thornton and colleagues computed the maximum likelihood evolutionary tree of cortisol-specific and aldosterone-specific receptors. They inferred the ancestral sequence, and 'resurrected' it, following Pauling and Zuckerkandl—synthesizing the protein, determining its specificity and solving its crystal structure. The computed and synthesized ancestral protein is activated by 11-deoxycorticosterone, aldosterone and, with lower affinity, cortisol.

Thornton and colleagues computed the evolutionary pathway from the computed ancestral protein. By synthesizing and assaying proteins with different substitutions, they identified two mutations as primarily reponsible for the loss of aldosterone affinity in tetrapod glucocorticosteroid receptors: S106P and L111Q. The L111Q substitution, by itself, has almost no effect on the affinity for any of the three ligands: aldosterone, deoxycorticosterone and cortisol. The S106P substitution, by itself, largely destroys affinity for all three. That L111Q is functional and S106P is not suggests the scenario that L111Q appeared first, followed by S106P. In this way, the evolutionary pathway passed through functional intermediates only. Other substitutions, with less dramatic effects, acted to 'tune' the affinities to their current value.

Molecular modelling allowed inference of the structural consequences in the computed-ancestral protein structure, of these two mutations, separately and together. In the hypothetical glucocorticoid receptor of the common ancestor of all jawed vertebrates, positions 106 and 111 appear in a loop adjacent to a helix. The mutation S106P on its own is predicted to reconform the loop, and the adjacent helix partially unwinds. However, a concomitant of these changes is to bring position 111 closer to the ligand, to the point where the L111Q mutation can form a hydrogen bond to the C17 hydroxyl of cortisol, stabilizing its binding. This hydroxyl is not present in aldosterone (or 11-deoxycorticosterone) so the hydrogen bond can form only to cortisol. Without the repositioning of position 111 by the S106P mutation, it is unlikely that the L111Q mutation, on its own, could form the hydrogen bond to enhance cortisol binding (see Plate VI).

Although some aspects of this investigation are hypothetical, there is reason to be fairly confident that they are correct. In revealing a detailed evolutionary pathway, one can see the mechanism whereby gene duplication followed by divergence produces differential ligand specificity.

The problem of varying rates of evolution

Suppose that the four species A, B, C and D have the phylogenetic tree:

This tree is consistent with the dissimilarity matrix:

	A	B	C	D
A	0	3	3	3
B		0	2	2
C			0	1
D				0

Suppose, however, that species D has changed very fast, although the phylogeny is unaltered. The dissimilarity matrix might then be observed to be:

	A	B	C	D
A	0	3	3	20
B		0	2	20
C			0	20
D				0

from which we would derive the incorrect phylogenetic tree:

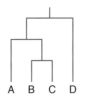

All the methods discussed here are subject to errors of this kind if the rates of evolutionary change vary along different branches of the tree. To test for varying rates, compare the species under consideration with an **outgroup**—a species more distantly related to all the species in question than any pair of them is to each other. For instance, if we are studying species of primates, a non-primate mammal such as the cow would be a suitable outgroup. If the rates of evolution among the primate species were constant, we should expect to observe approximately equal dissimilarity measures between all primate species and the cow. If this is not observed, the suggestion is that evolutionary rates have varied among the primates, and the character being used may well not provide the correct phylogenetic tree.

Are trees the correct way to present phylogenetic relationships?

In the classic model of evolution by descent and divergence, the biological process assures us that the relationship between species is a hierarchy. A tree structure is its natural and proper representation. The question of whether we can accurately and confidently determine the correct tree is a separate issue. However, in some cases, a tree structure does not adequately account for the data. This is observed particularly often in viral evolution, or in situations in which there has been a large amount of horizontal gene transfer.

H.-J. Bandelt and A.W.M. Dress developed a more general graphical clustering method, called **split decomposition,** that can better account for a set of distances than a tree structure. Here is a simple example:

Example 5.7 Alternative representations of similarity data

Bandelt and Dress* consider the following distance matrix:

	A	B	C	D	E	F	G
A	0	4	5	7	13	8	6
B	4	0	1	3	9	12	10
C	5	1	0	2	8	13	11
D	7	3	2	0	6	11	13
E	13	9	8	6	0	5	7
F	8	12	13	11	5	0	2
G	6	10	11	13	7	2	0

Alternative representations of similarity data *(continued)*

Application of the UPGMA method gives the following tree representation:

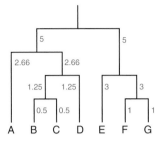

Note that the sum of the path labels between B and C is $0.5 + 0.5 = 1$, which is equal to the B–C distance in the matrix. However, the sum of the path labels between A and D is $2.66 + 2.66 + 1.25 = 6.57$ which is only approximately equal to 7. The tree does not account precisely for all the distance data.

Bandelt and Dress represent the distance matrix by a more complex network based on the split decomposition method:

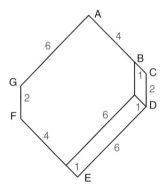

This is not a phylogenetic tree such as we are used to. But the sum of the indices of the edges linking any two nodes, along the shortest path, does reproduce the distance data. In this graph, the distance between A and D, along a shortest path—which might pass through B and C—is $4 + 1 + 2 = 7$, equal to the distance in the original data matrix. Observe that the graph contains some intermediary nodes that do not correspond to original data points.

From which representation do the clusters appear more naturally? Do you think that if the UPGMA graph were drawn so that the edge lengths were proportional to the distances indicated as the labels, the clustering would be more clear? (Try it!)

* Bandelt, H.-J. and Dress, A.W.M. (1992). Split decomposition: A new and useful approach to phylogenetic analysis of distance data. *Molecular Phylogenetics and Evolution* **1**, 242–252.

Computational considerations

Cladistic methods—maximum parsimony and maximum likelihood—are more accurate than simpler clustering methods such as UPGMA, but require large amounts of computer time if the number of species is appreciable. The total number of possible trees, which cladistic methods are committed to considering if they could, increases very rapidly with the number of species. As a result, in many cases of interest these methods can give only approximate answers, even with respect to their intrinsic assumptions.

Because calculated phylogenies are often approximations, it is important to try to test them. Methods include:

1. Comparison of phylogenies obtained from different characters describing the same set of taxa—are they consistent? If trees produced from different characters share a subtree, perhaps that portion of the phylogeny has been determined reliably and other portions have not.

2. Analysis of subsets of taxa should give the same answer—with respect to the subset—as appears within the full tree.

3. Formal statistical tests, involving rerunning the calculation on subsets of the original data, are known as **jackknifing** and **bootstrapping**:

   - Jackknifing is calculation with data sets sampled randomly from the original data. For phylogeny calculations from multiple sequence alignments, select different subsets of the positions in the alignment, and rerun the calculation. Finding that each subset gives the same phylogenetic tree lends it credibility. If each subset gives a different tree, none of them is trustworthy.
   - Bootstrapping is similar to jackknifing except that the positions chosen at random may include multiple copies of the same position, to form data sets of the same size as the original, to preserve statistical properties of the sampling.

4. If there are very long edges, consider seriously the possibility of unequal variation in evolutionary rate that may have perturbed the calculation. Introduce outgroup taxa to check.

Putting it all together

A fairly standard sequence of operations in bioinformatics involves:

1. Select a sequence of interest, and use PSI-BLAST or some equivalent tool to extract a set of similar sequences.

2. Perform a multiple alignment of the sequences retrieved.

3. From the multiple sequence alignment, derive and study the conservation patterns, and from the set of overall similarities and differences among the sequences, compute and draw a phylogenetic tree.

Numerous tools are available to facilitate the individual steps of the process, and to smooth the transition from each to the next. We have already mentioned the feature of the Sequence Retrieval System that feeds retrieved sequences into a multiple sequence alignment program such as CLUSTAL-W. CLUSTAL-W computes a tree based on the similarities of the input sequences, which can serve as input to one of a number of tree-drawing programs. PhyloGenie, by T. Frickey and A.N. Lupas[4] integrates and coordinates facilities for all these steps.

Web resources Phylogenetic trees
The taxonomic community has expended great effort to produce mature software. The PHYLIP package (PHYLogeny Inference Package) of J. Felsenstein is an integrated collection of many different techniques. The programs work on many different types of computers, and are freely distributed and easily obtained.

Summary of tools for phylogenetics; includes useful list of web sites, and general listing of phylogeny software:
http://evolution.genetics.washington.edu/phylip/software.html
and
Whelan, S., Liò, P. and Goldman, N. (2001). Molecular phylogenetics: State-of-the-art methods for looking into the past.
Trends in Genetics, **17**, 262–272.

Some multiple sequence alignment packages, such as CLUSTAL–W, provide facilities to launch a phylogenetic tree calculation from the alignments they produce.

---

online
resource
centre

Recommended reading

Altschul, S.F. and Koonin, E.V. (1998). Iterated profile searches with PSI-BLAST—a tool for discovery in protein databases. *Trends in Biochemical Sciences*, **23**, 445–447. [Description of one of the most important tools for database searching for sequence similarity.]

Altschul, S.F., Boguski, M.S., Gish, W. and Wootton, J.C. (1994). Issues in searching molecular sequence databases. *Nature Genetics*, **6**, 119–129. [General background to challenges in designing information-retrieval methods and interpreting the results.]

Eddy, S. (1996). Hidden Markov models. *Current Opinion in Structural Biology*, **6**, 361–365. [Readable introduction to an important mathematical technique providing powerful tools for detection of distantly related sequences, and protein fold recognition.]

Efron, B. and Gong, G. (1983). A leisurely look at the bootstrap, the jackknife, and cross-validation. *The American Statistician*, **37**, 36–48. [Classic paper on statistical methods for calibrating pattern recognition procedures.]

Penny, D., Hendy, M.D., Zimmer, E.A. and Hamby, R.K. (1990). Trees from sequences: panacea or Pandora's box? *Australian Systematic Botany*, **3**, 21–38. [Cautionary notes about determination of phylogenetic trees.]

---

4 Frickey, T. and Lupas, A.N. (2004). PhyloGenie: automated phylome generation and analysis. *Nucleic Acids Research*, **32**, 5231–5238; http://protevo.eb.tuebingen.mpg.de/download

Smith, T.F. (1999). The art of matchmaking: sequence alignment methods and their structural implications. *Structure*, **7**, R7–R12. [Explanation of different sequence alignment methods, and what can confidently be concluded from them about similarity of structure. Gives due emphasis to structure-based alignment as a 'gold standard'.]

Exercises, Problems and Weblems

Exercises

Exercise 5.1 What is the Hamming distance between the words DECLENSION and RECREATION?

Exercise 5.2 What is the Levenshtein distance between the words BIOINFORMATICS and CONFORMATION?

Exercise 5.3 The Levenshtein distance between the strings `agtcc` and `cgctca` is 3, consistent with the following alignment:

`ag-tcc`

`cgctca`

Provide a sequence of three edit operations that convert `agtcc` to `cgctca`.

Exercise 5.4 'I wasted time and now doth time waste me'. (a) First sketch the expected appearance of a dotplot of this character string against itself. (b) Then calculate the dotplot exactly, recording only character identities as dots in the matrix, and compare with (a).

Exercise 5.5 What values of window and threshold (see program, pages 248–9) could you use to eliminate the singletons in the DOROTHYHODGKIN dotplot, but retain the other matches shown?

Exercise 5.6 For each of the matrices (a) PAM250 and (b) BLOSUM62, which substitution is more probable, W ↔ F or H ↔ R?

Exercise 5.7 To what alignment does the path through the following dotplot correspond?

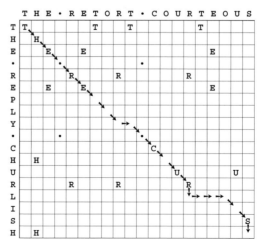

Exercise 5.8 In planning your trip from Malmö to Tromsø (see page 262), suppose that for personal reasons you wanted to include a visit to Uppsala. How could you adjust the costs of the segments to ensure that the minimal-cost route passed through Uppsala?

Exercise 5.9 How would you use a dotplot to pick up palindromic DNA sequences of the type that appear partly on each strand, as in the specificity sites of restriction endonucleases?

Exercise 5.10 Modify the PERL program on pages 248–9 that draws dotplots to accept sequences in FASTA format.

Exercise 5.11 To what value of P would a Z-score of 1 correspond in a normal distribution?

Exercise 5.12 For each of the alignments in Fig. 5.2, state whether it is in the twilight zone, more similar than the twilight zone or less similar than the twilight zone.

Exercise 5.13 Figure 5.2a shows the sequence alignment of papaya papain and kiwi fruit actinidin, and the corresponding dotplot. The sequence alignment shows two places at which one or more residues are deleted from the papain sequence, and one place at which a residue is deleted from the actinidin sequence. On a photocopy of Fig. 5.2a, indicate in the dotplot the approximate positions of these insertions and deletions.

Exercise 5.14 Suppose it were argued that randomizing a sequence is not an appropriate way to generate a control population for analysis of the statistical significance of pairwise sequence alignments, because natural sequences have non-random dipeptide or tripeptide frequencies. What improved way to generate a control population would you suggest?

Exercise 5.15 Comparisons of DNA sequences of homologous chromosomes in different people show that, on average, 1 of every 700 bp of non-coding DNA is different. 95% of the human genome is non-coding. Estimate the number of polymorphisms in the human genome, to give some idea of the number of potential DNA markers.

Exercise 5.16 Show the calculations that led to the entry with value 65 in Example 5.6. What is the significance of the observation that there are two arrows coming from it?

Exercise 5.17 The α-helix formed by residues 32–49 in E. coli thioredoxin is interrupted. On a photocopy of Fig. 5.6 indicate where this interruption appears. At what residue is this distortion likely to occur?

Exercise 5.18 (a) Using simple 'inventory' scoring, what hexapeptide gives the greatest possible value for a match to positions 25–30 in the thioredoxin scoring table (page 274). (b) Using a scoring scheme distributed among all 20 amino acids according to the BLOSUM62 matrix, compare the score of this hexapeptide with the score of the hexapeptide VDFSAE.

Exercise 5.19 (a) Make an inventory of the region from residues 90–95, similar to the table on page 274. What contribution would the following sequences aligned to these residues make to a simple profile score using inventories as weights? (b) ISSAVK. (c) FVGAKE.

Exercise 5.20 (a) Is the following pair of trees identical in topology?

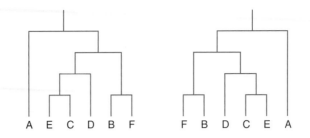

(b) Is the following pair of trees identical in topology?

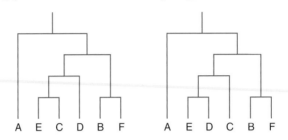

Exercise 5.21 Draw all possible rooted trees relating three taxa. How many are there?

Exercise 5.22 For the final graph in Example 5.6, how was the branch length 1.5 of the nodes joining the clusters {ATCC, ATGC} and {TTCG, TCGG} arrived at?

Exercise 5.23 Using the original matrix of distances, and the final tree, in Example 5.6, for each pair of species compare the original distance between them with the sum of the lengths of the paths joining them in the tree.

Exercise 5.24 Draw an example of a connected graph containing five nodes that is not a tree.

Exercise 5.25 If a completely connected graph is not a tree, what is the maximum number of nodes it can contain?

Exercise 5.26 Mitochondrial DNA sequences of European, African and Asian cattle suggest that European and African breeds are more closely related to each other than to Indian breeds. To exclude the appearance of this result as the spurious result of differential rates of evolution in the two lineages, suggest a reasonable choice of a species as an outgroup.

Exercise 5.27 Draw the final tree in Example 5.6 to scale, with sizes of the edges proportional to the assigned branch lengths.

Exercise 5.28 For the dynamic programming method for alignment of two sequences of length n, we noted that the execution *time* requirements scale as n^2. In a naive implementation of the algorithm, how would the storage *space* requirements scale with n: (a) If we want to determine an optimal alignment, so that traceback information must be stored? (b) If we want only the score and not an alignment, so that traceback need not be stored? [Note: subtle ways of implementing the algorithm substantially reduce the space requirements over the naive implementation.]

Problems

Problem 5.1 Draw a dotplot of the following sequence from the wheat dwarf virus genome: ttttcgtgagtgcgcggaggctttt against itself. In what respects is it not a perfect palindrome?

Problem 5.2 (a) How would you change the algorithm of the section on dynamic programming (starting on page 261) to find the optimal matches of a relatively short pattern $A = a_1 a_2 \cdots a_n$ in a long sequence $B = b_1 b_2 \cdots b_m$ with $n \ll m$. (No gap penalty in the regions of B that precede and follow the region matched by A.) This corresponds to motif matching as described in Chapter 1. (b) Redo the calculations of Example 5.5 for aligning the strings ggaatgg and $B = $ atg as a motif-matching problem, using the same scoring scheme: match = 0, mismatch = 20, *internal* gap initiation 25, gap extension 22. (c) How do the results that you get differ from those derived in the example?

Problem 5.3 At which residues in *E. coli* thioredoxin are there turns in the chain conformation that do *not* correspond to regions at or near which there are deletions in the multiple sequence alignment?

Problem 5.4 How could you modify the profile method to retain its ability to pick up non-mammalian thioredoxins if a large number of additional mammalian, closely related, sequences were added to the table. Consider (a) methods that attempt to remove redundancy by ignoring certain sequences, and (b) methods that retain all the sequences but include a weighting scheme to balance the representation of the closely related ones.

Problem 5.5 Write a PERL program to make profile inventories from a multiple sequence alignment, and to score the matching of query sequences using BLOSUM62. Assume that the query sequence has already been aligned before it is presented to the program.

Problem 5.6 (a) Write a PERL program to read in two character strings and report all matches of contiguous five-character regions. Test this on the strings

My.care.is.loss.of.care,.by.old.care.done,　　and
Your.care.is.gain.of.care,.by.new.care.won

(b) Develop this program to extend and combine the matches found, to the longest regions that contain perfectly matching 5-mers, without gaps, with no more than 25% mismatches overall.

Problem 5.7 Extend the previous problem by writing a PERL program to illustrate, in a form based on dotplots, the progress of a BLAST-type algorithm at the stages when it (a) detects all matching substrings of length 5, (b) extends them to maximal contiguous matches, (c) combines them to form a match with no more than k mismatches. You may make use of the PERL program for dotplots in the text.

Problem 5.8 Write a program to *animate* the progress of a BLAST-type algorithm as described in the previous problem. As background, look on the Web for examples of animation of string search algorithms. (This problem requires a relatively high level of experience with computing.)

Problem 5.9 Single-stranded RNAs, such as tRNA or rRNAs, adopt conformations containing *stem–loop* regions, in which a region of the chain loops back on itself to form a double-stranded helix from complementary base pairs, with antiparallel strands. How would a program that would detect palindromes be useful in analysing RNA sequences to detect regions capable of forming perfect (i.e. no mismatched bases) stem–loop structures?

Problem 5.10 Suppose that you have a pair of dice, one red and one green. Define a *state* of the pair of dice as the pair of numbers: the number appearing on the upper face of the red one followed by the number appearing on the upper face of the green one. Instead of rolling the dice, pass from state to state by tipping each of the dice by $90°$ in any direction with equal probability. Then a state in which 6 is up on one of the dice can by followed, with equal probability, by 2, 3, 4 or 5 up. (Dice are constructed so that the sum of the numbers on each of the three sets of opposite faces is 7. Therefore, the probability that 1 follows 6 is 0, because this would require a $180°$ rotation.) The probability of the sequence 6, 2, 6, 4 is $(1/4)^4 = 1/256$. The probability of generating the sequence 6, 2, 5, 4 is 0, because the transition $2 \to 5$ is not allowed, and the probability of the sequence 6, 6, 2, 3, 4 is zero because the system must change its state, so 6 cannot be followed by 6.

This procedure defines a first-order Markov process.

Write a program to answer the following questions: suppose the initial state has a 4 at the top of the red die and a 3 at the top of the green die. (a) What is the probability of another state in which the numbers add up to 7 appearing within 5 moves? (b) If the initial state is an 8, what is the probability that another 8 appears before a 7?

Problem 5.11 Show that any (undirected) graph that has either of the following properties must also have the other: (1) There is a unique path between any two nodes. (2) The graph contains no cycles.

Problem 5.12 How many paths are there altogether from Start to Finish in Fig. 5.11? Count them in each of the following ways:

(a) Brute force—write down all the possibilities. This is actually less of a mindless exercise than it appears. It demonstrates that it is really not so difficult to do as it first seems. Secondly, it shows how you will get to sense patterns as you do it.

(b) In Fig. 5.11, count the number of paths from Start to A and from A to Finish. Multiply these numbers together to get the total number of paths from Start to Finish that pass through A. Then count the number of paths from Start to B and from B to Finish. What is the relationship between these numbers? Multiply them together to get the total number of paths from Start to Finish that pass through B. Compute the total number of paths from Start to Finish as the sum of the number of paths from Start to Finish that pass through A, B, C and D.

(c) Recognize that to go from Start to Finish requires six steps, including exactly three left turns and three right turns (else you won't end up at the right place). Different choices of the order of right and left turns correspond to different paths. To count the number of paths, recognize that you need to decide only how many ways there are to choose the three steps at which you turn left (for then you must turn right

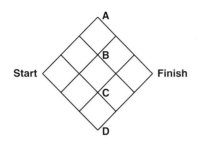

Fig. 5.11 Counting paths on a finite lattice.

at the other three steps). To assign three left turns to six steps, first we can choose one of six steps for one left turn, then one of the five remaining steps for the next left turn, and then one of the four remaining steps for the final left turn. However, the product of these numbers overcounts the number of possibilities, because it includes the same sets of steps assigned in different order. Each triple can arise in six different ways, and it is necessary to correct for this. The result is equal to the binomial coefficient $\begin{pmatrix} 6 \\ 3 \end{pmatrix} = 6!/(3!3!)$.

Problem 5.13 For the final tree in Example 5.10, derive possible ancestors at internal nodes chosen from a maximum parsimony criterion. Are there any ambiguities?

Problem 5.14 A convenient notation for trees uses nested parentheses to indicate the clusters. (a) Expand the following into a rooted tree: ((A(BC))D). (b) Write the parenthesis notation for the trees shown in Exercise 5.20.

Problem 5.15 Add a generous amount of comments to the PERL program for drawing dotplots and trees.

Problem 5.16 Write a PERL program for the UPGMA method of deriving a phylogenetic tree from a matrix of distances. You may use the program for tree drawing to produce graphical output.

Problem 5.17 For each pair of points in Example 5.7: (a) Calculate the sum of indices of the edges along the path joining the two points in the tree representation of the data. (b) Calculate the sum of indices of the edges along the shortest path joining the two points in the split-decomposition representation of the data. (c) What is the average value of the absolute value of the difference between the numbers computed in (a) and the distances in the data matrix? (d) Confirm that the values computed in part (b) precisely reproduce *all* values in the data matrix.

Problem 5.18 If you didn't draw the UPGMA graph with edge lengths scaled according to the values of the indices, do it now. Compare with the split-decomposition representation. Now, from which representation do the clusters appear more naturally?

Weblems

Weblem 5.1 Retrieve the gene sequence of mitochondrial ATPase subunit 6 from Atlantic hagfish (*Myxine glutinosa*). Draw a dotplot against the homologous gene from sea lamprey (*Petromyzon marinus*). Comment on the similarity observed and compare with the similarity between the lamprey and dogfish sequences shown on page 246.

Weblem 5.2 Submit the amino acid sequence of papaya papain to a BLAST search and to a PSI-BLAST search. Which of the homologues appearing in Fig. 5.2 are successfully detected by BLAST? Which by PSI-BLAST?

Weblem 5.3 Submit the amino acid sequence of papaya papain to a PSI-BLAST search (see previous Weblem). In the results for the match to human procathepsin L, indicate on a photocopy of the dotplot (Fig. 5.2b) the regions of local matches reported.

Weblem 5.4 Find structures of thioredoxins appearing in the alignment table in Plate V, from organisms other than *E. coli*. On a photocopy of the alignment table, indicate the regions of helix and strands of sheet as assigned in the Protein Data Bank entries, and compare with the helices and strands of *E. coli* thioredoxin.

Weblem 5.5 Align the amino acid sequence of papaya papain and the homologues shown in Fig. 5.2 using CLUSTAL-W or T-Coffee. Compare the results with the alignment table in Pfam based on Hidden Markov Models, and with the structural alignments in Fig. 5.2.

Weblem 5.6 Can PSI-BLAST identify the homology between immunoglobulin domains and domains of *Cellulomonas fimi* endogluconase C and *Streptococcus agalactiae* IgA receptor?

Weblem 5.7 (a) Can PSI-BLAST identify the relationship between *Klebsiella aerogenes* urease, *Pseudomonas diminuta* phosphotriesterase and mouse adenosine deaminase? (b) Compare the alignments of these three sequences produced by DALI or MUSTANG, and by CLUSTAL-W or T-Coffee.

Weblem 5.8 Which pair of species is more closely related (as measured by more recent time of divergence)? (a) human/bonobo, mouse/rat; (b) human/orangutan, whale/hippopotamus; (c) human/kangaroo, *Arabidopsis*/wheat.

Weblem 5.9 The growth hormones in most mammals have very similar amino acid sequences. (The growth hormones of the alpaca, dog, cat, horse, rabbit and elephant each differ from that of the pig at no more than three positions out of 191.) Human growth hormone is very different, differing at 62 positions. The evolution of growth hormone accelerated sharply in the line leading to humans. By retrieving and aligning growth hormone sequences from species closely related to humans and our ancestors, determine *where* in the evolutionary tree leading to humans the accelerated evolution of growth hormone took place.

The next series of weblems is designed to place the human species in its biological context by analysis of sequences from near and distant relatives, and to illustrate some of the variety of genetic information that has been used to investigate phylogenetic relationships.

Weblem 5.10 The living species most closely related to humans are apes and monkeys. Alu elements are a type of SINE (Short INterspersed Element) useful as species markers. Although part of the repetitive non-coding portion of the genome, some Alu elements function in gene regulation. On the basis of Alu elements that regulate the genes for parathyroid hormone, the haematopoietic cell-specific FcεRI-γ receptor, the central nervous system-specific nicotinic acetylcholine receptor α3 and the T cell-specific CD8α, derive a phylogenetic tree for human, chimpanzee, gorilla, orangutan, baboon, rhesus monkey and macaque monkey.

Weblem 5.11 Humans are primates, an order that we, apes and monkeys share with lemurs and tarsiers. On the basis of the gene contents and order of the β-globin gene clusters of human, a chimpanzee, an old-world monkey, a new-world monkey, a lemur and a tarsier, derive a phylogenetic tree of these groups.

Weblem 5.12 Primates are mammals, a class we share with marsupials and mono-tremes. Extant marsupials live primarily in Australia, except for the oppossum, found in North and South America. Extant monotremes are limited to two animals from Australia and New Guinea: the platypus and echidna. Collect the nuclear genes for mannose 6-phosphate/insulin-like growth factor II receptor from mammalian species including placentals, marsupials and monotremes. From them draw an evolutionary tree, indicating branch lengths. Are monotremes more closely related to placental mammals or to marsupials?

Weblem 5.13 Mammals are vertebrates, a subphylum that we share with fishes, sharks, birds and reptiles, amphibia and primitive jawless fishes (example: lampreys). For the coelacanth (*Latimeria chalumnae*), the great white shark (*Carcharodon carcharias*), skipjack tuna *(Katsuwonus pelamis)*, sea lamprey (*Petromyzon marinus*), frog (*Rana pipiens*), and Nile crocodile (*Crocodylus niloticus*), using sequences of cytochromes *c* and pancreatic ribonucleases, derive evolutionary trees of these species.

Weblem 5.14 Tetrapods are gnathostomes, a superclass that we share with fishes. The traditional view of fish → tetrapod evolution is that jawed vertebrates split into one group containing cartilaginous fishes, *Chondrichthyes*, including sharks and rays, and another containing both ray-finned fishes, *Actinopterygii*, including modern bony fishes such as cod and salmon; and lobe-finned fishes, *Sarcopterygii*, including coelacanths, lungfishes and tetrapods (see Fig. 5.12a). Test this hypothesis using at least 12 mitochondrial protein-coding genes from at least 30 species including sharks, lungfish, bony fishes, amphibia, reptiles, birds and mammals, using a lamprey as an outgroup. Of the three groups — cartilaginous fishes, bony fishes and tetrapods — which pair appears to be most closely related: bony fishes and tetrapods as in the traditional view, or a different pair? Consider specifically a phylogeny according to which the tetrapods split off first, and cartilaginous fishes, bony fishes and even lungfishes are sister taxa (see Fig. 5.12b).

Weblem 5.15 Vertebrates are chordates, a phylum that also includes lancelets (small fish-like marine animals; example, amphioxus), and jawless vertebrates (lacking a true vertebral column (example, lamprey). As in other organisms with bilateral symmetry (including insects), vertebrate HOX genes encode a family of DNA-binding proteins. The expression of these genes varies along the head-to-tail body axis, and controls the setting out of the body plan. Indeed there is an amazing mapping between the order of the genes on the chromosome, the order of their action along the body and the relative times during development of the onset of their activity.

Fig. 5.12 (a) Traditional view of gnathostome evolution: the first split was between cartilaginous fishes such as sharks and others; then tetrapods emerged from a group containing coelacanths, lungfishes and tetrapods split off. An alternative to be considered is that the split leading to the tetrapods was the earliest split. (b) Alternative tree, rooted using lamprey as an outgroup, in which the lineage leading to tetrapods split off first.

During the course of vertebrate evolution there have been large-scale genomic duplications, associated presumably with the development of greater complexity of body architecture, as presciently suggested by S. Ohno in 1970. The genomes of insects and amphioxus have a single HOX cluster. Zebrafish have seven HOX clusters, interpretable in terms of a series of duplications: $1 \to 2 \to 4 \to 8$ followed by loss of one to reduce $8 \to 7$.

Find the number of HOX clusters in the human and the lamprey, perform a multiple sequence alignment to assign correspondences among the individual genes and derive from the results a phylogenetic tree for amphioxus, lamprey, fishes and mammals.

Weblem 5.16 Chordates are deuterostomes (see page 23), a grouping we share with urochordates (example: sea squirts) hemicordates (example: amphioxus), and echinoderms (example: starfish). There are systematic differences between these four phyla in their mitochondrial genetic code. Determine, for examples of organisms in each phylum, the amino acids that correspond to the codons ATA and AGA. Derive from these results a phylogenetic tree of the four deuterostome phyla.

Structural bioinformatics and drug discovery

Learning goals

1. Understanding the concept of protein folding: the process by which the one-dimensional amino acid sequence encoded by a gene takes up a definite and biologically active three-dimensional conformation.

2. Recognizing that steric considerations severely limit the conformations of the polypeptide chain, with the Sasisekharan–Ramakrishnan–Ramachandran plot showing the allowed states of the mainchain.

3. Getting to know the 20 sidechains—the actors that play all the roles in all the proteins.

4. Understanding the hydrophobic effect and its implications for the structures and energetics of folded proteins.

5. Generalizing the ideas of sequence alignment to alignment of protein sequences by structural superposition.

6. Knowing the relationship between divergence of sequence and divergence of structure in protein evolution.

7. Becoming familiar with classification of protein folding patterns, as presented for example by the Structural Classification of Proteins (SCOP) database and web site.

8. Knowing some basic approaches to the prediction of protein structure from amino acid sequence, and the Critical Assessment of Structure Prediction (CASP) programmes.

9. Understanding the basis of Hidden Markov Models, the most powerful methods now available for deducing affinities between proteins on the basis of sequences.

10. Knowing the basic requirements of a successful drug, and understanding some approaches to drug discovery and design.

Introduction

To equip them to play their biological roles, proteins have three profound and fundamental properties:

1. For implementation of the metabolism, growth, architecture and regulation of cell and organism, proteins show a great variety of structures and functions.

2. For their synthesis, proteins show a unity, that permits all proteins to be synthesized by one apparatus, the ribosome. The implications for molecular biology are analogous to the impact of movable type on printing.

3. Proteins fold spontaneously to active native three-dimensional states. It is therefore sufficient to encode the amino acid sequence in terms of a one-dimensional sequence of nucleotides in a genome. Evolution takes advantage of the hereditary transmission of the information, and of the mutations that generate variants on which natural selection can act.

The great variety of three-dimensional structures and functions of proteins arises in molecules that share underlying common chemical features. Proteins are like strings of Christmas tree lights: Almost all proteins consist of a linear (i.e., unbranched) polymer mainchain with different amino acid sidechains attached at regular intervals (Fig. 1.6). The wire linking the string of lights corresponds to the repetitive mainchain or backbone, and the variable sequence of colours of the lights corresponds to the individuality of the sequence of sidechains.

The amino acid sequence of a protein is specified by the nucleotide sequence of a gene. The three-dimensional structures of protein molecules are determined, without further participation of nucleic acids, by the one-dimensional sequences of their amino acids. Proteins fold spontaneously to their native conformations.

How does the amino acid sequence encode the three-dimensional structure? Any possible folding of the mainchain places different residues in contact. The interactions of the sidechains and mainchain, with one another and with the solvent, and the restrictions placed on sidechain mobility, determine the relative stabilities of different conformations. This is a consequence of the second law of thermodynamics, which states that systems at constant temperature and pressure find an equilibrium state that is a compromise between comfort (low enthalpy, H) and freedom (high entropy, S), to give a minimum Gibbs free energy $G = H - TS$, in which T is the absolute temperature. (In human relationships, marriage is just such a compromise.)

Proteins have evolved so that one folding pattern of the mainchain is thermodynamically significantly better than other conformations. This is the native state. If we could calculate sufficiently accurately the energies and entropies of different conformations, and if we could computationally examine a large enough set of possible conformations to be sure of including the correct one, it would be possible consistently to predict protein structures from amino acid sequences on the basis of *a priori* physicochemical principles. There has been progress towards this goal but it is not yet achieved.

The mainchain of each protein in its native state describes a curve in space. We now know 50 000 protein structures. They show a great variety of folding patterns. The first problem in analysing these structures is one of presentation. Figure 6.1 illustrates, for the small protein acylphosphatase, the difficulty in interpreting a fully detailed, literal representation, and the kind of simplified pictures that computer programs produce to give us visual access to the material. An active cottage industry has produced many different simplified representations. A skilled molecular illustrator will combine them to show different parts of a structure in finely tuned degrees of detail.

The central frame of Fig. 6.1 shows the course through space of the mainchain of acylphosphatase. Two regions at the front of the picture have the form of helices—like classic barber poles—with their axes almost vertical in the orientation shown. Acyl-phosphatase also contains four strands of sheet. These too are approximately vertical in orientation. The four strands interact laterally to stabilize their assembly into a β-sheet. In the last frame, helices and strands are represented as 'icons': helices as cylinders and strands of sheet as large arrows. The top frame of Fig. 6.1, showing the most detailed representation of the structure, including mainchain and sidechains, demonstrates the importance of simplification in producing an intelligible picture of even a small protein.

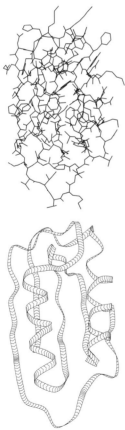
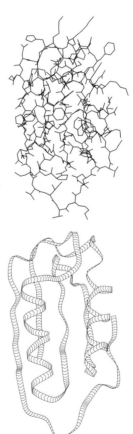

Fig. 6.1 (*See overleaf*)

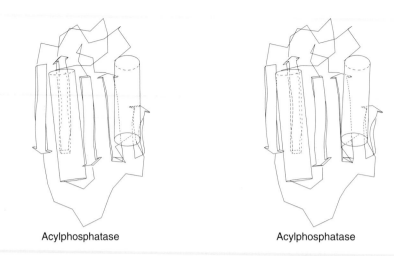

Acylphosphatase   Acylphosphatase

Fig. 6.1 Proteins are sufficiently complex structures that it has been necessary to develop specialized tools to present them. This figure shows a relatively small protein, acylphosphatase, at three different degrees of simplification. First: complete skeletal model; mainchain bolder than sidechains. Second: the course of the chain is represented by a smooth interpolated curve, the chevrons indicating the direction of the chain. Third: schematic diagram, in which cylinders represent helices and arrows represent strands of sheet. The solid objects in the picture are represented as 'translucent' by altering lines that pass behind them to broken lines. It is possible to superpose different representations visually by rotating the page $90°$ and viewing in stereo (but not for too long!)

Protein stability and folding

To form the native structure, the protein must optimize the interactions within and between residues, subject to constraints on the space curve traced out by the mainchain. Preferred conformations of the mainchain bias the folding pattern towards recurrent structural patterns: (1) helices, (2) extended regions that interact to form sheets and (3) several standard types of turns.

The Sasisekharan–Ramakrishnan–Ramachandran plot describes allowed mainchain conformations

To a good approximation, the mainchain conformation of each non-glycine residue is restricted to two discrete conformational states.

A fragment of the polypeptide chain common to all protein structures is shown in Fig. 6.2. Rotation is permitted around the N–Cα and Cα–C single bonds of all residues (with one exception: proline). The angles ϕ and ψ around these bonds, and the angle of rotation, ω, around the peptide bond, define the conformation of a residue. The peptide bond itself tends to be planar, with two allowed states: *trans*, $\omega \approx 180°$ and *cis*, $\omega \approx 0°$ Most peptide bonds in proteins are *trans*. *cis* peptides are rare, and most cases involve a proline residue. The sequence of ψ, ϕ and ω angles of all residues in a protein defines the backbone conformation.

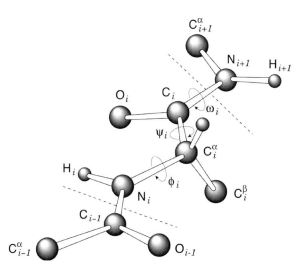

Fig. 6.2 Definition of conformational angles of the polypeptide backbone.

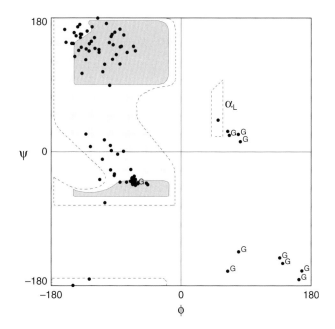

Fig. 6.3 A Sasisekharan–Ramakrishnan–Ramachandran plot of acylphosphatase (PDB code 2ACY). Note the clustering of residues in the α and β regions, and that most of the exceptions occur in glycine residues (labelled G).

The principle that two atoms cannot occupy the same space limits the values of conformational angles. The allowed ranges of ϕ and ψ, for $\omega = 180°$, fall into defined regions in a graph called a Sasisekharan-Ramakrishnan-Ramachandran plot—usually shortened to 'Ramachandran plot' (see Fig. 6.3). Solid lines in the figure delimit energetically preferred regions of ϕ and ψ; regions outside the broken lines are sterically disallowed. The conformations of most amino acids fall into either the α_R or β regions. Glycine has access to additional conformations. In particular it can form a left-handed helix: α_L. Figure 6.3 shows the typical distribution of residue conformations in a well-determined protein structure. Most residues fall in or near the allowed regions, although a few are forced by the folding into energetically less favourable states.

The allowed regions generate standard conformations. A stretch of consecutive residues in the α conformation (typically between 6 and 20 in native states of globular proteins) generates an α-helix. Repeating the β conformation generates an extended β-strand. Two or more β-strands can interact laterally to form β-sheets, as in acylphosphatase (Fig. 6.1). Helices and sheets are 'standard' or 'prefabricated' structural pieces that form components of the conformations of most proteins. They are stabilized by relatively weak interactions, **hydrogen bonds**, between mainchain atoms. In some fibrous proteins virtually all of the residues belong to one of these types of structure: wool contains α-helices; silk β-sheets. Amyloid fibrils, formed in disease states by many proteins, also contain extensive β-sheets.

Typical globular proteins contain several helix and/or sheet regions, connected by *turns.* Usually the ends of helix or strand regions appear on the surface of a domain of a protein structure. They are connected by turns, or loops: regions in which the chain alters direction to point back into the structure. Many but not all turns are short, surface-exposed regions that contain charged or polar residues.

How does the mainchain choose among the possible allowed conformations? What is unique about each protein is the sequence of its sidechains. Therefore, interactions involving sidechains must determine the mainchain conformation.

The sidechains

Sidechains offer the physicochemical versatility required to generate all the different folding patterns. The sidechains of the twenty amino acids vary in:

- **Size**   The smallest, glycine, consists of only a hydrogen atom; one of the largest, phenylalanine, contains a benzene ring.

- **Electric charge**   Some sidechains bear a net positive or negative charge at normal pH. Asp and Glu are negatively charged. Lys, Arg and often His are positively charged. Charged residues of opposite sign can form attractive pairwise interactions called *salt bridges.*

- **Polarity**   Some sidechains are polar; they can form hydrogen bonds to other polar sidechains, or to the mainchain, or to water. Other sidechains are electrically neutral. Some of these contain chemical groups related to ordinary hydrocarbons such as methane or benzene. Because of the thermodynamically unfavourable interaction of hydrocarbons with water, these are called 'hydrophobic' residues. Congregation of hydrophobic residues in protein interiors, predicted by W.J. Kauzmann before the first protein structures were determined, is an important contribution to protein stability. This effect is analogous to the formation of droplets of oil in salad dressing.

- **Shape and rigidity**   The overall shape of a sidechain depends on its chemical structure and on its degrees of internal conformational freedom.

The hydrophobic effect

The difference among the different amino acid sidechains in their preferences for aqueous or oil-like environments is one of the governing principles of protein structure.

What is the hydrophobic effect? Phase separation in oil–water mixtures—for instance, salad dressing—is one common example; another is that gases (unlike most solids) are less soluble in water as the temperature increases. Readers with whistling tea kettles will have heard low levels of sound prior to proper boiling—this occurs when dissolved air comes out of solution as the water is heated.

What is the origin of the hydrophobic effect? Cold water is a highly structured liquid. It contains many hydrogen bonds, which account for its high heat of vaporization and low density. But water is even more highly ordered around solutes than in the pure liquid. Methane dissolved in water—it is only slightly soluble, but soluble enough to study—is surrounded by a cage of water molecules called a clathrate complex. As a result, dissolving methane in water makes the solvent even *more* ordered, lowering the entropy. The natural tendency toward states of higher entropy inhibits the dissolving of methane in water. This is why methane and other hydrocarbons are only very slightly water-soluble. The solubilities of non-polar gases decrease upon heating—from an already small value in cold water—because as the temperature increases entropy plays an even more important role in determining the equilibrium state.

The hydrophobic effect in aqueous solutions of simple non-polar solutes was well known to physical chemists when W.J. Kauzmann, in 1959, recognized its importance for protein structure.

The non-polar side chains of proteins are similar to oil-like solutes. Their interaction with water is unfavourable. Kauzmann predicted that they would be sequestered in protein interiors, away from the solvent. This *oil-drop model of protein interiors* was confirmed by the X-ray crystal structures of globular proteins. We now also recognize the importance of high packing densities in protein interiors, and that it is better to regard the interior of a folded protein as more like a crystal than like an organic liquid. But the hydrophobic effect has lost none of its significance.

As a consequence of the hydrophobic effect, charged residues are largely excluded from protein interiors; in rare cases they form internal salt bridges, or gain or lose protons to form neutral states. Obviously, the backbone must traverse the interiors of the protein, and carries with it the polar N and O atoms, which interact with other polar mainchain atoms and with polar side chains such as threonine or asparagine. Thus the interior is not completely oil-like. Conversely, the surface of a protein is not exclusively charged or polar. About half the residues on the surface of a protein are non-polar.

Protein stability and denaturation

What are the chemical forces that stabilize native protein structures? What is the process by which a protein folds from an ensemble of denatured conformations to a unique native state?

To address these questions, biochemists have studied the denaturation of proteins in response to heat, or to increasing concentrations of urea or guanidinium hydrochloride (commonly used denaturants). Some measurements are *static*—determination of the amount of native and denatured states at equilibrium under different conditions, or the heat released at points along the transition. Others are *kinetic*—measurement of rates of folding or unfolding, or identification of structures that appear transiently during the process.

One important message from these studies is that proteins are only *marginally stable*. The native state of globular proteins is typically only 20–60 kJ mol^{-1} (5–15 kcal mol^{-1}) more stable than the denatured state. This is the equivalent of about one or two water–water hydrogen bonds.

Precisely why proteins have marginal stability is unclear. Some people believe that it facilitates protein turnover. Others suggest that proteins are as stable as they need to be so 'why bother' (less informally: there is no selective advantage in) further optimizing the stabilizing interactions. We do know that the interactions that stabilize native proteins are capable of producing protein structures with much higher stabilities.

Suppose you are a globular protein in aqueous solution, and you want to achieve a stable native state. Your major problem is the great loss of conformational freedom, relative to the large ensemble of denatured states, that is exacted from you in adopting a unique conformation. This entails a large reduction in entropy, which is thermodynamically unfavourable. One way in which you can compensate is to form a compact globular state, burying many residues in the interior away from contact with water. The release of water from interaction with the non-polar atoms of the protein produces a compensating favourable *increase* in entropy arising from the *hydrophobic effect* (see Box).

That's fine, but now you discover that to form the compact state you have buried many polar atoms, including but not limited to mainchain nitrogen and carbonyl oxygens. In the denatured state, these atoms make hydrogen bonds to water. When buried in the interior, their hydrogen-bonding potential must somehow be satisfied. (Don't forget: one or two uncompensated hydrogen bonds and you've blown it; your native state would be unstable.) A fairly general-purpose solution that satisfies mainchain hydrogen-bonding potential is to form helices or sheets.

There is a bonus: formation of secondary structure also ensures that the main-chain is in a stereochemically acceptable conformation, as limited by the Sasisekharan–Ramakrishnan–Ramachandran plot. Residues in α-helices are all in the α conformation; residues in strands of β-sheet are all in the β conformation.

How do you decide which regions should form helices or strands? Enthalpically, helix and sheet are reasonably similar for most residues. However, entropically, some sidechains are more hindered in helices than in strands (Ile, Val, Thr); these prefer strands. Such effects bias the formation of secondary structures. Specific sequences providing sidechain–mainchain hydrogen bonds form **helix caps**, governing where α-helices begin and end.

How compact is the globular state required to be? You could achieve exclusion of water from your interior by fairly loose packing—as long as no channel is larger than 1.4 Å in radius—the size of a water molecule (see Box). But the closer together you can squeeze your atoms, the better advantage you can take of Van der Waals forces: general forces of attraction between atoms that give matter its general cohesion. Protein interiors are densely packed: the fitting together of the sidechains is like a solved jigsaw puzzle. However, the puzzle pieces (the residues) are deformable, so the folding process is more complicated than the rigid matching of pieces in ordinary jigsaw puzzles.

For a more in depth discussion of protein folding, see *Introduction to Protein Science*, chapter 5.

Accessible and buried surface area

In a pioneering development of computational analysis of protein structures, B.D. Lee and F.M. Richards defined and measured the solvent-accessible surface areas of proteins.

The accessible surface area of a protein or protein complex is the area swept out by a water molecule (modelled as a sphere 1.4 Å in radius) rolling around the outside the protein. The accessible surface includes nooks and crannies in the protein surface that are larger than 2.8 Å wide, but smooths over finer wrinkles. Accessible surface area calculations rationalize the hydrophobic contribution to the thermodynamics of protein folding and interactions. Observed regularities include:

1. C. Chothia established a basic calibration: each $Å^2$ of buried surface area contributes 6 J (25 cal) of free energy of stabilization.

2. The accessible surface area (ASA) of monomeric proteins of up to about 300 residues varies as the 2/3 power of the molecular weight M: ASA $= 11.1M^{\frac{2}{3}}$.

3. The formation of oligomeric proteins from monomers buries an additional 1000–5000 $Å^2$ of surface. Lower values characterize proteins for which the monomer structure is stable in isolation; higher values characterize proteins in which association must stabilize the structure of the monomers as well as the complex.

4. The nature of the accessible and buried area in native protein structures. The average solvent-accessible surface of monomeric proteins—the protein *exterior*—is ~58% non-polar (hydrophobic), ~29% polar and ~13% charged. The average buried surface of monomeric proteins—the protein *interior*—is ~60% non-polar (hydrophobic), ~33% polar and ~7% charged. Many people expect the large *buried* hydrophobic surface but are surprised at how large the *exterior* hydrophobic area is. In fact, the main 'take-home message' about the difference between the surface and the interior is that proteins rarely bury charged groups.

In summary, you have to find a conformation of the chain that simultaneously solves all the following problems:

1. All residues must have stereochemically allowed conformations. This applies to both the mainchain and the sidechains. Steric collisions would raise the energy of the conformation and render it unstable.

2. Buried polar atoms must be hydrogen-bonded to other buried polar atoms. If you miss out a few hydrogen bonds, the protein will prefer to form the denatured state in order to allow these unsatisfied polar atoms to hydrogen-bond to solvent.

3. Enough hydrophobic surface must be buried, and the interior must be sufficiently densely packed, to provide thermodynamic stability.

For most proteins, there is a unique solution to all these problems, defining the native state. Other proteins change conformation when they bind ligands, or pass through metastable states, as part of their mechanisms of function.

The fact that one conformation of a protein—the native state—has substantially greater stability than others is complex but not mysterious. It is a question of optimizing the available interactions, and selecting sequences for which this optimum is unique and substantially lower than others. For most regions the local structure is determined by local interactions. Therefore, if the native state were not unique, there would have to be more than one way to fit a given set of pieces together. Given the chain constraints, it is easy for evolution to avoid this.

Protein folding

Suppose again that you are a protein, and that you are denatured. Now that you understand how your native state is stabilized, how would you go about finding it? Clearly you can't try all conformations—many years ago C. Levinthal calculated that a simple conformational search, using reasonable assumptions about speeds of internal rotations, would require much too much time to finish. Two circumstances conspire to make the *process* by which proteins fold to their native states a mysterious one.

First is the fact that proteins are only marginally stable. This implies that any quasi-stable intermediate in protein folding must be even less stable, else the folding process would get trapped in the intermediates. Indeed, for many proteins, measurements of fractions of molecules in native and denatured states, as a function of temperature or denaturant concentration, imply simple two-state Native ↔ Denatured equilibria in which undetectably few molecules are anything but native or denatured. This confirms that there are no intermediates with more than marginal stability. But this makes it difficult to follow the folding transition structurally.

The second circumstance that makes protein folding mysterious is that the denatured state is so heterogeneous that in the absence of stable intermediates there is no convenient way to visualize the complete pathway.

Contrast protein folding with two other types of structure formation:

1. In assembling do-it-yourself furniture, one passes through a succession of well-defined intermediate states. First one screws A to B in the native-like conformation.

The structure of the A–B fragment is determined and stabilized purely by the interactions between A and B. Were it not for gravity, a stable A–B intermediate would be formed. But proteins don't have the luxury of forming stable intermediates.

2. In assembling an arch from its voussoirs, the structure as a whole has no stability until the keystone is inserted. Only the completed arch has independent stability, there are no stable intermediates, and the only way to assemble the structure is by using scaffolding which is subsequently removed. But proteins don't have the luxury of using external scaffolding.

What proteins have to do is to work with unstable intermediates—like do-it-yourself furniture in the *presence* of gravity—and to get the job finished before the intermediates fall apart, or else keep reforming them and trying again.

Identification of transient structure during protein folding can be achieved experimentally by isotope exchange measurements. Prepare a sample of denatured protein in which all hydrogen atoms are replaced by deuterium. (It is possible to separate signals from H and D in NMR experiments.) At various times during refolding, in separate experiments, expose the sample to a pulse of protons. After the native state is formed, detect where in the structure D↔H exchange occurred and when. Such studies justify the model that many proteins fold by initial formation of a 'molten globule' containing some native secondary structure, but without the tertiary structural interactions that lock the molecule into its final conformation. This is followed by a hierarchical condensation to form supersecondary structure, etc., leading eventually to accretion of the native state. For most proteins, there is no evidence for non-native structures as intermediates along productive folding pathways, although non-native structures—such as incorrect proline isomers—can divert and thereby slow down the folding process.

The conclusion is that structures of local regions are determined primarily by local interactions, and, although these interactions may be inadequate to stabilize local regions to the point where they can be isolated, they are good enough to provide a low-energy pathway for structure assembly.

Most of what we understand about the process of protein folding comes from measurements on purified proteins in dilute solution. Within cells, protein folding is made more difficult by the intermolecular crowding. What is worse, misfolded proteins have an enhanced tendency to aggregate. Protein aggregates are implicated in many diseases, including spongiform encephalopathies and Alzheimer disease. Cells contain **chaperones** to catalyse proper folding and reduce aggregation.

Applications of hydrophobicity

Using a **hydrophobicity scale** that assigns a value to each amino acid, we can plot the variation of hydrophobicity along the sequence of a protein. This is called a **hydrophobicity profile.** Hydrophobicity profiles have been used to predict the positions of turns between elements of secondary structure, exposed and buried residues, membrane-spanning segments and antigenic sites.

Example 6.1 Use of hydrophobicity profiles to predict the positions of turns between helices and strands of sheet

Figure 6.4a shows the hydrophobicity profile of hen egg white lysozyme. It has pronounced minima at the following residues: 17, 44, 70, 93 and 117. Figure 6.4b shows the structure of hen egg white lysozyme, from which it is possible to check the correlation between turns in the structure and the positions of the minima in the hydrophobicity profile.

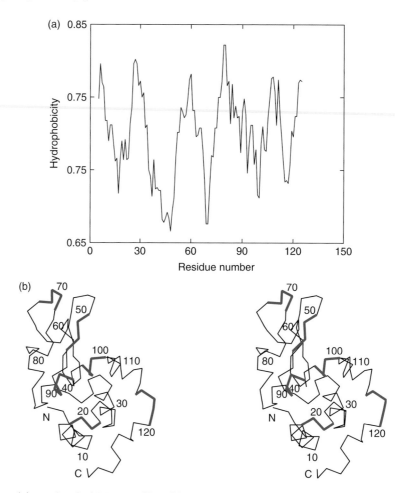

Fig. 6.4 (a) Hydrophobicity profile of hen egg white lysozyme. (Produced using the Primary Structure Analysis tools available through http://www.expasy.org) (b) Structure of hen egg white lysozyme. Regions corresponding to minima in the hydrophobicity plot are shown in bold blue lines.

Four of the major minima in the hydrophobicity profile appear at or near the positions of turns. Another minimum occurs in a surface-exposed region, but in the structure this one corresponds to a strand of a β-sheet rather than to a turn. One of the minima is within a helix. Conversely, many of the turns do not correspond to pronounced minima in the hydrophobicity plot. Hydrophobicity profiles provide useful information, but do not unambiguously predict all turns in a protein structure.

Example 6.2 **The helical wheel**

O.B. Ptitsyn observed that α-helices in globular proteins often have a 'hydrophobic face' turned inwards towards the protein interior, and a 'hydrophilic face' turned outwards towards the solvent. Each residue in an α-helix appears at a position 100° around the circumference helix from its predecessor. Therefore, to achieve Ptitsyn's effect, the sequence of residues should alternate between hydrophobic and hydrophilic with a periodicity of approximately four.

To check this relationship, the residues can be projected onto a plane perpendicular to a helix axis, a diagram called a **helical wheel**. This example shows the sequence of an α-helix of sperm whale myoglobin. Charged and polar residues appear in bold.

The helix has a hydrophobic face—which points to the inside of the structure—and a hydrophilic face—which points outside. In this picture, the hydrophilic face is at the bottom of the diagram. From such a pattern of hydrophobicity we can predict whether a region of an amino acid sequence is likely to form an α-helix in the native protein structure.

The next Box shows a PERL program to draw helical wheels.

A PERL program to draw helical wheels

```
#!/usr/bin/perl
#helwheel.pl---draw helical wheel
#usage: echo DVAGHGQDILIRLFKSH | helwheel.prl > output.ps
# or    echo 20DVAGHGQDILIRLFKSH | helwheel.prl > output.ps
#       the numerical prefix sets the first residue number

# The output of this program is in PostScript (TM),
#       a general-purpose graphical language

# The next section prints a header for the PostScript file
```

A PERL program to draw helical wheels *(continued)*

```perl
print <<EOF;
%!PS-Adobe-
%%BoundingBox: (atend)
%1 0 0 setrgbcolor
%newpath
%37.5 161 moveto 557.5 161 lineto 557.5 681 lineto 37.5 681 lineto
%closepath stroke
297.5 421. translate 2 setlinewidth 1 setlinecap
/Helvetica findfont 20 scalefont setfont 0 0 moveto
EOF


#  Define fonts to associate with each amino acid

$font{"G"} = "Helvetica";      $font{"A"} = "Helvetica";       $font{"S"} = "Helvetica";
$font{"T"} = "Helvetica";      $font{"C"} = "Helvetica";       $font{"V"} = "Helvetica";
$font{"I"} = "Helvetica";      $font{"L"} = "Helvetica";       $font{"F"} = "Helvetica";
$font{"Y"} = "Helvetica";      $font{"P"} = "Helvetica";       $font{"M"} = "Helvetica";
$font{"W"} = "Helvetica";      $font{"H"} = "Helvetica-Bold";  $font{"N"} = "Helvetica-Bold";
$font{"Q"} = "Helvetica-Bold"; $font{"D"} = "Helvetica-Bold";  $font{"E"} = "Helvetica-Bold";
$font{"K"} = "Helvetica-Bold"; $font{"R"} = "Helvetica-Bold";

$_= <>;                                 #  read line of input
chop();$_ =~ s/\s//g;                   #  remove terminal carriage return and blanks

if ($_ =~ s/^(\d+)//)                   #  if input begins with integer
    {$resno = $1;}                      #  extract it as initial residue number
else {$resno = 1}                       #  if not, set initial residue number = 1

$radius = 50;                           #  initialize values for radius,
$x = 0; $y = -50; $theta = -90;         #  x, y and angle theta

#  print light gray spiral arc as succession of line segments, 10 per residue

$npoints = 10*(length($_) - 1);

print "0.8 0.8 0.8 setrgbcolor\n";      #  set colour to light gray
print "newpath\n";                      #  draw spiral arc
printf("%8.3f %8.3f moveto\n",$x,$y);
foreach $d (1 .. $npoints) {            #  10 points per residue
    $theta += 10; $radius += 0.6;       #  increase radius and theta
    $x = $radius*cos($theta*0.01747737); #  calculate new value of x
    $y = $radius*sin($theta*0.01747737); #     and y
    printf("%8.3f %8.3f lineto\n",$x,$y);
}
print "stroke\n";

#  print residues and residue numbers

$radius = 50;                           #  reinitialize values for radius,
$x = 0; $y = -50; $theta = -90;         #  x, y and angle theta
print "0 setgray\n";                    #  set colour to black

foreach (split ("",$_)) {               #  loop over characters from input line
    print "/$font{$_} findfont ";       #  set font appropriate
    print "20 scalefont setfont\n";     #  for this amino acid
    printf("%8.3f %8.3f moveto\n",$x,$y); #  move to current point
    print " ($resno$_) stringwidth";    #  adjust position to centre residue
    print " pop -0.5 mul -7 rmoveto\n"; #     identification on point on spiral
    print " ($resno$_) show\n";         #  print residue number and id
    print "% $theta $resno$_\n";
    $theta += 100; $radius += 6;        #  set new values of angle, radius
    $x = $radius*cos($theta*0.01747737); #  compute new values of x
    $y = $radius*sin($theta*0.01747737); #     and y
    $resno++;                           #  increase residue number
}
```

→

A PERL program to draw helical wheels *(continued)*

```
print "showpage\n";                        #  postscript signals to
print "%%BoundingBox:";                     #  print
$xl = 297.5 - 1.05*$radius;                 #    x
$xr = 297.5 + 1.05*$radius;                 #     and
$yb = 421.  - 1.05*$radius;                 #      y
$yt = 421.  + 1.05*$radius;                 #       limits
printf("%8.3f %8.3f %8.3f %8.3f\n",$xl,$xr,$yb,$yt);

print "%%EOF\n";                            #  and wind up
```

Coiled-coil proteins

Helical wheel diagrams also illuminate coiled-coil structures, in which two α-helical regions wind around each other.

Proteins containing coiled-coils are known among structural proteins such as α-keratin, and also in a variety of globular proteins associated with a number of functions prominently including transcription regulation. The pitch is 14.7 nm. The chains go in opposite directions. The JUN–DNA complex is a typical example (see Fig. 6.5).

Such coiled-coil domains contain a signature pattern in their amino acid sequences. They show **heptad repeats**—seven-residue patterns—containing positions denoted

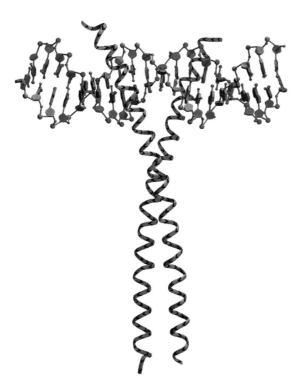

Fig. 6.5 Coiled-coil BZIP domain from proto-oncogene c-jun bound to cAMP response element (CRE) DNA target.

a, *b*, *c*, *d*, *e*, *f* and *g*, of which the first and fourth positions—*a* and *d*—are usually hydrophobic. The sequences of coiled-coil regions frequently contain a motif called the 'leucine zipper' because of the leucine repeats every seven residues. Here is the sequence of the coiled-coil region of GCN4, with the heptads demarcated and the hydrophobic positions indicated by asterisks:

```
abcdefg abcdefg abcdefg abcdefg
 *   *    *  *     *   *     *  *
```

R│MKQLEDK│VEELLSK│NYHLENE│VARLKKL│VG

The distribution of leucine residues (blue) is clear on a helical wheel:

Prediction of transmembrane proteins and signal peptides

Many proteins are designed to sit within membranes. Membrane proteins mediate the exchange of matter, energy and information between cell interiors, organelles and surroundings. Examples of membrane protein functions include energy transduction, via the generation or release of concentration gradients across cell or organelle membranes; and signal reception and transmission.

It is estimated that in the human genome, 30% of genes encode membrane proteins. Approximately 70% of known targets of drugs are membrane proteins. Given that membrane proteins are so common, it is important to have reliable tools for their identification. Relatively few membrane protein structures have been experimentally determined. This places a greater burden on computational tools for sequence analysis, to identify and characterize them.

Among their adaptations, membrane proteins contain regions of mostly non-polar residues, that interact with the organic layer. Thereby they differ from water-soluble proteins that show biases towards a hydrophobic interior and a hydrophilic exterior. Many membrane proteins contain a set of seven consecutive α-helices that traverse the membrane, oriented approximately perpendicular to the plane of the membrane. These helices are connected by loops that protrude into the aqueous surroundings. A second class of membrane protein structures contains a β-barrel. Transmembrane helices are typically 15–30 residues long. Although enriched in hydrophobic residues, they contain some polar sidechains, usually in interfaces between helices packed together in the structure.

A useful clue to the orientation of the helices across the membrane is the 'positive-inside rule'. The loops between helices lie either entirely inside or entirely outside the cell or organelle. Those inside contain a preponderance of positively charged residues.

Example 6.3 Detection of transmembrane helical segments

Bacteriorhodopsin was the first solved structure to show the configuration of seven helices traversing a membrane, connected by loops (see Fig. 6.6).

Residues within the membrane-spanning segments are almost exclusively hydrophobic, because the entire helix is embedded in a non-aqueous medium. They are separated by regions containing polar amino acids.

An unfiltered hydrophobicity plot of the amino acid sequence of *H. salinarum* bacteriorhodopsin shows seven regions of maxima, corresponding to the seven transmembrane helices (positions indicated by horizontal bars).

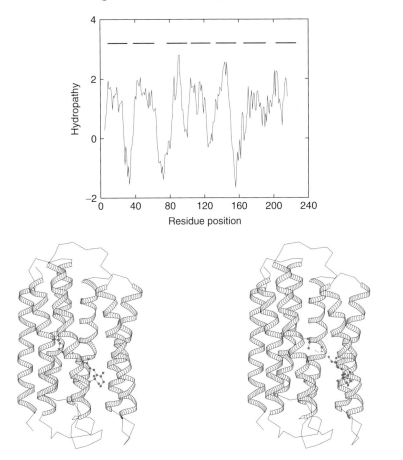

Fig. 6.6 Bacteriorhodopsin from the bacterium *Halobacterium salinarum* (formerly *Halobacterium halobium*) [2BRD] viewed in the plane of the membrane. The ligand shown in ball-and-stick representation is the chromophore, retinal.

These applications of hydrophobicity are typical of early approaches to sequence–structure correlation. Using some specific physical idea about proteins, they describe, or infer, structural properties of one protein at at time, from one sequence at a time. Modern developments have demonstrated the predictive power of multiple sequence alignments. Indeed, a program to predict coiled-coil regions, by C.D.D. Barry in 1982, was the first application of position-specific scoring matrices (see Chapter 5). The best current methods for predicting coiled coils and transmembrane domains make use of Hidden Markov Models (see page 278).

Superposition of structures, and structural alignments

Some aspects of sequence analysis carry over fairly directly into structural analysis, some must be generalized, and others have no analogues at all.

As in the case of sequences, a fundamental question in analysing structures is to devise and compute a measure of similarity. If two molecules have very similar structures, we can imagine superposing them so that corresponding points are as close together as possible. Then the average distance between corresponding points is a measure of the structural similarity. In practice it is conventional to report the root-mean-square (r.m.s.) deviation of the corresponding atoms:

$$\text{r.m.s. deviation} = \sqrt{\sum d_i^2/n}$$

where d_i is the distance between the ith pair of atoms after optimal fitting, and n is the number of points.

This assumes that we have pre-specified the correspondence between the points.

If the correspondence is not known, we must first determine it and only then calculate the r.m.s. deviation of the alignable substructures. If each point corresponds to an atom representing the successive residues of a protein or nucleic acid structure (the Cα atoms of proteins or the phosphorus atoms of nucleic acids), the problem is literally a question of alignment (= assignment of residue–residue correspondences) (see Box: *Determination of similarily and alignment in computational chemistry*). Indeed, determination of residue–residue correspondences via structural superposition of two or more proteins is a powerful method of sequence alignment. Because structure tends to diverge more conservatively than sequence during evolution, structure alignment is a more powerful method than pairwise sequence alignment for detecting homology and aligning the sequences of distantly related proteins.

Example 6.4 Structural alignment of γ-chymotrypsin and *Staphylococcus aureus* epidermolytic toxin A

Chymotrypsin and *S. aureus* epidermolytic toxin A are both members of the chymotrypsin family of proteinases. Figure 6.7 shows a structural superposition of PDB entries 8GCH (γ-chymotrypsin) (black) and 1AGJ (*S. aureus* epidermolytic toxin A) (blue). The molecules share the common chymotrypsin family serine proteinase folding pattern, and the Ser-His-Asp catalytic triad (thicker lines). A sequence alignment derived from the superposition follows:

```
8gch  CGVPAIQPVLIVNG----------------------------EEAVP--GS----WPWQVSLQ-DKTG
1agj  -------------EVSAEEIKKHEEKWNKYYGVNAFNLPKELFSKVDEKDR-QKYPYNTIGNVFVK-G-

8gch  FH--FCGGSLINE-NWVVTAAHC-GV-T---T-SDVVVAGEFDQG---SSSEKI--QKLKIAKVFK-NS-
1agj  --QTSATGVLIG-KNTVLTNRHIAK-FANGDPSKVSFRPSI-NTDDNGNT-E-TPYGEYEVKEILQEP-F

8gch  KYNSLTINNDITLLKLST-----AAS--FSQTVSAVCLPSASD--DFAAGTTCVTTGWG-LTRYNTPD-R
1agj  GAG-----VDLALIRLKPDQNGVSL-GDK---ISPAKIGT---SNDLKDGDKLELIGYPFDH----KVNQ

9gch  LQQASLPLL-SNTNCKKYWGTKIKDAM--ICAGASGV-SSCMGDSGGPLVCKKNGAWTLVGIVSWGSSTC
1agj  MHRSEIELTTLS--------------RGLRYY----GFTVPGNSGSGIFNSN---GELVGIHSSK----

8gch  STST---------PGVYARVTA-LVNWVQQTLAAN-
1agj  ----VSHLDREHQINYGVGIGNYVKRIINEKN---E
```

The resemblance between these two sequences is well within the 'twilight zone'. It could not be derived correctly from standard pairwise alignment of the two sequences alone.

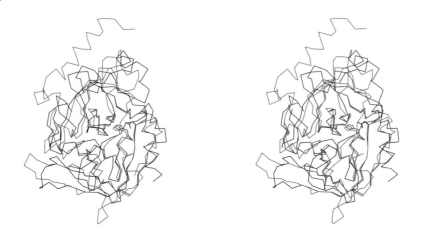

Fig. 6.7 Structural superposition of γ-chymotrypsin [8GCH] (black) and *S. aureus* epidermolytic toxin A [1AGJ] (blue). The sidechains of the catalytic triads are shown. Observe that the region around the active site is the best-conserved part of the protein.

Determination of similarity and alignment in computational chemistry

Three related problems have applications in molecular biology.

1. Similarity of two sets of atoms with known correspondences:

$$p_i \longleftrightarrow q_i, i = 1, \ldots N.$$

   The analogue, for sequences, is the Hamming distance: mismatches only.

2. Similarity of two sets of atoms with unknown correspondences, but for which the molecular structure—specifically the linear order of the residues—restricts the possiblities. In the case of proteins or nucleic acids we are limited to correspondences in which we retain the order along the chain:

$$p_{i(k)} \longleftrightarrow q_{j(k)}, k = 1, \ldots K \leq N, M$$

   with the constraint that: if $k_1 > k_2$, then it is necessary that $i(k_1) > i(k_2)$ and $j(k_1) > j(k_2)$. This can be thought of as corresponding to the Levenshtein distance, or to sequence alignment with gaps. The result of such a calculation is a structural alignment of parts or all of the sequences.

3. Similarities between two sets of atoms with unknown correspondence, with no restrictions on the correspondence:

$$p_{i(k)} \longleftrightarrow q_{j(k)}$$

   This problem arises in the following important case: suppose two (or more) molecules have similar biological effects, such as a common pharmacological activity. It is often the case that the structures share a common constellation of a relatively small subset of their atoms that are responsible for the biological activity. These atoms are called a **pharmacophore**. The problem is to identify them: to do so it is useful to be able to find the maximal subsets of atoms from the two molecules that have a similar structure.

DALI and MUSTANG

As proteins evolve, their structures change. Among the subtle details that evolution has strongly tended to conserve are the patterns of contacts between residues. That is, if two residues are in contact in one protein, the residues aligned with these two in a related protein are also likely to be in contact. This is true even in very distant homologues, and even if the residues involved change in size. Mutations that change the sizes of packed buried residues produce adjustments in packing of the helices and sheets against one another.

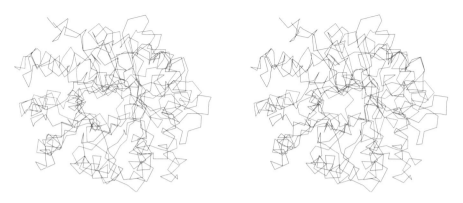

Fig. 6.8 The regions of common fold, as determined by the program DALI by L. Holm and C. Sander, in the TIM-barrel proteins mouse adenosine deaminase [1FKX] (black) and *Pseudomonas diminuta* phosphotriesterase [1PTA] (blue). In the alignment shown in this figure, the sequences have only 13% identical residues — closer to midnight than to the twilight zone.

L. Holm and C. Sander applied these observations to the problem of structural alignment of proteins. If the inter-residue contact pattern is preserved in distantly related proteins, then it should be possible to *identify* distantly related proteins by detecting conserved contact patterns.

Computationally, one makes matrices of contact patterns in two proteins (this is very easy), and then seeks the maximal matching submatrices (this is hard). Using carefully chosen approximations, Holm and Sander wrote an efficient program called DALI (for Distance matrix ALIgnment) that is now in common use for identifying proteins with folding patterns similar to that of a query structure. The program runs fast enough to carry out routine screens of the entire Protein Data Bank for structures similar to a newly determined structure, and even to perform a classification of protein domain structures from an all-against-all comparison. Holm and Sander have found several unexpected similarities not detectable at the level of pairwise sequence alignment.

An example of DALI's 'reach' into recognition of very distant structural similarities is its identification of the relationship between mouse adenosine deaminase, *Klebsiella aerogenes* urease and *Pseudomonas diminuta* phosphotriesterase (see Fig. 6.8).

DALI is available on the Web. You can submit coordinates to the site http://www2.ebi.ac.uk/dali/, and receive the set of similar structures and their alignments with the query.

MUSTANG, written by A.S. Konagurthu, is a development of DALI's distance-matrix approach to *multiple* structural alignment (http://www.cs.mu.oz.au/~arun/mustang/).

Evolution of protein structures

Included in the 50 000 protein structures now known are several families in which the molecules maintain the same basic folding pattern over ranges of sequence similarity from near identity down to well below 20% conservation. The serine proteinases

(γ-chymotrypsin and *S. aureus* epidermolytic toxin A, Fig. 6.7) and the adenosine deaminase—phosphotriesterase family (Fig. 6.8) are examples.

The general response to mutation is structural change. It is characteristic of biological systems that the objects we observe to have a certain form arose by evolution from related objects with similar but not identical form. They must, therefore, be robust, having the freedom to tolerate some variation. We can take advantage of this robustness in our analysis: by identifying and comparing related objects, we can distinguish variable and conserved features, and thereby determine what is crucial to structure and function.

Natural variations in families of homologous proteins that retain a common function reveal how structures accommodate changes in amino acid sequence. Surface residues not involved in function are usually free to mutate. Loops on the surface can often accommodate changes by local refolding. Mutations that change the volumes of buried residues generally do not change the conformations of individual helices or sheets, but produce distortions of their spatial assembly. The nature of the forces that stabilize protein structures sets general limitations on these conformational changes. Particular constraints derived from function vary from case to case.

Families of related proteins tend to retain common folding patterns. However, although the general folding pattern is preserved, there are distortions which increase as the amino acid sequences progressively diverge. These distortions are not uniformly distributed throughout the structure. Usually, a large central **core** of the structure retains the same qualititative fold, and other parts of the structure change conformation more radically. As a simple analogy, consider the letters B and R. As structures, they have a common core which corresponds to the letter P. Outside the common core they differ: at the bottom right B has a loop and R has a diagonal stroke.

Figure 6.9 compares spinach plastocyanin and cucumber stellacyanin.

Systematic studies of the structural differences between pairs of related proteins have defined a quantitative relationship between the divergence of the amino acid sequences of the core of a family of structures and the divergence of structure. As the sequence diverges, there are progressively increasing distortions in the mainchain conformation, and the fraction of the residues in the core usually decreases. Until the fraction of identical residues in the sequence drops below about 40–50%, these

For other illustrations of structural comparisons of homologous proteins, and discussion of classification schemes, see chapter 4 of *Introduction to Protein Architecture.* That book contains a large number of pictures of protein structures suitable for browsing, for any reader interested in exploring the stunning variety of protein folding patterns.

(a)

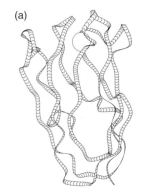 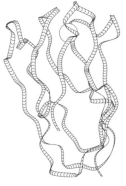

Fig. 6.9

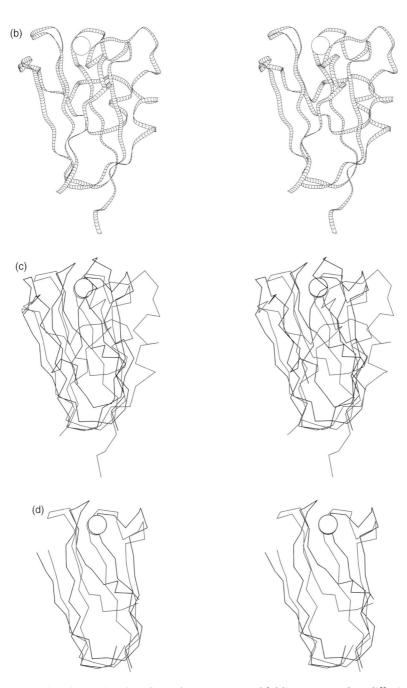

Fig. 6.9 Two related proteins that share the same general folding pattern, but differ in details. Circles represent copper ions. (a) Spinach plastocyanin [1AG6], (b) cucumber stellacyanin [1JER], (c, d) superpositions, showing (c) the entire structures, and (d) only the well-fitting core. The main secondary structural elements of these proteins are two β-sheets packed face-to-face. It is seen in the superposition that several strands of β-sheet are conserved but displaced, and that the helix at the right of cucumber stellacyanin has no counterpart in the spinach plastocyanin structure. Even the relatively well-fitting core shows the conservation of folding topology but nevertheless reveals considerable distortion.

effects are relatively modest. Almost all the structure remains in the core, and the deformation of the mainchain atoms is on average no more than 1.0 Å. With increasing sequence divergence, in most cases some regions refold entirely, reducing the size of the core, and the distortions of the residues remaining within the core increase in magnitude.

A correlation between the divergence of sequence and structure applies to all families of proteins. Figure 6.10a shows the changes in structure of the core, expressed as the r.m.s. deviation of the mainchain atoms after optimal superposition; plotted against the sequence divergence, expressed as the percentage conserved amino acids of the core after optimal alignment. The points correspond to pairs of homologous proteins from many related families. (Those at 100% residue identity are proteins for which the structure was determined in two or more crystal environments, and the deviations show that crystal packing forces—and, to a lesser extent, solvent and temperature—can slightly modify the conformation of the proteins.) Figure 6.10b shows the changes in the fraction of residues in the core as a function of sequence divergence. The fraction of residues in the cores of distantly related proteins can vary widely: in some cases the fraction of residues in the core remains high, in others it can drop to below 50% of the structure.

Classifications of protein structures

Organization of protein structures according to the folding pattern imposes a very useful logical structure on the entries in the Protein Data Bank. It affords a basis for structure-oriented information retrieval. Several databases derived from the PDB are built around classifications of protein structures. They offer useful features for exploring the protein structure world, including: search for keyword or sequence, navigation

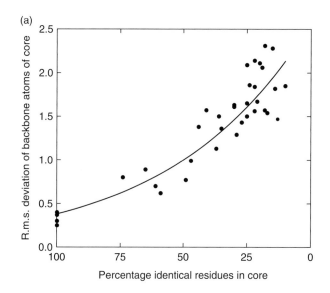

Fig. 6.10

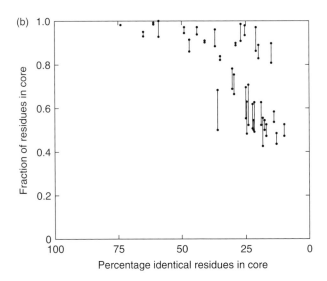

Fig. 6.10 Relationships between divergence of amino acid sequence and three-dimensional structure of the core, in evolving proteins. (a) Variation of r.m.s. deviation of the core with the percentage of identical residues in the core. (b) Variation of size of the core with the percentage of identical residues in the core. This figure shows results calculated for 32 pairs of homologous proteins of a variety of structural types. [Adapted from Chothia, C. and Lesk, A.M. (1986). Relationship between the divergence of sequence and structure in proteins. *The EMBO Journal*, **5**, 823–826.]

among similar structures at various levels of the classification hierarchy, presentation of structure pictures, probing the databank for structures similar to a new structure, and links to other sites. These databases include SCOP (Structural Classification of Proteins), CATH (Class, Architecture, Topology, Homologous superfamily), FSSP/DDD (Fold classification based on Structure–Structure alignment of Proteins/Dali Domain Dictionary) and CE (The Combinatorial Extension Method).

SCOP

SCOP, by A.G. Murzin, L. Lo Conte, B.G. Ailey, S.E. Brenner, T.J.P. Hubbard and C. Chothia, organizes protein structures in a hierarchy according to evolutionary origin and structural similarity. At the lowest level of the SCOP hierarchy are individual **domains**, extracted from the Protein Data Bank entries. Sets of domains are grouped into **families** of homologues, for which the similarities in structure, sequence and sometimes function imply a common evolutionary origin. Groups of families containing proteins of similar structure and function, but for which the evidence for an evolutionary relationship is suggestive but not compelling, form **superfamilies**. Superfamilies that share a common folding topology, for at least a large central portion of the structure, are grouped as **folds**. Finally, each fold group falls into one of the general **classes**. The major classes in SCOP are α, β, $\alpha + \beta$, α/β and miscellaneous 'small proteins', which often have little secondary structure and are held together by disulphide bridges or ligands.

The Box shows the SCOP classification of flavodoxin from *Clostridium beijerinckii* (Plate VII). For illustrations of the degree of similarities of proteins grouped together at different levels of the hierarchy, and discussion of other classification schemes, see *Introduction to Protein Architecture: The Structural Biology of Proteins,* chapter 4.

SCOP classification of flavodoxin from *Clostridium beijerinckii*

1. Root SCOP
2. Class α and β proteins (α/β)

   Mainly parallel β-sheets (β-α-β units)
3. Fold Flavodoxin-like

   Three layers, α/β/α; parallel β-sheet of five strands, order 21345
4. Superfamily Flavoproteins
5. Family Flavodoxin-related. Binds FMN (flavin mononucleotide)
6. Protein Flavodoxin
7. Species *Clostridium beijerinckii*

The SCOP release of November 2007 contained 34 494 PDB entries, split into 97 178 domains. The distribution of entries at different levels of the hierarchy is:

Class	Number of families	Number of superfamilies	Number of folds
All α proteins	772	459	259
All β proteins	679	331	165
α/β proteins	736	232	141
α + β proteins	897	488	334
Multidomain proteins	74	53	53
Membrane and cell surface proteins	104	92	50
Small proteins	202	122	85
Total	3464	1777	1086

Numerous other web sites offering classifications of protein structures are indexed at: http://www.bioscience.org/urllists/protdb.htm

Protein structure prediction and modelling

The observation that each protein folds spontaneously into a unique three-dimensional native conformation implies that nature has an algorithm for predicting protein structure from amino acid sequence. Some attempts to understand this algorithm are based solely on general physical principles; others are empirical, based on observations of the known amino acid sequences and protein structures. A proof of our understanding would be the ability to reproduce the algorithm in a computer program that could predict protein structure from amino acid sequence.

Overview of modelling methods

Nature has an algorithm that computes protein native structure from amino acid sequence. All the information needed to do this computation is contained in the sequence itself: proteins don't need to look things up in databases. We do. Many of the most effective methods for protein structure prediction make use of known structures of homologous proteins. Indeed, the degree of sequence similarity between a protein of unknown structure and its nearest homologue of known structure controls what we can achieve in prediction of the unknown structure, and dictates what methods to use.

Generally speaking:

1. If a protein of unknown structure has homologues of known structure with ≥ 40% identical residues in an optimal alignment, homology modelling methods are likely to produce a nearly complete structural model. The quality of the model is likely to be good enough to interpret the protein's function. (The higher the sequence similarity, the more accurate the model.) Mature software for homology modelling is available.

2. If no homologue of known structure has sequence similarity to the unknown with ≥40% residue identity, it may still be possible to assign a general folding pattern to the protein of unknown structure. It should be possible to predict its secondary structure with ~70–80% accuracy on a residue by residue basis. Many servers will apply a variety of methods to a submitted sequence.

3. If no homologue of known structure is recognizable from the sequences, the last recourse is to use a prediction method general enough to handle novel folds. Such methods include both *a priori* and knowledge-based approaches. At present, the program ROBETTA, by D. Baker and colleagues, is the most effective tool for protein structure prediction whenever homology modelling is not applicable.

A priori and empirical methods

Many attempts to predict protein structure from basic physical principles alone try to reproduce the interatomic interactions in proteins, to define a computable energy associated with any conformation. Computationally, the problem of protein structure prediction then becomes a task of finding the global minimum of this conformational energy function. So far this approach has not succeeded, partly because of the inadequacy of the energy function, partly because the minimization algorithms tend to get trapped in local minima and partly because the calculations require more computer resources than are available.

Other *a priori* approaches to structure prediction are based on attempts to simplify the problem, to capture somehow the essentials.

There is a spectrum of approaches, distributed between two extremes:

1. Establish the *most detailed and accurate* model of the interatomic interactions within a protein itself and between protein and solvent. Apply molecular dynamics to simulate the motion of the system starting with a denatured conformation—perhaps the extended chain—and ending with something in the vicinity of the native state. The idea is that the physics of the problem is fairly well understood, down to the detailed microscopic level. The challenges are computational—how to simulate the system for long enough to attain the native state.

2. Establish the *least detailed* and *least accurate* model that can give the correct answer. If one could identify the essentials, great computational power might not be needed. The idea is that the physics of the problem is *not* well understood, *except* in microscopic detail. Of course everyone accepts the principles of mechanics and thermodynamics, but much of the detail is irrelevant and unilluminating, and this is a crucial part of the picture. Proteins just aren't that fussy: at the melting temperature half the molecules are in the native state, despite the very great alteration in the relative strengths of various terms in the free energy relative to normal physiological conditions. This argues for a great robustness in the determinants of structure, which is difficult to capture in detailed calculations, or to explain even if they were captured. It argues for a distinction between determinants of structure and determinants of stability, also difficult to explain from detailed calculations. Also, many proteins with substantial sequence differences fold to very similar native states. However, in some cases there are large perturbations of the folding pathway.

The field contains many scientists widely scattered between these end-points, linked by a certain creative tension.

It is partly a question of choosing goals. To go beyond a prediction of the native state, to account for trajectories, transition states, intermediates, if any, and melting temperatures, a detailed simulation may well be necessary. If one wants a perspicuous and satisfying explanation of how amino acid sequence determines protein structure, then even a successful fully detailed calculation may not provide it.

A proponent of molecular dynamics might argue that (1) from a successful fully detailed calculation, one could generate a series of simplified models by making approximations that keep the broad picture intact, and (2) a simplified model that works is—by virtue of its success—interesting, but may be unrealistic, or, even if realistic, incomplete. After all, there may be many simplified models, all of which work, but which do not agree on what is essential. The reason for suspecting that this may be true is the observation that folded proteins solve many problems at once—stereochemistry, packing, hydrogen bonding, entropy compensation: any spoke may lead to the hub.

The alternative to *a priori* methods are approaches based on assembling clues to the structure of a target sequence by finding similarities to known structures. These empirical or 'knowledge-based' methods have become very powerful.

We are coming closer and closer to saturating the set of possible folds with known structures. This is the stated goal of **structural genomics** projects (see Box). Once we have a complete set of folds and sequences, and powerful methods for relating them, empirical methods will provide pragmatic solutions of many problems. What will be the effect of this on attempts to predict protein structure *a priori*? The intellectual appeal of the problem will still be there—nature folds proteins without searching databases. Moreover, some methods may not merely identify the native conformation, but illuminate folding pathways. But it is unlikely that the problem will continue to command interest of the same intensity, and support of the same largesse, once a pragmatic solution has been found.

Structural genomics

In analogy with full-genome sequencing projects, structural genomics has the commitment to deliver the structures of the complete protein repertoire. X-ray crystallographic and NMR experiments will solve a 'dense set' of proteins, such that all proteins are within homology-modelling range of one or more known experimental structures. More so than genomic sequencing projects, structural genomics projects combine results from different organisms. The human proteome is of course of special interest, as are proteins unique to infectious microorganisms.

The goals of structural genomics have become feasible partly by advances in experimental techniques, which make high-throughput structure determination possible; and partly by advances in our understanding of protein structures, which define reasonable general goals for the experimental work, and suggest specific targets.

The theory and practice of homology modelling suggests that at least 30% sequence identity between target and some experimental structure is necessary. This means that experimental structure determinations will be required for an exemplar of every sequence family, including many that share the same basic folding pattern. Experiment will have to deliver the structures of on the order of 10 000 domains. In the year 2007, 8127 structures were deposited in the PDB, so the throughput rate is adequate.

$\longrightarrow$

Structural genomics *(continued)*

Methods of bioinformatics can help select targets for experimental structure determination that offer the highest pay-off in terms of useful information. Goals of target selection include:

- Elimination of redundant targets—proteins too similar to known structures.

- Identification of sequences with undetectable similarity to proteins of known structure.

- Identification of sequences with similarity only to proteins of unknown function, or

- Proteins of unknown structure with 'interesting' functions; for example, human proteins implicated in disease, or bacterial proteins implicated in antibiotic resistance

- Proteins with properties favourable for structure determination—likely to be soluble—contain methionine (which facilitates solving the phase problem of X-ray crystallography)

The machinery for carrying out the modelling is already up and running. MODBASE http://modbase.compbio.ucsf.edu/ and the SWISS-MODEL Repository http://swissmodel.expasy.org/repository/ collect homology models of proteins of known sequence.

Methods for prediction of protein structure from amino acid sequence include:

- **Attempts to predict secondary structure** without attempting to assemble these regions in three dimensions. The results are lists of regions of the sequence predicted to form α-helices and regions predicted to form strands of β-sheet.

- **Homology modelling** Prediction of the three-dimensional structure of a protein from the known structures of one or more related proteins. The results are a complete coordinate set for mainchain and sidechains, intended to be a high-quality model of the structure, comparable with at least a low-resolution experimental structure.

- **Fold recognition** Given a library of known structures, determine which of them shares a folding pattern with a query protein of known sequence but unknown structure. If the folding pattern of the target protein does not occur in the library, such a method should recognize this. The results are a nomination of a known structure that has the same fold as the query protein, or a statement that no protein in the library has the same fold as the query protein.

- **Prediction of novel folds,** either by *a priori* or knowledge-based methods. The results are a complete coordinate set for at least the mainchain and sometimes the sidechains also. The model is intended to have the correct folding pattern, but would not be expected to be comparable in quality with an experimental structure.

D. Jones has likened the distinction between *a priori* modelling and fold recognition to the difference between an essay and a multiple-choice question in an exam.

Critical Assessment of Structure Prediction (CASP)

CASP organizes blind tests of protein structure predictions, in which participating crystallographers and NMR spectroscopists make public the amino acid sequences of the proteins they are investigating, and agree to keep the experimental structure secret until predictors have had a chance to submit their models. CASP runs on a 2-year cycle. Every 2 years the sequences are published in the spring, and predictions are due in the autumn. At the end of the year a gala meeting brings the predictors together to discuss the current results and to gauge progress.

The CASP program was introduced briefly in Chapter 1.

Protein structure predictions in CASP have fallen into three main categories: (1) comparative modelling—in effect homology modelling, (2) fold recognition and (3) modelling of novel folds:

CASP category	Nature of target
Comparative modelling	Close homologues of known structure are available; homology modelling methods are applicable
Fold recognition	Structures with similar folds are available, but no relative sufficiently close for homology modelling; the challenge is to identify structures with similar topology
Novel fold	No structure with same folding pattern known; requires either a genuine *a priori* method or a knowledge-based method that can combine features of several structures

Three groups of assessors, one for each category, compare the predicted and experimental structures, and judge the predictions. Speakers at the end-of-year meeting include the organizers, the assessors, and selected predictors, including those who have been particularly successful, or who have an interesting novel method to present.

The lastest CASP programme took place in 2006. There were 104 targets. In all categories, 207 groups of predictors submitted a total of 63 717 models. This was approximately equal to twice the number of entries in the PDB at the time!

Many predictions are prepared by groups of researchers who study the results generated by their computer programs, and select and edit them before submission. In addition, the target sequences are sent to web servers that return predictions without human intervention. The CAFASP: Critical Assessment of Fully Automated Structure Prediction programme monitors the quality of these predictions. It is thereby possible to determine to what extent successful procedures could be made fully automatic. There are three challenges:

Human against protein	CASP
Computer against protein	CAFASP
Human against computer	CASP versus CAFASP

A separate programme of blind tests of prediction evaluates methods for predicting protein–protein interactions, or 'docking'. This is CAPRI: Critical Assessment of Predicted Interactions. CAPRI also held an assessment meeting in 2006.

Structure predictions at recent CASP programmes have showed continued improvements. For the most part, progress has been incremental rather than spectacular, with one notable exception: David Baker's group predicted and refined the structure of a small (70-residue) protein from *Thermus thermophilus*, producing a model that deviated by 1.59 Å from the X-ray structure! Indeed, improvements in knowledge-based methods originally developed for prediction of novel folds threaten to supersede traditional methods for fold recognition, such as threading, that make explicit reference to libraries of complete structures.

Results at CAPRI show that complexes between partners that do not undergo major conformational changes can now be predicted from the structures of the components. Large conformational changes upon complex formation still present difficulties. However, progress could be seen in at least one case, the trimeric TBE envelope protein.

For both CASP and CAPRI, the best results are very impressive. An observer of this scene once commented some years ago that protein structure prediction had advanced to the point that 'failure can no longer be guaranteed'. Things are now somewhat better than that. However, consistency in quality of prediction is still the challenge.

Secondary structure prediction

It seems obvious that: (1) it should be easier to predict secondary structure than tertiary structure, and (2) to predict tertiary structure a sensible way to proceed would be first to predict the helices and strands of sheet and then to assemble them. Whether or not these propositions are correct, many people have believed in, and acted upon, them. Given the amino acid sequence of a protein of unknown structure, they produce **secondary structure predictions**, the assignment of regions in the sequence as helices or strands of sheet.

To assess the quality of a secondary structure prediction, classify the residues in the experimental three-dimensional structure into three categories (helix = H, strand = E (extended), and other = −). The percentage of residues predicted correctly is denoted Q3. At the 2000 CASP programme, the PROF server by B. Rost achieved a good prediction of a domain from the *Thermus aquaticus* mismatch repair protein MutS. The value of Q3 for Rost's prediction is 81%:

```
                           10        20        30        40        50
                            |         |         |         |         |
Amino acid sequence ALVEDPPLKVSEGGLIREGYDPDLDALRAAHREGVAYFLELEERERERTG
Prediction          HH------EEE---HHHHHHHHHH-HHHHHHHHHHHHHHH-
Experiment          -E-------E---HHHHHHHHHHHHHHHHHHHHHHHHHHHHH-

                           60        70        80        90       100
                            |         |         |         |         |
Amino acid sequence IPTLKVGYNAVFGYYLEVTRPYYERVPKEYRPVQTLKDRQRYTLPEMKEK
Prediction          -EEEEEEEEEEEEEEEEE------EEEEEEEE-EEEE-HHHHHH
Experiment          --EEEEE--EEEEEEEEHHHHHH---EEEEE--EEEEE-HHHHHH

                           110       120
                            |         |
```

```
Amino acid sequence  EREVYRLEALIRRREEEVFLEVRERAKRQ
Prediction           HHHHHHHHHHHHHHHHHHHHHHHHHHHH-
Experiment           HHHHHHHHHHHHHHHHHHHHHHHHHHHH-
```

Figure 6.11 shows the experimental structure, with the *predicted* secondary structures distinguished. Except for a short 3_{10} helix, the secondary structural elements are predicted correctly apart from some minor discrepancies in the positions at which they start and end. (Other scoring schemes that check for *segment overlap* are less sensitive to end effects.) The quality of this result is very high but not exceptionally rare. This target was classified as being of *medium* difficulty by the CASP4 assessors. At present, PROF is running at an average accuracy of Q3 $\simeq$77%. Other secondary structure prediction methods are also doing comparably well.

The most powerful methods of secondary structure prediction are based on **artificial neural networks**.

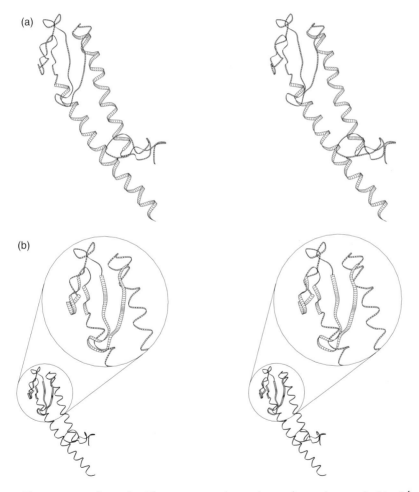

Fig. 6.11 The structure from the *Thermus aquaticus* mismatch repair protein MutS [1EWQ]. (a) The regions predicted by the PROF server of Rost to be helical are shown as wider ribbons. The prediction missed only a short 3_{10} helix, at the top left of the picture. (b) The regions predicted to be in strands are shown as wider ribbons.

Artificial neural networks

Artificial neural networks are a class of general computational structures based loosely on the anatomy and physiology of biological nervous systems. They have been applied successfully to a wide variety of pattern recognition, classification and decision problems.

A single neuron, in the computational scheme, is a node in a directed graph, with one or more entering connections designated as input, and a single leaving connection called the output:

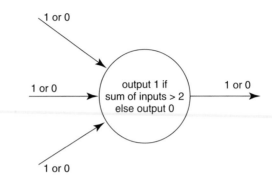

In the physiological metaphor, one says that the neuron 'fired' if the output is 1, and that the neuron 'didn't fire' if the output is 0. Simulated neurons can differ in the number of input and output connections, and in the formula for deciding whether to fire (see Box).

To form a network, assemble several neurons and connect the outputs of some to the inputs of others. Some nodes contain connections that provide input to the entire network; some deliver output information from the network to the outside world; and others, that do not interact directly with the outside, are called **hidden layers**.

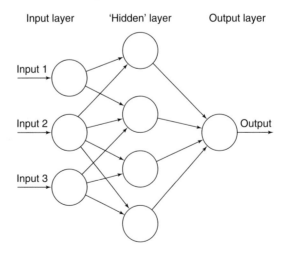

An unlimited degree of complexity is available by assembling and connecting neurons, and by varying the strengths of the connections. That is, instead of taking a simple sum of inputs, $i_1 + i_2 + i_3$, take a weighted sum—for instance, $10i_1 + 5i_2 + i_3$ —which would make the neuron most sensitive to input 1 and least sensitive to input 3. Biologically, this may correspond to changing the strengths of synapses.

Logic of neural networks

For a single neuron, a discrete decision process governing the output has a geometric interpretation in terms of lines and planes. The neuron in the following figure has two inputs. If we interpret the inputs as the coordinates of a point (x, y) in the plane, the neuron 'decides' on which side of a line the input point lies. The output will be 1 if and only if $x + y < 2$; that is, if the point is below and to the left of the line $x + y = 2$.

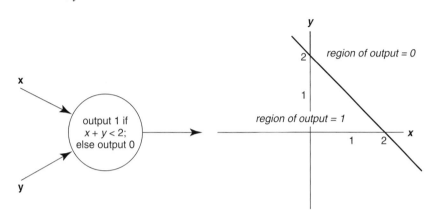

A *neural network* is specified by the topology of its connections, and the weights and decision formulas of its nodes. A network can make more complex decisions than a single neuron. Thus, if one neuron with two inputs can decide on which side of a line a point lies, three neurons can select points that lie within a triangle:

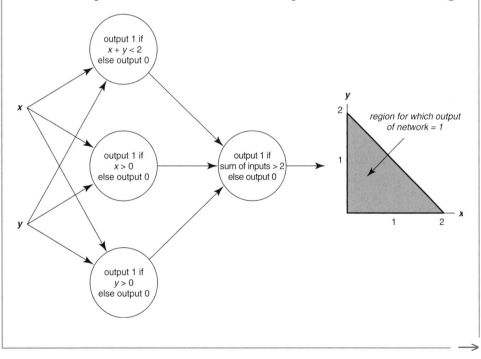

→

Logic of neural networks *(continued)*

Neural networks are more powerful and robust if the output is a smoothly varying function of the inputs. Such networks can perform more general kinds of computations and are better at pattern recognition. Also, for training the network it is useful if the output is a differentiable function of the parameters. To this end, a sharp threshold function for the output of a neuron is replaced by a smoothed-out step, or sigmoidal, function:

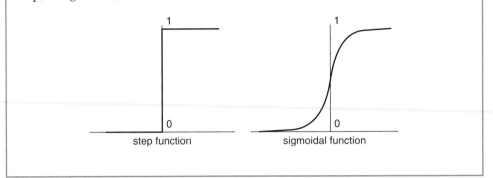

A property of a neural network that gives it great power is that the weights may be regarded as *variables,* and a calculation or learning process may determine the weights appropriate for a particular decision or pattern identifier. To train a network, feed the system sets of sample input for which the desired output is known, and compare the output with the correct answer. If the observed output differs from the desired one, adjust the parameters. The topology of the network remains invariant during the training process, although of course setting a weight to 0 has the effect of detaching an input.

The type of neural network that has been applied to secondary structure prediction is shown in Fig. 6.12.

A major advance in secondary structure prediction occurred with the application of evolutionary information, the recognition that multiple sequence alignment tables contain much more information than individual sequences. The conservation of secondary structure among related proteins means that the sequence–structure correlations are much more robust when a family as a whole is taken into account. Most neural network-based methods for secondary structure prediction now feed the input layer not simply with the identities of the amino acid at successive positions, but with a profile derived from a multiple sequence alignment.

It has also proved useful to run two neural networks in tandem, to make use of observed correlations among conformations of residues at neighbouring positions. Predictions of the states of several successive residues by one network similar to the one shown in Fig. 6.12 are combined by a second network into a final prediction.

A test of the maturity of a prediction method is whether it can be made fully automatic (see page 337). Some computational methods require human intervention and editing of results. Others, including PROF, the system that predicted the secondary structure of MutS, are fully automatic.

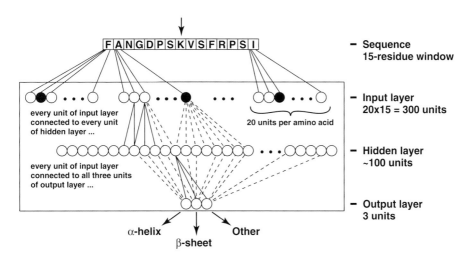

Fig. 6.12 A neural network applicable to secondary structure prediction contains three layers:

- The input layer sees a sliding 15-residue window in the sequence. That is, it treats a 15-residue region, predicts the secondary structure of the central residue (marked by an arrow, at the top) and then moves the window one residue along the amino acid sequence and repeats the process. To each of the 15 residues in the current window there correspond 20 nodes in the input layer of the network, one of which will be triggered according to the amino acid in that position.

- A hidden layer of ~100 units connects the input with the output. Each node of the hidden layer is connected to *all* input and output units; not all the connections are shown.

- The output layer consists of only three nodes, that signify prediction that the central residue in the window be in a helix, strand or other conformation.

Homology modelling

Model building by homology is a useful technique for predicting the structure of a target protein of known sequence, when the target protein is related to at least one other protein of known sequence *and* structure. If the proteins are closely related, the known protein structures—called the parents—can serve as the basis for a model of the target. Although the quality of the model will depend on the degree of similarity of the sequences, it is possible to estimate this quality before experimental testing (see Fig. 6.10). In consequence, knowing how good a model is necessary for the intended application permits intelligent prediction of the probable success of the exercise.

Steps in homology modelling are:

1. Align the amino acid sequences of the target and the protein or proteins of known structure. It will generally be observed that insertions and deletions lie in the loop regions between helices and sheets.

2. Determine mainchain segments to represent the regions containing insertions or deletions. Stitching these regions into the mainchain of the known protein creates a model for the complete mainchain of the target protein.

3. Replace the sidechains of residues that have been mutated. For residues that have not mutated, retain the sidechain conformation. Residues that have mutated tend to keep the same sidechain conformational angles, and could be modelled on this basis. However, computational methods are now available to search over possible combinations of sidechain conformations.

4. Examine the model—both by eye and by programs—to detect any serious collisions between atoms. Relieve these collisions, as far as possible, by manual manipulations.

5. Refine the model by limited energy minimization. The role of this step is to fix up the exact geometrical relationships at places where regions of mainchain have been joined together, and to allow the sidechains to wriggle around a bit to place themselves in comfortable positions. The effect is really only cosmetic—energy refinement will not fix serious errors in such a model.

To a great extent, this procedure produces 'what you get for free' in that it defines the model of the protein of unknown structure by making minimal changes to its known relatives. In some cases, it is possible to make substantial improvements. A rule of thumb (referring again to Fig. 6.10) is that if two or more sequences have at least 40–50% identical amino acids in an optimal alignment of their sequences, the procedure described will produce a model of sufficient accuracy to be useful for many applications. If the sequences are more distantly related, neither the procedure described nor any other currently available method will produce a model, correct in detail to atomic resolution, of the target protein from the structure of its relative.

In most families of proteins the structures contain relatively constant regions and more variable ones. A single parent structure will permit reasonable modelling of the conserved portion of the target protein, but may fail to produce a satisfactory model of the variable portion. From only one target and one parent sequence, it will not be easy even to predict which are the variable and constant regions. A more favourable situation occurs when several related proteins of known structure provide a basis for modelling a target protein. These reveal the regions of constant and variable structure in the family. The observed distribution of structural variability among the parents dictates an appropriate distribution of constraints to be applied to the model.

SWISS-MODEL hosts a web site that will accept the amino acid sequence of a target protein, determine whether a suitable parent or parents for homology modelling exist, and, if so, deliver a set of coordinates for the target. SWISS-MODEL was developed by T. Schwede, M.C. Peitsch and N. Guex, now at The Geneva Biomedical Research Institute.

An example of the automatic prediction by SWISS-MODEL is the prediction of the structure of a neurotoxin from red scorpion (*Buthus tamulus*) from the known structure of the neurotoxin from the related North African yellow scorpion (*Androctonus australis hector*). These two proteins have 52% identical residues in their sequence alignment. With such a close degree of similarity, it is not surprising that the model fits the experimental result very closely, even with respect to the sidechain conformation (Fig. 6.13).

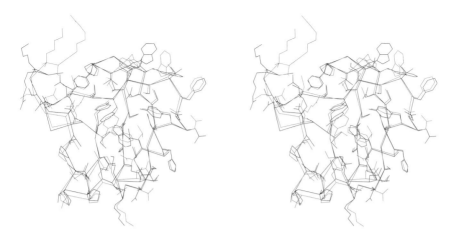

Fig. 6.13 SWISS-MODEL predicts the structure of red scorpion neurotoxin [1DQ7] (blue) from a closely related protein [1PTX] (black). The prediction was done *automatically*. Observe that most of the buried sidechains have not mutated, and have very similar conformations. Some sidechains on the surface have different conformations, and the mainchain of the C-terminus is in a different position (upper left). Not shown is a network of disulphide bridges, which constrain the structure. However, a model of this high quality would be expected, for two such closely related proteins, even without the extra constraints.

Fold recognition

Searching a sequence database for a probe sequence and searching a structure database with a probe structure are problems with known solutions. The mixed problems—probing a sequence database with a structure, or a structure database with a sequence, are less straightforward. They require a method for evaluating the compatibility of a given sequence with a given folding pattern.

The goal is to abstract the essence of a set of sequences or structures. Other proteins that share the pattern are expected to adopt similar structures.

Three-dimensional profiles

We have discussed patterns and profiles derived from multiple sequence alignments and their application to detection of distant homologues. One way to take advantage of available structural information to improve the power of these methods is a type of profile derived from the available sequences *and* structures of a family of proteins.

J.U. Bowie, R. Lüthy and D. Eisenberg analysed the *environments* of each position in known protein structures and related them to a set of preferences of the 20 amino acids for these structural contexts.

Given a protein structure, classify the environment of each amino acid in three separate categories:

1. Its mainchain hydrogen-bonding interactions, i.e. its secondary structure.

2. The extent to which it is buried within or on the surface of the protein structure.

3. The polar/non-polar nature of its environment.

The secondary structure may be one of three possibilities: *helix*, *sheet* or *other*. A sidechain is considered buried if the accessible surface area is less than 40 Å², partially buried if the accessible surface area is between 40 and 114 Å², and exposed if the accessible surface area is greater than 114 Å². The fraction of sidechain area covered by polar atoms is measured. The authors define six classes on the basis of accessibility and polarity of the surroundings. Sidechains in each of these six classes may have any of three types of secondary structure assignment: helix, sheet or other. This gives a total of 18 classes.

Assigning each sidechain to one of 18 categories makes it possible to write a coded description of a protein structure as a message in an alphabet of 18 letters, called a **three-dimensional structure profile.** Algorithms developed for sequence searches can thereby be applied to 'sequences' of encoded structures. For example, one could try to align two distantly-related proteins by aligning their three-dimensional structure profiles rather than their amino acid sequences. The three-dimensional profile method translates protein structures into one-dimensional probe (or probe-able) objects that do not explicitly retain either the sequence or structure of the molecules from which they were derived.

Next, how can one relate the three-dimensional structure profile to the set of known protein folding patterns? It is clear that some amino acids will be unhappy in certain kinds of sites; for example, a charged sidechain would prefer not to be buried in an entirely non-polar environment. Other preferences are not so clear-cut, and it is necessary to derive a preference table from a statistical survey of a library of well-refined protein structures.

Suppose now that we are given a sequence and want to evaluate the likelihood that it takes up, say, the globin fold. From the three-dimensional structure profile of the known sperm whale myoglobin structure we know the environment class of each position of the sequence. We must consider all possible alignments of the sequence of the protein of unknown structure with the three-dimensional structure profile of myoglobin. Consider a particular alignment, and suppose that the residue in the unknown sequence that corresponds to the first residue of myoglobin is phenylalanine. The environment class in the three-dimensional structure profile of the first residue of sperm whale myoglobin is: exposed, no secondary structure. One can score the probability of finding phenylalanine in this structural environment class from the table of preferences of particular amino acids for this three-dimensional structure profile class. (The fact that the first residue of the sperm whale myoglobin sequence is actually valine is not used, and indeed that information is not directly accessible to the algorithm. Sperm whale myoglobin is represented only by the sequence of environment classes of its residues, and the preference table is averaged over proteins with many different folding patterns.) Extension of this calculation to all positions and to all possible alignments (not allowing gaps within regions of secondary structure) gives a score that measures how well the given unknown sequence fits the sperm whale myoglobin profile.

A particular advantage of this method is that it can be automated, with a new sequence being scored against every three-dimensional profile in the library of known folds, in essentially the same way as a new sequence is routinely screened against a library of known amino-acid sequences.

Use of three-dimensional profiles to assess the quality of structures

The three-dimensional profile derived from a structure depends only very indirectly on the amino acid sequence. It is therefore meaningful to ask not only whether it is possible to identify other amino acid sequences compatible with the given fold, but whether the score of a three-dimensional profile for its own parent sequence is a measure of the compatibility of the sequence with the structure. Naturally, if real sequences did not generally appear to be compatible with their own structures, credibility in the method as capturing a valid connection between sequence and structure would be severely impaired. Two interesting results are observed: (1) Protein structures determined correctly do fit their own profiles well, although other, related, proteins, may give *higher* scores. The profile is abstracting properties of the family, not of individual sequences. (2) When a sequence does *not* match a profile computed from an experimental structure of that protein, there is likely to have been an error in the structure determination. The positions in the profile that do not match can identify the regions of error.

Threading

Threading is a method for fold recognition. Given a library of known structures, and a sequence of a query protein of unknown structure, does the query protein share a folding pattern with any of the known structures? The fold library could include some or all of the Protein Data Bank, or even hypothetical folds.

The basic idea of threading is to build many rough models of the query protein, based on each of the known structures and using different possible alignments of the sequences of the known and unknown proteins. This systematic exploration of the many possible alignments gives threading its name: imagine trying out all alignments by pulling the query sequence gently through the three-dimensional framework of any known structure. Gaps must be allowed in the alignments, but if the thread is thought of as being sufficiently elastic the metaphor of threading survives.

Both threading and homology modelling deal with the three-dimensional structure induced by an alignment of the query sequence with known structures of homologues. Homology modelling focuses on one set of alignments and the goal is a very detailed model. Threading explores many alignments and deals with only rough models usually not even constructed explicitly:

Homology modelling	Threading
First, identify homologues	Try all possible folds
Then, determine optimal alignment	Try many possible alignments
Optimize one model	Evaluate many rough models

Successful fold recognition by threading requires:

1. A method to score the models, so that we can select the best one.

2. A method for calibrating the scores, so that we can decide whether the best-scoring model is likely to be correct.

Several approaches to scoring have been tried. One of the most effective is based on empirical patterns of residue neighbours, as derived from known structures. First, we observe the distribution of inter-residue distances in known protein structures, for all 20×20 pairs of residue types. For each pair, derive a probability distribution, as a function of the separation in space, and in the amino acid sequence. For instance for the pair Leu–Ile, consider every Leu and Ile residue in known structures, and, for each Leu–Ile pair, record the distance between their Cβ atoms, and the difference in their positions in the sequence. Collecting these statistics permits estimation of how well the distributions observed in a model agree with the distributions in known structures.

The Boltzmann equation relates probabilities and energies. Usual applications of the Boltzmann equation start from an energy function and predict a probability distribution. (A standard example is the prediction of the density of the atmosphere as a function of altitude from the gravitational potential energy function of the air molecules.) For threading, one turns this on its head, and *derives* an energy function *from* the probability distribution. This energy function is then used to score threading models.

For each structure in the fold library, the procedure finds the assignment of residues that produces the lowest energy score. Although this is an alignment problem, the non-local interactions mean that it can't be solved by dynamic programming.

Fold recognition at CASP

The best methods for fold recognition are consistently effective. These include, but are not limited to methods based on threading.

Figures 6.14 and 6.15 show a prediction by A.G. Murzin, and another prediction by Bonneau, Tsai, Ruczinski and Baker, of targets from the 2000 CASP programme, both proteins of unknown function from *H. influenzae*.

Prediction of coiled-coils by Hidden Markov Models

Approaches to prediction of coiled-coil regions in proteins include:

* Profile methods using running windows (PCOILS)

* Profile methods, running windows, with residue correlations (PairCoil2)

* Hidden Markov Models (MARCOIL)

MARCOIL gave the best overall performance in controlled tests.

MARCOIL uses an HMM trained on a database containing nine classes of proteins:

* Tropomyosins   * Myosins   * Intermediate filaments
* Dyneins   * Kinesins   * Laminins
* SNARE proteins   * Transcription factors   * Others

Submitting to MARCOIL the chicken proto-oncogene protein c-fos, and selecting default parameters:

Some of the most powerful sequence → structure prediction methods involve HMMs (see chapter 5).

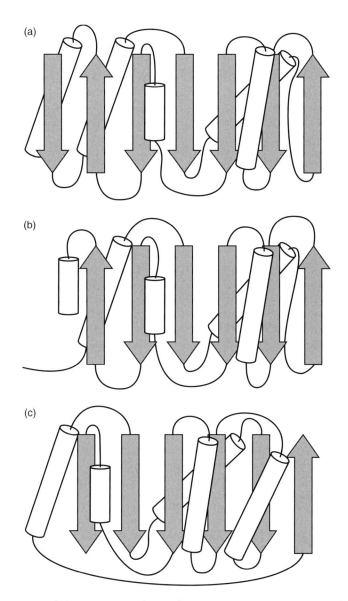

Fig. 6.14 Prediction of the structure of *H. influenzae,* hypothetical protein. (a) The folding pattern of the target. (b) Prediction by A.G. Murzin. (c) Folding pattern of the closest homologue of known structure: an *N*-ethylmaleimide-sensitive fusion protein involved in vesicular transport (PDB entry 1NSF). The topology of Murzin's prediction is closer to the target than that of the closest single parent.

```
>P11939|FOS_CHICK Proto-oncogene protein c-fos - Gallus gallus (Chicken).
MMYQGFAGEYEAPSSRCSSASPAGDSLTYYPSPADSFSSMGSPVNSQDFCTDLAVSSANF
VPTVTAISTSPDLQWLVQPTLISSVAPSQNRGHPYGVPAPAPPAAYSRPAVLKAPGGRGQ
SIGRRGKVEQLSPEEEEKRRIRRERNKMAAAKCRNRRRELTDTLQAETDQLEEEKSALQA
EIANLLKEKEKLEFILAAHRPACKMPEELRFSEELAAATALDLGAPSPAAAEEEAFALPLM
TEAPPAVPPKEPSGSGLELKAEPFDELLFSAGPREASRSVPDMDLPGASSFYASDWEPLG
AGSGGELEPLCTPVVTCTPCPSTYTSTFVFTYPEADAFPSCAAAHRKGSSSNEPSSDSLS
SPTLLAL
```

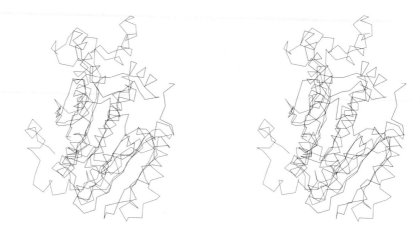

Fig. 6.15 Prediction by Bonneau, Tsai, Ruczinski and Baker of another hypothetical protein from *H. influenzae*, based on glycine *N*-methyltransferase [1XVA]. Experimental structure, black; prediction, blue. Note that much of the prediction superposes well on the experimental structure, and that the parts that do not superpose well have similar local structures but improper orientation and packing against the main body of the protein.

the program returned the prediction shown in Fig. 6.16. The program is quite confident that the protein contains a coiled-coil domain, approximately between residues 125 and 200.

Prediction of transmembrane helices and signal sequences by Hidden Markov Models

Some fold-recognition procedures strive for sufficient generality to identify all known domain structures. Others are specialized to particular types of folds. The best algorithms for prediction of transmembrane helices and coiled-coils make use of HMMs.

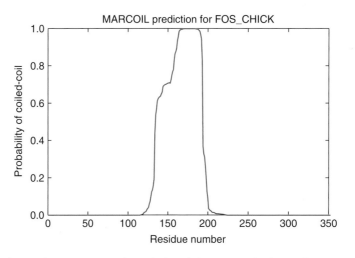

Fig. 6.16 Prediction by MARCOIL of a coiled-coil domain in chicken c-fos.

A simple approach to prediction of membrane proteins involves looking for amino acid segments 15–30 residues in length that are rich in hydrophobic residues. However, signal peptides also contain hydrophobic helices: the signal sequence typically comprises a positively charged n-region, followed by a helical hydrophobic h-region, followed by a polar c-region. Therefore, methods for recognizing transmembrane helices in amino acid sequences tend to pick up the h-regions of signal peptides as false positives. Methods for recognizing signal peptides in amino acid sequences tend to pick up transmembrane helices as false positives.

Käll, Krogh and Sonnhammer trained HMMs to test simultaneously for transmembrane helices and signal peptides. The goals are to find both at the same time, to discriminate between them in the results, and to predict not only the positions of the transmembrane helices but the locations—cytoplasmic or interior—of the loops. They called their method *PHOBIUS*.

PHOBIUS is the most successful algorithm currently available for recognizing signal peptides and helical transmembrane proteins, and for predicting the orientation of the transmembrane segments. PHOBIUS is capable of distinguishing h-domains of signal peptides from transmembrane helices: the number of false classifications of signal peptides was 3.9%, and the number of false classifications of transmembrane helices was 7.7%. These results represent a substantial improvement over previous methods. It is interesting that addressing the two problems at once proved to be more successful than treating them separately.

Web resources Membrane proteins

PHOBIUS (L. Käll, A. Krogh and E. Sonnhammer)
http://phobius.cgb.ki.se/

PHDhtm (B. Rost)
http://www.predictprotein.org

Membrane Protein EXplorer (S. White)
http://blanco.biomol.uci.edu/mpex/

The Membrane Protein Data Bank (P. Raman, V. Cherezov and M. Caffrey)
http://www.mpdb.ul.ie/

**online
resource
centre**

Conformational energy calculations and molecular dynamics

A protein is a collection of atoms. The interactions between the atoms create a unique state of maximum stability. Find it, that's all!

The computational difficulties in this approach arise because (1) the model of the interatomic interactions is not complete or exact, and (2) even if the model were exact we face an optimization problem in a large number of variables, involving non-linearities in the objective function and the constraints, creating a very rough energy surface with many local minima. Like a golf course with many bunkers, such problems are very difficult.

The interactions between atoms in a molecule can be divided into:

1. Primary chemical bonds—strong interactions between atoms that must be close together in space. These are regarded as a fixed set of interactions that are not broken or formed when the conformation of a protein changes, but are equally consistent with a large number of conformations.

2. Weaker interactions that depend on the conformation of the chain. These can be significant in some conformations and not in others—they affect sets of atoms that are brought into proximity by different folds of the chain.

The conformation of a protein can be specified by giving the list of atoms in the structure, their coordinates and the set of primary chemical bonds between them (this can be read off, with only slight ambiguity, from the amino acid sequence). Terms used in the evaluation of the energy of a conformation typically include:

- **Bond stretching:** $\sum_{\text{bonds}} K_r(\mathbf{r} - r_0)^2$ Here r_0 is the equilibrium interatomic separation and K_r is the force constant for stretching the bond. r_0 and K_r depend on the type of chemical bond.

- **Bond angle bend:** $\sum_{\text{angles}} K_\theta(\theta - \theta_0)^2$ For any atom i that is chemically bonded to two (or more) other atoms j and k, the angle j–i–k has an equilibrium value θ_0 and a force constant for bending K_θ.

- Other terms to enforce proper stereochemistry penalize deviations from planarity of certain groups, or enforce correct chirality (handedness) at certain centres.

- **Torsion angle:** $\sum_{\text{dihedrals}} \frac{1}{2}V_n[1 + \cos n\phi]$ For any four connected atoms: i bonded to j bonded to k bonded to l, the energy barrier to rotation of atom l with respect to atom i around the $j - k$ bond is given by a periodic potential. V_n is the height of the barrier to internal rotation; n barriers are encountered during a full $360°$ rotation. (For instance, for ethane $n = 3$.) The mainchain conformational angles ϕ, ψ and ω are examples of torsional rotations (see Fig. 6.2).

- **Van der Waals interactions:** $A_{ij}R_{ij}^{-12} - B_{ij}R_{ij}^{-6}$ For each pair of non-bonded atoms i and j, separated by a distance R_{ij}, the first term accounts for a short-range repulsion and the second term for a long-range attraction between them. The parameters A and B depend on atom type.

- **Hydrogen bonds:** $C_{ij}R_{ij}^{-12} - D_{ij}R_{ij}^{-10}$ The hydrogen bond is a weak chemical/electrostatic interaction between two polar atoms. Its strength depends on distance and also on the bond angle. This approximate hydrogen bond potential does not explicitly reflect the angular dependence of hydrogen bond strength; other potentials attempt to account for hydrogen bond geometry more accurately.

- **Electrostatics:** $Q_i Q_j/(\epsilon R_{ij})$ For each pair of charged atoms i and j, Q_i and Q_j are the effective charges on the atoms, R_{ij} is the distance between them, and ϵ is the dielectric 'constant'. This formula applies only approximately to media that are not infinite and isotropic, including proteins.

- Solvent Interactions with the solvent, water and co-solutes, such as salts and sugars, are crucial for the thermodynamics of protein structures. Attempts to model the solvent as a continuous medium, characterized primarily by a dielectric constant, are approximations. With the increase in available computer power, it is now possible to include solvent explicitly, simulating the motion of a protein in a box of water molecules.

There are numerous sets of conformational energy potentials of this or closely related forms, and a great deal of effort has gone into the tuning of parameter sets. The energy of a conformation is computed by summing these terms over all bonded and non-bonded atoms.

The potential functions satisfy necessary but not sufficient conditions for successful structure prediction. One test is to take the right answer—an experimentally determined protein structure—as a starting conformation, and minimize the energy starting from there. Most high-quality energy functions produce a minimized conformation that is about 1 Å (r.m.s. deviation) away from the starting model. This can be thought of as a measure of the resolution of the force field. Another test has been to take deliberately misfolded proteins and minimize their conformational energies, to see whether the energy value of the local minimum in the vicinity of the correct fold is significantly lower than that of the local minimum in the vicinity of an incorrect fold. Such tests reveal that multiple local minima cannot be reliably distinguished from the correct one on the basis of calculated conformational energies.

Attempts to predict the conformation of a protein by minimization of the conformational energy have so far not provided a general method for predicting protein structure from amino acid sequence. Molecular dynamics offers a way to overcome the problems of getting trapped in local minima, and of the absence of a good static model for protein–solvent interactions. In molecular dynamics calculations, the protein plus explicit solvent molecules are treated—via the force field—by classical Newtonian mechanics. It is true that this permits exploration of a much larger sector of phase space. However, as an *a priori* method of structure prediction, it has still not succeeded consistently. However, these are calculations that are extremely computationally intensive and here, perhaps more than anywhere else in this field, advances deriving from the increased power of processors will have an effect.

Is lack of computational power the only reason for lack of success in prediction of protein structure by simulation of the folding pathway? There have been several attempts to apply 'brute-force', including the IBM Blue Gene supercomputer project; and the distributed computing approach of FoldingHome, which makes use of contributions of computer power from over a million participating CPUs. (A similar approach has been applied to drug design.) An IBM group folded a 20-residue peptide from a fully extended conformation to a state within ~1.5 Å r.m.s. deviation of the native state.

Scaling of resource requirements for molecular dynamics calculations

Fully detailed molecular dynamics calculations perform a series of individual time steps of duration 10^{-15}s($= 1$ fs). The computer time required for an individual time step scales approximately as $N \ln N$, where N is the length of the protein.

$\longrightarrow$

$\longrightarrow$

Scaling of resource requirements for molecular dynamics calculations *(continued)*

The time required for a protein to fold depends on a number of features, but, for purposes of a 'back-of-the-envelope' calculation it varies with the length N of the protein as $\sim N^{\frac{2}{3}}$.

Therefore, the total computer resources required to fold a protein may be expected to vary approximately as $N^{\frac{5}{3}} \ln N$. This means that if it takes 3 months (of uninterrupted time on a supercomputer running flat out) to fold a protein of length N, it would be expected to require 1.5 years, on the same system, to fold up a protein of length $3N$ residues.

In the meantime, molecular dynamics, if supplemented by experimental data, regularly makes extremely important contributions to structure determinations by both X-ray crystallography (usually) and NMR (always). How is molecular dynamics integrated into the process of structure determination? For any conformation, one can measure the consistency of the model with the experimental data. In the case of crystallography, the experimental data are the absolute values of the Fourier transform of the electron density of the molecule. In the case of NMR, the experimental data provide constraints on the distances between certain pairs of residues. But in both X-ray crystallography and NMR, the experimental data underdetermine the protein structure. To solve a structure one must seek a set of coordinates that minimizes a combination of the deviation from the experimental data and the conformational energy. Molecular dynamics is successful at determining such coordinate sets: the dynamics provides adequate coverage of conformation space, and the bias derived from the experimental data channels the calculation quite effectively towards the correct structure.

ROSETTA

ROSETTA is a program by D. Baker and colleagues that predicts protein structure from amino acid sequence by assimilating information from known structures. At recent CASP programmes, ROSETTA has shown consistent success on targets in both the Fold Recognition and Novel Fold categories. At present, it leads the field by several lengths.

ROSETTA predicts a protein structure by first generating structures of fragments using known structures, and then combining them. First, for each contiguous region of three and nine residues, instances of that sequence and related sequences are identified in proteins of known structure. For fragments this small, there is no assumption of homology to the target protein. The distribution of conformations of the fragments serves as a model for the distribution of possible conformations of the corresponding fragments of the target structure.

ROSETTA explores the possible combinations of fragments using Monte Carlo calculations (see Box, page 356). The energy function has terms reflecting compactness, paired β-sheets and burial of hydrophobic residues. The procedure carries out 1000 independent simulations, with starting structures chosen from the fragment conformation distribution pattern generated previously. The structures that result from these simulations are clustered, and the centres of the largest clusters presented as predictions of the target structure. The idea is that a structure that emerges many times from independent simulations it is likely to have favourable features.

Figure 6.17 shows successful predictions by ROSETTA of two targets from the 2000 CASP programme.

ROBETTA (http://robetta.bakerlab.org) is a web server designed to integrate and implement the best of the protein structure prediction tools. The central pipeline of the software involves first the parsing of a submitted amino acid sequence of a protein of unknown structure into putative domains. Then homology modelling techniques are applied to those domains for which suitable parents of known structure exist, and the *de novo* methods developed by Baker and co-workers are applied to other domains. In addition, the user will receive the results of other prediction methods based on software developed outside the ROBETTA group. These include, for example, predictions of secondary structure, coiled-coils and transmembrane helices.

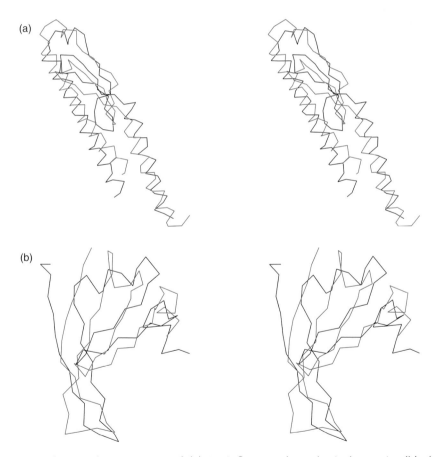

Fig. 6.17 Predictions by ROSETTA of (a) *H. influenzae*, hypothetical protein, (b) the N-terminal half of domain 1 of human DNA repair protein Xrcc4. (b) shows a selected substructure containing the N-terminal 55 out of 116 residues. Experimental structures, solid lines; predicted structures, broken lines.

Monte Carlo algorithms

Monte Carlo algorithms are used very widely in protein structure calculations, to explore conformations efficiently, and in many other optimization problems, to search for the minimum of a complicated function. Simple minimization methods based on moving 'downhill' in energy fail, because the calculation gets trapped in a local minimum far from the native state.

In general, **Monte Carlo methods** make use of random numbers to solve problems for which it is difficult to calculate the answer exactly. The name was invented by J. von Neumann, referring to the applications of random number generators in the famous gambling casino.

To apply Monte Carlo techniques to find the minimum of a function of many variables—for instance, the minimum energy of a protein as a function of the variables that define its conformation—suppose that the configuration of the system is specified by the variables $\mathbf{x}$, and that for any values of these variables, we can calculate the energy of the conformation, $\mathcal{E}(\mathbf{x})$. ($\mathbf{x}$ stands for a whole set of variables—perhaps the set of atomic coordinates of a protein, or the mainchain and sidechain torsion angles.)

Then the Metropolis procedure (invented in 1953, allegedly at a dinner party in Los Alamos) prescribes:

1. Generate a random set of values of $\mathbf{x}$, to provide the starting conformation. Calculate the energy of this conformation, $\mathcal{E} = \mathcal{E}(\mathbf{x})$.

2. Perturb the variables: $\mathbf{x} \rightarrow \mathbf{x}'$, to generate a neighbouring conformation.

3. Calculate the energy of the new conformation, $\mathcal{E}(\mathbf{x}')$.

4. Decide whether to *accept* the step: to move $\mathbf{x} \rightarrow \mathbf{x}'$, or to stay at $\mathbf{x}$ and try a different perturbation:
   (a) If the energy has decreased: $\mathcal{E} = \mathcal{E}(\mathbf{x}) > \mathcal{E}(\mathbf{x}')$, i.e. the step went *downhill*, always accept it. The perturbed conformation becomes the new current conformation: set $\mathbf{x}' \rightarrow \mathbf{x}$ and $\mathcal{E} = \mathcal{E}(\mathbf{x}')$.
   (b) If the energy has increased or stayed the same, i.e. $\mathcal{E}(\mathbf{x}) \leq \mathcal{E}(\mathbf{x}')$—in other words the step goes *uphill*—*sometimes* accept the new conformation. If $\Delta = \mathcal{E}(\mathbf{x}') - \mathcal{E}(\mathbf{x})$, accept the step with a probability $\exp[-\Delta/(kT)]$, where k is Boltzmann's constant, and T is an effective temperature.

5. Return to step 2.

It is step 4b that is the ingenious one. It has the potential to get over barriers, out of traps in local minima. The effective temperature, T, controls the chance that an uphill move will be accepted. T is not the physical temperature at which we wish to predict the protein conformation, but simply a numerical parameter that controls the calculation.

For any temperature, the higher the uphill energy difference, the less likely that the step will be accepted. For any value of $\mathcal{E}$, if T is low, then $\mathcal{E}(\mathbf{x})/(kT)$ will be high, and $\exp[-\mathcal{E}(\mathbf{x})/(kT)]$ will be relatively low. If T is high, then $\mathcal{E}(\mathbf{x})/(kT)$ will be low,

$\longrightarrow$

Monte Carlo algorithms (continued)

and $\exp[-\mathcal{E}(\mathbf{x})/(kT)]$ will be relatively high. The higher the temperature, the more probable the acceptance of an uphill move.

This relatively simple idea has proved extremely effective, with successful applications including but by no means limited to protein structure calculations.

Simulated annealing is a development of Monte Carlo calculations in which T varies—first it is set high to allow efficient exploration of conformations, then it is reduced to drop the system into a low-energy state.

LINUS

LINUS (= Local Independently Nucleated Units of Structure) is a program for prediction of protein structure from amino acid sequence, by G.D. Rose and R. Srinivasan. It is a completely *a priori* procedure, making no explicit reference to any known structures or sequence–structure relationships. LINUS folds the polypeptide chain in a *hierarchical* fashion, first producing structures of short segments, and then assembling them into progressively larger fragments.

An insight underlying LINUS is that the structures of local regions of a protein—short segments of residues consecutive in the sequence—are controlled by local interactions within these segments. During natural protein folding, each segment will preferentially sample its most favourable conformations. However, these preferred conformations of local regions, even the one that will ultimately be adopted in the native state, are below the threshold of stability. Local structure will form transiently and break up many times before a suitable interacting partner stabilizes it. But in the computer one is free to anticipate the results. In a LINUS simulation, favourable structures of local fragments, as determined by their frequent recurrence during the simulation, transmit their preferred conformations as biases that influence subsequent steps. The procedure applies the principle of a rachet to direct the calculation along productive lines.

LINUS begins by building the polypeptide from the sequence as an extended chain. The simulation proceeds by perturbing the conformations of a succession of randomly chosen three-residue segments, and evaluating the energies of the results. Structures with steric clashes are rejected out of hand; other energetic contributions are evaluated only in terms of local interactions. A Monte Carlo procedure (see Box) is used to decide whether to accept a perturbed structure or revert to its predecessor. LINUS performs a large number of such steps. It periodically samples the conformations of the residues, to accumulate statistics of structural preferences.

Subsequent stages in the simulation assemble local regions into larger fragments, using the conformational biases of the smaller regions to guide the process. The window within the sequence controlling the range of interactions is progressively opened, from short local regions, to larger ones, and ultimately to the entire protein.

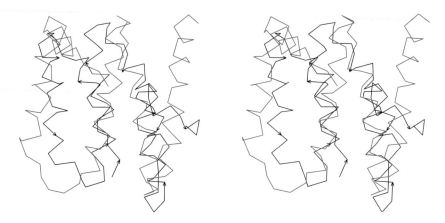

Fig. 6.18 A LINUS prediction of the C-terminal domain of rat endoplasmic reticulum protein ERp29. Experimental structure, black; prediction, blue.

The LINUS representation of the protein folding process is realistic in essential respects, although approximate. All non-hydrogen atoms of a protein are modelled, but the energy function is approximate and the dynamics simplified. The energy function captures the ideas of: (1) steric repulsion preventing overlap of atoms, (2) clustering of buried hydrophobic residues, (3) hydrogen bonding and (4) salt bridges.

LINUS is generally successful in getting correct structures of small fragments (size between supersecondary structure and domain), and in some cases can assemble them into the right global structure. Figure 6.18 shows the LINUS prediction of the C-terminal domain of rat endoplasmic reticulum protein ERp29, one of the targets of the 2000 CASP programme.

Assignment of protein structures to genomes

A genome sequence is the complete statement of a potential life. Assignment of structures to gene products is a first step in understanding how organisms implement their genomic information.

We want to understand the structures of the molecules encoded in a genome, their individual activities and interactions, and the organization of these activities and interactions in space and time during the lifetime of the organism. We want to understand the relationships among the molecules encoded in the genome of one individual, and their relationships to those of other individuals and other species.

For individual proteins, knowing their structure is essential for understanding the mechanism of their function and interactions. For entire organisms, knowing the structures tells us how the repertoire of possible protein folds is called upon, and how it is distributed among different functional categories in different species.

For interspecies comparisons, protein structures can reveal relationships invisible in highly diverged sequences.

Several methods have been applied to structure assignment:

- Experimental structure determination The best way of all!

- Detection of homology in sequences Sophisticated sequence comparison methods such as PSI–BLAST or HMMs can identify relationships between proteins, both within an organism and between species. If the structure of any homologue is known experimentally, at least the general fold of the family can be inferred.

- Fold-recognition methods can assign folds to some proteins even in the absence of evidence for homology.

- Specialized techniques detect *membrane proteins,* and *coiled-coils.*

The results of structure assignments provide partial inventories of proteins in the different genomes, and, for the subset of proteins with sufficiently close relatives of known structure, detailed three-dimensional models. The degree of coverage of assignments is changing very fast, primarily because of the rapid growth of sequence and structural data. The Table contains a current scorecard.

Species	Number of sequences	Structures assigned	%
E. coli	4289	916	21
M. jannaschii	1773	262	14
S. cerevisiae	6289	1109	17
D. melanogaster	13 687	2990	21

From: GeneQuiz, http://jura.ebi.ac.uk:8765/ext-genequiz/.

What do these results tell us about the usage of the potential protein repertoire? A comparison of folding patterns of proteins deduced from the genomes of an archaeon *Methanococcus jannaschii,* a bacterium *Hamophilus influenzae* and a eukaryote *Saccharomyces cerevisiae* revealed that out of a total of 148 folds, 45 were common to all three species — and, by implication, probably common to most forms of life. The archaeon, *M. jannaschii,* had the fewest unshared folds (see Fig. 6.19).

An inventory of the structures common to all three species showed that the five most common folding patterns of domains, are: (1) the P-loop-containing NTP hydrolase fold, (2) the NAD-binding domain, (3) the TIM-barrel fold, (4) the flavodoxin fold, and (5) the thiamin-binding fold. Plate VIII shows the structure and a simplified schematic diagram of the topology of the first of these (see also Weblems 6.3 and 6.4). All are of the α/β type.

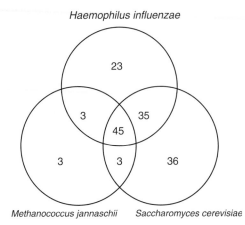

Fig. 6.19 Shared protein folds in an archaeon *Methanococcus jannaschii*, a bacterium *Haemophilus influenzae*, and a eukaryote *Saccharomyces cerevisiae*. (After: Gerstein, M. (1997), A structural census of genomes: comparing bacterial, eukaryotic, and archaeal genomes in terms of protein structure. *Journal of Molecular Biology*, **274**, 562–576.)

Prediction of protein function

The cascade of inference should ideally flow from sequence → structure → function. However, although we can be confident that similar amino acid sequences will produce similar protein structures, the relationship between structure and function is more complex. Proteins of similar structure and even of similar sequence can be recruited for very different functions. Very widely diverged proteins may retain similar functions. Moreover, just as many different sequences are compatible with the same structure, proteins with different folds can carry out the same function (see Fig. 6.20).

As proteins evolve they may:

♦ retain function and specificity

♦ retain function but alter specificity

♦ change to a related function, or a similar function in a different metabolic context, or

♦ change to a completely unrelated function

People often ask: How much must a protein sequence or structure change before the function changes? The answer is: some proteins have multiple functions, so they need not change at all!

♦ In the duck, an active lactate dehydrogenase and an enolase serve as crystallins in the eye lens, although they do not encounter the substrates *in situ*. In other cases, crystallins are closely related to enzymes, but some divergence has already occurred, with loss of catalytic activity. This proves that the enzymatic activity is not necessary in the eye lens.

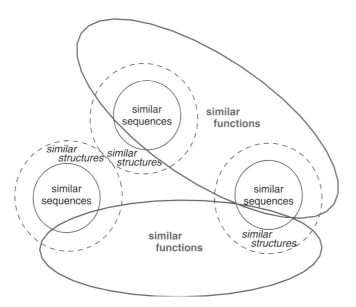

Fig. 6.20 Relationships among sequence, structure and function:

- *Similar sequences can be relied on to produce similar protein structures*, with divergence in structure increasing progressively with the divergence in sequence.

- Conversely, *similar structures are often found with very different sequences.* In many cases the relationships in a family of proteins can be detected *only* in the structures, the sequences having diverged beyond the point of our being able to detect the underlying common features.

- *Similar sequences and structures often produce proteins with similar functions*, but exceptions abound.

- Conversely, *similar functions are often carried out by non-homologous proteins with dissimilar structures*; examples include the different families of proteinases, sugar kinases and lysyl-tRNA synthetases.

- A protein from *E. coli*, called Do or DegP or HtrA, acts as a chaperone (catalysing protein folding) at low temperatures, but at 42° C turns into a proteinase. The rationale seems to be: under normal conditions or moderate heat stress, the goal is to salvage proteins that are having difficulty folding; under more severe heat stress, when salvage is impossible, the goal is to recycle them.

- We have already mentioned the *E. coli* enzyme lipoate dehydrogenase that is *also* an essential subunit of pyruvate dehydrogenase, 2-oxoglutarate dehydrogenase and the glycine cleavage complex.

These examples of structure–function relationships are at the extreme end of a wide spectrum.

One problem is that it is not easy to define the idea of difference in function quantitatively. When are two different functions more similar to each other than two other different functions? In some cases altered function may conceal similarity of mechanism. For example, the enolase superfamily contains several homologous enzymes that catalyse different reactions with shared mechanistic features. This group includes enolase itself, mandelate racemase, muconate lactonizing enzyme I and D-glucarate dehydratase. Each acts by abstracting an α proton from a carboxylic acid to form an enolate intermediate. The subsequent reaction pathway, and the nature of the product, vary from enzyme to enzyme. These proteins have very similar overall structures, a variant of the TIM-barrel fold. Different residues in the active site produce enzymes that catalyse different reactions.

Several authors have devised measures of differences between protein functions based on the Enzyme Commission and Gene Ontology classification.

Divergence of function: orthologues and paralogues

The family of chymotrypsin-like serine proteinases includes closely related enzymes in which function is conserved, and widely diverged homologues that have developed novel functions.

Evolutionary relationships among proteins: homologues, orthologues and paralogues

- Proteins are homologous *if and only if* they are descended from a common ancestor.

- Homologues in different species, descended from a single ancestral protein, are *orthologues*.

- Homologues in the same species, arising from gene duplication, are *paralogues*. Their descendants are also paralogues. After gene duplication, one of the resulting pairs of proteins can continue to provide its customary function, releasing the other to diverge, to develop new functions. Therefore, inferences of function from homology are more secure for orthologues than for paralogues.

Trypsin, a digestive enzyme in mammals, catalyses the hydrolysis of peptide bonds adjacent to a positively charged residues, Arg or Lys. (A **specificity pocket**, a surface cleft in the active site, is complementary in shape and charge distribution to the sidechain of the residue adjacent to the scissile bond.) Enzymes with similar sequence, structure, function and specificity exist in many species, including human, cow, Atlantic salmon and even *Streptomyces griseus* (Fig. 6.21). The similarity of the *S. griseus* enzyme to vertebrate trypsins suggests a lateral gene transfer. For the three vertebrate enzymes, each pair of sequences has ≥64% identical residues in the alignment, and the bacterial homologue has ≥30% identical residues with the others; all have very similar structures. These enzymes are called *orthologues*—homologous proteins in different species. (Other bacterial homologues are very different in sequence.)

TRYPSIN

```
                10        20        30        40        50        60        70        80
                |         |         |         |         |         |         |         |
Human           IVGGYNCEENSVPYQVSLNSGYHFCGGSLINEQWVVSAGHCYKSR---IQVRLGEHNIEVLEGNEQFINAAKIIRHPQYD
Cow             IVGGYTCGANTVPYQVSLNSGYHFCGGSLINSQWVVSAAHCYKSG---IQVRLGEDNINVVEGNEQFISASKSIVHPSYN
Atlantic salmon IVGGYECKAYSQAHQVSLNSGYHFCGGSLVNENWVVSAAHCYKSR---VEVRLGEHNIKVTEGSEQFISSSRVIRHPNYS
S. griseus      VVGGTRAAQGEFPFMVRLSMG---CGGALYAQDIVLTAAHCVSGSGNNTSITATGGVVDLQSGAAVKVRSTKVLQAPGYN

                iVGGy c        p qVsLnsGyhfCGGsL  n  wVvsA HCyks      vrlge ni v eG eqfi   k i hp y


                90       100       110       120       130       140       150       160
                |         |         |         |         |         |         |         |
Human           RKTLNNDIMLIKLSSRAVINARVSTISLPTAPPATGTKCLISGWGNTASSGADYPDELQCLDAPVLSQAKCEASYP-GKI
Cow             SNTLNNDIMLIKLKSAASLNSRVASISLPTSCASAGTQCLISGWGNTKSSGTSYPDVLKCLKAPILSDSSCKSAYP-GQI
Atlantic salmon SYNIDNDIMLIKLSKPATLNTYVQPVALPTSCAPAGTMCTVSGWGNTMSSTADS-NKLQCLNIPILSYSDCNNSYP-GMI
S. griseus      --GTGKDWALIKLAQPINQ----PTLKIATTTAYNQGTFTVAGWGANREGGSQQRYLLKAN-VPFVSDAACRSAYGNELV

                nDimLIKL    a  n  v    lpT      gt c  sGWGnt ssg       L cl  P 1S    C  Yp g i


                170       180       190       200       210       220       230
                |         |         |         |         |         |         |
Human           TSNMFCVGFLE-GGKDSCQGDSGGPVVCNG-----QLQGVVSWGDGCAQKNKPGVYTKVYNYVKWIKNTIAANS
Cow             TSNMFCAGYLE-GGKDSCQGDSGGPVVCSG-----KLQGIVSWGSGCAQKNKPGVYTKVCNYVSWIKQTIASN-
Atlantic salmon TNAMFCAGYLE-GGKDSCQGDSGGPVVCNG-----ELQGVVSWGYGCAEPGNPGVYAKVCIFNDWLTSTMASY-
S. griseus      ANEEICAGYPDTGGVDTCQGDSGGPMFRKDNADEWIQVGIVSWGYGCARPGYPGVYTEVSTFASAIASAARTL-

                t  mfC G le GGkDsCQGDSGGPvvc g       lqG VSWG GCA      PGVYtkV    wi     t a
```

Fig. 6.21 Alignment of sequences of trypsins from human, cow, Atlantic salmon and *Streptomyces griseus*. In the lines under the blocks, uppercase letters indicate absolutely conserved residues and lowercase letters indicate residues conserved in three of the four sequences (in most but not all cases *S. griseus* is the exception).

Evolution has also created related enzymes in the *same* species with different specificities. Chymotrypsin and pancreatic elastase are other digestive enzymes that, like trypsin, cleave peptide bonds, but next to different residues: chymotrypsin cleaves adjacent to large flat hydrophobic residues (Phe, Trp) and elastase cleaves adjacent to small residues (Ala). The change in specificity is effected by mutations of residues in the specificity pocket. Another homologue, leukocyte elastase (the object of database searching in Chapter 4), is essential for phagocytosis and defence against infection. Under certain conditions it is responsible for lung damage leading to emphysema.

Homologous proteins in the same species are called *paralogues*. Trypsin, chymotrypsin and pancreatic elastase function in digestion of food. Another set of paralogues mediates the blood coagulation cascade. Although all are proteinases, the requirements for activation and control are very different for digestion and blood coagulation, and the families have diverged and become specialized for these respective roles.

Some homologues of trypsin have developed entirely new functions:

♦ Haptoglobin is a chymotrypsin homologue that has lost its proteolytic activity. It acts as a chaperone, preventing unwanted aggregation of proteins. Haptoglobin forms a tight complex with haemoglobin fragments released from erythrocytes, with several useful effects including preventing the loss of iron.

♦ The serine proteinase of rhinovirus has developed a separate, independent function, of forming the initiation complex in RNA synthesis, using residues on the opposite side of the molecule from the active site for proteolysis. This is not a modification of an active site—it is the creation of a new one.

♦ Subunits homologous to serine proteinases appear in plasminogen-related growth factors. The role of these subunits in growth factor activity is not yet known, but it cannot be a proteolytic function because essential catalytic residues have been lost.

- An antifreeze glycoprotein in Antarctic fish is homologous to chymotrypsin.
- The insect 'immune' protein scolexin is a distant homologue of serine proteinases that induces coagulation of haemolymph in response to infection.

In the chymotrypsin family we see a retention of structure with similar functions in closely related proteins, and progressive divergence of function in some but not all distantly related ones.

The message is that the overall folding pattern of a protein is an unreliable guide to predicting function, especially for very distant homologues. For correct prediction of function in distantly related proteins it is necessary to focus on the active site. For example:

- J.F. Bazan and R. Fletterick and, independently, P. Argos, G. Kamer, M.J. Nicklin and E. Wimmer, recognized that viral 3C proteinases are chymotrypsin homologues, despite the fact that the serine of the catalytic triad is changed to cysteine.
- W.R. Taylor and L. Pearl recognized the distant homology between retroviral and aspartic proteinases from conserved Asp, Thr and Gly residues.

Like motif libraries such as PROSITE, such approaches go directly from signature patterns of active site residues in the sequence to conserved function, even in the absence of an experimental structure.

In focusing on the active site, there is opportunity to use methods similar to those used in drug design, to predict ligands that might bind to the proteins. These would be putative substrates. It will be important to make use of other experimental information available, such as tissue distribution patterns of expression, and catalogues of proteins that interact. Attempts to measure function directly, for instance by means of 'gene knockouts', will sometimes provide an answer, but are unproductive if the knockout phenotype is lethal or if there are multiple proteins that share a function.

It seems likely that the contribution of bioinformatics to prediction of protein function from sequence and structure will not be a simple algorithm that provides an unambiguous answer. (In contrast there is reasonable hope that there will someday be a program that will predict structure from sequence.) More reasonable aims are to suggest productive experiments and to contribute to the interpretation of the results. These are not unworthy goals.

Drug discovery and development

It is a sobering experience to ask a classroom full of students how many would be alive today without at least one course of drug therapy during a serious illness. (This ignores diseases escaped entirely, through vaccination.) Or to ask the students how many of their surviving grandparents would be leading lives of greatly reduced quality without regular treatment with drugs. The answers are eloquent. They engender fear of the new antibiotic-resistant strains of infectious microorganisms. It is necessary to develop new drugs, which, in combination with genomic information that can improve their specificity and reduce side effects, will extend and improve our lives.

However, it is not easy to be a drug. For a chemical compound to qualify as a drug, it must be:

1. safe

2. effective

3. stable—both chemically and metabolically

4. deliverable—the drug must be absorbed and make its way to its site of action

5. available—by isolation from natural sources or by synthesis

6. novel, i.e. patentable

Steps in the development of a new drug are summarized in the Box. The process involves scientific research, clinical testing to prove safety and efficacy, and very important economic and legal aspects involving patent protection and estimation of returns on the very high required investment.

Steps in the development of a new drug

1. Understanding the biological nature and symptoms of a disease. Is it caused by
   - an infectious agent—bacterium, virus, other?
   - a poison of non-biological origin?
   - a mutant protein in the patient?

2. Developing an assay. Given a candidate drug, can you test it by:
   - its effect on the growth of a microorganism?
   - its effect on cells grown in tissue culture?
   - its effect on animals that suffer the disease or an analogue?
   - its binding to a known protein target?

3. Is an effective agent from a natural source known from folklore? If so, go to 6.

4. Identify a specific molecular target, usually a protein. Determine its structure experimentally or by model building.

5. Get a general idea of what kind of molecule would fit the site on the target. Is there a known substrate or inhibitor?

6. Identification of a *lead compound*: any chemical that shows the desired biological activity to *any* measurable extent. A lead compound is a bridgehead; finding lead compounds and subsequently modifying them are quite different kinds of activities.

→

→ ──

Steps in the development of a new drug *(continued)*

7. Development of the lead compound: extensive study of variants of the compound, with the goal of building in all the desired properties and enhancing the biological activity.

8. Preclinical testing, *in vitro* and with animals, to prove effectiveness and safety. At this point the drug may be patented. (In principle, one wants to delay patenting as long as possible because of the finite lifetime of the patent. Many lengthy steps still remain before the drug can be sold.)

9. In the USA: submission of an Investigational New Drug Application to the Federal Drug Administration. This is followed by three phases of clinical trials.

10. Phase I clinical trials. Test the compound for safety on healthy volunteers. Determine how the body deals with the drug—how it is absorbed, distributed, metabolized, excreted. The results suggest a safe dosage range.

11. Phase II clinical trials. Test the compound for efficacy against a disease on approximately 200 volunteer patients. Does it cure the disease or alleviate symptoms? Calibrate the dosage.

12. Phase III clinical trials. Test approximately 2000 patients, to demonstrate conclusively that the compound is better than the best known treatment. These are randomized double-blind tests, either against a placebo or against a currently used drug. These trials are very expensive; it is not uncommon to kill a project before embarking on this step, if the phase II trials expose side effects or unsatisfactory efficacy.

13. File a New Drug Application with the FDA, containing supporting data proving safety and efficacy. FDA approval allows the drug to be sold. Only now can the drug bring in income.

14. Phase IV studies, subsequent to FDA approval and marketing, involve continued monitoring of the effects of the drug, reflecting the wider experience in its use. New side effects may turn up in some classes of patients, leading to restrictions on the use of the drug, or even possibly its recall.

To develop a drug, first you must choose a target disease. You will want to study what is known about its possible causes, its symptoms, its genetics, its epidemiology, its relationship to other diseases—human and animal—and all known treatments. Assuming that the potential utility of a drug justifies the major time, expense and effort required for its development, you are now ready to begin.

You must develop a suitable assay with which to detect success in the initial phase. If a known protein is the target, binding can be measured directly. A potential antibacterial drug can be tested by its effect on growth of the pathogen. Some compounds might be tested for effects on eukaryotic cells grown in tissue culture. If a laboratory animal is susceptible to the disease, compounds can be tested on animal subjects. However, compounds may have different effects on animals and humans (see page 174).

The lead compound

A goal in the early stages of drug development is identification of one or more **lead compounds.** A lead compound is any substance that shows the biological activity you seek. It demonstrates that a compound exists that possesses at least some of the desired properties.

There are a number of ways to find lead compounds:

1. Serendipity: penicillin is the classic example.

2. Survey of natural sources. 'Grind and find' is the medicinal chemist's motto. Sometimes traditional remedies point to a source of active compounds. For example, digitalis was isolated from leaves of the foxglove, which had been used for congestive heart failure. (Why not just continue to use the traditional remedy? Isolation of the active principle makes it possible to regulate dosage, and to explore variants.)

3. Study of what is known about substrates, inhibitors and the mechanism of action of a protein implicated in a disease, and select potentially active compounds from these properties.

4. Drugs effective against similar diseases.

5. Large-scale screening. Techniques of combinatorial chemistry permit parallel testing of large sets of related compounds. A special technique applicable to polypeptides is phage display.

6. Occasionally, from side effects of existing drugs. Minoxidil (2,4-diamino-6-piperidino-pyrimidine-3-oxide), originally designed as an antihypertensive, was found to induce hair growth. Viagra, originally developed as a heart medicine, is another example.

7. Screening. The US National Cancer Institute has screened tens of thousands of compounds. (Screening of variants is also very important *after* a lead compound has been found.)

8. Computer screening and *ab initio* computer design.

Discovery of a lead compound triggers other kinds of research activities. Many variants of the lead compound must be tested to improve its effectiveness, and to build in other essential properties. For instance, a compound that binds to its target is no good as a drug unless it can get there. Deliverability of a drug to a target within the body requires the capacity to be absorbed and transmitted. It requires metabolic stability. It requires the proper solubility profile—a drug must be sufficiently water-soluble to be absorbed, but not so soluble that it is excreted immediately; it must (in most cases) be sufficiently lipid-soluble to get across membranes, but not so lipid-soluble that it is merely taken up by fat stores.

Improving on the lead compound: quantitative structure-activity relationships (QSAR)

For any compound with pharmacological activity, similar compounds typically exhibit related activity but vary in potency and specificity. Starting with a lead compound, chemists must survey large numbers of related molecules to optimize the desired pharmacological properties. To search systematically, it would be very useful to understand how the variation in structural and physicochemical features in the family of molecules is correlated with pharmacological properties. The problem is that there are very many possible descriptors for characterizing molecules. These include structural features such as the nature and distribution of substituents; experimental features such as solubility in aqueous and organic solvents, or dipole moments; and computed features such as charges on individual atoms.

Quantitative structure–activity relationships (QSAR) provide methods for predicting the pharmacological activity of a set of compounds from the relationship between molecular features and pharmacological activity, based on test cases. The method was developed by C. Hansch and colleagues in the 1960s, and has been of very widespread use.

C. Hansch, J. McClarin, T. Klein and R. Langridge applied QSAR methods to study inhibitors of carbonic anhydrase. Carbonic anhydrase is an enzyme that catalyses the reaction $CO_2 + H_2O \rightleftharpoons H^+ + HCO_3^-$. Clinical applications of carbonic anhydrase inhibitors include diuretics, treatment of high interocular pressure in glaucoma by supressing secretion of aqueous humour (the fluid within the eye), and antiepileptic agents. High-altitude climbers take carbonic anhydrase inhibitors for relief of symptoms of acute mountain sickness.

Measurements of carbonic anhydrase binding of 29 phenylsulphonamides:

(X stands for a set of substituents on the ring that are variable in both structure and position) showed that the binding constant was related to the Hammett electronic substituent constant σ, a measure of the electron-withdrawing or -donating strength of the substituent; the octanol–water partition coefficient P of the unionized form of the ligand; and the location (ortho or meta) of the substitution:

$$\log K \approx 1.55\sigma + 0.65 \log P - 2.07 I_1 + 3.28 I_2 + 6.94$$

in which K = binding constant, $I_1 = 1$ if X is *meta* and 0 otherwise, and $I_2 = 1$ if X is *ortho* and 0 otherwise. The substituents X were of the form alkyl, or -COO-alkyl, or -CONH-alkyl.

This type of correlation has two implications:

1. A large number of compounds can be screened in the computer and those predicted to be the best can then be tested experimentally.

2. It is possible to visualize the binding site from analysis of the parameters:

   ♦ The positive coefficient of σ, implying that electron-withdrawing substituents are favoured, suggests that the ionized form of the SO_2NH_2 moiety binds to the Zn^{2+} ion in the carbonic anhydrase active site.

- The positive coefficient of $\log P$ suggests a hydrophobic interaction between the protein and ligand.

- The negative coefficients of I_1 and I_2 suggest steric clashes with substituents in the *meta* or *ortho* positions.

Structures of ligated carbonic anhydrase confirm these conclusions (see Web-lem 6.10).

Bioinformatics in drug discovery and development

Computing and information retrieval contribute to several steps in drug discovery and development projects. These include: target identification; design, analysis and enhancement of ligands; and selection and *in silico* screening of libraries. Information systems are also important in the organization of the theoretical predictions, the experimental designs and analysis of the data. D. Searls has called the intimate interplay between theory and experiment 'wet–dry cycles'.

Target selection

To develop a drug against a disease, it is necessary to select a protein linked to the disease in a way that suggests that it would be therapeutically useful to affect its function or expression. New high-throughput data sources, particularly of genome sequences and protein expression patterns, provide a rich source of material for identifying potential drug targets. **Differential genomics and proteomics**, the comparisons of healthy and diseased humans or animals, can pinpoint which particular protein is missing, dysfunctional, improperly regulated or expressed only in affected cells. Comparisons between antibiotic-resistant and susceptible strains of bacteria can elucidate the mechanism of resistance. Information about protein–protein complexes makes it possible to target not just a single protein, but a specific protein–protein interaction.

Knowledge of prokaryotic and viral genomes supports identification of targets for drugs against infectious disease. Of particular interest are metabolic pathways specific to microorganisms, and the proteins that participate in them. A drug affecting such a target is less likely to interact with a human homologue with consequent side effects. Proteins with sequences similar across bacterial clades offer the possibility of broad-spectrum antibiotics. Conversely, gene duplications warn of potential redundant functions, with concomitant insensitivity to inactivation of the target. Knowledge of the relative speed of evolution of different proteins, including horizontal gene transfer rates, indicates the expected stability of a therapy against development of resistant strains.

Commitment to a target by a large pharmaceutical company involves a very heavy investment of resources. The profit expected to flow from a successful drug exerts a very important influence on the choice of targets actively pursued. Analysis of the history of drugs that currently yield high profits suggests that prediction of economic returns is not a very precise science. Now, even generously supported bioinformatics efforts are much less expensive than laboratory work. The possibility that calculations will improve predictions and enhance profit is behind the espousal of bioinformatics by the pharmaceutical industry, in addition to the purely scientific contributions of bioinformatics to drug discovery. This contribution to economic forecasting is especially important when a company considers high-risk projects, such as those aimed at developing a drug against a new class of targets. Such projects must compete with lower-risk activities such as trying to improve on a competitor's success.

Prediction of a lead compound

Methods for predicting ligands suitable as lead compounds for drug discovery can be divided into inductive and deductive approaches.

Inductive methods depend on correlations between known affinities of some test set of compounds, and molecular features characterizing entire libraries of potential ligands. These features include structural properties such as size, geometry, charge distributions and specific functional groups including hydrogen-bond donors and acceptors. They include general 'drug-like' qualities such as solubility in aqueous and organic solvents, easy route of administration, appropriate distribution in body tissues and metabolic turnover rate. The relevant characteristics of compounds are compiled into a **feature vector** used to compare the overall match between compounds of known affinity and a complete library. The requirements for organization, encoding, storage and searching of information about small molecules has created a new field, **chemoinformatics**, which complements bioinformatics in applications to drug discovery.

Deductive methods are applicable if the binding site on the target protein is known or can be inferred. However, because binding affinity and specificity are only two requirements for a lead compound—admittedly essential ones—it is necessary to combine deductive methods with the correlation to desirable properties as in the purely inductive approach. Binding assays on purified systems give little idea of the behaviour of a compound as a drug in its biological context. Bioinformatics has a contribution to make in integrating the information available from molecular and cell biology, and physiology and pharmacology, to help bridge the gap between *in vitro* experiments and *in vivo* therapeutic activities.

Molecular modelling in drug discovery

A central problem in drug discovery is the identification of a compound that will bind tightly and specifically to a target protein. Tight binding is necessary for efficacy at low concentrations. Specificity is necessary to minimize side effects.

If the structure of the target is known from experiment, it is possible to apply molecular modelling directly to ligand design. If the structure of the target is unknown, a picture of the binding site must be created from indirect evidence, and ligand design is correspondingly more difficult.

Goals of molecular modelling applied to drug design include:

♦ Ideally: suggestion of a lead compound that already shows reasonable affinity and specificity. This is a rare achievement.

♦ Analysis of compounds known to bind to the target. Understanding the important interactions serves as a guide to design and testing of potential ligands, and for selecting structural features to build into combinatorial synthesis of libraries. In the

---

Medicinal chemists apply an equivalent of the duck test: only if it walks like a drug, swims like a drug and quacks like a drug, then maybe it will be a drug.

Ligand design without the target structure is like trying to catch a bank robber from eyewitness descriptions; ligand design to a target of known structure is like trying to catch the bank robber from a clear image on a CCTV videotape.

case of antibacterial or antiviral projects, a model of the protein–ligand complex can give some idea of how easy it would be for the pathogen to develop resistance by mutations that lower the affinity.

- Pharmacophore identification is the identification of common substructures of many compounds that share a pharmacological activity, or at least that bind to the same site on a protein. The hypothesis is that there is some common constellation of atoms within the structures that is responsible. The computational problem of extracting the pharmacophore from a set of compounds is similar to that of structural alignment of a set of homologous proteins. Although typical ligands are much smaller than proteins, the combinatorial problems are more severe because one has lost the linear ordering of the residues in proteins (see page 326). Inferred pharmacophore properties are integrated with QSAR methods to filter libraries of compounds for candidate ligands.

- *In silicio* screening: predicting of affinities, even qualitatively, suggests candidate ligands from a library of chemical structures. The results can either be used for setting priorities in experimental tests, or can be integrated into broader approaches to computer screening of libraries on the basis of features correlated with favourable chemical and pharmacological properties. Many readers will be aware of the harnessing of screensavers worldwide to search for potential drugs http://www.chem.ox.ac.uk/curecancer.html At present, 2.6 million computers have joined this project. They have contributed a cumulative total of over 320 000 years of CPU power.

See Box, Docking: prediction of ligand geometry and affinity

- Lead compound improvement: once a compound is identified that binds to a target protein, albeit with low affinity and specificity, interactive modelling can suggest modifications that are expected to enhance the fit. Synthesis and testing of compounds predicted to show enhanced affinity, and even solution of crystal structures of their complexes, can guide the search for improved compounds. The modelling is usually coupled with combinatorial chemistry and experimental library screening.

Docking: prediction of ligand geometry and affinity

Docking is prediction of ligand binding. In includes prediction both of binding of small molecules to proteins, and of protein–protein binding. The goals of docking are (1) to identify the binding site on the protein, and determine the position and orientation of the ligand, and (2) to estimate the affinity.

1. **Identification of the mode of binding** Docking of small molecules to proteins requires matching of the ligand to a site on a protein of known structure. The binding site may be known in advance, or it may be necessary to try many different modes of apposition of the ligand and protein to predict the optimal binding site.

$\longrightarrow$

→

Docking: prediction of ligand geometry and affinity *(continued)*

The basis for docking is the identification of complementarity in size, shape and distribution of charge, polarity and potential for hydrophobic and hydrogen-bonding interactions. A complication is the possibility of flexibility in both partners. Small organic molecules containing many single bonds have a high degree of conformational flexibility. (Drug designers love structures with rings and bridges.) Many proteins show conformational changes upon binding ligands. Therefore, the experimental structure of an unligated protein cannot be assumed to serve as a rigid target for docking. However, allowing for flexibility complicates docking calculations substantially.

Water molecules at interfaces present another difficulty. They can contribute to the surface complementarity, and provide bridging hydrogen bonds.

2. **Estimation of affinity** It is difficult to estimate absolute affinities. However, comparative docking can provide useful information about *relative* affinities. A suitable scoring function, that can predict the ranking of different ligands in approximate order of affinity, allows selectivity, and setting of priorities, in experimental testing. Such scoring schemes can be *ab initio*—based on the kinds of force fields described on page 352—or empirical. Conversely, comparative docking of one ligand to many proteins can predict the specificity of the interaction.

Compare:

Docking calculation	Information provided
1 ligand–1 protein	Mode of binding, estimate of affinity
Many ligands–1 protein	Ranking of affinities of a series of potential ligands
1 ligand–many proteins	Prediction of specificity

Docking and scoring are important steps in the filter between a total potential library and testing at the bench. A typical narrowing of the funnel might run as follows:

Overall library size	10^{12} compounds
After general filters	10^5
Docking	10^4
Scoring	10^3
Visual	10–100 for experimental testing

Two case studies illustrate the range of chemical and molecular biological techniques involved in drug development, and show some interesting similarities and contrasts. They concern well-known families of analgesic drugs—colloquially, 'painkillers'—typified by morphine and aspirin. The the two groups of compounds have different mechanisms of actions, different potencies and different spectra of side effects.

Case Study 6.1 Development of analgesic drugs based on morphine*

Morphine and codeine are natural alkaloids contained in the latex of the opium poppy (*Papaver somniferum*) (Fig. 6.22). The pharmacological effects have been known since antiquity. Modern chemistry has explored and developed many variants. Heroin was synthesized in 1874 (Fig. 6.22). More hydrophobic than the natural compounds, heroin traverses the blood–brain barrier more readily, giving it a more rapid onset of action.

   Both codeine and heroin are metabolized to produce morphine, the active form. Codeine is therefore a natural example of a *prodrug*, an inactive agent that is converted to an active one. The conversion depends on a cytochrome, CYP2D6, which is absent in 5–10% of Caucasians and 1–3% of Afro-Americans and Asians.

   Morphine and codeine have been applied in medicine and surgery as analgesics, i.e. drugs to relieve severe pain. Side effects include passivity and euphoria, and physical dependence and addiction. Drug developers have therefore long sought a compound that would relieve pain without the harmful side effects. Of course there was no guarantee that this would be possible.

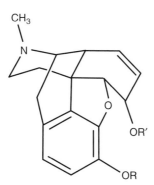

Fig. 6.22 Morphine, codeine and heroin have structures differing only in substituents at two positions:

Compound	R	R'
Morphine	—H	—H
Codeine	—CH$_3$	—H
Heroin	—COCH$_3$	—COCH$_3$

$\longrightarrow$

Case Study 6.1 *(continued)*

Synthetic variants of morphine allow correlation of biological effects with chemical structure.

One approach is to try to simplify the structure. The goals are (1) to infer the minimal pharmacophore required for activity, and (2) if possible, to dissect the parts of the structure that relieve pain, away from those causing addiction. Morphine, codeine and heroin are rigid compounds containing five fused rings. Levorphanol differs from morphine by loss of the bridging oxygen (i.e. removal of the tetrahydrofuran ring) and one of the hydroxyl groups (Fig. 6.23). It is a more potent analgesic than morphine but still addictive. Benzomorphan, cyclazocine and pentazocine break the cyclohexene ring (Fig. 6.24). The addictive effects of these compounds are less than those of morphine and levorphanol. Demerol, which opens the cyclohexene ring, and methadone, which has *no* fused rings, retain analgesic activity, sharing even smaller common substructures with morphine

From these structures, one can infer the pharmacophore shown in Fig. 6.25.

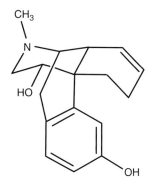

Fig. 6.23 The structure of levorphanol.

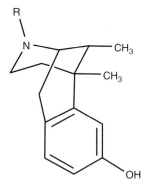

Fig. 6.24 The structures of benzomorphan: R = CH$_3$, cyclazocine: R = CH$_2$-cp (cp = cyclopropane); pentazocine: R = CH$_2$CH=C(CH$_3$)$_2$.

Case Study 6.1 *(continued)*

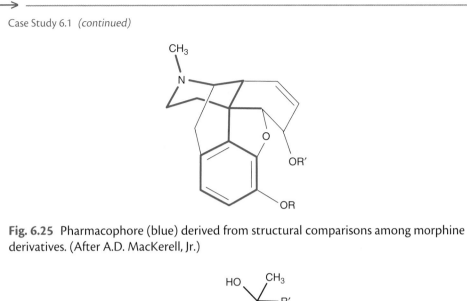

Fig. 6.25 Pharmacophore (blue) derived from structural comparisons among morphine derivatives. (After A.D. MacKerell, Jr.)

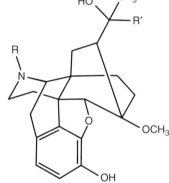

Fig. 6.26 The structures of etorphine: R = CH$_3$, R$'$ = C$_3$H$_7$; buprenorphine: R = CH$_2$-cp (cp = cyclopropane), R$'$ = t−butyl.

   In contrast to simplifying the molecule to identify a pharmacophore, attempts to enhance specificity have retained the pharmacophore but made the molecule more complex. Some success has been achieved. Etorphine and buprenorphine, discovered in the 1960s, are far more powerful analgesics than morphine (etorphine is used for sedation of large animals), and have lower addictive potential (see Fig. 6.26). Indeed, the most important clinical use of buprenorphine is in treatment of drug addiction, rather than in analgesia.

   This exploration of variants went on before the natural receptors were identified. We now know that the natural targets of action of morphine and related molecules are receptors for endogenous peptides called endorphins. These include:

β-endorphin	YGGFMTSEKSQTPLVTLFKNAIIKNAYKKGE
dynorphin	YGGFLRRIRPKLKWDNQ

and their cleavage products:

Leu-enkephalin	YGGFL
Met-enkephalin	YGGFM

$\longrightarrow$

Case Study 6.1 *(continued)*

Morphine is therefore a natural **peptidomimetic**, a non-peptide that shares a structure and activity with a peptide.

Several classes of receptors are known, including μ, κ, and δ types, and a recently discovered fourth type, called ORL-1 (ORL = opiate receptor-like). They are G-protein-coupled receptors, similar in structure to bacteriorhodopsin (see Fig. 6.6). Their sequences are about 50–70% identical at the residue level. Different ligands—natural and synthetic—have differential affinity for different receptors, and different kinetics of binding and dissociation. The natural targets of morphine are μ receptors. It is thought that μ receptors tend to be more involved in physical dependence and addiction than κ receptors, although this statement of the situation is extremely oversimplified. Nevertheless the suggestion is that an approach to producing a drug that provides analgesia with reduced side effects is to look at the *distribution* of affinities of compounds for the different types of receptor.

*Coop, A. and MacKerell, A.D., Jr. (2002). The future of opioid analgesics. *American Journal of Pharmaceutical Education*, **66**, 153–156.

Case Study 6.2 Computer-aided drug design: specific inhibitors of prostaglandin cyclooxygenase 2

Prostaglandins are a family of natural compounds that mediate a wide variety of physiological processes. Pharmacological applications include the use of prostaglandins themselves, and, conversely, drugs that block prostaglandin synthesis. Prostaglandin E_2 (dinoprostone) is used in obstetrics to induce labour. Aspirin, ibruprofen, acetaminophen (paracetamol) and other **non-steroidal anti-inflammatory drugs** (NSAIDs) are effective against arthritis and related diseases (see Box, page 378). They achieve this effect by inhibiting enzymes in the pathway of prostaglandin synthesis, specifically, prostaglandin cyclooxygenases. A well-known side effect of aspirin is bleeding from the walls of the stomach. This occurs because prostaglandins (the production of which aspirin inhibits) suppress acid secretions by the stomach and promote formation of a mucus coating protecting the stomach lining.

Aspirin and other NSAIDs inhibit *two* closely related prostaglandin cyclooxygenases, called COX-1 and COX-2. (Unfortunately the same abbreviations are used for cytochrome oxidases 1 and 2.) COX-1 is expressed constitutively in the stomach lining. COX-2 is inducible, and upregulated in response to inflammation. This suggests that a drug that would inhibit COX-2 but not COX-1 would retain the desired activity of NSAIDs but reduce unwanted side effects.

The amino acid sequences and crystal structures of COX-1 and COX-2 are known. (These proteins have 65% sequence identity.) Figure 6.27 shows part of the structure of COX-1, acetylated by the aspirin analogue 2-bromoacetoxybenzoic acid (aspirin brominated on the methyl group of the acetyl moiety). The salicylate moiety binds

$\longrightarrow$

Case Study 6.2 *(continued)*

nearby. The effect is to block the entrance to the active site. Most NSAIDs bind but do not covalently modify the enzyme.

Figure 6.28 shows the same figure with the corresponding region of COX-2 superposed. Can you see regions of structural difference, that could be clues to the design of selective drugs? Figure 6.29 shows the region of COX-2 with the selective inhibitor SC-558 (1-phenylsulphonamide-3-trifluoromethyl-5-parabromophenylpyrazole, made by Searle). From Fig. 6.30 we can see why SC-558 cannot inhibit COX-1. There would be steric clashes with the isoleucine sidechain, which corresponds to a valine in COX-2.

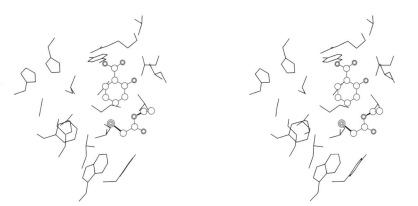

Fig. 6.27 The binding site in COX-1 for an aspirin analogue, 2-bromoacetoxybenzoic acid. The ligand has reacted with the protein, transferring the bromoacetyl group to the sidechain of serine 530. The protein is shown in skeletal representation, in black. The aspirin analogue is shown in ball-and-stick representation, in blue.

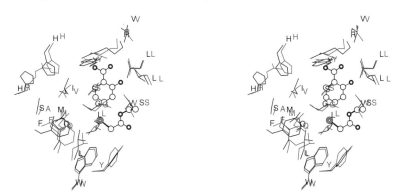

Fig. 6.28 The binding site in COX-1 for an aspirin analogue, 2-bromoacetoxybenzoic acid, in black, and the homologous residues of COX-2, in blue. Can you see what unoccupied space exists in the site that could accomodate a larger ligand? Can you see any sequence differences that might be exploited to design an inhibitor that would bind to COX-2 (blue) but *not* to COX-1 (black)?

Case Study 6.2 *(continued)*

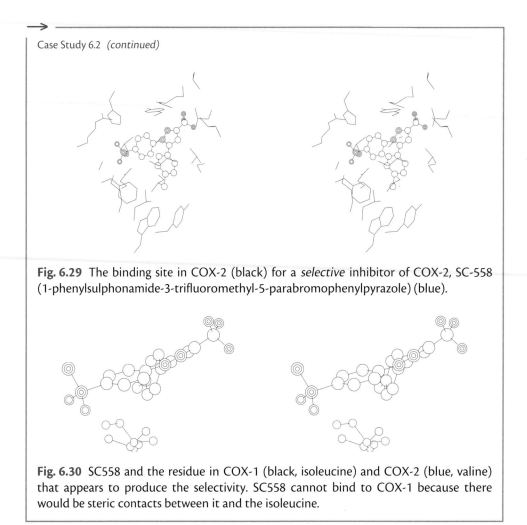

Fig. 6.29 The binding site in COX-2 (black) for a *selective* inhibitor of COX-2, SC-558 (1-phenylsulphonamide-3-trifluoromethyl-5-parabromophenylpyrazole) (blue).

Fig. 6.30 SC558 and the residue in COX-1 (black, isoleucine) and COX-2 (blue, valine) that appears to produce the selectivity. SC558 cannot bind to COX-1 because there would be steric contacts between it and the isoleucine.

Aspirin

Aspirin is one of the oldest of folk remedies and newest of scientific ones. Hippocrates noted the effectiveness of preparations of willow leaves or bark to assuage pain and reduce fever. The active ingredient, salicin, was purified in 1828, and synthesized in 1859 by Kolbe. The mechanism of its action was unknown, and indeed remained unknown until, in the 1970s, J. Vane and colleagues discovered that aspirin acts by blocking prostaglandin synthesis. Not knowing the mechanism of action was never an impediment to its use.

A century ago, sodium salicylate was used in the treatment of arthritis. Because stomach irritation was a serious side effect, F. Hoffman sought to reduce the compound's acidity by forming acetylsalicylic acid, or aspirin. Aspirin was the

→

Aspirin *(continued)*

first synthetic drug, which launched the modern pharmaceutical industry. (The name salicin comes from the Latin name for willow, *salix*, and the name aspirin comes from 'a' for acetyl and 'spir' from the *spirea* plant, another natural source of salicin.)

Aspirin has the effect of reducing fever, and giving relief from aches and pains. In high doses it is effective against arthritis. Aspirin is also used for prevention and treatment of heart attacks and strokes. The applications to cardiovascular disease depend on inhibition of blood clotting by suppressing prostaglandin control over platelet clumping. The many applications of aspirin reflect the many physiological processes that involve prostaglandins.

Aspirin's many uses

Small doses	Medium doses	Large doses
Interferes with blood clotting	Fever/pain	Reduces pain and inflammation of arthritis and related diseases

Recommended reading

Protein folding

Baldwin, R.L. and Rose, G.D. (1999). Is protein folding hierarchic? I. Local structure and peptide folding. II. Folding intermediates and transition states. *Trends in Biochemical Science* **24**, 26–32; 77–83.

Rose, G.D., Fleming, P.J., Banavar, J.R. and Maritan, A. (2006). A backbone-based theory of protein folding. *Proceedings of the National Academy of Sciences, USA*, **103**, 16623–16633. [Discussions of current thinking about protein stability and folding.]

Structure alignment and sequence–structure relationships

Holm, L. and Sander. C. (1995). Dali: a network tool for protein structure comparison. *Trends in Biochemical Science*, **20**, 478–480. [Describes DALI and its applications to structural alignment.]

Connections among sequences, structures and functions.

Das, R., Junker, J., Greenbaum, D. and Gerstein, M.B. (2001). Global perspectives on proteins: comparing genomes in terms of folds, pathways and beyond. *The Pharmacogenomics Journal*, **1**, 115–125.

Galperin, M.Y. and Koonin, E.V. (2003). *Sequence–Evolution–Function/Computational Approaches in Comparative Genomics*. Kluwer, Boston, MA.

State of the art in homology modelling and its application in structural genomics.

Guex, N., Diemand, A. and Peitsch, M.C. (1999). Protein modelling for all. *Trends in Biochemical Science*, **24**, 364–367; Schwede, T., Kopp, J., Guex, N. and Peitsch, M.C. (2003). SWISS–MODEL: an automated protein homology-modeling server. *Nucleic Acids Research*, **31**, 3381–3385. [Descriptions of SWISS–MODEL.]

Peitsch, M.C., Schwede, T. and Guex, N. (2000). Automated protein modelling—the proteome in 3D. *Pharmacogenomics*, **1**, 257–266. [What it will take to complete the structural genomics problem.]

Marti-Renom, M.A., Stuart, A.C., Fiser, A., Sánchez, R., Melo, F. and Šali, A. (2000). Comparative protein structure modeling of genes and genomes. *Annual Review of Biophysics and Biomolecular Structure*, **29**, 291–325.

Pieper, U., Eswar, N., Braberg, H., Madhusudhan, M.S., Davis, F.P., Stuart, A.C., Mirkovic, N., Rossi, A., Marti-Renom, M.A., Fiser, A., Webb, B., Greenblatt, D., Huang, C.C., Ferrin, T.E. and Šali, A. (2004). MODBASE, a database of annotated comparative protein structure models, and associated resources. *Nucleic Acids Research*, **32**, D217–D222.

Tramontano, A. (2004). Integral and differential form of the protein folding problem. *Physics of Life Reviews*, **1**, 103–127.

Tramontano, A. (2006). *Protein Structure Prediction: Concepts and Applications.* Wiley-VCH, Weinheim, Baden-Württemberg, Germany.

Other protein structure prediction methods

Bonneau, R. and Baker, D. (2001). *Ab initio* protein structure prediction: progress and prospects. *Annual Review of Biophysics and Biomolecular Structure*, **30**, 173–189. [Recent review of structure prediction methods by authors of the most successful of them.]

Structure-based drug discovery

Abraham, D.J. (2007). Structure-based drug design—a historical perspective and the future. In *Comprehensive Medicinal Chemistry II.* Mason, J.S., volume ed.; Triggle, D. and Taylor, J. eds. in chief. Vol. 4, 1–22. Elsevier, Amsterdam.

Introduction to molecular visualization

Lesk, A.M., Bernstein, H.J. and Bernstein, F.C. (2008). Molecular graphics in stuctural biology. In *Computational Structural Biology.* Peitsch, M. and Schwede, T., eds. World Scientific Publishing.

Classics still well worth reading

Kauzmann, W. (1959). Some factors in the interpretation of protein denaturation. *Advances in Protein Chemistry*, **14**, 1–63.

Richards, F.M. (1977). Areas, volumes, packing and protein structure. *Annual Review of Biophysics and Bioengineering*, **6**, 151–176.

Chothia, C. (1984). Principles that determine the structure of proteins. *Annual Review of Biochemistry*, **53**, 537–572.

Richards, F.M. (1991). The protein folding problem. *Scientific American* **264 (1)**, 54–57; 60–63.

Exercises, Problems and Weblems

Exercises

Exercise 6.1 The heat of sublimation of ice = 51 kJ mol^{-1} at the freezing point. In the solid state, each molecule of H_2O makes two hydrogen bonds. What is the energy of a single water–water hydrogen bond?

Exercise 6.2 Which pairs are orthologues, which are paralogues and which are neither?

(a) Human haemoglobin α and human haemoglobin β.

(b) Human haemoglobin α and horse haemoglobin α.

(c) Human haemoglobin α and horse haemoglobin β.

(d) Human haemoglobin α and human haemoglobin γ.

(e) The proteinases human chymotrypsin and human thrombin.

(f) The proteinases human chymotrypsin and kiwi fruit actinidin.

Exercise 6.3 On a photocopy of Plate VIII, indicate the locations in the structure that correspond to X, Y and Z in the following diagram.

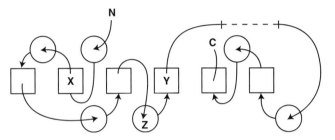

Exercise 6.4 On a photocopy of Fig. 6.11a, highlight the region of the 3_{10} helix that was not predicted to be helical.

Exercise 6.5 Which of the following shows the correct topology—correct strand order in the sequence, and orientation—of the β-sheet in Fig. 6.11b?

(a) ↑ ↑ ↑ ↑   (b) ↑ ↓ ↑ ↓   (c) ↑ ↑ ↓ ↑
    1 2 3 4   3 4 2 1   1 3 2 4

Exercise 6.6 On a photocopy of Fig. 6.9a, indicate with highlighters of two different colours the strands that form the two β-sheets.

Exercise 6.7 In the structure prediction of the *H. influenzae* hypothetical protein, Fig. 6.14: (a) What are the differences in folding pattern between the target protein and the experimental parent? (b) What are the differences in folding pattern between the prediction by A.G. Murzin and the target? (c) What are the differences in folding pattern between the prediction by A.G. Murzin and the experimental parent? In what respects is Murzin's prediction a better representation of the folding pattern than the experimental parent?

Exercise 6.8 Draw the chemical structures of aspirin and 2-bromoacetoxy benzoic acid.

Exercise 6.9 Many proteins from pathogens have human homologues. Suppose you had a method for comparing the determinants of specificity in the binding sites of two homologous proteins. How could you use this method to select propitious targets for drug design?

Exercise 6.10 In the neural networks illustrated on page 341, how many parameters —variable weights and thresholds—are available to adjust, assuming a linear decision procedure?

Exercise 6.11 What is the geometrical interpretation of a neuron that accepts two inputs x and y and 'fires' if and only if $x + 2y \geq 2$?

Exercise 6.12 Sketch a neuron with two inputs x and y, each of which may have any numerical value, that will emit 1 if and only if the value of the first input is greater than or equal to that of the second. What is the geometric interpretation of this neuron?

Exercise 6.13 Which of the following compounds would you expect to have the higher affinity for carbonic anhydrase?

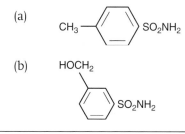

Problems

Problem 6.1 In the table of aligned sequences of ETS domains (see Problem 1.1): (a) Which are the most similar and most distant members of the family? (b) Suppose that an experimental structure is known only for the first sequence. For which others would you expect to be able to build a model with an overall r.m.s. deviation of ≤ 1.0 Å for 90% or more of the residues?

Problem 6.2 Sketch a network that accepts eight inputs, each of which has value 0 or 1, with the interpretation that the eight inputs correspond to the residues in a sequence of eight amino acids, and that the value of the ith input is 0 if the ith residue is hydrophilic and 1 if the ith residue is hydrophobic. The network should output 1 if the pattern appears helical—for simplicity demand that it be PPHHPPHH where H = hydrophobic (uncharged) and P = polar or charged—and 0 otherwise.

Problem 6.3 Write a more reasonable set of patterns to identify helices from the hydrophobic/hydrophilic character of the residues in a 10-residue sequence. Your patterns might include 'wild cards'—positions that could be either hydrophobic or hydrophilic, or correlations between different positions. Generalize the previous problem by sketching neural networks to detect these more complex patterns.

Problem 6.4 We, and computers, can do logic with arithmetic. Define: 1 = TRUE and 0 = FALSE. Sketch simulated neurons with two inputs, each of which can have only the values 0 or 1, and a linear decision process for firing, for which (a) the output is the logical AND of the inputs and (b) the output is the logical OR of the two inputs. (c) What is the simplest neural network, with each neuron having a linear decision process for firing, that produces as its output the EXCLUSIVE OR of the two inputs (the exclusive or is true if either one of the inputs is true, false if neither or both inputs are true). Can this be done with a single layer? If not, what is the minimum number of layers in the network required?

Problem 6.5 Modify the PERL program for drawing helical wheels (pages 319–20) so that different amino acids appear in different colours, as follows: GAST, cyan; CVILFYPMW, green; HNQ, magenta, DE, red; and KR blue.

Problem 6.6 Hydrophobic cluster analysis. Suppose a region of a protein forms an α-helix. To represent its surface, imagine winding the sequence into an α-helix (even if in fact it forms a strand of sheet or loop in the native structure). Then 'ink' the surface of the helix, and roll it onto a sheet of paper, to print the names of the residues. By rolling the helix over twice, all surfaces are visible.

From such a diagram, hydrophobic patches on surfaces of helices can be identified. In this way it is possible to try to predict which regions of the sequence actually form helices in the native structure. Comparisons of hydrophobic clusters can also be used to detect distant relationships.

Write a PERL program to produce such diagrams.

Problem 6.7 In the 2000 Critical Assessment of Structure Prediction (CASP4), one of the targets in the category for which no similar fold was known was the N-terminal domain of the human DNA end-joining protein Xrcc4, residues 1–116.

The secondary structure prediction by B. Rost, using the method PROF: profile-based neural network prediction, is as follows (An H under a residue means that the residue is predicted to be in a Helix, an E means that the residue is predicted to be in an Extended conformation, or strand, and a dash means Other):

```
                 1         2         3         4         5         6
                 0         0         0         0         0         0
Sequence     MERKISRIHLVSEPSITHFLQVSWEKTLESGFVITLTDGHSAWTGTVSESEISQEADDMA
Prediction   ---EEEEEEE----HHHHHH-HHHHHHH--EEEEEE-------EE---HHHHHHHHHHHH

                                       1         1
                 7         8         9 0         1
                 0         0         0 0         0
Sequence     MEKGKYVGELRKALLSGAGPADVYTFNFSKESCYFFFEKNLKDVSFRLGSFNLEKV
Prediction   HHH-HHHHHHHHHHHH-----EEEEEE-----EEEEE------EEEE-----HHHH
```

The experimental structure of this domain, released after the predictions were submitted (PDB entry [1FU1]) is shown here:

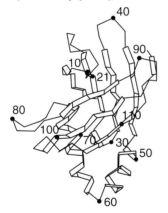
HUMAN XRCC4 [1fu1] domain1

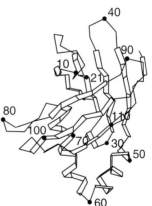
HUMAN XRCC4 [1fu1] domain1

The secondary structure assignments from the PDB entry are:

Secondary structure	Residue ranges
Helix	27–29, 49–59, 62–75
Sheet 1	2–8, 18–24, 31–37, 42–48, 114–115
Sheet 2	84–88, 95–101, 104–111

(a) Calculate the value of Q3, the percentage of residues correctly assigned to helix (H), strand (E) and other (−).

(b) On a photocopy of the picture of XRCC4, highlight, in separate colours, the regions *predicted* to be in helices and strands.

(c) From the result of (b): how many predicted helices overlap with helices in the experimental structure? How many strands overlap with strands in the experimental structure?

Problem 6.8 In CASP4 the group of Bonneau, Tsai, Ruczinski and Baker made a prediction of the full three-dimensional structure of protein XRCC4, residues 1–116. The secondary structure prediction derived from their model is as follows (H = helix, E = strand (extended), − = other):

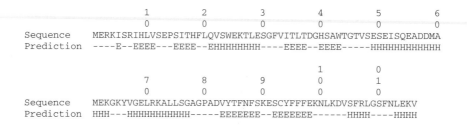

```
                 1         2         3         4         5         6
                 0         0         0         0         0         0
Sequence      MERKISRIHLVSEPSITHFLQVSWEKTLESGFVITLTDGHSAWTGTVSESEISQEADDMA
Prediction    ----E--EEEE---EEEE--EHHHHHHHH----EEEE--EEEE-----HHHHHHHHHHHH

                           1         0
                 7         8         9         0         1
                 0         0         0         0         0
Sequence      MEKGKYVGELRKALLSGAGPADVYTFNFSKESCYFFFEKNLKDVSFRLGSFNLEKV
Prediction    HHH---HHHHHHHHHHH-----EEEEEEE--EEEEEEE------HHHH----HHHH
```

(a) What is the value of Q3 for this prediction? (b) In this case, which method gives the better results, as measured by Q3, for the prediction of secondary structure: the neural network that produces only a secondary structure prediction, or a prediction of the full three-dimensional structure.

Problem 6.9 Write a PERL program that implements the neural network shown on page 341, bottom.

Problem 6.10 Suppose that you are trying to evaluate, using a threading approach, whether a sequence of length M is likely to have the folding pattern of a protein of known structure of length N > M. (a) How many different possible alignments of the sequences are possible. (b) Suppose that half the residues of the known protein form α-helices, and no gaps within helical regions are permitted. How many different possible alignments of the sequences are now possible? (c) How many alignments are there, under each of these assumptions, if N = 200 and M = 150?

Problem 6.11 Write a PERL program to calculate approximate values of π by a Monte Carlo method, as follows: the square in the plane with corners at (0, 0), (1, 0), (0, 1) and (1, 1) has area 1. Compute a series of *pairs* of random numbers (x, y) in the range [0, 1] to generate points distributed at random in this square. Count the number of points that lie within a circle of radius 0.5 inscribed in the square. The ratio of the number of points that fall within the circle to the total number of points = the ratio of the area of the circle to the area of the square = π/4.

Determine the average relationship between the number of points chosen and the number of correct digits in the calculated value of π. Estimate the number of points required to determine π correctly to 50 decimal places.

Problem 6.12 To convert the output of a neuron from a step function to a smooth function (see page 342), one can replace a statement of the form 'Let X be some weighted sum of the inputs; then output 1 if $X > 0$, else output 0' to 'Let X be some weighted sum of the inputs; then output $1/(1 + e^{-X})$.' (a) Verify that as $X \rightarrow -\infty, 1/(1 + e^{-X}) \rightarrow 0$, as $X \rightarrow +\infty, 1/(1 + e^{-X}) \rightarrow 1$, and that if $X = 0, 1/(1 + e^{-X}) = 0.5$. (b) Suppose the network for determining whether a point lies within a triangle (page 341) is so altered, so that the output of each neuron is described by the smooth function $1/(1 + e^{-X})$ rather than a step function, and that a point is considered inside the accepted area if the output of the network is > 0.5. Write a PERL program to determine what area is then defined.

Problem 6.13 The pollen antigen from Western ragweed *Ambrosia psilostachya* (SWISS-PROT ID MPA5A_AMBPS) is a 77-residue protein with the sequence:

```
MNNEKNVSFEFIGSTDEVDEIKLLPCAWAGNVCGEKRAYC
CSDPGRYCPWQVVCYESSEICSQKCGKMRMNVTKNTI
```

A BLAST search in the non-redundant protein sequence databank produced the following hits:

```
                                                            Score      E
Sequences producing significant alignments:                (Bits)   Value

sp|P43174|MP5A_AMBPS  Pollen allergen Amb p 5a precursor (Amb ...    142    8e-33
gb|AAA20067.1|  Amb p V allergen                                    140    2e-32
sp|P43175|MPA5B_AMBPS  Pollen allergen Amb p 5b precursor (Amb...    116    3e-25
gb|AAA20066.1|  Amb p V allergen                                    115    5e-25
gb|AAA20068.1|  Amb p V allergen                                    115    1e-24
sp|P02878|MPA5_AMBEL  Pollen allergen Amb a 5 (Amb a V) (Allergen   81.3   2e-14
sp|P10414|MPAT5_AMBTR  Pollen allergen Amb t 5 precursor (Amb ...   42.4   0.008
```

The first six 'hits' have E-values substantially less than 1.0. These proteins can be confidently taken to be homologous to the probe sequence.

The last 'hit', with an E-value of 0.008, is a likely homologue, a pollen antigen from a closely related plant: ragweed pollen allergen from giant ragweed *Ambrosia trifida* (SWISS-PROT ID MPAT5_AMBTR). Although the similarity of the sequences is above Doolittle's 'twilight zone', the E-value suggests that there is almost a 1% chance of finding a sequence, with this degree of similarity to the probe sequence, at random.

What can we do to try to confirm a true relationship? The structure of the mature form of the *A. trifida* protein, corresponding to the C-terminal 40 residues of that sequence, is known (PDB entry 1BBG). In the full alignment of the sequences, uppercase letters indicate the portion of the sequence that corresponds to the mature protein, and which appears in the structure; and the letter B underneath the blocks indicates the residues buried within the structure (computed from coordinate set 1BBG):

```
P43174|MPA5A_AMBPS  mnne---------knvsfefigstdevdeikllP--CAWAGNVCGEKRAYCCSDPGRYCP 49
P10414|MPAT5_AMBTR  mknifmltlfiliitstikaigstnevdeikqeDDGLCYEGTNCGKVGKYCCSPIGKYC- 59
                    *:*          . :::  ****:******     .: *. **:    **** *:**
                                                                      B   BB

P43174|MPA5A_AMBPS  WQVVCYESSEICSQKCGkmrmnvtknti 77
P10414|MPAT5_AMBTR  ---VCYDSKAICNKNCT----------- 73
                       ***:*. **.::*
                         B
```

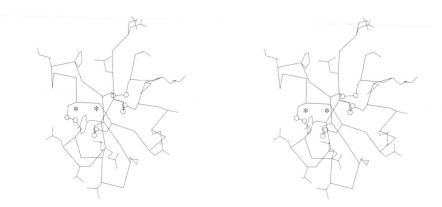

Fig. 6.31 *Ambrosia trifida* pollen antigen. Sidechains shown are those that differ from pollen antigen of *A. psilostachya*

These two sequences share the same residue at 28 positions. From the structure, the following pairs of cysteines form disulphide bridges: 5–35, 11–26, 18–28, 19–39.

Figure 6.31 shows the structure of the mature fragment of the Giant ragweed (*A. trifida*) antigen, including the putative disulphide bridges. Sidechains corresponding to positions of mutations are shown in blue. The site of the insertion in the *A. psilostachya* sequence is marked by an asterisk.

(a) Does the overall extent of sequence similarity suggest that the proteins are homologous? (b) On a photocopy of Fig. 6.31 mark the N- and C- termini. (c) On a photocopy of Fig. 6.31 write next to the sidechain of each mutated residue the one-letter code of the amino acid that appears in the parent sequence. (d) Is the site of insertion in a loop between two elements of secondary structure? (e) Consider each of the mutations. Which are easy to reconcile with a conservation of structure and which are difficult to reconcile with a conservation of structure? (f) Was MODBASE able to construct a model of the parent sequence? (This will require checking a web site.)

Weblems

Weblem 6.1 The bacterium *Pseudomonas fluorescens* and the fungus *Curvularia inaequalis* each possess a chloroperoxidase, an enzyme that catalyzes halogenation reactions. Do these enzymes have the same folding pattern?

Weblem 6.2 Check the prediction of transmembrane helical segments in bacteriorhodopsin from the secondary structure assignments in the experimental structure in the Protein Data Bank.

Weblem 6.3 Plate VIII showed the structure of a thiamin-binding domain, identified by M. Gerstein as one of the five most common folding patterns appearing in archaea, bacteria and eukarya. Using facilities available in SCOP, draw pictures of the four other structures.

Weblem 6.4 Using either the results of Weblem 6.3, or pictures available in *Introduction to Protein Architecture: The Structural Biology of Proteins*, draw simplified topology diagrams analogous to the one in Plate VIII for the other four structures.

Weblem 6.5 Does the human θ_1 globin gene encode an active globin? Or is it really a pseudogene? Send the amino acid sequence of human θ_1 globin to SWISS-MODEL, including a request for a WhatCheck report on the result. What can you conclude from the result about the status of human θ_1 globin?

Weblem 6.6 Compare the number of entries in SCOP, in different categories, listed on page 332, with the number that SCOP contains now.

Weblem 6.7 Align the sequences of human neutrophil elastase and *C. elegans* elastase. (a) In the optimal alignment, how many identical residues are there? (b) Would it be reasonable to build a model of *C. elegans* elastase starting from the structure of human neutrophil elastase?

Weblem 6.8 Align the sequences of γ-chymotrypsin and *S. aureus* epidermolytic toxin A using pairwise sequence alignment methods. Compare the result with the structural alignment shown in the text.

Weblem 6.9 S. Chakravarty and K.K. Kannan solved the structures of carbonic anhydrase with a benzenesulphonamide ligand (Protein Data Bank entry 1CZM.) Draw pictures of the binding site showing the nature of the interactions between the protein and ligands. Describe the nature of the interactions in terms of the conclusions drawn from the QSAR analysis.

Weblem 6.10 MARCOIL predicts with high confidence that residues 164–191 of chicken c-fos form a coiled-coil. Residues in coiled-coil structures are assignable to positions abcdefg in successive seven-residue oligopeptides (see page 322). What assignment of residues 164–191 in chicken c-fos does MARCOIL predict?

Weblem 6.11 Submit the sequence of chicken c-fos to MARCOIL http://www.isrec.isb-sib.ch/webmarcoil/webmarcoilC1.html and COILS http://www.ch.embnet.org/software/coils/COILS_doc.html. Compare the predictions on a residue-by-residue basis. For how many residues do the predictions agree (to within 10% probability)?

Weblem 6.12 Chicken c-fos was used as an example of coiled-coil prediction by MARCOIL. In order to get some idea of how challenging the prediction was:

(a) Identify the protein of known structure with amino acid sequence most similar to that of chicken c-fos.

(b) Determine from inspection of the structure the residues in that structure that form a coiled-coil.

(c) Draw a dotplot of the two sequences. Indicate on the dotplot the regions in the protein of known structure that form a coiled-coil, and the regions in chicken c-fos that are predicted by MARCOIL to form a coiled-coil.

(d) Align the two sequences and report the percentage identical residues in the two complete sequences.

(e) Report the percentage identical residues in the region that forms a coiled-coil in the protein of known structure.

Weblem 6.13 M. Vihinen curates a database of elastase mutants, ELA2base http://bioinf.uta.fi/ELA2base/ [Thusberg, J. and Vihinen, M. (2006). Bioinformatic analysis of protein structure–function relationships: case study of leukocyte elastase (ELA2) missense mutations. *Human Mutation*, **27**, 1230–1243. J. Moult, P. Yue, E. Melamud and M. Zuhl created a database to evaluate the effects on protein structure of human SNPs http://www.snps3d.org See also Yue, P. and Moult, J. (2006). Identification and analysis of deleterious human SNPs. *Journal of Molecular Biology*, **356**, 1263–1274.]

(a) For the single amino acid replacements in ELA2base, draw a picture of the human neutrophil elastase structure [1HNE] showing the positions of the residues. How would you describe the distribution in the structure of the disease-causing substitutions? (b) Submit each single amino acid replacement in ELA2base to SNPS3d, and tabulate the scores returned. (c) Can you correlate the scores with the distribution of residues in the structure? (d) Can you correlate the scores with the severity of the symptoms? (e) The following elastase mutations are associated with cyclic neutropenia: A32V, L177F, R191Q. The following elastase mutations are associated with severe congenital neutropenia (non-cyclic): P110L, V72M, S97L. Plot these residues on the structure. Can you see a systematic difference between the locations within the structure of the two different phenotypes? The residue numbers in this part (e) refer to the following sequence:

```
  1   MTLGRRLACL FLACVLPALL LGGTALASEI VGGRRARPHA WPFMVSLQLR    50
 51   GGHFCGATLI APNFVMSAAH CVANVNVRAV RVVLGAHNLS RREPTRQVFA   100
101   VQRIFENGYD PVNLLNDIVI LQLNGSATIN ANVQVAQLPA QGRRLGNGVQ   150
151   CLAMGWGLLG RNRGIASVLQ ELNVTVVTSL CRRSNVCTLV RGRQAGVCFG   200
201   DSGSPLVCNG LIHGIASFVR GGCASGLYPD AFAPVAQFVN WIDSIIQRSE   250
251   DNPCPHPRDP DPASRTH    267
```

Weblem 6.14 The structure of the SH3 domain from the cytoskeletal protein α-spectrin from chicken has been determined both by X-ray crystallography [1SHG] and by NMR [1AEY]. The crystallographic result contains one set of coordinates, containing residues 6–62. The NMR result contains 15 sets of coordinates, each containing residues 5–62. For every pair of structures (15 NMR structures and one X-ray crystal structure), compute and tabulate the root-mean-square deviation of (a) the Cα atoms, (b) all mainchain (N, Cα, C and O) atoms and (c) all atoms. (d) Describe your results. Do you see any systematic differences between X-ray and NMR structures?

Proteomics and systems biology

Learning goals

1. Understanding the goals of proteomics—the measurement of amounts and distributions of proteins within a cell or organism.

2. Becoming familiar with the data derivable from microarrays and their application to inferring and interpreting similarities and differences in gene expression patterns. Understanding the relationship between typical 'raw' microarray data (see for instance Plate IX) and the gene expression table.

3. Understanding the applications of mass spectrometry to analysis of mixtures of proteins, to partial protein sequencing and to high-throughput nucleic acid sequencing and searching for variant genetic sequences.

4. Appreciating a trend towards a new point of view: the theme of systems biology is integration.

5. Recognizing the distinction between physical and logical networks.

6. Understanding the general features of graphs, including the distinction between undirected, directed and labelled graphs. Understanding the representation of networks by graphs.

7. Appreciating the different possible kinds of dynamic states of networks.

8. Understanding the characteristics of protein–protein and protein–nucleic acid complexes.

9. Understanding the structure and some of the building blocks of regulatory networks.

Proteomics is the study of the distribution and interactions of proteins in time and space in a cell or organism. High-throughput experimental methods of data analysis, including microarray analysis and mass spectrometry, are giving us a large-scale picture of the protein economy in living things.

The goal of systems biology is the synthesis of proteomic, genomic and other data into an integrated picture of the structure, dynamics, logistics and, ultimately, the logic of living things. For any protein in a cell, a systems biologist will combine study of the protein, its gene, the molecules that control its expression or its activity once expressed, and the set of other proteins with which it interacts. A systems biologist will assemble into a metabolic network the chemical reactions catalysed by the enzymes of a cell, and assemble into control networks the mechanisms that regulate their activities and expression.

DNA microarrays

See Box:
Summary of applications
page 54–5.

DNA microarrays analyse the mRNAs in a cell, to reveal the expression patterns of proteins; or genomic DNAs, to reveal absent or mutated genes.

1. **For an integrated characterization of cellular activity, we want to determine what proteins are present, where and in what amounts** To determine the expression pattern of a cell's genes, we measure the relative amounts of many different mRNAs. Hybridization is an accurate and sensitive way to detect whether any particular nucleic acid sequence is present. The key to high-throughput analysis is to run many hybridization experiments in parallel.

2. **Knowing the human genome sequence can help to identify genes associated with propensities to diseases** Some diseases, such as cystic fibrosis, arise from mutations in single genes. For these, isolating a region by classical genetic mapping can lead to pinpointing the lesion. Other diseases, such as asthma, depend on interactions among many genes, with environmental factors as complications. To understand the aetiology of multifactorial diseases requires the ability to determine and analyse multiple expression patterns of genes, which may be distributed around different chromosomes.

DNA microarrays, or DNA chips, are devices for checking a sample *simultaneously* for the presence of many sequences.

The basic idea is this: to detect whether one oligonucleotide has a particular known sequence, test whether it can bind to an oligonucleotide with the complementary sequence (a 'one-to-one' test). To detect the presence or absence of a query oligonucleotide in a mixture, spread the mixture out, and test each component of the mixture for binding to the oligonucleotide complementary to the query (a 'many-to-one' test). This is a northern or Southern blot. To detect the presence or absence of *many* oligonucleotides in a mixture, synthesize a set of oligonucleotides, one complementary to each sequence of the query list, and test each component of the mixture for binding to each member of the set of complementary oligonucleotides (a 'many-to-many' test). Microarrays provide an efficient, high-throughput way of carrying out these tests in parallel.

To achieve parallel hybridization analysis, a large number of DNA oligomers are affixed to known locations on a rigid support, in a regular two-dimensional array. The mixture to be analysed is prepared with fluorescent tags, to permit detection of the hybrids. After exposing the array to the mixture, each element of the array to which some component of the mixture has become attached bears the tag. Because we know the sequence of the oligomeric probe in each spot in the array, measurement of the *positions* of the probes identifies their sequences. This analyses the components present in the sample.

DNA microarrays are distributed on a small wafer of glass or nylon, typically 2 cm square. Oligonucleotides are attached in a square array, at densities between 10 000 and 250 000 positions per cm^2. The spot size may be as small as ~150 μ in diameter. The grid is typically a few centimetres across. A **yeast chip** contains 6000 oligonucleotides, covering all known genes of *Saccharomyces cerevisiae*. A DNA array, or DNA chip, may contain 400 000 probe oligomers. Note that this is larger than the total number of genes even in higher organisms (excluding immunoglobulin genes).

To analyse a mixture, expose it to the microarray under conditions that promote hybridization, then wash away any loose probe. To compare two sets of oligonucleotides, tag the samples with differently coloured fluorophores (Plate IX). Scanning the array collects the data in computer-readable form.

Different types of chips designed for different investigations differ in the types of DNA immobilized.

1. In an **expression chip**, the immobilized oligonucleotides are cDNA samples, typically 20–80 bp long, derived from mRNAs of known genes. The target sample might be a mixture of mRNAs from normal or diseased tissue.

2. In **genomic hybridization**, one looks for gains or losses of genes or changes in copy number. The target sequences, fixed on the chip, are large pieces of genomic DNA, from known chromosomal locations, typically 500–5000 bp long. The probe mixtures contain genomic DNA from normal or disease states. For instance, some types of cancer arise from chromosome deletions, which can be identified by microarrays.

3. In **mutation microarray analysis** one looks for patterns of single-nucleotide polymorphisms (SNPs).

> The immobilized material on the chip is the *probe*. The sample tested is the *target*.

Microarray data are quantitative but imprecise

Microarrays are capable of comparing concentrations of probe oligonucleotides. This allows investigation of responses to changed conditions. However, the precision is low. Moreover, mRNA levels, detected by the array, do not always quantitatively reflect protein levels. Indeed, usually mRNAs are reverse-transcribed into more stable cDNAs for microarray analysis; the yields in this step may also be non-uniform. Microarray data are therefore semi-quantitative, in that distinction between presence and absence is possible, determination of relative levels of expression in a controlled experiment is more difficult, and measurement of absolute expression levels is beyond the capacity of current microarray techniques.

Web resources   Microarray databases

Microarrays provide another high-throughput stream of data production in bioinformatics. A standard called MIAME (Minimum Information About a Microarray Experiment) describes the contents and format of the information to be recorded in the experiment, and deposited.

Major publicly available microarray databases include:

The European Bioinformatics Institute hosts a database, ArrayExpress:
http://www.ebi.ac.uk/arrayexpress/

The US National Center for Biotechnology Information hosts the Gene Expression Omnibus database:
http://www.ncbi.nlm.nih.gov/geo/

The Stanford Microarray Database:
http://genome-www5.stanford.edu/MicroArray/SMD/

A listing of microarray databases for plants appears in:
http://www.univ-montp2.fr/~plant_arrays/databases.html

For additional lists, see:
Butte, A. (2002). The use and analysis of microarray data. *Nature Reviews Drug Discovery*, **1**, 951–960.
Penkett, C.J. and Bähler, J. (2004). Getting the most from public microarray data. *European Pharmaceutical Review*, **1**, 8–17.

Analysis of microarray data

The raw data of a microarray experiment are displayed as an image, in which the colour and intensity of the fluorescence reflect the extent of hybridization to alternative probes. The two sets of probes are tagged with red and green fluorophores. If only one probe hybridizes, the spot appears red; if only the other probe hybridizes, the spot appears green. If both hybridize, the colour of the corresponding spot appears red + green = yellow (see Plate IX).

The initial goal of data processing is a **gene expression table.** This is a matrix in which the rows correspond to different genes, and the columns to different samples. Different spots in a microarray pattern such as that shown in Plate IX correspond to different genes. For each gene, results from different sets of samples appear in the red or green channel, respectively (or neither, or both). There is extensive redundancy in the oligonucleotides in a microarray—each gene may be represented by several spots, corresponding to different regions of the gene sequence; inclusion of controls with a deliberate mismatch allows data verification. Typically one gene may correspond to ~30–40 spots.

The samples may vary according to experimental conditions and/or physiological states, or they may be extracted from different individuals, or different tissues or developmental stages.

The process of data reduction to produce the gene expression matrix involves many technical details of image processing, checking internal controls, dealing with missing data, selecting reliable measurements, and putting the results of different arrays on consistent scales. The derived gene expression table indicates *relative* expression levels. A change in expression levels of a gene between two samples by a factor $\geq 1.5\text{--}2$ is generally considered significant.

Extraction of reliable biological information from a gene expression table is not straightforward. Despite extensive internal controls, there is considerable noise in the experimental technique. In many cases, variability is inherent within the samples themselves. Microorganisms can be cloned; animals can be inbred to a comparable degree of homogeneity. However, experiments using RNA from human sources—for example, a set of patients suffering from a disease and a corresponding set of healthy controls—are at the mercy of the large individual variations that humans present. Indeed, inbred animals, and even apparently identical eukaryotic tissue culture samples, show extensive variability.

Another intrinsic disadvantage—and a severe one—in interpreting gene expression data is the fact that the number of genes is much larger than the number of samples. Computationally, we are trying to understand the relationship of a space of very many variables (the genes) to a space of observations (the phenotype), from only a few measured points (the samples). The sparsity of the observations does not give us anywhere near adequate coverage. Statistical methods bear a heavy burden in the analysis to give us confidence in the significance of our conclusions.

Two general approaches to the analysis of a gene expression matrix involve (1) comparisons focused on the genes, i.e. comparing distributions of expression patterns of different genes by comparing *rows* in the expression matrix, or (2) comparisons focused on samples, i.e. comparing expression profiles of different samples by comparing *columns* of the expression matrix:

1. **Comparisons focused on genes: how do gene expression patterns vary among the different samples?** Suppose a gene is known to be involved in a disease, or to a change in physiological state in response to changed conditions. Other genes co-expressed with the known gene may participate in related processes contributing to the disease or the change in state. More generally, if two rows (two genes) of the gene expression matrix show similar expression patterns across the samples, this suggests a common pattern of regulation, and possibly some relationship between their functions, including but not limited to a possible physical interaction.

2. **Comparisons focused on samples: how do samples differ in their gene expression patterns?** A consistent set of differences among the samples may characterize the classes which the samples represent. If the samples are from different controlled groups (e.g., diseased and healthy animals), do samples from different groups show consistently different expression patterns? If so, given a novel sample, we can assign it to its proper class on the basis of its observed gene expression pattern.

How then do we measure the similarity of different rows or columns? Each row or column of the expression matrix can be considered as a vector, in a space of many dimensions. The row-vectors (a row corresponds to a gene), each entry of which refers to the same gene in different samples, has as many elements as there are samples. The column-vectors (a column corresponds to a sample), each entry of which refers to a different gene in a single sample, have as many elements as there are genes reported. It is possible to calculate the 'angle' between different row-vectors, or between different column-vectors, to provide a measure of their similarities. It is then natural to ask whether subsets of the points form natural clusters—points with high mutual similarity—characterizing either sets of genes or sets of samples.

Depending on the origin of the samples, what is already known about them and what we want to learn, data analysis can proceed in different ways.

1. The simplest case is a carefully controlled study, using two different sets of samples *of known characteristics*. For instance, the samples might be taken from bacteria grown in the presence or absence of a drug, from juvenile or adult fruit flies, or from healthy humans and patients with a disease. We can focus on the question, what differences in gene expression pattern characterize the two states? Can we design a classification rule such that, given another sample, we can assign it to its proper class? This would be applicable in diagnosis of disease. For instance, determination of the subtype of a leukaemia permits more accurate treatment and prognosis. Subject to the availability of adequate data, such an approach can be extended to systems of more than two classes.

   Computationally, training such a classification algorithm is called 'supervised learning'. The expression pattern of each sample is given by a vector corresponding to a single column of the matrix. This corresponds to a point in a many-dimensional space—as many dimensions as there are genes. In favourable cases, the points may fall in separated regions of space. Then a scientist, or a computer program, will be able to draw a boundary between them. In other cases, separation of classes may be more difficult. Consider the distribution of football players during a match. At the start of play, a line drawn across the midfield separates the teams; that is, the midfield line divides the field into two regions, each region containing exclusively the players of one of the teams. During play, the teams become commingled, and it is impossible to draw a single line that divides the field into regions that separate the teams.

2. In a different experimental situation, we might not be able to *pre-assign* different samples to different categories. Instead, we hope to extract the classification of samples from the analysis. The goal is to cluster the data to *identify* classes of samples and the differences between the genes that characterize them.

   Many clustering algorithms have been applied to microarray data, including those that try to work out simultaneously both the number of clusters and the boundaries between them. All algorithms must face the difficulty arising from the sparsity of sampling of the very high-dimensional space of the measurement. Sometimes it is possible to simplify the problem by identifying a small number of combinations of genes that account for a large portion of the variability. This is called **reduction of dimensionality** (see Box, and compare with discussion of odour classification by neural networks, Chapter 3).

Reduction of dimensionality

The distribution of gene expression data in a space of a large number of dimensions means that (1) coverage of the space with a limited number of samples is sparse and (2) it is difficult to visualize the distribution of sample points. In some cases, the distribution may depend primarily on fewer equivalent variables, and it is very advantageous to find them and transform the data accordingly.

A simple example illustrates the basic idea. Consider a distribution of groups of people picnicking on a beach. Represent the position of each person by the x, y and z coordinates of the tip of his or her nose. Take the x-axis to be parallel to the shoreline the y-axis perpendicular to the shoreline, and the z-axis vertical. Obviously height is irrelevant: this is really a two-dimensional, not a three-dimensional, distribution. To cluster the people into groups (perhaps families, or surfing clubs) the x and y coordinates carry all the significant data, and the z-coordinate carries only irrelevant information, such as the heights of the people and whether or not they are standing up or sitting on the sand. In this case, to reduce the dimensionality from three to two we need only ignore the z-coordinate. (Indeed, if the tide comes in and the beach area becomes narrower, the dimension along the shoreline carries the bulk of the information and the dimensionality could be further reduced from two to one.)

Now suppose that groups of people are climbing a vertical rock face rising parallel to the shoreline. This also is really a two-dimensional, not a three-dimensional, distribution, but in this case it is the x and z coordinates that carry the information.

In more complex cases, reduction in dimension requires more than simply picking coordinates to ignore. Suppose the people are distributed on a ski slope. To reduce the distribution from three to two dimensions, we could not simply ignore a coordinate, but would have to *project* the data, onto the oblique plane parallel to the slope. This idea of *projection* of the data onto a lower-dimensional space that contains the important components of the variation, is the key to the methods.

Practical problems of data analysis are harder than these simple illustrations. For one thing, the starting dimensions are much higher than three and the reduction in dimensionality is potentially much greater. For another, it is not obvious how to achieve the dimensionality reduction because we don't have the easily visualizable picture of the physical space and the distribution of people on a beach, rock face or ski slope.

Nevertheless, the questions to be answered remain: along what directions should we project the data to retain the largest discrimination using the fewest

$\longrightarrow$

→

Reduction of dimensionality (continued)

dimensions? Mathematical methods known as principal component analysis (PCA) using the singular value decomposition (SVD) can solve this problem. These methods automatically select a new coordinate system that best represents the variability of the data along the fewest axes, and, for each new coordinate axis, the calculation gives a measure of the contribution of that coordinate to accounting for the overall variability of the data.

Although two dimensions may well not contain all important components of the variation, we can always pick the best two-dimensional projection and plot the result on a graph; this has the immense advantage of allowing scientists to stare at the data and think about them. (Three-dimensional distributions can also be represented visually, with somewhat greater difficulty.)

Case Study 7.1 Interpretation of microarray data — regulation of genes by BRCA1 and implications for the role of BRCA1 dysfunction or silencing in tumour development

The *BRCA1* gene encodes a tumour suppressor. It is mutated in approximately 90% of patients with familial predisposition to breast and ovarian cancer. A single defective *BRCA1* allele is sufficient to increase risk, for in any cell the normal copy of the gene may be lost, or, in a small fraction of cases, rendered inactive by promotor methylation.

BRCA1 is an 1863-residue protein. It has an N-terminal ring finger domain, followed by a predicted helical coiled-coil region, followed by two tandem BRCT domains, that bind other proteins, and also regulate transcription. (BRCT abbreviates BRCA C-terminal domain.)

BRCA1 interacts with many other proteins to form functional complexes and is thereby involved in several different activities, including:

- Sensing and signalling of lesions in DNA BRCA1 responds to several types of DNA damage — for instance double-strand breaks — and activates repair mechanisms appropriate to each.

- Preserving chromosome structure Chromosome integrity may suffer *as a consequence of* inaccurate repair of DNA damage. These functions are related.

- Mediating checkpoint tests at points in the cell cycle, in part at least by regulating transcription of genes encoding proteins involved in checkpoint enforcement.

A unifying idea about *BRCA1* is that the protein encoded mediates responses to DNA damage by eliciting repair mechanisms and, in case repair is unsuccessful,

→

Case Study 7.1 *(continued)*

checkpoint mechanisms that stop cells with unrepaired damage from propagating. Loss of BRCA1 function leads to the accumulation of damaged DNA in cells, enhancing the chances of transition to a cancerous state.

The variety and complexity of the processes involving BRCA1 make it difficult to sort out the detailed mechanism of its relationship to cancer:

1. Is tumour formation a direct consequence of loss of one or more functions of BRCA1 and its interacting partners? If so, which one(s)?

2. What is the importance of transcriptional regulation—of *BRCA1* by products of other genes, of other genes by BRCA1, or both? To what extent do changing expression patterns involving *BRCA1* lead *indirectly* to tumorigenesis? We shall see that the distinction between direct and indirect effects is not really a hard and fast one: BRCA1 binds directly to some of the proteins the expression of which it regulates.

3. DNA repair mechanisms are common to many types of cells. Why does BRCA1 dysfunction or silencing specifically lead to increased risk of cancers of the breast and ovary (and other epithelial tissues, including pancreas and prostate)?

One function of BRCA1 is control over transcription. In order to investigate the regulatory context of the relationship of *BRCA1* to cancer risk, Welcsh *et al.** used microarray analysis to compare the expression patterns of genes in cells producing high and low levels of BRCA1, using a cell line in which BRCA1 expression was selectively inducible. The chip used for detection of the response contained oligonucleotides representing ~6800 human genes. (Note that this is a relatively small fraction of the total human proteome.)

The results implicated 373 genes, differentially expressed by significant and reproducible amounts in response to higher levels of BRCA1 expression. Standing out among these were 57 upregulated genes and 15 downregulated genes, for which expression levels changed by factors ≥ 2. These candidates for involvement in functions of BRCA1 relevant to tumorigenesis were checked for differential expression in cancer tissues from patients and normal controls.

Clustering the gene expression matrix shows the clear distinction between up- and downregulated genes, and gives an impression of the variability among replicates. Many of the proteins encoded by upregulated genes are hormone receptors and structural proteins. Many of the proteins encoded by downregulated genes are involved in DNA replication and translation.

See Plate X.

Notable among the genes identified in the study are the following:

1. Consistent with the tissue specific appearance of tumours as a result of BRCA1 dysfunction, some of the genes with altered expression patterns are involved in estrogen-mediated control pathways, suggesting a possible link to the tissue-specificity enigma. The set of proteins implicated includes cyclin D1 and myc, which are upregulated by lower levels of BRCA1. (For comparison with the

→

Case Study 7.1 *(continued)*

clinical setting, low levels of *BRCA1* expression correspond to patients with reduced or absent BRCA1 function, i.e. the high-risk group; and high levels analogous to normal controls. However, the experiments of Welcsh *et al.* did not try to reproduce actual endogenous BRCA1 expression levels observed in patients and normal counterparts.) Cyclin D1 and myc are observed to be overexpressed in 20% of breast cancers, consistent with their repression by functional BRCA1.

2. Conversely, JAK and STAT proteins are downregulated by decreased levels of BRCA1. These proteins are implicated as growth inhibitors in control pathways that govern proliferation, differentiation, apoptosis and transformation. Loss of BRCA1 activity would be expected to reduce JAK1 and STAT1 levels, promoting cellular proliferation and reducing apoptosis. This is consistent with the observation that *Stat1*-null mice develop tumours more readily than normals.

The relationships detected by Welcsh *et al.* are part of the cell's control network. However, some of the products of genes regulated by BRCA1 are also known to be involved in formation of functional complexes with BRCA1. For instance, the product of *myc*—a potent oncogene—binds to BRCA1, suggesting a direct inhibition of myc by BRCA1. Thus reduced BRCA1 levels would have the dual effect of reducing the inhibition of myc through binding, and increasing the expression of *myc* through loss of transcriptional repression.

Thus, *myc* is linked to BRCA1 through both physical and regulatory interactions. We shall see in a later section that the idea of two parallel interaction networks in cells—physical interactions and regulatory interactions—is a useful distinction. However, it is one that is difficult to maintain in a system such as BRCA1 function in which the two are so closely intertwined.

* Welcsh, P.L., Lee, M.K., Gonzalez-Hernandez R.M., Black, D.J., Mahadevappa, M., Swisher, E.M., Warrington, J.A. and King, M.-C. (2002) BRCA1 transcriptionally regulates genes involved in breast tumorigenesis. *Proceedings of the National Academy of Sciences, USA*, **99**, 7560–7565.

Mass spectrometry

Mass spectrometry is a physical technique that characterizes molecules by measurements of the masses of their ions. Investigations of large-scale expression patterns of proteins require methods that give high throughput rates as well as fine accuracy and precision. Mass spectrometry achieves this, which has stimulated its development into a mature technology in widespread use. Applications to molecular biology include:

♦ Rapid identification of the components of a complex mixture of proteins.

♦ Sequencing of proteins and nucleic acids, including high-throughput genomic sequencing, and surveying populations for genetic variability.

◆ Analysis of post-translational modifications, or substitutions relative to an expected sequence.

Identification of components of a complex mixture

First the components are separated by electrophoresis. Then the isolated proteins are digested by trypsin to produce peptide fragments with relative molecular masses of about 800–4000 (Fig. 7.1). Trypsin cleaves proteins after Lys and Arg residues. Given a typical amino acid composition, a protein of 500 residues yields about 50 tryptic fragments. The mass spectrometer measures the masses of the fragments with very high accuracy (Fig. 7.2). The list of fragment masses, called the **peptide mass**

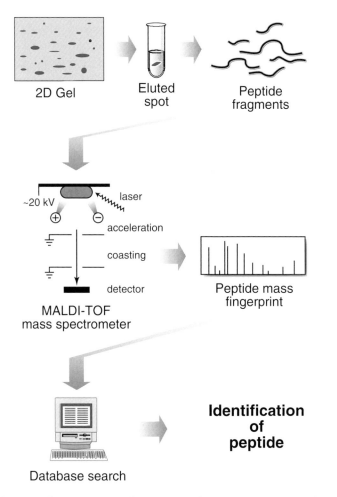

Fig. 7.1 Identification of components of a mixture of proteins by elution of individual spots, digestion and fingerprinting of the peptide fragments by MALDI–TOF (matrix-assisted laser desorption ionization-time-of-flight) mass spectrometry, followed by looking up the set of fragment masses in a database.

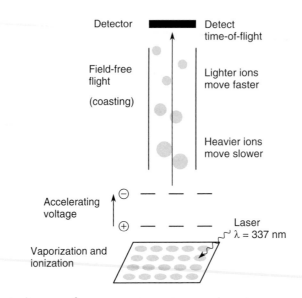

Fig. 7.2 Schematic diagram of a mass spectrometry experiment.

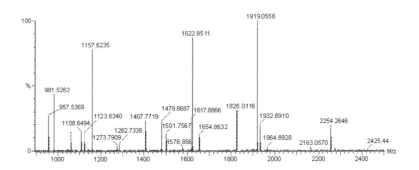

Fig. 7.3 Mass spectrum of a tryptic digest. Of the 21 highest peaks, 15 match expected tryptic peptides of the 39 kDa subunit of cow mitochondrial complex I. This easily suffices for a positive identification. (Figure courtesy of Dr I. Fearnley.)

fingerprint, characterizes the protein (Fig. 7.3). Searching a database of fragment masses identifies the unknown sample.

Construction of a database of fragment masses is a simple calculation from the amino acid sequences of known proteins, translations of open reading frames in genomes or (at a pinch) of segments from EST libraries. The fragments correspond to segments cut by trypsin at lysine and arginine residues, and the masses of the amino acids are known. (Note that trypsin doesn't cleave Lys–Pro peptide bonds, and may also fail to cleave Arg–Pro peptide bonds.)

Web resources	**Protein identification from peptide mass fingerprints**
MS-Fit	http://prospector.ucsf.edu/prospector/4.0.8/html/msfit.htm
Mascot	http://www.matrixscience.com/home.html
Aldente	http://ca.expasy.org/tools/aldente/
ProFound	http://prowl.rockefeller.edu/prowl-cgi/profound.exe

A recent source of lists of web sites for proteomics is the two-part review: Palagi, P.M., Hernandez, P., Walther, D. and Appel, R.D. (2006). Proteome informatics I: Bioinformatics tools for processing experimental data. *Proteomics*, **6**, 5435–5444, and Lisacek, F., Cohen-Boulakia, S. and Appel, R.D. (2006). Proteome informatics II: Bioinformatics for comparative proteomics. *Proteomics*, **6**, 5445–5466.

Mass spectrometry is sensitive and fast. Peptide mass fingerprinting can identify proteins in sub-picomole quantities. Measurement of fragment masses to better than 0.1 mass units is quite good enough to resolve isotopic mixtures. It is a high-throughput method, capable of processing 100 spots/day (though sample preparation time is longer). However, there are limitations. Only proteins of known sequence can be identified from peptide mass fingerprints, because only their predicted fragment masses are included in the databases. (As with other fingerprinting methods, it would be possible to show that two proteins from different samples are likely to be the same, even if no identification is possible.) Also, post-translational modifications interfere with the method because they alter the masses of the fragments.

The results shown in Fig. 7.3 are from an experiment in which the molecular masses of the ions were determined from their time-of-flight (TOF) over a known distance, as illustrated in Figs 7.1 and 7.2. The operation of the spectrometer involves these steps:

1. Production of the sample in an ionized form in the vapour phase.

2. Acceleration of the ions in an electric field. Each ion emerges with a velocity proportional to its charge/mass ratio.

3. Passage of the ions into a field-free region, where they 'coast'.

4. Detection of the times of arrival of the ions. The 'TOF' indicates the mass-to-charge ratio of the ions.

5. The result of the measurements is a trace showing the flux as a function of the mass-to-charge ratio of the ions detected.

Proteins being fairly delicate objects, it has been challenging to vaporize and ionize them without damage. Two 'soft-ionization' methods that solve this problem are:

1. **The matrix-assisted laser desorption ionization (MALDI)**, in which the sample is introduced into the spectrometer in dry form, mixed with a substrate or matrix that moderates the delivery of energy. A laser pulse, absorbed initially by the matrix, vaporizes and ionizes the protein. The MALDI-TOF combination, that produced the results shown in Fig. 7.3, is a common experimental configuration.

2. The **electrospray ionization (ESI)** method starts with the sample in liquid form. Spraying it through a small capillary with an electric field at the tip creates an aerosol of highly charged droplets. As the solvent evaporates, the droplets contract, bringing the charges closer together and increasing the repulsive forces between them. Eventually the droplets explode into smaller droplets, each with less total charge. This process repeats, creating ions, which may be multiply charged, devoid of solvent. These ions are transferred into the high vacuum region of the mass spectrometer.

Because the sample is initially in liquid form, ESI lends itself to automation in which a mixture of tryptic peptides passes through a high-performance liquid chromatograph (HPLC) into the mass spectrometer directly.

Protein sequencing by mass spectrometry

Fragmentation of a peptide produces a mixture of ions. Conditions under which cleavage occurs primarily at peptide bonds yield series of ions differing by the masses of single amino acids (Fig. 7.4). The amino acid sequence of the peptide is therefore deducible from analysis of the mass spectrum (Fig. 7.5), subject to ambiguities: Leu and Ile have the same mass and cannot be distinguished, and Lys and Gln have almost the same mass and usually cannot be distinguished. Discrepancies from the masses of standard amino acids signal post-translational modifications. In practice, the sequence of about 5–10 amino acids can be determined from a peptide of length <20–30 residues.

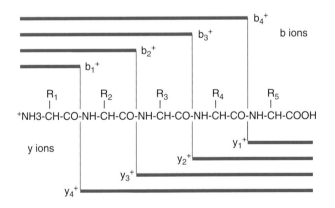

Fig. 7.4 Fragments produced by peptide bond cleavage of a short peptide. b ions contain the N-terminus; y ions contain the C-terminus. The difference in mass between successive b ions or successive y ions is the mass of a single residue, from which the peptide sequence can be determined. Two ambiguities remain: Leu and Ile have the same mass and cannot be distinguished, and Lys and Gln have almost the same mass and usually cannot be distinguished. In collision-induced dissociation, bond breakage can be largely limited to peptide linkages by keeping to low-energy impacts. Higher energy collisions can fragment sidechains, occasionally useful to distinguish Leu/Ile and Lys/Gln.

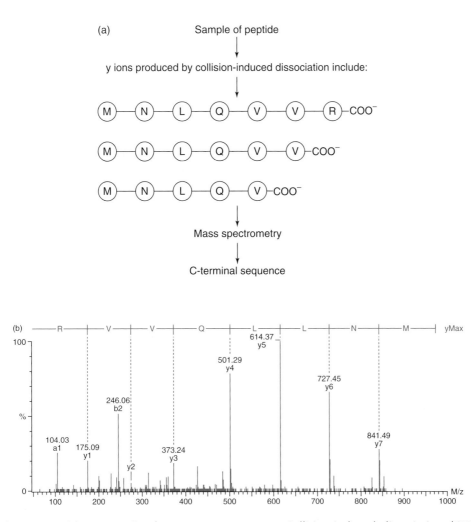

Fig. 7.5 Peptide sequencing by mass spectrometry. Collision-induced dissociation (CID) produces a mixture of ions. (a) The mixture contains a series of ions, differing by the masses of successive amino acids in the sequence. In CID the ions are not produced in sequence as suggested by this list, but the mass-spectral measurement automatically sorts them in order of their mass/charge ratio. (b) Mass spectrum of fragments suitable for C-terminal sequence determination. The greater stability of y ions over b ions in fragments produced from tryptic digests simplifies the interpretation of the spectrum. The mass differences between successive y ion peaks are equal to the individual residue masses of successive amino acids in the sequence. Because y ions contain the C-terminus, the y ion peak of smallest mass contains the C-terminal residue, etc., and therefore the sequence comes out 'in reverse'. The two leucine residues in this sequence could not be distinguished from isoleucine in this experiment. From Carroll, J., Fearnley, I.M., Shannon, R.J., Hirst, J. and Walker, J.E. (2003). Analysis of the subunit composition of complex I from bovine heart mitochondria. *Molecular and Cellular Proteomics*, **2**, 117–126 (Supplementary figure S138).

In current practice, the fragments are produced *in situ*: first the peptide is vaporized, then it is fragmented by collision-induced dissociation (CID) with argon gas. This approach requires two mass analysers, operating in tandem in the same instrument (called MS/MS). The vaporized sample first passes through one mass analyser, to separate an ion of interest. The selected ion passes into the collision cell where impacts with argon atoms excite and fragment it. By keeping the energy of impact low, the fragmentation can be limited largely to peptide bond breakage (Fig. 7.4). The second mass analyser determines the masses of the fragments.

Masses of amino acid residues, standard isotopes					
Gly	57.02146	Ala	71.03711	Ser	87.03203
Pro	97.05276	Val	99.06841	Thr	101.04768
Cys	103.00919	Leu	113.08406	Ile	113.08406
Asn	114.04293	Asp	115.02694	Gln	128.05858
Lys	128.09496	Glu	129.04259	Met	131.04049
His	137.05891	Phe	147.06841	Arg	156.10111
Tyr	163.06333	Trp	186.07931		

Measuring deuterium exchange in proteins

If a protein is exposed to D_2O, mobile hydrogen atoms will exchange with deuterium at rates dependent on the protein conformation. By exposing proteins to D_2O for variable amounts of time, mass spectrometry can give a conformational map of the protein. Applied to native proteins, the results give information about the structure. Using pulses of exposure, the method can give information about intermediates in folding.

Genome sequence analysis by mass spectrometry

Mass spectrometry of nucleic acids provides a very precise and high-throughput technique for quantitative analysis of DNA and RNA sequences in individuals and in populations. Its advantages include:

- **High precision** The standard deviation of typical mass spectral concentration measurement replicates is ~3%, compared with ~200% for microarray measurements.

- **More data per sample** A mass spectrum contains many peaks rather than a single value. This allows analysis of mixtures; and permits 'multiplexing', or simultaneous analysis of features of a set of mixed samples.

- **High specificity and sensitivity** Very small sample sizes are required. PCR amplification can be pushed to very high gain, as there is little risk of mistaking a contaminant for a true sample amplicon. In fact, it is possible to determine sequences from individual cells or even single DNA strands.

To prepare for the measurement, samples are treated by gene-specific PCR amplification by allele-specific primer extension, to produce single-stranded oligonucleotides.

Products are purified and embedded in a matrix suitable for MALDI vaporization and mass analysis. No hybridization step is required for detection. Assembly of many subjects on an array allows for automation of data collection. (Throughput rates can reach 10^5 spectra per instrument per day.)

The typical relative molecular mass of an oligonucleotide measured is ~6000, corresponding to about 20 bases. Under conditions where the amplified products of different alleles contain different numbers of bases, the mass difference is ≥ 300, a very large difference relative to the accuracy of mass spectrometry. In fact, it is feasible to pick up a single-base substitution in oligonucleotides of the same length, or even the methylation of a gPc site. For nucleotide substitutions, the mass differences between bases range from ± 9 for t $\leftrightarrow$ a to ± 40 for c $\leftrightarrow$ g.

Applications include:

1. **Measurement of allele frequencies in populations, or detection of alleles in individuals, by identification of SNPs** For population studies, samples from several individuals in the selected groups can be pooled, and genotype frequencies measured to about 3% accuracy. Several SNPs can be determined from a single spectrum. Such studies have impact on a wide variety of fields, including anthropology, agriculture, and forensics, but medical applications are the major driving force. For example, controlled comparisons between healthy populations and those predisposed to a disease can identify genetic factors of clinical importance.

2. **Characterization of individual genotypes** A selection of 100 000 SNPs offers about three polymorphisms per gene, enough for fairly thorough characterization of the protein-coding portion of an individual person's genome. Determination of one individual's SNP profile is achievable using one instrument for 1 day. Clinical applications include: (a) diagnosis, based on systematic differences, between healthy individuals and those with a disease, previously established from controlled population studies, and (b) pharmacogenomics, to distinguish patients who will benefit from treatment with a drug from those who will not benefit or even risk severe side effects.

3. **Measurement of individual haplotypes** Haplotypes are local combinations of genetic polymorphisms that tend to be co-inherited (see Chapter 2). Haplotypes simplify the search for phenotype–genotype correlations, because they reduce the number of variables with which to characterize the genotype. Mass-spectrometric methods based on amplifying regions around SNPs in a sample containing a single DNA molecule provide an accurate and high-throughput method of individual haplotype determination.

4. **Measurement of gene expression levels on an absolute scale with a precision of ~3%.** This is achieved by spiking the RNA extracted from a sample with a known amount of a related oligoribonucleotide, and measuring the relative amounts of signal from the calibrating oligo and the natural ones.

5. **Non-invasive prenatal diagnosis based on the small amount of foetal DNA that leaks into maternal blood** Because of the 95–99% maternal DNA background, only paternal contributions to the foetus can be identified. However, the technique is

sensitive enough to detect the *SRY* gene, demonstrating that the foetus is male, or other paternal alleles that may be useful in diagnosing genetic abnormalities. It should be emphasized that the use of only a *maternal blood sample* avoids the significant risks of an invasive procedure to sample amniotic fluid.

6. **Genomic sequencing** Mass spectrometry has the potential to compete in accuracy and throughput with gel-based methods for large-scale DNA sequence determination.

Case Study 7.2 Application of combined genomic, proteomic and structural methods to antibiotic resistance in tuberculosis

Tuberculosis is an infectious disease caused by *Mycobacterium tuberculosis*. Despite development of vaccines and drugs, it remains a potent killer. Tuberculosis is the most common cause of death from infectious disease, claiming about 2 million victims per year. Of the 9 million new cases per year (estimated by the World Health Organization), 80% occur in developing countries in Asia and sub-Saharan Africa. HIV infection, also more prevalent in developing countries, exacerbates the mortality of tuberculosis infection by lowering resistance.

Our bodies' front-line defences against most bacterial infections include macrophages, cells of the immune system that engulf bacteria and attack them with a variety of chemical and biochemical agents. *Mycobacterium tuberculosis*, exceptionally, is adapted to survive *within* the macrophage. Part of its adaptation is structural: cells of *M. tuberculosis* and close relatives surround themselves with a waxy coat. The low permeability of the coat shields them from the inhospitable environment within the macrophage, including low pH and oxidative stress. The bacteria also make substantial changes to gene expression patterns, to adapt their physiological state to these surroundings.

After several decades of decline following the development of effective drugs, the incidence of tuberculosis began to increase in the mid-1980s. One reason is emergence of resistant strains.

A primary drug used in prevention and therapy of tuberculosis is isoniazid (isonicotinic acid hydrazide). Isoniazid attacks *M. tuberculosis* by interfering with synthesis of its cell wall, without which the bacterium cannot survive. Targets of isoniazid include an NADH-dependent enoyl acyl carrier protein reductase (InhA), and a β-keto-acyl ACP synthase (KasA). These enzymes participate in synthesis of mycolic acids, major components of the cell wall.

Isoniazid must be converted to an active form after absorption by the bacterial cell. The enzyme that effects the conversion, KatG, is a natural suspect for involvement in resistance. Its natural function is to detoxify peroxides.

Several methods have been applied to elucidate the adaptations responsible for isoniazid resistance:

1. Changes in gene expression patterns were detected using microarrays.

→

Case Study 7.2 (continued)

2. Genes that change expression were sequenced in susceptible and resistant strains, and mutations observed.

3. The crystal structure of isoniazid bound to InhA has been determined.

1. Changes in gene expression patterns

Wilson and colleagues* examined susceptible and resistant strains of *Mycobacterium tuberculosis* at times up to 8 hours of exposure to isoniazid (Plate XI). The array included almost all ORFs identified in the *Mycobacterium tuberculosis* genome. Although biochemical studies had already implicated some proteins in resistance, a general screen was carried out in order to identify as many drug targets as possible.

Exposure to isoniazid greatly enhanced the transcription of two classes of genes. One set is involved in cell wall synthesis, including an operon-like cluster encoding components of a fatty acid synthase complex (FAS-II). Additional genes, including one encoding a subunit of alkyl hydroxyperoxide reductase (AhpC), that handle oxidative stress, were also upregulated. The logic of the experiment is that the treated cells are recognizing the effects of the drug, and feedback mechanisms are acting to try to compensate for reduced activities by enhanced expression.

2. Mutations conferring resistance to isoniazid

On the basis of the changed expression profiles, Ramaswamy *et al.* (2003) sequenced a total of 2.6 Mb from 124 *Mycobacterium tuberculosis* isolates.[†] These include mutations in KatG that impede activation of isoniazid, and mutations in InhA to escape inhibition by the activated form.

Note that because oxidative stress is part of the host's natural defence to infection, simple knockout of KatG could be a dangerous strategy for the bacterium. Ideally the bacterium would reduce the activity of the enzyme in isoniazid activation but retain activity against small peroxides. In this way it would reduce susceptibility to the drug while maintaining its general fitness in the environment within the macrophage. Precisely this balance is achieved by the the most common KatG mutation in resistant strains, is S315T.

The most common mutation in InhA is S94A. The inhibitory effectiveness of activated isoniazid is reduced in this modified protein.

3. Crystallography

Rozwarski *et al.* (1998) solved the structure of the complex between the activated form of isoniazid and InhA (Fig. 7.6). The drug is covalently attached to the nicotinamide ring of NAD, bound to the active site of InhA. The sidechain of S94 is also shown. In the inhibitory complex, the protein binds the NAD-activated isoniazid adduct. The compling of these molecules can occur *only* on the enzyme (in solution activated isoniazid and NADH do not react).

How does the S94A mutant achieve resistance? In the absence of inhibitor, the enzyme can bind either substrate first and then cofactor, or cofactor first and then substrate. Because the substrate occupies the same site on the enzyme as the inhibitor, only if cofactor is bound first can an inhibitory complex form. Two pathways

The genome of *Mycobacterium tuberculosis* is ∼4.4 Mb long and contains ∼4000 genes.

Case Study 7.2 *(continued)*

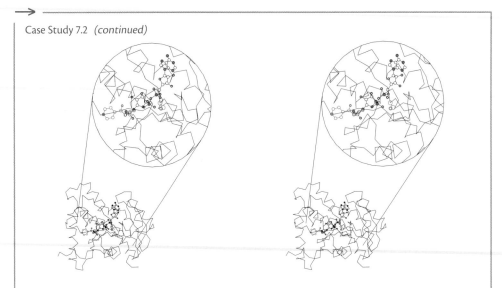

Fig 7.6 Structure of long fatty acid chain enoyl-acp reductase (InhA) in complex with inhibitor [1ZID]. The ligand is an adduct of activated isoniazid and NADH. Shown in blue are the isoniazid moiety of the inhibitor (centre of blown-up circle), and the sidechain of Ser94 (left in blown-up circle). The mutation S94A contributes to isoniazid resistance. (See Rozwarski, D.A., Grant, G.A., Barton, D.H., Jacobs, W.R., Jr. and Sacchettini, J.C. (1998). Modification of the NADH of the isoniazid target (InhA) from *Mycobacterium tuberculosis. Science,* **279**, 98–102.)

are possible, the first leading exclusively to product, the other producing an inhibitory complex (E = enzyme, S = substrate, C = cofactor and I = inhibitor):

```
If substrate binds first:

(1a) E + S → ES   + C → ESC → E + C + P

(1b) E + S → ES   + I  ✕

If cofactor binds first:

(2a) E + C → EC   + S → ESC → E + C + P

(2b) E + C → EC   + I → ESI =   inhibitory
                                  complex
```

If substrate binds first (1a and 1b), the inhibitory complex cannot form. If cofactor binds first (2a and 2b), a stable inhibitory complex may form, taking the enzyme out of the game permanently.

The S94A mutation reduces the affinity of the enzyme for NADH. This enhances the substrate-bound-first pathway (1a and 1b), lowering the amount of inhibitory complex produced (2b), and also enhancing the dissociation rate of the inhibitory complex.

It is also possible that S94A and other mutations reduce the affinity of the adduct.

→

Case Study 7.2 *(continued)*

Research and development of anti-tuberculosis drugs is a continuing challenge. This example shows the effectiveness of coordinated application of many different techniques.

* Wilson, M., DeRisi, J., Kristensen, H.H., Imboden, P., Rane, S., Brown, P.O. and Schoolnik, G.K. (1999). Exploring drug-induced alterations in gene expression in *Mycobacterium tuberculosis* by microarray hybridization. *Proceeding of the National Academy of Sciences, USA*, **96**, 12833–12838.
† Ramaswamy, S.V., Reich, R., Dou, S.J., Jasperse, L., Pan, X., Wanger, A., Quitugua, T. and Graviss, E.A. (2003). Single nucleotide polymorphisms in genes associated with isoniazid resistance in *Mycobacterium tuberculosis*. *Antimicrobial Agents and Chemotherapy*, **47**, 1241–1250.

Systems biology

Proteins are social animals, and life depends on their interactions. Because individual proteins have specialized functions, control mechanisms are required to integrate their activities. The right amount of the right protein must function in the right place at the right time. Failure of control mechanisms can lead to disease and even death.

Under unchanging environmental conditions, an organism's biochemical systems must be stable. Under changing conditions, the system must be robust, accommodating both neutral and stressful perturbations. Over longer periods of time, processes must have their rates altered, or even be switched on and off. This regulation includes short-term adjustments, for instance in the stages of the cell cycle, or responses to external stimuli such as changes in the composition or levels of nutrients or oxygen. Longer-term regulatory activities include control over developmental stages during the entire lifetime of an organism.

Metabolism is the flow of molecules and energy through pathways of chemical reactions. Of course many metabolic reactions involve proteins and nucleic acids, as well as small compounds such as amino acids and sugars. The full panoply of metabolic reactions forms complex traffic patterns. Some patterns are linear pathways, such as the multistep synthesis of tryptophan from chorismate. Others form closed loops, such as the tricarboxylic acid (Krebs) cycle. Moreover, the pathways interlock densely. The structure of the totality of metabolic pathways—its connectivity or topology—and its activity patterns, can be analysed in terms of a mathematical apparatus dealing with graphs and flows and throughputs.

To control metabolic flow patterns, **regulatory pathways** connect proteins and metabolite concentrations. The structure and dynamics of the regulatory pathways are different from those of the metabolic pathways. Corresponding to the succession of enzymatic transformations in metabolism, a regulatory pathway is an assembly of signalling cascades.

Systems biology describes metabolic and regulatory interactions in terms of **interaction networks**.

Systems biology focuses on the integration and control of gene and protein activity.

Two parallel networks: physical and logical

In cells, the two interaction networks operate in parallel: (1) a **physical network** of protein–protein and protein–nucleic acid complexes, and (2) a **logical network** of control cascades. Metabolic pathways partake of both: many but not all metabolic pathways are mediated by physical protein-protein interactions and regulated by logical interactions.

Examples of purely physical interactions include the assembly of photosynthetic reaction centres, complexes of proteins and cofactors that convert light to chemical energy; assemblies of collagen in connective tissue; and the ribosome. Examples of logical interactions *not* mediated entirely by direct physical interaction between proteins include feedback loops in which the increase in concentration of a product of a metabolic pathway inhibits an enzyme catalysing one of the early steps in the pathway, or the secretion of a small molecule as a signal to other cells ('fire and forget' mode). In these cases the logical interaction is transmitted by a diffusing small molecule. Many other examples appear in the very common theme of regulation of gene expression. A transcription factor, binding to DNA, may never interact physically with the proteins the expression of which it controls.

The allosteric change in haemoglobin is an example of simultaneous physical and logical interaction: the subunits of haemoglobin respond to changes in oxygen levels by a conformational change that alters oxygen affinity. Another example is the transmission of a signal from the surface of a cell across the membrane to the interior by dimerization of a receptor. This can be the initial trigger of a process that ultimately affects gene expression. Not all links of this process need involve protein–protein interactions; some may be mediated by diffusion of small molecules such as cyclic AMP.

Even though certain protein–protein and protein–nucleic acid complexes participate in both physical and logical networks, the two networks remain distinct in terms of their logic and their biological function, and it is useful to keep the distinction in mind, *especially* when considering proteins that participate in both.

See Box:
Cell–cell communication in microorganisms: quorum sensing

Cell–cell communication in microorganisms: quorum sensing

Control mechanisms *not* involving direct protein–protein interactions mediate intercellular signalling in microorganisms. *Vibrio fischeri* is a marine bacterium that can adopt alternative physiological states in which bioluminescence is active or inactive. (Literally a 'light switch'.) The organism can live free in seawater, or colonize the light organs of certain species of fish or squid. It is bioluminescent only when growing within the animal.

The bacteria respond to the local density of cells; a form of communication called **quorum sensing**. In *V. fischeri*, quorum sensing is mediated by secretion and detection of a small signalling molecule, N-(3-oxohexanoyl)-homoserine lactone. Related species use other N-acyl homoserine lactones, abbreviated AHL. AHL can diffuse freely out of the cells in which it is synthesized. Within the light organs,

$\longrightarrow$

$\longrightarrow$

Cell–cell communication in microorganisms: quorum sensing *(continued)*

culture densities can reach $10^{10} - 10^{11}$ cells/ml, and the AHL concentration can exceed the threshold of about 5–10 nM for triggering the physiological switch.

Bacterial genes *LuxI* and *LuxR* govern the regulation. The product of *LuxI* is involved in the synthesis of AHL. The *LuxR* gene product contains a membrane-bound domain, that detects the AHL signal, and a transcriptional activator domain. *LuxR* activates an operon that includes (1) genes for synthesis of luciferase (the enzyme responsible for the bioluminescence), and (2) *LuxI*, expression of which synthesizes additional AHL, amplifying the signal and sharpening the transition.

The host also senses the bacteria: the light organs of squid grown in sterile salt water do not develop properly. This appears to be a reaction to the intensity of luminescence, rather than to the concentration of AHL. For the animal, the luminescence contributes to camouflage: disguise from predators at lower depths, by blending with illumination from the sky. The masking of shadows is a natural form of 'make-up'. (The bioluminescence also regularly surprises diners in seafood restaurants, who jump to the conclusion that their glowing dinner is of extra-terrestrial origin. However, most bioluminescent bacteria are harmless, although some strains of the related *Vibrio* species, *V. cholerae*, the causative agent of cholera, are weakly bioluminescent. In fact the virulence of *V. cholerae* is also under the control of quorum sensing, by a related mechanism.)

Protein complexes and aggregates

The basis of our understanding of how life within a cell is organized and regulated is the set of protein–protein and protein–nucleic acid interactions. The development of high-throughput methods for detecting interactions has been a focus of recent interest.

Interacting proteins and nucleic acids span a range of structures and functions:

- Simple dimers or oligomers in which the monomers appear to function independently.

- Oligomers with functional 'cross-talk', including ligand-induced dimerization of receptors, and allosteric proteins such as haemoglobin, phosphofructokinase and asparate carbamoyltransferase.

- Large fibrous proteins such as actin or keratin.

- Non-fibrous structural aggregates such as viral capsids.

- Large aggregates with dynamic properties such as F1-ATPase, pyruvate kinase, the GroEL–GroES chaperonin and the proteasome.

- Protein–nucleic acid complexes, including ribosomes, nucleosomes, transcription regulation complexes, splicing and repair particles, and viruses. In many cases initial binding is followed by recruitment of additional proteins to form large complexes.

◆ Many proteins, whether monomeric or oligomeric, function by interacting with other proteins. These include all enzymes with protein substrates, and many antibodies, inhibitors and regulatory proteins.

◆ Protein interactions are frequently associated with disease, as misfolded or mutant proteins are prone to aggregation (see Box).

Diseases associated with protein aggregates

Disease	Aggregating protein	Comment
Sickle-cell anaemia	Deoxyhaemoglobin–S	Mutation creates hydrophobic patch on surface
Classical amyloidoses	Immunoglobulin light chains, transthyretin and many others	Extracellular fibrillar deposits
Emphysema associated with Z-antitrypsin	Mutant α_1-antitrypsin	Destabilization of structure facilitates aggregation
Huntington	Altered huntingtin	One of several polyglutamine repeat diseases
Parkinson	α-Synuclein	Found in Lewy bodies
Alzheimer	Aβ, τ	Aβ = 40–42 residue fragment
Spongiform encephalopathies	Prion proteins	Infectious, despite containing no nucleic acid

---

Properties of protein–protein complexes

Stoichiometry — what is the composition of the complex?

Stable oligomeric proteins may contain many copies of one protein, or combine different ones. Among aggregates of a single protein, complexes containing *odd* numbers of molecules are less common than those containing even numbers. Oligomers (complexes containing a few copies of the same protein — dimers, trimers, . . .) usually show symmetry. For instance, insulin forms a hexamer with three-fold and two-fold axes.

Some prokaryotic proteins containing identical subunits are homologous to eukaryotic proteins containing related but non-identical subunits, arising by gene duplication and divergence. The proteasome is an example. Some viruses achieve diversity *without* duplication, by combining proteins with the same sequence but different conformations.

Protein complexes vary widely in the numbers and variety of molecules they contain. Some complexes contain only a few proteins, but others are very large; for example, pyruvate dehydrogenase contains hundreds of subunits, and some viral capsids contain thousands.

Many very large aggregrates have clinical importance, including bovine spongiform encephalopathy (BSE, or 'mad cow disease'), Alzheimer and Huntington disease. Amyloidoses are diseases characterized by extracellular fibrillar deposits, usually with a common crossed-β-sheet structure. They arise from a variety of causes, including destabilizing mutations, overproduction of a protein and inadequate clearance in renal failure. Misfolded proteins are more prone to aggregate, and mutated proteins are more prone to misfold. Large local concentrations, such as can occur in myelomas that overproduce immunoglobulin light chains, also aggravate the threat of aggregation.

See Box: *Diseases associated with protein aggregation.*

Affinity — how stable is the complex?

A common index of the affinity of a complex is the **dissociation constant**, K_D, the equilibrium constant for the *reverse* of the binding reaction:

$$\text{Protein–Ligand} = \text{Protein} + \text{Ligand} \qquad K_D = \frac{[\text{P}][\text{L}]}{[\text{PL}]}$$

[P], [L] and [PL] denote the numerical values of the concentrations of protein, ligand, and protein-ligand complex, respectively, expressed in mol l^{-1}. The lower the K_D, the tighter the binding. K_D corresponds to the concentration of free ligand at which half the proteins bind ligand and half are free: [P] = [PL].

The Michaelis constant of an enzyme is the dissociation constant of the enzyme-substrate complex.

The K_D is related to the Gibbs free energy change of dissociation by the relationship:

$$\text{PL} = \text{P} + \text{L} \qquad \Delta G^\circ = \Delta H^\circ - T\Delta S^\circ = -RT \ln K_D$$

Dissociation constants of protein–ligand complexes span a wide range:

Biological context	Ligand	Typical K_D
Allosteric activator	Monovalent ion	$10^{-4} - 10^{-2}$
Coenzyme binding	NAD, for instance	$10^{-7} - 10^{-4}$
Antigen–antibody complexes	Various	$10^{-4} - 10^{-16}$
Thrombin inhibitor	Hirudin	5×10^{-14}
Trypsin inhibitor	Bovine pancreatic trypsin inhibitor	10^{-14}
Streptavidin	Biotin	10^{-15}

Structural studies have elucidated several important features of the interactions between soluble proteins, that contribute to affinity:

◆ **What holds the proteins together?** Burial of the hydrophobic surface, hydrogen bonds and salt bridges.

◆ **Do proteins change conformation upon formation of complexes?** In some cases they do. In these cases the interaction energy has to 'pay for' the conformational change, and the interface tends to be larger.

◆ **What determines specificity?** Complementarity of the occluding surfaces, in shape, hydrogen-bonding potential and charge distribution. Prediction of protein complexes from the structures of the partners is the **docking problem**. Reliable solution of this problem, together with progress in structural genomics, would permit *in silicio* screening of proteomes for interacting partners.

Kinetics of formation and break-up, average lifetime

The dissociation constant of a complex indicates the fraction of time that the components spend in the bound state and the fraction of time in which they are unbound. But the **average lifetime** of the bound state can vary without affecting K_D. Defining individual rate constants for association and dissociation, k_{on} and k_{off}, the dissociation constant is equal to their ratio:

$$P + L \underset{k_{off}}{\overset{k_{on}}{\rightleftarrows}} [PL] \qquad K_D = k_{off}/k_{on}$$

A short average lifetime, corresponding to large values of both k_{off} and k_{on}, or a long average lifetime, corresponding to small values of both k_{off} and k_{on}, can produce the same K_D. Lifetimes are important: if you want to purify a complex, it is important that its average lifetime is longer than the duration of the isolation procedure! Conversely, if a protein–protein complex is to mediate transmission of a signal, a short lifetime provides a natural 'reset mechanism' to preclude the signal's being locked 'on' for too long.

The 'on rate' is limited by diffusion rates. Under ordinary conditions $k_{on} \leq 10^{-9} M\,s^{-1}$. k_{on} may be considerably smaller if, for example, a conformational change is required for binding. Typical k_{on} values are $10^{-6} - 10^{-7} M^{-1} s^{-1}$, and typical lifetimes $\sim$1 s.

How are complexes organized in three dimensions?

When two proteins form a complex, each leaves a 'footprint' on the surface of the other, defining the portion of the surface involved in the interaction. If two proteins interact using the *same* surface on both, the complex is **closed**. If two proteins interact through *different* surfaces, the complex is **open**. The significance is that a closed complex does not allow additional proteins to bind with the same interaction. An open complex, in which the surface of potential interaction is not occluded, can grow by accretion of additional subunits. Thus, open but not closed complexes are compatible with formation of repetitive aggregates.

Do proteins change conformation upon complex formation?

Some protein complexes form by the coming together of rigid subunits. The subunits in these complexes have the same structure in the complex that they have separately.

Other protein complexes involve structural changes upon complex formation. These include complexes of subunits that are not stable separately.

Features of protein–protein interfaces

♦ **Burial of protein surface** The surface buried by formation of a complex is the difference between the accessible surface area (ASA) of the complex and the sum of the ASAs of the components separately.

A typical protein–protein interface might involve 22 residues, and 90 atoms, of which 20% would be mainchain atoms, and an occasional water molecule. A histogram of surface area buried in binary protein complexes shows a peak centred at 1600 $Å^2$.

The minimum buried surface for stability of a protein–protein complex is about 1000 $Å^2$. Complexes that bury >2000 $Å^2$ tend to involve conformational changes upon complex formation.

Each $Å^2$ of hydrophobic surface buried contributes about 105 J to the free energy of stabilization.

♦ **The composition of the interface** The chemical character of protein–protein interfaces is intermediate between that of the surfaces and interiors of monomeric globular proteins. Interfaces are enriched in neutral polar atoms at the expense of charged atoms. The amino acid composition of interfaces are enriched in aromatic residues—His, Phe, Tyr, Trp—relative to remaining exposed surface. There is a lesser degree of enrichment in aliphatic sidechains—Leu, Ile, Val, Met—and Arg (but, surprisingly, not Lys).

♦ **Complementarity** of interfaces is responsible for specificity. Complementarity involves both good packing at the occluding surfaces and proper juxtaposition of hydrogen-bonded and charged atoms. Typically there is one hydrogen bond per 170 $Å^2$ of interface area. Isolated water molecules occupy sites in many interfaces. Typically there is one fixed water molecule per 100 $Å^2$ of interface.

Protein interaction networks

The units from which interaction networks are assembled are:

♦ For physical networks, a protein–protein or protein–nucleic acid complex.

♦ For logical networks, a dynamic connection in which the activity of a process is affected by a change in external conditions, or by the activity of another process.

Most experiments reveal only pairwise interactions. The challenges are to integrate pairwise interactions into a network and then to study the structure and dynamics of the system.

Many techniques detect physical interactions directly. These include:

- **X-ray and NMR structure determinations** not only can identify the components of the complex, but can reveal how they interact, and whether conformational changes occur upon binding.

- **Two-hybrid screening systems** Transcriptional activators such as Gal4 contain a DNA-binding domain and an activation domain. Suppose these two domains are separated, and one test protein is fused to the DNA-binding domain and a second test protein is fused to the activation domain. Then a reporter protein will be expressed only if the components of the activator are brought together by formation of a complex between two test proteins. High-throughput methods allow parallel screening of a 'bait' protein for interaction with a large number of potential 'prey' proteins. (See Box: *Protein interaction networks determined by two-hybrid screening systems*.)

- **Chemical cross-linking** fixes complexes so that they can be isolated. Subsequent proteolytic digestion and mass spectrometry permits identification of the components.

- **Co-immunoprecipitation** An antibody raised to a 'bait' protein binds the bait together with any other 'prey' proteins that interact with it. The interacting proteins can be purified and analysed, for instance by western blotting, or mass spectrometry.

- **Chromatin immunoprecipitation** identifies DNA sequences that bind proteins. Treatment with formaldehyde cross-links proteins and DNA, fixing the complexes that exist within a cell. Then, isolation of the chromatin and breaking the DNA into small fragments allows separation of proteins by binding to specific antibodies, carrying the DNA sequences along with them. Reversal of the cross-link followed by sequencing of the DNA identifies the specific DNA sequence to which each protein binds.

- **Phage display** Genes for a large number of proteins are individually fused to the gene for a phage coat protein, to create a population of phage each of which carries copies of one of the extra proteins exposed on its surface. Affinity purification against an immobilized 'bait' protein selects phage displaying potential 'prey' proteins. DNA extracted from the interacting phages reveals the amino acid sequences of these proteins.

- **Surface plasmon resonance** analyses the reflection of light from a gold surface to which a protein has been attached. The signal changes if a ligand binds to the immobilized protein. (The method detects localized changes in the refractive index of the medium adjacent to the gold surface. This is related to the mass being immobilized.)

- **Fluorescence resonance energy transfer** If two proteins are tagged by different chromophores, transfer of excitation energy can be observed over distances up to about 60 Å.

Other methods provide complementary information:

- **Domain recombination networks** Many eukaryotic proteins contain multiple domains. A feature of eukaryotic evolution is that a domain may appear in different proteins with different partners. In some cases, proteins in a bacterial operon cata-lysing successive steps in a metabolic pathway are fused into a single multidomain protein in eukarya. The domains of the eukaryotic protein are individually homo-logous to the separate bacterial proteins. (Examples of proteins fused in eukarya and separate in prokaryotes are also known.)

It is possible to create a network by defining an interaction between two protein domains whenever homologues of the two domains appear in the same protein. This is evidence for some functional link between the domains, even in species where the domains appear in separate proteins.

◆ Co-expression patterns Clustering of microarray data identifies proteins with common expression patterns. They may have the same tissue distribution, or be up- or downregulated in parallel in different physiological states. This is also suggestive evidence that they share some functional link. In the response of *M. tuberculosis* to isoniazid (page 406), genes for the fatty acid synthesis complex are coordinately upregulated. They are on an operon-like gene cluster, and in fact these proteins do form a physical complex. On the other hand, alkyl hydroperoxidase (AHPC) is also upregulated in response to isoniazid. AHPC acts to relieve oxidative stress. There is no evidence that it physically interacts with the fatty acid synthesis complex, or that it mediates a metabolic transformation coupled to fatty acid synthesis. It is a second component of the response to isoniazid.

◆ Phylogenetic distribution patterns The **phylogenetic profile** of a protein is the set of organisms in which it and its homologues appear. Proteins in a common structural complex or pathway are functionally linked and expected to co-evolve. Therefore, proteins that share a phylogenetic profile are likely to have a functional link, or at least to have a common subcellular origin. There need be no sequence or structural similarity between the proteins that share a phylogenetic distribution pattern. A welcome feature of this method is that it derives information about the function of a protein from its relationship to *non-homologous* proteins.

Protein interactions detected by two-hybrid screening systems*

	H. pylori	*S. cerevisiae*	*C. elegans*	*D. melanogaster*
Total proteome size	1576	5585	33 469	13 843
Proteins tested	732	987 / 790	1415	4685
Interactions detected	1465	936 / 800	2131	4876

The two sets of numbers for yeast are the results of independent investigations.

* From: Aloy, P. and Russell, R.B. (2004). Ten thousand interactions for the molecular biologist. *Nature Biotechnology*, **22**, 1317–1321.

There are many ways to link proteins, including direct physical protein–protein interactions, two-hybrid complementarity, domain recombination, coexpression patterns, phylogenetic profiles, etc. Each provides a basis for a protein interaction network. The networks formed by combining each set of interactions are different, although they overlap, to a greater or lesser extent. They give different views of the kinds of relationships between proteins that exist in cells. It is possible to form a more comprehensive network by combining different types of interactions. For instance, the DIP database http://dip.doe-mbi.ucla.edu/ is a curated collection of experimentally determined protein–protein interactions. It contains data about 56 186 interactions between 19 490 proteins from 161 organisms.

Plate XII shows a portion of an interaction network of yeast proteins, based on sets of proteins that have been found together in solved structures.

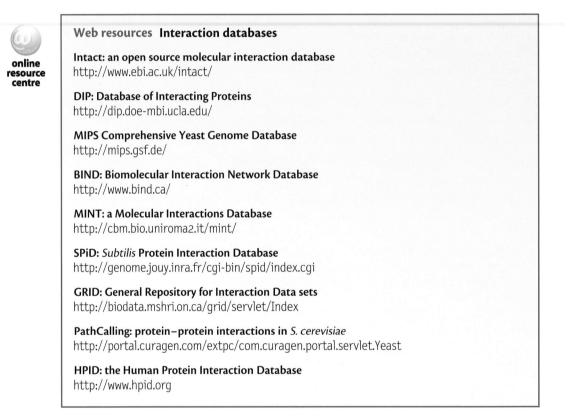

online resource centre

Web resources   Interaction databases

Intact: an open source molecular interaction database
http://www.ebi.ac.uk/intact/

DIP: Database of Interacting Proteins
http://dip.doe-mbi.ucla.edu/

MIPS Comprehensive Yeast Genome Database
http://mips.gsf.de/

BIND: Biomolecular Interaction Network Database
http://www.bind.ca/

MINT: a Molecular Interactions Database
http://cbm.bio.uniroma2.it/mint/

SPiD: *Subtilis* Protein Interaction Database
http://genome.jouy.inra.fr/cgi-bin/spid/index.cgi

GRID: General Repository for Interaction Data sets
http://biodata.mshri.on.ca/grid/servlet/Index

PathCalling: protein–protein interactions in *S. cerevisiae*
http://portal.curagen.com/extpc/com.curagen.portal.servlet.Yeast

HPID: the Human Protein Interaction Database
http://www.hpid.org

Case Study 7.3   Components of the primosome assembly in *Bacillus subtilis.*

The first step in DNA replication in *Bacillus subtilis* is the binding of initiator proteins to specific DNA sequences that serve as origins of replication. These then recruit a nucleoprotein complex called the primosome. A major component of the primosome is DnaC, a hexameric replicative helicase.*

$\longrightarrow$

Case Study 7.3 *(continued)*

It is believed that steps in the process include:

1. Binding of an initiator protein, DnaA or PriA, to an appropriate single-stranded DNA sequence.

2. Other proteins—DnaB, DnaC and DnaI—are recruited. DnaB and DnaI are regulators of DnaC activity.

3. DnaC is loaded onto the single-stranded DNA, forming a hexameric assembly.

4. DnaG is recruited to prime DNA synthesis.

Scientists at the *Institut National de la Récherche Agronomique* maintain a database of the protein interaction network of *B. subtilis*[†] (see http://genome.jouy.inra.fr/cgi-bin/spid/index.cgi).

Figure 7.7 shows a small fragment the network, limited to immediate neighbours of DnaC.

The web site is active: clicking on a node either *adds* the interaction partners of the node to the graph, or *replaces* the graph with another centred on the selected protein. By adding partners, one can look at more extended neighbourhoods of DnaC. By replacing the graph, one can walk through the network. (See Weblem 7.6.)

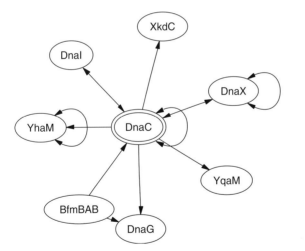

Fig. 7.7 DnaC and proteins that interact directly with it. Arrows linking partners point from 'bait' to 'prey'; bidirectional arrows indicate cases where the interaction was detected in reciprocal experiments. In the original web site, the arrows are colour coded according to the nature of the evidence for the interaction. Reproduced by permission.

---

[*] Be aware that the nomenclature of these proteins differs between *E. coli* and *B. subtilis*.

[†] Hoebeke, M., Chiapello, H., Noirot, P. and Bessières, P. (2001). SPiD: a *subtilis* protein interaction database. *Bioinformatics*, **17**, 1209–1212; Noirot-Gros, M.F., Dervyn, E., Wu, L.J., Mervelet, P., Errington, J., Erlich, S.D. and Noirot, P. (2002). An expanded view of bacterial DNA replication. *Proceedings of the National Academy of Sciences, USA*, **99**, 8342–8347.

Networks and graphs

See Box: *The idea of a graph.*

In the abstract, networks have the form of graphs.

The routes between cities on the map of Sweden in Fig. 5.4 is a network represented by a graph, similar to those appearing in systems biology. Each city is a node. The thick lines joining them indicate routes. Other examples familiar to many readers are the map of the London Underground,[1] and maps of the tube (subway) systems of other cities (see Box: *Examples of graphs*). Each station is a node of the graph, and edges correspond to tracks connecting the stations. The modern London underground map shows the *topology* of the network; it does not quantitatively represent the geography of the area. An early map, from 1925, did maintain geographic accuracy.[2] This was possible when the system was simpler than it is now. Some of the maps now posted in the Paris Métro are fairly accurate geographically. *Considered as networks, a geographically accurate map and a simplified map with the same topology correspond to the same graph.*

The idea of a graph

- Mathematically, a graph consists of a set of vertices V and a set of edges E.

- Each edge is specified by a pair of vertices.

- In a **directed graph** the edges are **ordered** pairs of vertices.

- In a **labelled graph** there is a value associated with each edge. (A directed graph is a special case of a labelled graph: consider the arrowheads as labels.)

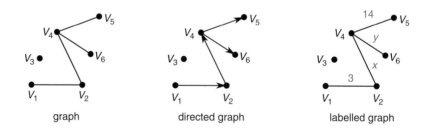

| graph | directed graph | labelled graph |

An undirected unlabelled graph specifies the connectivity of a network but not the distances between vertices (the topology but not the geometry, as in the modern London Underground map). Labels on the edges can indicate distances. For example, some phylogenetic trees indicate only the topology of the ancestry. Others indicate quantitatively the amount of divergence between species.

$\longrightarrow$

---

1 http://www.transportforlondon.gov.uk/tfl/tube_map.shtml or http://www.afn.org/~alplatt/tube.html Exercises 7.11 and 7.12, Problem 7.5 and Weblem 7.4 also make use of this map.
2 http://www.ltmuseum.co.uk/collections/posters_b.html

→

The idea of a graph *(continued)*

Phylogenetic trees are often drawn with the lengths of the branches indicating the time since the last common ancestor. This is a pictorial device for labelling the edges.

Some graphs do not correspond to physical structures, and in any event edge labels need not reflect geometry in the usual sense. For example, the links in a network of metabolic pathways might be labelled to reflect flow patterns.

The London Underground network is connected, in that there is a path between any two stations. Many questions familiar to commuters are shared in the analysis of biological networks; for example: what are the paths connecting Station A and Station B? Regarding different lines as subnetworks, how easy is it to transfer from one to another; that is, what is the nature of the patterns of connectivity? In case of failure of one or more links, is the network robust—does it remain fully connected?

Examples of graphs

- Sets of people who have met each other
- Electricity distribution systems
- Phylogenetic trees
- Metabolic pathways
- Chemical bonding patterns in molecules
- Citation patterns in scientific literature
- The World Wide Web

Robustness and redundancy

Biological systems need to be robust, both for survival of individuals under stress and for the plasticity required for evolution. In yeast, for example, single gene knockouts of over 80% of the ~6200 open reading frames are survivable injuries.

In principle, networks can achieve robustness through redundancy. The most direct mechanism is simple **substitutional redundancy**: if two proteins are each capable of doing a job, knock out one and the other takes over. In the London Underground, this would correspond to a second line running over the same route. For instance, when the Circle Line is not running, passengers travelling between Paddington and King's Cross stations can travel by the District and the Hammersmith & City lines running on the same tracks.

In cells, some genes have closely related homologues resulting from gene duplication, and some of these contribute to substitutional redundancy. For example, in establishing mouse models for diabetes it appeared that mice and rats (but not humans) have two similar but non-allelic insulin genes. However, substitutional redundancy requires equivalence not only of function but also of control of expression. In the mouse, knocking out either insulin gene leads to compensatory increased expression of the other, resulting in a normal phenotype.

Equivalent expression patterns are more probable among duplicated genes than among unrelated ones. For example, *E. coli* contains two fructose-1,6-bisphosphate aldolases. One, expressed only in the presence of special nutrients, is non-essential under normal growth conditions. However, the other is essential. In this case functional redundancy does *not* provide robustness. These two enzymes are probably homologous, but they are distant relatives, not the product of a recent gene duplication. One is a member of a family of fructose-1,6-bisphosphate aldolases typical of bacteria and eukarya, and the other is a member of another family that occurs in archaea. *E. coli* is unusual in containing both.

An alternative mechanism of network robustness is **distributed redundancy**: the same effect achieved through different routes. In normal *E. coli* approximately two-thirds of the NADPH produced in metabolism arises via the pentose phosphate shunt, which requires the enzyme glucose-6-phosphate dehydrogenase. Knocking out the gene for this enzyme leads to metabolic shifts, after which increased NADH produced by the tricarboxylic acid cycle is converted to NADPH by a transhydrogenase reaction. The growth rate of the knockout strain is comparable with that of the parent.

Connectivity in networks

If V_A and V_Z are vertices in a graph, a **path** from V_A to V_Z is a series of vertices: $V_A, V_B, V_C, \ldots V_Z$, such that an edge in the graph connects each successive pair of vertices. The number of vertices in the chain is called the **length** of the path. A **cycle** is a path of length >2 in a non-directed graph for which the initial and final end-points are the same, but in which no intermediate link is repeated.

A graph that contains a path between any two vertices is called **connected**. Alternatively, a graph may split into several connected components. The graph in the Box on page 420 contains two connected components, one containing five vertices and one containing only one vertex. (In the extreme, a graph could contain many vertices but no edges at all.) It is often useful to determine the *shortest path* between any two nodes, and to characterize a network by the distribution of path lengths. The phrase 'Six Degrees of Separation'—the title of a play by John Guare, made into a film—refers to the assertion (attributed originally to Marconi) that if the people in the world are vertices of a graph and the graph contains an edge whenever two people know each other, then the graph is connected, and there is a path between any two vertices with length ≤ 6.

A **tree** is a special form of graph. A tree is a connected graph containing only one path between each pair of vertices. A hierarchy is a tree: examples include military chains of command, and Linnaean taxonomy. Note that some family trees are not trees in the mathematical sense; examples are plentiful in the Royal families of Europe. A tree cannot contain a cycle: if it did, there would be two paths from the initial point (= the final point) to each intermediate point. In the graph in the Box on page

The Enzyme Commission classification of protein function is a tree; the Gene Ontology classification is not.

420, the subgraph consisting of vertices V_1, V_2, V_4, V_5 and V_6 is a tree. Adding an edge from V_1 to V_5 would create an alternative path from V_1 to V_5, and the cycle $V_1 \rightarrow V_2 \rightarrow V_4 \rightarrow V_5 \rightarrow V_1$; the graph is no longer a tree.

The **density of connections** = the mean number of edges per vertex, characterizes the structure of a graph. A fully connected graph of N vertices has $N - 1$ connections per vertex; a graph with no edges has 0. Nervous systems of higher animals achieve their power not only by containing large number of neurons but also by high connectivities.

In some systems there are limits on numbers of connections: For many human societies, in the graph in which individuals are the vertices, and edges link people married to each other, each node has connectivity 0 or 1. In hydrocarbons, the graphs in which carbon and hydrogen atoms are the vertices and edges link atoms bonded to each other, each node has ≤ 4 connections. In other networks, connectivities follow observable regularities (see Box: '*Small-world*' *networks*). For instance, the World Wide Web can be considered as a directed graph. Individual documents are the nodes, and hyperlinks are the edges. It is observed that the distribution of incoming and outgoing links follow power laws: $P(k)$ = probability of k edges k^{-q}, where $q = 2.1$ for incoming links, and $q = 2.45$ for outgoing links.

'Small-world' networks

Many observed networks, including biological networks, the World Wide Web and electric power distribution grids, have the characteristics of high clustering and short path lengths. They include relatively few nodes with very large numbers of connections, called 'hubs', and many that contain few connections. These combine to produce short path lengths between all nodes. From this feature they are called 'small-world' networks. Such networks tend to be fairly robust—staying connected after failure of random nodes. Failure of a hub would be disastrous, but is unlikely, because there are so few of them.

Many networks, notably the World Wide Web, are continuously adding nodes. The connectivity distribution tends to remain fairly constant as the network grows. These are called 'scale-free' networks.

The density of connections is very important in defining the properties of a network. For instance, the interactions that spread disease among humans and/or animals form a network. Whether a disease will cause an epidemic depends not only on the ease of transmission in any particular interaction, but also on the density of connections. As the density of connections—the rate of interactions—increases, the system can exhibit a *qualitative* change in behaviour, analogous to a phase change in physical chemistry, from a situation in which the disease remains under control to an epidemic spreading through an entire population. The classic approach of 'quarantine'—isolating people for 40 days—works by cutting down the degree of connectivity of the disease transmission network. Note that a carrier who shows no symptoms—'Typhoid Mary' was a classic case—serves as a hub of the disease transmission network.

Mary Mallon (1869–1938) presented the following unfortunate combination of features: (1) she was infected with typhoid, (2) she did not show symptoms and (3) she worked for many families as a cook.

Two historical epidemics associated with wars demonstrate the distinction between topology and geometry in network connectivity. In the early years of The Peloponnesian War, Athens suffered a severe epidemic. (Despite Thucydides' detailed description of the symptoms, the disease has not been definitively identified, but was probably bubonic plague.) A factor contributing to its transmission was the crowding of people into the city from the more militarily vulnerable surrounding countryside. After the First World War, an epidemic of influenza killed an estimated 20 million people, more than died in the war itself. Long-distance travel by soldiers returning from the war helped spread the disease. Any epidemic needs an infectious agent, and a high density of routes of transmission. These examples show that the controlling factor is the density of the *connections* and not the density of the people.

A change in behaviour analogous to the transition to an epidemic appears in nuclear fission. In a sample of Uranium–235, decaying nuclei produce neutrons that can trigger fission of other atoms. If the sample is small, so many secondary neutrons are lost through the surface that the sample remains stable. Above a critical mass, enough neutrons are captured within the sample to create a chain reaction. If the atoms are vertices of a graph, and the edges are the trajectories of neutrons from one atom to another, the change in behaviour can be seen as the effect of increasing the connectivity of a network. (The background to Michael Frayn's recent popular play, *Copenhagen*, involves the attempts, before and during the Second World War, to estimate the size of the critical mass, in order to determine whether nuclear explosions would be feasible.)

Dynamics, stability and robustness

An unlabelled, undirected graph gives a *static* structure of the topology of a network. For our molecular interaction networks, this may be an adequate description of many of the physical interactions.

For some networks, such as metabolic pathways or patterns of traffic in cities, the *dynamics* of the system depends on the transmission capacities of the individual links. These capacities can be indicated as labels of the edges of the graph. This allows modelling of patterns of flow through the network. Examples include route planning, in travel or deliveries. Note that the shortest path may well not give optimal throughput. In many cities, taxi drivers are exquisitely sensitive—and insensitively garrulous—about optimal traffic paths.

In molecular biology, metabolic pathways and signal transduction cascades are networks that lend themselves to pathway and flow analysis. Even optimal sequence alignment by dynamic programming (see Chapter 5) involves determining the optimal path through an edit graph.

Although much is known about the mechanisms of individual elements of control and signalling pathways, understanding their integration is a subject of current research. For instance, the idea that healthy cells and organisms are in stable states is certainly no more than an approximation (and in most cases a gross idealization). The description of the actual dynamic state of the metabolic and regulatory networks is a very delicate problem. Understanding *how* cells achieve even an apparent approximation to stability is also quite tricky. It is likely that great redundancy of control processes lies at its basis. Regulation is based on the resultant of many individual control mechanisms—here a short feedback loop, there a multistep cascade. Somehow the independent actions of all the individual signals combine to achieve an overall, integrated result. It is like the operation of the 'invisible hand' that, according to Adam Smith, coordinates individual behaviour into the regulation of national economies.

Several types of dynamic states of a network are possible (see Box):

- Equilibrium
- Steady state
- States that vary periodically
- Unfolding of developmental programmes
- Chaotic states
- Runaway or divergence
- Shutdown

States of a network of processes

- At **equilibrium** one or more forward and reverse processes occur at compensating rates, to leave the amounts of different substances unchanging:

$$A \rightleftharpoons B$$

Chemical equilibria are generally self-adjusting upon changes in conditions, or in concentrations of reactants or products.

- A **steady state** will exist if the total rate of processes that produce a substance is the same as the total rate of processes that consume it. For instance, the two-step conversion

$$A \longrightarrow B \longrightarrow C$$

could maintain the amount of B constant, provided that the rate of production of B (the process $A \rightarrow B$) is the same as the rate of its consumption (the process $B \rightarrow C$). The net effect would be to convert A to C.

A cyclic process could maintain a steady state in all its components:

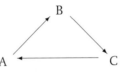

A steady state in such a cyclic process with all reactions proceeding in one direction is very different from an equilibrium state. Nevertheless, in some

$\longrightarrow$

→ _____

States of a network of processes *(continued)*

cases, it is still true that altering external conditions produces a shift to another, neighbouring, steady state.

- **States that vary periodically** appear in the regulation of the cell cycle, circadian rhythms, and seasonal changes such as annual patterns of breeding in animals and flowering in plants. Circadian and seasonal cycles have their origins in the regular progressions of the day and year, but have evolved a certain degree of internalization.

- Many equilibrium and some steady-state conditions are **stable**, in the sense that concentrations of most metabolites are changing slowly if at all, and the system is robust to small changes in external conditions. The alternative is a **chaotic state**, in which small changes in conditions can cause very large responses. Weather is a chaotic system: the meteorologist Lorenz asked, 'Does the flap of a butterfly's wings in Brazil set off a tornado in Texas?' In a carefully regulated system, chaos is usually well worth avoiding, and it is likely that life has evolved to damp down the responses to the kinds of fluctuations that might give rise to it. Chaotic dynamics does sometimes produce approximations to stable states—these are called **strange attractors**. Understanding stability in dynamic systems subject to changing environmental stimuli is an important topic, but beyond the scope of this book.

- **Unfolding of developmental programmes** occurs over the course of the lifetime of the cell or organism. Many developmental events are relatively independent of external conditions, and are controlled primarily by internal regulation of gene expression patterns.

- **Runaway or divergence**. Breakdown in control over cellular proliferation leads to unconstrained growth, in cancer.

- **Shutdown** is part of the picture. Apoptosis is the programmed death of a cell, as part of normal developmental processes, or in response to damage that could threaten the organism, such as DNA strand breaks. Breakdown of mechanisms of apoptosis—for instance, mutations in protein p53—is an important cause of cancer.

The background of systems biology: sources of ideas

Several related ideas are important in coping with the static and dynamic aspects of systems biology. These include **complexity**, **entropy**, **randomness**, **redundancy**, **robustness**, **predictablility** and **chaos**. We deal with these in our daily lives, but without the need to define them precisely and quantitatively. How well do we really understand these concepts? What are the relationships among them? And how can they be used to illuminate biology in general and systems biology in particular?

Complexity of sequences

The simplest complex object in biology is a sequence. We have all heard of random sequences, and probably agree that the more random the sequence the more complex it is. For example, genomic sequences contain 'low-complexity' regions. In the human genome, such regions include simple repeats, or microsatellites, or regions of highly skewed nucleotide composition such as AT-rich or GC-rich regions, or polypurine and polypyrimidine stretches. Are these regions more, or less, random than a region containing a gene that encodes a specific protein? How can such properties of sequences be measured?

Take a sequence of characters:

AGTCTCTA..., or AATAAAAATAAA..., or ABZXUVJFLT....

What determines the amount of information needed to specify the next character in each sequence? Less information is required if the set of possible characters—A, T, G, C—is very small, or if the distribution is very skewed—AATAAAAATAAA—than if the set is very large and the ratio of different characters is more even.

How can we make this quantitative? Genomic sequences are limited to characters A, T, G and C. To identify each symbol it is enough to ask two 'yes-or-no' questions. For instance,

> Question 1: Is it a purine (or a pyrimidine)? (Purine implies it's A or G.)
> Question 2: Is it 6-amino (or 6-keto)? (6-amino implies it's A or C.)

Knowing the answer to these two questions is enough for us to identify one of the four bases uniquely.

Representing yes = 1 and no = 0, each 'yes-or-no' question provides one binary digit, or **1 bit** of information. We could encode each nucleotide of a genome sequence as a **2-bit** binary string.

To identify a character of the ordinary alphabet–*abcd...z*—requires *more* than two yes–no questions. It is therefore reasonable to think that a character string of full text is more complex than a genomic sequence of the same length containing only characters A, T, G and C.

Questions of how much information is needed to specify an amino acid appear in the genetic code itself. How many nucleotides are required to encode 20 amino acids? If each position in a gene can contain one of four nucleotides, then there are only 16 possible dinucleotides—not enough. So, if the same number of nucleotides is to be required for each amino acid, there must be at least three nucleotides per codon, as observed. Because there are only 20 amino acids, the triplet code contains redundancy.

Shannon's definition of entropy

In 1948, C.E. Shannon introduced the concept of **entropy** into information theory, as part of his analysis of signal transmission. Suppose a text contains symbols with

Why not make do with fewer amino acids? If 15 amino acids (plus a STOP signal)—not unreasonable—would suffice, then from the information point of view a doublet code would be possible. However, a two-base codon–two-base anticodon interaction would probably not have adequate stability.

relative probability p_i. Shannon's measure of entropy is:

$$H = -\sum_i p_i \log_2 p_i$$

The entropy H can be interpreted as the minimum average number of bits per symbol required to transmit the sequence.

For example, for a genomic sequence with equimolar base composition, $p_G = p_C = p_G = p_C = 0.25$,

$$H = -\sum_i p_i \log_2 p_i$$

$$= -[0.25 \log_2 0.25 + 0.25 \log_2 0.25 + 0.25 \log_2 0.25 + 0.25 \log_2 0.25]$$

$$= 2$$

(Note that $\log_2 0.25 = \log_2 \frac{1}{4} = -2$.)

The result $H = 2$ for the gene sequence with equimolar base composition recovers our informal result that 2 bits, or two 'yes or no' questions, are required. For a sequence limited to *two* equiprobable characters A and T: $p_A = p_T = 0.5$, $H = -[0.5 \log_2 0.5 + 0.5 \log_2 0.5] = 1$. This also makes sense because, knowing that the only choices are A and T, we can decide which it is with *one* 'yes-or-no' question, or 1 bit.

Suppose that a sequence is known to have the skewed nucleotide composition: $p_A = p_T = 0.42$, and $p_G = p_C = 0.08$, Then

$$H = -[0.42 \log_2 0.42 + 0.42 \log_2 0.42 + 0.08 \log_2 0.08 + 0.08 \log_2 0.08] = 1.63$$

What is the significance of the fact that the value $H = 1.63$ is less than 2? It suggests that we might be able to encode the sequence with fewer than 2 bits/character, on average.

The Morse code for telegraphy took such advantage of unequal letter distribution frequencies to encode common letters with short sequences and uncommon letters with longer one. For instance E = dot (length one) and J = dot–dash–dash–dash (length four). Note that to take advantage of entropy values lower than those corresponding to equal distributions of characters requires variable-length encoding. Huffman devised an algorithm for assigning length-optimal codes to symbols knowing their relative probabilities.

It would be difficult to devise a Morse code for single nucleotides because the fact that we can easily encode them with no more than 2 bits doesn't give us much room to play with; after all, we can't easily subdivide a single bit. However, consider encoding a genome sequence at the trinucleotide level. Assume that there is no bias in trinucleotide frequencies other than that expected from the mononucleotide frequencies (i.e. $p_{ATC} = p_A \times p_T \times p_C$, etc.) There are 64 trinucleotides to encode. Six bits per triplet obviously suffice, but for the skewed distribution $p_A = p_T = 0.42$, $p_G = p_C = 0.08$, $H = 4.9$. We could encode the sequence using 5 bits per trinucleotide instead of 6.

The uncertainty in each transmitted symbol is not complete—it is more likely to be A or T than G or C. In principle, we can use this knowledge to improve the coding efficiency.

Conversely, looking at distributions of oligonucleotides (dinucleotides, triplets, etc.) is a useful way to detect biologically significant patterns. Codon usage patterns in protein-coding regions are examples. Some algorithms for gene identification make use of biases, in coding regions, of frequencies of hexanucleotides.

Although the actual genetic code does not achieve the theoretical efficiency that entropy calculations suggest, and indeed there does not even seem to be selection for reduction in the size of non-viral genomes, it is clear that the redundancy in the genetic code has biological significance. Many single-base mutations are silent. Conservative mutations allow proteins to evolve with small non-lethal changes that, cumulatively, can achieve large changes in structure and function. And of course the redundancy in having two copies of the genetic information in two strands of DNA is used to detect and correct errors in replication and translation, and to repair DNA damage.

Randomness of sequences

The Shannon entropy of sequences is related to the idea of randomness, another concept that we know from everyday life without worrying too much about exactly what it means. A.N. Kolmogorov defined, as a quantitative measure of the randomness of a sequence of numbers, *the length of the shortest computer program that can reproduce the sequence*. Thus the sequence $0, 0, 0, 0, 0, 0, 0 \ldots$ is far from random, as it is the output of the very short program:

Step 1:	print 0.
Step 2:	go back to step 1.

Periodic sequences, such as: Monday, Tuesday, Wednesday, Thursday, Friday, Saturday, Sunday, Monday, . . . are also of low complexity. In contrast, *a truly random sequence has no description shorter than the sequence itself.*

The relationship between complexity, randomness and compressibility

One way to shorten the specification of a non-random sequence is to compress it. We all use compression algorithms on our files to save disk space. If a sequence is truly random, in the sense of Kolmogorov, it cannot be compressed. By definition, non-random sequences *can* be compressed.

One basic principle of compression is that: *if you can predict what is coming next, you can compress effectively.*

The reason that sequences such as $0, 0, 0, 0, \ldots$ and Monday, Tuesday, Wednesday, Thursday, Friday, Saturday, Sunday, Monday, . . . are so effectively compressible is that it is simple to decide what the successor of any element is. Even sequences for which it is not possible to decide unambiguously what the next element is can be compressed if some indications are available. It is not even necessary that the rules be supplied 'up front' as they can be for sequences such as $0, 0, 0, 0, \ldots,$ and Monday, Tuesday, Wednesday, Thursday, Friday, Saturday, Sunday, Monday, . . . The rules and statistics

Shannon entropy is linked with thermodynamic entropy through the general notion of disorder or randomness. The relationship has been explored by physicists, including J.C. Maxwell and L. Szilard, in their discussions of 'Maxwell's demon', and by E.T. Jaynes.

of prediction of a successor can be generated on the fly from the incoming data. The rule, 'the weather tomorrow is likely to be the same as the weather today' would—in most places—be good enough for effective compression of a series of daily weather reports.

Putting together these considerations suggests a general idea that *the harder it is to predict the contents of a data set from a subset of the data, the more complex the data set is.*

The relationships among complexity, predictability and compressibility, which we have so far described for character strings, apply to the static structures of other types of objects, including images, three-dimensional structures and—especially—networks. Indeed, most types of biological data can be regarded as networks. For instance, a nucleotide sequence is equivalent to a network in which the individual bases are the nodes, and each base is connected by a directed edge pointing to the next base. That's a perfectly proper graph! Conversely, recognizing that sequences are networks can usefully lead us to ask: can we define analogues of sequence alignment for more general networks? (Yes, we can.)

Complexity of other types of biological data

A paradox in applying Kolmogorov's ideas to protein structures is that the shortest representation of a protein structure is an amino-acid or DNA sequence!

Many types of biological data are not sequences. These include *static* data, such as structures, gene expression patterns measured with microarrays, and regulatory networks; and *dynamic* data, describing processes.

For static data, generalizations of Kolmogorov's approach are suitable for defining complexity. The description of the complexity of a *process* is a matter of more delicacy.

Computational complexity

Perhaps the best-developed area of analysis of complexity of *processes* comes from studies of the complexities of computational problems.

An algorithm in computer science defines a process for solving a computational problem. For some problems, the execution time required to solve it is directly proportional to the size of the problem. These are said to be of order $\mathcal{O}(N)$ (read 'Oh-N'). For instance, searching for a number in an *unsorted* table requires an execution time proportional to the length N of the table, $\mathcal{O}(N)$. For some problems, the execution time increases only as $N \log N$. Sorting a list is an $\mathcal{O}(N \log N)$ problem. For some problems, the execution time increases as a power N^2, N^3, ... The alignment by dynamic programming (see Chapter 5) of two sequences, both of length N, by dynamic programming is an $\mathcal{O}(N^2)$ problem. These are called **polynomial time problems**. Still other problems have even greater time demands. Enumerating all subsets of a set containing N members is $\mathcal{O}(2^N)$.

Computer scientists define the complexity of a problem in terms of the dependence of execution time on problem size (see Box: *Classes P and NP*).

In principle, constraints of computational complexity apply to biological systems much as to any other kind of computer. Computational complexity describes the complexity of the *problem*, not the complexity of the device that solves it. However, classical computational complexity theory applies to computers that execute programs *sequentially.* Biological computers do lots of parallel processing. This allows them to solve problems of substantial complexity. For instance, the regulatory activities that biological systems carry out are complicated non-linear optimization calculations. Prediction of protein structure from amino acid sequence is an example. Another, described by Sydney Brenner, is the growth of bacteria in heavy water. Changing from H_2O to D_2O has the effect of changing the kinetic constants of many enzymatic reactions. After a relatively short period, cells readjust and resume activity and growth.

Classes P and NP

A problem that can be solved in polynomial time is said to be in **class P**. $\mathcal{O}(N \log N)$ algorithms are faster than $\mathcal{O}(N^2)$, and are therefore in class P.

   Suppose on the other hand that the optimal algorithm to solve a problem has order worse than polynomial—for instance it might have exponential order $\mathcal{O}(2^N)$—but that if you *propose* a solution it can be *checked* in polynomial time. Such a problem is said to be of **class NP**. (NP does *not* stand for non-polynomial, but for non-deterministic polynomial, referring to a different model for the computation. Don't worry about this technical distinction.)

   Consider the problem of sorting a list of numbers into order. That is, given a series of N numbers: 2, 1, 7, 5, 8, 4, 3, …; an algorithm must produce as output the numbers rearranged into order: 1, 2, 3, 4, 5, 7, 8, …. Whatever the order of the optimal algorithm that *solves* the problem, an algorithm to *verify* that 1, 2, 3, 4, 5, 7, 8, … is a solution (or that 1, 8, 7, 2, 4, 5, 3, …, is *not* a solution) can run in time linear in the length of the list. It is necessary only to check that each number is equal to or greater than its predecessor, which can be done by looking at each element of the list once. Therefore, sorting a list of numbers into order is a problem in class NP. [Sorting happens also to be in class P; sorting algorithms are known with order $\mathcal{O}(N \log N)$.]

   For many problems, we don't know whether any polynomial time algorithm exists.

NP-complete problems. Does P = NP?

Many NP problems have equivalent complexities, in the sense that if a polynomial algorithm were discovered for one, it could be applied to solve others. The set of NP-complete problems is the set of NP problems, such that if we could solve any one of them in polynomial time, we would be able to solve all of them in polynomial time. In other words, the discovery of a polynomial time algorithm for *any* problem known to be NP complete would cause the classes P and NP-complete to coalesce. But are there *any* NP problems that are NOT in class P? This is the famous unsolved conjecture of computer science: Does P = NP? (see Fig. 7.8).

$\longrightarrow$

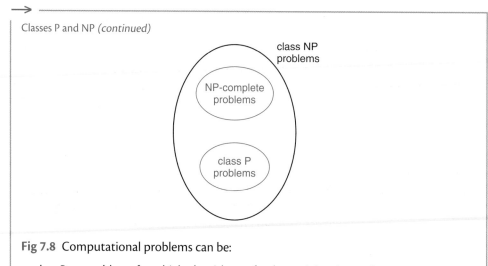

Classes P and NP *(continued)*

Fig 7.8 Computational problems can be:

- class P = problems for which algorithms of polynomial order are known
- class NP = problems, for which proposed answers can be checked in polynomial time
- NP-complete = a set of problems for which no algorithms of asymptotic polynomial time order are known but which are reducible to one another in the sense that the discovery of an algorithm of asymptotic polynomial time order for one of them (proving it to be of class P) would show that all NP-complete problems are of class P, or

If P = NP, class P would expand to fill the entire class NP set.

Static and dynamic complexity

One dimension of complexity is time. Is it possible to distinguish static from dynamic complexity? If we could define and measure the static complexity of a system, this would provide an approach to dynamic complexity: we could ask how the static complexity of a system changes with time.

For example, a program that sorts a list of numbers into order may proceed through a series of steps in which the numbers appear ordered to progressively greater extents. The entropy, or the randomness, of the list may steadily decrease. This provides an important connection between *complexity of static data* and *complexity of process*. We can collect the historical records of a process, and treat them as a succession of cases of static data. We can apply ideas of predictablity and complexity of structures to these historical records, to give insight into the changes in complexity of the system during the process.

For real physical processes, changes in complexity over time appear to be governed by some general rules. If you stop someone on the street, they might well say that in closed systems the laws of thermodynamics *require* that structural complexity always increases in natural processes. Other passers-by might say that the solar system is structurally complex but, ignoring tidal effects, dynamically simple. Will these statements hold up to rigorous analysis?

Within classical Newtonian mechanics, we could base an analysis of dynamic complexity on the definition and description of the trajectories of a system of particles.

From the Kolmogorov point of view, the initial positions and velocities of the particles, knowledge of the forces between them, and Newton's laws of motion together provide a concise description of the dynamics of such a system.

However, even within the framework of classical dynamics, this concise description can break down in the case of chaotic states. In chaotic states, very small changes in the initial conditions can lead to very large changes in the ensuing trajectories. Prediction of the dynamics requires very precise statement of the initial conditions, and very precise knowledge of the forces. Specification of the information required to describe the dynamics cannot in these cases be concise. Chaos is an extreme form of dynamic complexity.

Another way to look at this is directly relevant to systems biology: the dynamics of non-chaotic systems are robust to small changes in initial conditions. The dynamics of chaotic systems are not robust to small changes in initial conditions.

> The original meaning of the word chaos (from the Greek word for vast empty void) suggests a structural significance, but modern physics, since Maxwell, has given it a dynamic one.

Chaos and predictability

The discovery of the laws of mechanics in the seventeenth century—Newton's *Principia* was published in 1687—gave rise to the hope that the dynamics of the solar system in particular (and much if not all of the universe in general) was predictable. Laplace expressed the view that:

If we can imagine a consciousness great enough to know the exact locations and velocities of all the objects in the universe at the present instant, as well as all forces, then there could be no secrets from this consciousness. It could calculate anything about the past or future from the laws of cause and effect.

Leaving aside philosophical questions of the implications about free will and responsibility, there are also issues of computability. How much information do we really need, and how accurately do we need it, to predict the dynamics of the solar system? the weather? the universe? In chaotic systems, accurate prediction of the dynamic development requires unachievably accurate knowledge of the initial conditions. (At the atomic level Heisenberg's uncertainty principle killed off Laplace's hope of perfect determinism.)

It is true that in classical mechanics, even chaotic systems are subject to Poincaré's recurrence principle: any system of particles held at fixed total energy will eventually return arbitrarily closely to any set of initial positions and velocities. (What rescues the second law of thermodynamics is that the closer the reapproach demanded, the longer the time required; that is, the rarer the fluctuations that achieve the recurrence.) However, knowing that the configuration will recur does not simplify the calculation of the trajectories of the particles.

Through unpredictability, chaotic dynamics is associated with complexity. However, chaotic dynamics is not entirely incompatible with order and even the 'spontaneous' generation of order. In governing the time course of evolution of a system, chaotic dynamics does sometimes produce stable states or approximations to stable states—these are called **attractors.** Sometimes these are unique points; in other cases they are periodic and/or localized states. There have been examples of apparent generation of order in model systems evolving 'at the edge of chaos'.

There are even examples of static or structural order in chaotic systems. Many sequences associated with chaotic behaviour have a **fractal** structure. This means that if an object is dissected into parts, the parts have a structure similar to that of the whole (as well as to one another). B. Mandelbrot has produced many beautiful images. This self-similarity at different scales implies that if we know part of such a structure we can predict a larger segment of it. This should recall the idea that predictability should permit compressibility, and effectively reduce complexity. Indeed, such internal structural relationships have been applied to compression. Fractal image compression is an effective tool for reducing the sizes of images, to a form from which the recovered image is not exactly the same as the starting image but perceptually equivalent.

Fractal structures in biology include branching patterns of plants, and of the circulatory systems of vertebrates. At the molecular level, the storage polysaccharide glycogen has features of a fractal structure.

Alignment of metabolic pathways

Metabolic pathways provide interesting examples of the generalization of ideas of alignment from sequences to more general networks.

Alignment of two or more character strings is the assignment of correspondences between positions in the strings, usually preserving the relative order. The constraint that relative order must be conserved means that:

is an allowable alignment, but

is not. The concept of alignment, including the relative-order constraint, carries over fairly directly to protein structures, because of the linear chemistry of the polypeptide chain: a structural alignment is still a correspondence between the amino acid *sequences*, despite an appeal to three-dimensional data to determine it.

However, many objects of interest in bioinformatics have a fundamentally non-linear structure. These include the most general networks, such as sets of regulatory interactions among transcription factors. How does the concept of alignment generalize?

Metabolic pathways are an interesting example. Some present themselves as linear sequences; others are higher dimensional.

Comparing linear metabolic pathways

Many linear metabolic pathways are extractable from general metabolic networks. In principle, alignment of linear metabolic pathways is directly analogous to alignment of any other sequences. The extension to alignment of non-linear metabolic pathways takes us off the familiar track.

How we characterize steps in metabolic pathways depends on the kinds of questions we want to explore.

In its simplest form, a metabolic pathway is a sequence of metabolites. Associated with each step, in each organism, is an enzyme. In many cases, the steps of the pathway are catalysed in different species (or different tissues) by homologous enzymes. But not always. It may therefore be of interest to include, in characterizing the steps of a pathway, information about the enzyme, in addition to the substrate and product. Associated with each enzyme is also a gene. In some cases, for example the tryptophan synthesis pathway in *E. coli,* the genes for successive steps of the pathway are collinear in the genome with the steps of the pathway (see Fig. 2.1). Alignment methods can detect this. The results can be useful in assigning function to proteins in newly sequenced genomes (see page 216).

Phylogenetic trees from metabolic pathways

Chorismate synthesis illustrates conservation on a metabolic pathway between archaea and bacteria, with loss of homology in one of the enzymes. Metabolic pathways themselves may differ among species. Some pathways appear in limited sets of taxa: Capture of sunlight energy by photosynthesis occurs only in higher plants, and some algae and prokaryotes; only certain prokaryotes fix nitrogen; only certain archaea form methane from carbon dioxide and hydrogen.

Some pathways have common and divergent segments: Yeast under anaerobic conditions converts glucose to pyruvate to ethanol. *Lactobacillus* shares the pathway from glucose to pyruvate, but converts pyruvate to lactate. A novel catalytic activity required during evolution of pathways usually arises by recruitment of an enzyme with a related mechanism—after gene duplication or horizontal gene transfer—followed by evolution of specificity.

In other cases, one transformation can follow different routes. Yeast, *Lactobacillus,* and many other species, including humans, convert glucose to pyruvate by glycolysis. Some bacteria use an alternative route from glucose to pyruvate—the Entner-Doudoroff pathway—with few intermediates in common with glycolysis. Many organisms have both pathways.

Metabolic pathways characterize species. Just as we construct phylogenetic trees from similarities in DNA and protein sequences, metabolic pathways can also serve as a basis of classification. Many phylogenetic trees based on metabolic pathways are similar in topology to corresponding sequence-based ones. They do not reflect differences in individual enzymes, such as the non-homologous archaeal/bacterial shikimate kinases. This confers some immunity to effects of horizontal gene transfer. However, phylogenetic trees based on metabolic pathways may reflect convergent evolution—common adaptations of distantly-related organisms to similar environments. Convergent evolution disturbs the link between hierarchies of similarity and evolutionary relationships.

Genome sequences and metabolic pathway repertoires are two complementary aspects of the individuality of species.

In studies of evolution of metabolic pathways it also is useful to associate cofactors with reactions. Well known to biochemistry students is the succinyl-CoA synthetase reaction, converting succinyl-CoA to succinate in the Krebs cycle. The reaction is coupled to phosphorylation of GDP in mammals and ADP in bacteria and plants.

Some differences in pathways between organisms are common knowledge. A vitamin is by definition *not* the product of a metabolic pathway. Humans and other primates require a diet containing vitamin C, because we cannot synthesize it. Most animals can synthesize vitamin C. All those that cannot, lack the enzyme L-gulano-γ-lactone oxidase, the enzyme that catalyses the last step in the pathway, the conversion of L-gluonate to vitamin C. From the point of view of alignment of metabolic sequences, the pathway in humans is truncated, relative to that of animals such as the mouse that are competent to synthesize vitamin C. In primates there is a deletion of a large component of the gene for L-gulano-γ-lactone oxidase.

Similar considerations apply to catabolic pathways. The end-product of purine metabolism—the form in which nitrogen is excreted—differ among animals in different phyla (see Fig. 7.9). Most mammals degrade purines to allantoin, produced from uric acid by urate oxidase. Primates (and Dalmatian dogs) lack functional urate oxidase, and consequently excrete its substrate, uric acid.

The much lower solubility of uric acid relative to allantoin creates clinical problems in humans including kidney stones and gout. The drug allopurinol inhibits xanthine oxidase, the enzyme that converts hypoxanthine → xanthine → uric acid. The precursors, hypoxanthine and xanthine, are more soluble than uric acid, and are cleared much faster by the kidneys. Moreover, in a mixture of hypoxanthine, xanthine and uric acid, each solute has *independent* solubility. Therefore, formation of a precipitate is less likely from a mixed solution of hypoxanthine, xanthine and uric acid, than from a solution of the same total concentration of uric acid alone.

The enzyme hypoxanthine-guanine phosphoribosyltransferase (HGPRT) recovers degraded purines for nucleic acid synthesis. It converts hypoxanthine and guanine to AMP and GMP. Absence of HGPRT activity causes a build-up of uric acid, associated with the Lesch–Nyhan syndrome, an inherited metabolic disease. Gout and kidney stones are common symptoms, together with mental retardation and behavioural syndromes including uncontrollable lip and finger biting.

> The Lesch–Nyhan syndrome was the first unambiguous correlation of a biochemical defect with a psychological abnormality.

Comparing non-linear metabolic pathways: the pentose phosphate pathway and the Calvin–Benson cycle

The pentose phosphate pathway, and the Calvin–Benson cycle in photosynthesis, are two metabolic pathways involving transformations of sugars.

Metabolism of glucose can proceed through glycolysis and the Krebs cycle, to couple glucose oxidation to production of ATP. The pentose phosphate pathway is an alternative, that produces NADPH and ribose-5-phosphate. A cell that needs reducing power or ribose-5-phosphate for nucleic acid synthesis will divert some of its glucose metabolism through the pentose phosphate pathway. Several intermediates in the pentose phosphate pathway can be shuttled back into glycolysis.

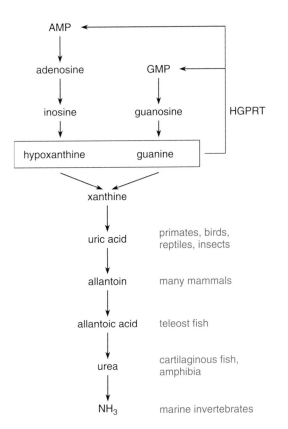

Fig. 7.9 Purine degradation in animals. In marine invertebrates, purines are broken down all the way to ammonia, which is excreted. Other phyla have truncated the purine degradation pathway to different extents, excreting urea, allantoic acid, allantoin or uric acid. The relative insolubility of uric acid causes gout and kidney stones, in humans; hyperuricaemia is treatable by allopurinol, which inhibits the conversion of xanthine to uric acid.

An alternative to excretion is conversion of hypoxanthine and guanine to nucleotides, for nucleic acid synthesis, by the enzyme hypoxanthine-guanine phosphoribosyl transferase (HGPRT). Mutations in HGPRT can lead to accumulations of uric acid, causing Lesch–Nyhan syndrome.

The Calvin–Benson cycle is the route of CO_2 fixation in photosynthesis. The enzyme ribulose-1,5-bisphosphate carboxylase (RUBISCO) couples CO_2 to ribulose-1,5-bisphosphate to form an intermediate that breaks down spontaneously to two molecules of glyceraldehyde-3-phosphate. Of every six molecules of glyceralde-3-phosphate produced, five are used to reconstitute three molecules of ribulose-1,5-bisphosphate, and the sixth harvested for energy.

The pentose phosphate pathway and the Calvin–Benson cycle share many intermediates. Several intermediates link each pathway with 'mainstream' glycolysis.

A. Sillero, V.A. Selivanov and M. Cascante presented three-dimensional diagrams of these two metabolic subnetworks, that brings out the similarities more clearly than standard two-dimensional textbook presentations do (see Fig. 7.10).

Five 3-carbon molecules → three 5-carbon molecules.

Regulatory networks

Regulatory networks pervade living processes. Control interactions are organized into linear signal transduction cascades, and reticulated into control networks.

Any individual regulatory action requires (1) a stimulus, (2) transmission of a signal to a target, (3) a response and (4) a 'reset' mechanism to restore the resting state (see Fig. 7.11). Many regulatory actions are mediated by protein–protein complexes. Transient complexes are common in regulation, as dissociation provides a natural reset mechanism.

Some stimuli arise from genetic programmes. Some regulatory events are responses to current internal metabolite concentrations. Others originate outside the cell: a signal detected by surface receptors is transmitted across the membrane to an intracellular target.

Control may be exerted:

- 'In the field': by several mechanisms, such as: inhibitors; dimerization; ligand-induced conformational changes including but not limited to allosteric effects; GDP–GTP exchange or kinase–phosphorylase switches; and differential turnover rates.

- 'At headquarters': through control over gene expression.

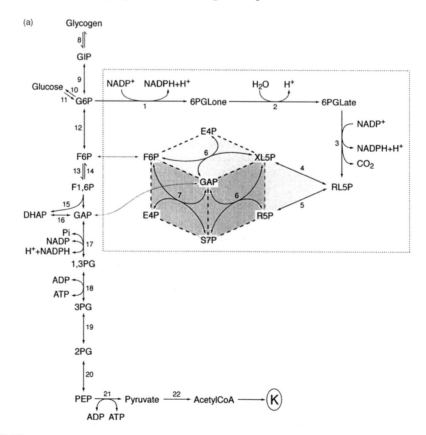

Fig. 7.10

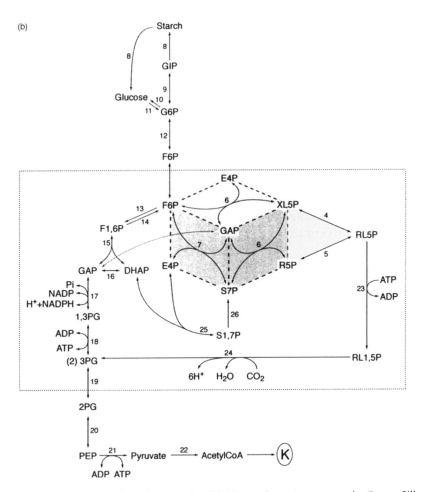

Fig. 7.10 (a) The pentose phosphate cycle. (b) The Calvin–Benson cycle. From: Sillero, A., Selivanov, V.A. and Cascante, M. (2006). Pentose phosphate and Calvin cycles: similarities and three-dimensional views. *Biochemistry and Molecular Biology Education*, **34**, 275–277.

In both (a) and (b), numbers in the figure correspond to the following enzymes: 1, glucose-6-P dehydrogenase; 2, gluconolactonase; 3, 6-P-gluconate dehydrogenase; 4, ribulose-5-P 3-epimerase; 5, ribulose-5-P-isomerase; 6, transketolase; 7, transaldolase; 8, enzymes acting in the interconversion of glucose 1-P and glycogen; 9, phosphoglucomutase; 10, glucose-6-phosphatase; 11, hexokinase; 12, phosphoglucose isomerase; 13, 6-phosphofructokinase; 14, fructose-1,6-bisphosphatase; 15, aldolase; 16, triosephosphate isomerase; 17, glyceraldehyde-3-P dehydrogenase; 18, phosphoglycerate kinase; 19, phosphoglycerate mutase; 20, enolase; 21, pyruvate kinase; 22, pyruvate dehydrogenase. K represents the Krebs cycle.

In (b), 23, phosphoribulose kinase; 24, RUBISCO; 25, transaldolase, 26, sedoheptulose-1,7-bisphosphatase. Copyright 2006. International Union of Biochemistry and Molecular Biology. Reprinted with permission of John Wiley & Sons, Inc.

One signal can trigger many responses. Each response may be stimulatory—increasing an activity—or inhibitory—decreasing an activity. Transmission of signals may damp out stimuli or amplify them. There are ample opportunities for complexity, opportunities of which cells have taken extensive advantage.

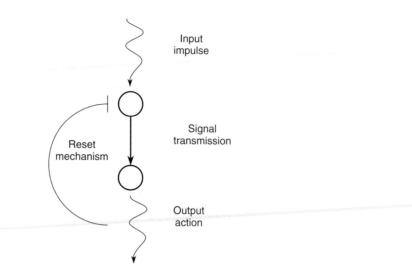

Fig. 7.11 The elementary step in a regulatory network. An input impulse is received by a node, which transmits a signal to a downstream node, causing an output action. This is followed by reset of the upstream node to its inactive state. Combination of such elementary diagrams gives rise to the complex regulatory networks in biology.

G-protein-coupled receptors (GPCRs) illustrate the components of signal transduction. Recall that GPCRs contain seven transmembrane helices, with a binding site for triggering ligands on the extracellular side, and a binding site for the downstream recipient of the signal, a heterotrimeric G-protein on the intracellular side.

G-proteins consist of three subunits: $G_\alpha G_\beta G_\gamma$. G_α and G_γ are anchored to the membrane. In the resting, inactive state, G_α binds GDP. An activated GPCR binds to a specific G-protein, and catalyses GDP–GTP exchange in the G_α subunit. This destabilizes the trimer, dissociating G_α:

$$G_\alpha(GDP)G_\beta G_\gamma \rightleftharpoons G_\alpha(GTP) + G_\beta G_\gamma$$

The separated components, G_α and $G_\beta G_\gamma$, activate downstream targets, such as adenylyl cyclase.

A single activated GPCR can interact successively with many G-protein molecules, amplifying the signal. It is therefore essential to turn the signal off, after it has had its effect. Mutations that render a GPCR constitutively active cause a number of diseases, the symptoms emerging from a war between the rogue receptor and the feedback mechanisms that are unequal to the task of restraining its effects. Different GPCRs have different mechanisms for restoring the resting state. Rhodopsin, for example, is inactivated by cleavage of the isomerized chromophore.

The activity of the heterotrimeric G-proteins is turned off by the GTPase activity of $G\alpha$, converting $G_\alpha(GTP) \rightarrow G_\alpha(GDP)$. $G_\alpha(GDP)$ does not bind to its receptors—shutting down that pathway of signal transmission. Instead, $G_\alpha(GDP)$ rebinds the $G_\beta G\gamma$ subunits. This resets the system.

Signal transduction and transcriptional control

The signal transduction network exerts control *'in the field'*, by a variety of mechanisms including: inhibitors; dimerization; ligand-induced conformational changes including but not limited to allosteric effects; GDP–GTP exchange or kinase–phosphorylase switches; and differential turnover rates. This component acts quickly, on sub-second timescales. The transcriptional regulatory network exerts control *'at headquarters'*, through control over gene expression. This component is slower, acting on a timescale of minutes.

General characteristics of all control pathways include:

* a single signal can trigger a single response or many responses

* a single response can be controlled by a single signal or influenced by many signals

* each response may be stimulatory—increasing an activity—or inhibitory—decreasing an activity

* transmission of signals may damp out stimuli or amplify them

Structures of regulatory networks

Think of control, or regulatory networks, as assemblies of *activities*. Although mediated in part by physical assemblies of macromolecules—protein–protein and protein–nucleic acid complexes—regulatory networks:

1. **Tend to be unidirectional** A transcription activator may stimulate the expression of a metabolic enzyme, but the enzyme may not be involved directly in regulating the expression of the transcription factor.

stimulatory interaction

2. **Have a logical component** It is not enough to describe the connectivity of a regulatory network. Any regulatory action may stimulate or repress the activity of its target. If two interactions combine to activate a target, activation may require *both* stimuli (logical 'and') or *either* stimulus may suffice (logical 'or').

inhibitory interaction

3. **Produce dynamic patterns** Signals may produce combinations of effects with specified time courses. Cell cycle regulation is a classic example.

autoregulatory interaction

The structure of a regulatory network can be described by a graph, in which edges indicate steps in pathways of control. Regulatory networks are directed graphs: the influence of vertex A on vertex B is expressed by a directed edge connecting A and B. An edge directed from vertex A to vertex B is called *an outgoing connection* from A and *an incoming connection* to B. Conventionally, an arrow indicates a stimulatory interaction, and a T-symbol indicates an inhibitory interaction. An edge connecting a vertex to itself indicates autoregulation. A double-headed arrow indicates reciprocal stimulation of two nodes; note that this is *not* the same as an undirected edge.

reciprocal interaction

Structural biology of regulatory networks

Any regulatory interaction involves one or more proteins and nucleic acids. Examples of regulatory mechanisms include a protein binding a ligand, undergoing chemical modification such as phosphorylation/dephosphorylation, changing conformation, or all of the above. X-ray crystallography and NMR spectroscopy have elucidated some of the general mechanisms underlying control processes.

Many molecules involved in regulation are multidomain proteins. A domain is a segment of a protein that has independent stability and can appear in conjunction with different partners through evolutionary recombination. A multidomain protein contains a linear sequence of domains each of which is relatively free to interact with other molecules. Assembly of a protein from domains therefore permits the joining into one molecule of a set of functions. 'Mixing and matching' of domains gives evolution access to a wide variety of functional combinations (see Fig. 2.4).

One important feature of regulatory proteins is recognition. An **interaction domain** is a part of a protein that confers specificity in ligation of a partner. Regulatory proteins contain a limited number of types of interaction domains, which have diverged to form large families with different individual specificities. For instance, the human genome contains 115 SH2 domains and 253 SH3 domains. (*Src*-Homology domains SH2 and SH3 are named for their homologies to domains of the src family of cytoplasmic tyrosine kinases.) Many individual interaction domains even interact with different partners as they participate in successive steps of a control cascade. Initial interactions may also trigger recruitment of additional proteins to form large regulatory complexes.

Many interaction domains are sensitive to the state of post-translational modification of their ligands, for instance binding preferentially to states of a ligand in which specific tyrosines, serines or threonines are phosphorylated. These and other post-translational modifications function as switches, turning on or interrupting/resetting a signalling cascade.

Protein–protein complex formation allows a cell to detect a signal molecule in the external medium, and report its arrival to the cell interior, without the signal molecule itself ever needing to enter the cell. Many receptors use an ingenious dimerization mechanism: the receptor has external, transmembrane and internal segments. An external ligand binds to *two* molecules of receptor. The juxtaposition of the external portions brings the internal portions together also, because they are tethered to the external regions by the transmembrane segments. Interaction between the interior segments triggers a conformational change that activates a process such as phosphorylation of a protein. This may initiate a signal transduction cascade that can amplify the original stimulus.

Figure 7.12 shows types of interaction domain complexes with ligands, including binding of peptides (which may be attached to proteins), protein–protein complexes, extracellular dimer formation upon binding a hormone, and a protein–nucleic acid complex.

Understanding the mechanism of regulation will require the structures of large protein–protein and protein–nucleic acid complexes. The sizes of many of the large complexes challenge the limits of NMR spectroscopy. X-ray diffraction has had some major successes, but is at the mercy of being able to grow adequate crystals. Cryo-electron microscopy is another approach to structure determination of larger assemblies.

Electron microscopy of specimens at liquid nitrogen temperatures has revealed structures in the range $M_r = 500\,000$ to 4×10^8, 100–1500 Å in diameter. These results do not achieve atomic resolution. However, if the structures of individual components of a complex are known to high resolution from X-ray diffraction or NMR spectroscopy, the component structures can be fitted into the low-resolution structure determined by electron microscopy, to produce a detailed model of the entire assembly.

A limitation that remains is the difficulty of determining structures of transient complexes, or of systems showing substantial conformational changes upon assembly. The situation is shared with much of current molecular biology: we are coming to grips with static structures of increasing size, but awaiting the development of methods for treating the dynamics.

The genetic switch of bacteriophage λ

A discussion of the *Lac* operon appears in *Introduction to Genomics*, chapter 7.

Two classic control systems in biology are well understood at the molecular level: the *E. coli Lac* operon, and the lytic/lysogenic switch in bacteriophage λ. These are also the simplest examples of developmental pathways.

- The *Lac* operon is a set of genes appearing in tandem on the genome of *E. coli*, that are jointly regulated in reponse to the presence of lactose and glucose in the medium.

(a)

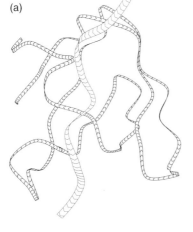
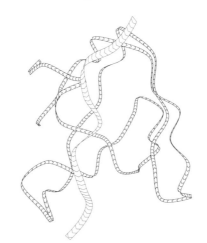

(b)

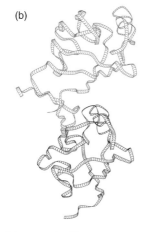

Fig. 7.12 (*See overleaf*)

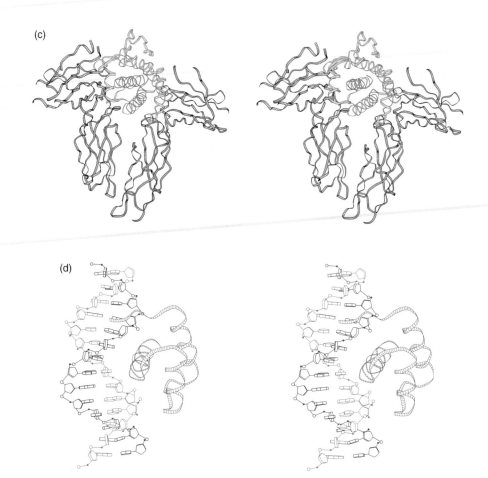

Fig. 7.12 Types of interactions involved in regulatory signalling. (a) Binding of a peptide by an SH3 domain [1CKA]. SH3 domains are common constituents of regulatory proteins. Functions of SH3 domains include signal transduction, protein and vesicle trafficking, cytoskeletal organization, cell polarization and organelle biosynthesis. (b) Domain–domain interaction: PDZ domains in syntrophin (black) and neuronal nitric oxide synthase (blue) [1QAV]. (c) Binding of a molecule of human growth hormone (blue) to two molecules of the external segment of the human growth hormone receptor (black) (d) The homeodomain antennapedia–DNA complex [9ANT]. Homeodomains are highly conserved eukaryotic proteins, active in control of animal development. They regulate homeotic genes; that is, genes that specify locations of body parts. Antennapedia is a *Drosophila* protein responsible for initiating leg development. The earliest mutations found in antennapedia produced ectopic legs at the positions of, and instead of, antennae. Loss-of-function mutations convert legs into antennae. As with many DNA-binding proteins, an α-helix binds in the major groove of DNA.

♦ Phage λ can adopt an active or passive lifestyle, effected by alternative gene expression profiles. It is probably the simplest form of life that makes a decision.

λ is a bacteriophage, a virus that infects *E. coli* (see Fig. 7.13). The mature virion contains an icosahedral head, that encapsulates the viral DNA; and a tail, that recognizes

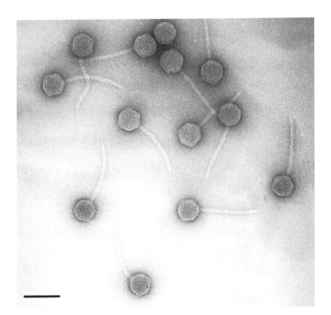

Fig. 7.13 Bacteriophage λ. Bar at lower left indicates 100 nm. (Picture courtesy Professor R.B. Inman, University of Wisconsin. From: ICTVdB — The Universal Virus Database, version 4. http://www.ncbi.nlm.nih.gov/ICTVdb/ICTVdB/)

and attaches to the host, and functions as a syringe to inject the viral DNA. The virion contains ~15 different proteins. The genome is a single molecule of double-stranded DNA 58 402 bp long, containing 50 genes, organized into seven operons. As in bacteria, an operon is a set of successive genes under coordinated transcriptional control.

After attaching to an *E. coli* cell and injecting its DNA, the phage may follow either of two paths:

- In the **lytic** state, replication and intracellular reconstitution of daughter phage particles is followed, in about 45 minutes, by rupture of the host cell and release of the ~100 progeny. The expression patterns of several distinct sets of genes are under control of a developmental programme during this process.

- In the **lysogenic** state, the phage DNA becomes integrated into the bacterial genome, to form a prophage. Only one phage gene is expressed: the cI protein, which acts as a repressor to inhibit the expression of phage genes responsible for initiating viral multiplication, thereby maintaining the lysogenic state.

Here we shall focus on the subset of the λ regulatory network involved in the switch between lytic and lysogenic states.

Given a healthy host population, the lytic state perpetuates itself as progeny viruses infect additional bacterial cells.

The lysogenic state of the phage is stable under normal conditions. The viral DNA, integrated into the bacterial genome, replicates with the bacterial DNA. This creates a population of infected bacteria. Although the virus does not reproduce completely to form intact progeny phage, the viral DNA *is* replicated, as a passenger in the dividing cells.

Sleeping Beauty can be awakened, by damage, such as UV radiation, that threatens the host cell. The virus resumes active replication, in order to escape conditions that endanger its host. Of course this *ensures* that the host cell will not survive.

The strategy of the virus is to take advantage of a thriving host population to reproduce lytically, but to adopt lysogeny to get through 'lean' periods.

What are the characteristics of the switch, that must be implemented by DNA–protein interactions?

Here we are interested more in the logic of the system than in the details of the molecular biology.

M. Ptashne and colleagues, and a large community of virologists, have clarified the molecular biology of phage λ in very great detail.

- The states of the switch must be mutually incompatible. Each state must repress the other.

- Under *constant conditions*, each state must be self-maintaining. In other words, not only is the choice of one or the other commitment enforced; once selected, the chosen state persists until conditions change.

- In response to *changing conditions*, it must be possible to move from one state to the other. We expect to find a simple trigger that leads to a complex cascade of consequences.

To implement this logic, the system has the following variables at its disposal:

- **DNA sequences**: in particular the sequences at sites of promotors and operators (see page 216).

- **Local flexibility in DNA structure**: the ability of the DNA to form loops, bridged by interacting proteins bound to sites distant in the DNA sequence.

- **DNA–protein interactions**: relative affinities of different proteins for different sites, including RNA polymerase and regulatory proteins.

- **Interactions among DNA-protein complexes**:

  - positive cooperativity: enhancement of binding by stabilizing protein–protein interactions on the DNA
  - negative cooperativity (or anticooperativity)—especially the blocking, by binding of one protein, of the binding site of another.

Sites of protein–DNA interactions in transcription control

A **promoter** is a site on DNA—typically ~60 bp in prokaryotes—near the beginning of a gene. It binds RNA polymerase, required for initiation of transcription. RNA polymerase is a bacterial enzyme. However, part of the developmental programme of lytic phage λ takes place through the modification of the bacterial polymerase by viral proteins, to alter its response to termination signals. The result is to extend the region of transcription of viral genes in successive stages of the lytic cycle.

An **operator** is a site on DNA that binds regulatory proteins. A repressor, or negative regulator, blocks the site where RNA polymerase binds to the operator, preventing transcription. A positive regulator interacts with RNA polymerase to enhance its binding affinity for a promoter.

Which proteins will bind DNA, and where, depends on the:

◆ Intrinsic affinity of different sites for different proteins

◆ Cooperativity of protein binding

◆ Competition of proteins for sites.

Availability of a site may be denied by occlusion, caused by binding of another protein at or near the site. Conversely, favourable interaction with another protein bound at a neighbouring site may enhance affinity (cooperative binding). These effects may involve interactions among regulatory proteins, or of regulatory proteins with RNA polymerase.

Principle: a DNA-binding protein will choose the available site for which it has the highest affinity.

The materials

1. Proteins

- **RNA polymerase**, the enzyme that transcribes DNA into RNA. RNA polymerase binds to available promoter sites.

- **cro**, a transcription regulator that inhibits synthesis of cI.

- **cI**, or *repressor,* a transcription regulator that inhibits expression of cro, and regulates its own expression.

2. Sites on the phage DNA

- The accessibility of two adjacent promoters controls the transcription of cI and cro.

- Three operator sites, one within each promoter and a third overlapping both, that are binding sites for cro and cI. Figure 7.14 shows the layout of promoter and operator sites on the DNA, and the two mutually exclusive states in which cro or cI are expressed.

The operation of the system depends on the relative values of the affinities of different operator sites for different proteins and for different combinations of proteins

Protein(s)	Relative affinity
RNA polymerase itself	cro promoter > cI promoter
cro	OR3 > OR2 ≃ OR1
cI	OR1 ≃ OR2 > OR3
2 × cI	OR1 + OR2 high, cooperative binding
2 × cro	No cooperative binding
cro + RNA polymerase	cro promoter > 0, cI promoter = 0
cI + RNA polymerase	cI promoter > 0, cro promoter = 0
High concentration of cI + RNA polymerase	cI promoter = 0, cro promoter = 0

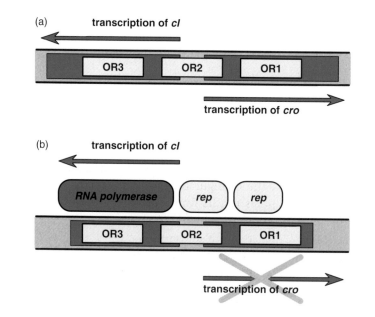

(a) transcription of *cl*

OR3   OR2   OR1

transcription of *cro*

(b) transcription of *cl*

RNA polymerase   rep   rep

OR3   OR2   OR1

transcription of *cro*

Fig. 7.14

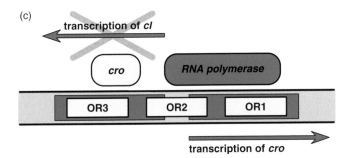

(c)

transcription of *cl*

cro   RNA polymerase

OR3   OR2   OR1

transcription of *cro*

Fig. 7.14 (a) Region of the phage λ genome containing promoters for *cro* and *cl*. (b) In the lytic state, *cro* is expressed and *cl* is off. (c) In the lysogenic state, *cro* is off and *cl* (encoding repressor) is expressed. OR1, OR2 and OR3 are operator sites—binding sites for regulatory proteins. Each is about 15–20 bp long, or roughly two turns of DNA. OR1 overlaps the *cro* promoter, OR3 overlaps the *cl* promotor, and OR2 overlaps both.
The relative affinities of *cro* and *cl* for the operator sites:

Protein	Relative affinity	Effect
cro	OR3 > OR1 = OR2	Binding of cro to OR3 prevents cl synthesis
cl	OR1 ≃ OR2 > OR3	Binding of 2 × cl to OR1/OR2 prevents cro synthesis

create alternative states:

State	cro	cl
Lytic	on	off
Lysogenic	off	on

This scheme—in particular the relative affinities of cI and and cro for the different operator sites—explains the configurations of Fig. 7.14(b) and (c). The mutual incompatibility of the two states results from binding of transcriptional regulatory proteins to the operators, repressing one of the two genes and enhancing expression of the other. Binding of cI, preferentially and cooperatively to OR1 and OR2, turns off transcription of *cro* and, through favourable interaction with RNA polymerase on the DNA, stimulates transcription of *cI*. At higher concentrations of cI, after titration of the OR1 and OR2 sites, cI will bind to OR3, turning off *cl* transcription. This acts to regulate the ambient cI concentration.

The cI concentration—about 100 molecules per cell—is high enough to prevent lytic infection by phage λ of an *E. coli* cell already containing lysogenized λ.

The diagram shows the logical relationships between these two components. High concentrations of cro inhibit its further expression. The combination of both stimulatory and repressive links from cI to itself signifies the regulation of cI concentration: the autostimulatory link is active at low cI concentration and the autorepressive link is active at high cI concentration. The phage protein cII also activates expression of cI but using a different promoter.

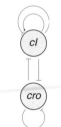

cI binds as a tetramer to OR1/OR2. There is an additional set of promoters and operators OL1, OL2 and OL3 ~2.3 kb from OR1, OR2 and OR3. A tetramer of cI, bound to OR1/OR2 and another tetramer of cI bound to OL1/OL2 can form an octamer, enhancing the affinity for DNA. To do this the DNA must loop around to allow apposition of the two tetramers.

How to 'throw' the switch

- To change from the lysogenic to the lytic state: UV irradiation or other hindrance to DNA replication causes bacterial protein RecA to cleave cI. This frees the OR1/OR2 sites to bind RNA polymerase, to express *cro*. Cro has its highest affinity for the OR3 site, turning off synthesis of cI. As the concentration of cro builds up, it also binds to OR1 and OR2, turning off its own expression. Expression of cro also initiates a cascade of events that effect the transition to the lytic state.

- The switch from lytic to lysogenic state can occur only upon infection, of necessity by a phage that has emerged from a lytic event. The phage may either remain lytic (the default) or become lysogenic.

The choice appears to be determined primarily by the concentration of a phage protein cII. cII activates transcription of:

- cI, the repressor (but via a different operator from that shown in Fig. 7.14)

- int, a protein required for integration of phage DNA into the bacterial genome

- an antisense RNA that prevents the viral modification of bacterial RNA polymerase, which shuts down the lytic programme

The transition to lysogeny requires build up of the concentration of cII. The concentrations of cII and cIII appear to depend primarily on the dose of viral DNA. About 1% of cells infected by one virus become lysogenic. About 50% of cells infected simultaneously by two or more viruses become lysogenic.

A high multiplicity of infection implies a low ratio of bacteria to phage. For the phage to remain lytic under these conditions would threaten to deplete the population of bacteria, to the point where progeny phase could not find hosts to infect. Similar considerations rationalize the greater frequency of lysogenation if the host cell is starved.

A bacterial protease HflB destroys cII, and a viral protein cIII inhibits HflB. Readers may wonder why the cell would synthesize a molecule that promotes lysis, and why the phage would synthesize one that promotes lysogeny. Both the bacterium and the phage appear to be acting to *reduce* the number of their own immediate progeny. However, there may be long-term benefits for the populations as a whole. In a population of lysogenized bacteria, occasionally a cell spontaneously goes lytic. The progeny phage do not damage the bacterial population—remember that lysogeny confers 'immunity' to phage infection—but may protect the bacterial population against competition with foreign invading susceptible bacteria. Sociobiologists may see this as an example of altruism.

It is interesting that the lytic/lysogenic choice of phage λ can be extracted as a fairly simple subset of a far more complicated regulatory network.

The genetic regulatory network of *Saccharomyces cerevisiae*

A recent study of transcription regulation in yeast treated a network containing 3562 genes, corresponding to approximately half the known proteome of *S. cerevisiae*.[3] The genes included 142 that encode transcription regulators, and 3420 that encode target genes exclusive of transcription regulators. There are 7074 known regulatory interactions among these genes, including effects of regulators on one another, and of regulators on non-regulatory targets.

Analysis of the overall network architecture reveals that:

- The distribution of incoming connections to target genes has a mean value of 2.1, and is distributed exponentially. Most target genes receive direct input from about two transcriptional regulators. The probability that a gene is controlled by k transcription regulators, $k = 1, 2, \ldots$, is proportional to $e^{-\alpha k}$, with $\alpha = 0.8$.

- The distribution of outgoing connections has a mean value of 49.8, and obeys a power law. The probability that a given transcriptional regulator controls k genes is proportional to $k^{-\beta}$, with $\beta = 0.6$. Power-law behaviour characterizes topologies in which a few nodes—the 'hubs'—have many connections, and many nodes have few. In regulatory networks, hubs tend to be fairly far upstream, forming important foci of regulation with far-reaching control.

- The average number of intermediate nodes in a minimal path between a transcriptional regulator and a target gene is 4.7. The maximal number of intermediate nodes in a path between two nodes is 12.

3 Luscombe, N.M., Babu, M.M., Yu, H., Snyder, M., Teichmann, S.A. and Gerstein, M. (2004). Genomic analysis of regulatory network dynamics reveals large topological changes. *Nature*, **431**, 308–312.

◆ The clustering coefficient of a node is a measure of the degree of local connectivity within a network. If all neighbours of a node are connected to one another, the clustering coefficient of the node = 1. If no pair of neighbours of a node is connected to each other, the clustering coefficient of the node = 0. The mean clustering coefficient, averaged over all nodes, is a measure of the overall density of the network. For the yeast transcriptional regulatory network, the mean clustering coefficient is 0.11.

Figure 7.15 is a cartoon-like sketch of a fragment of such a network, indicating rather loosely some of its general features. Nodes are divided into **transcriptional regulators**, shown as circles, and **target genes**, shown as squares. Target genes are distinguished by having no output connections. There is extensive inter-regulation among the transcription factors, to a much higher density of interconnections than can intelligibly be shown in this diagram. Think of an active network of transcription factors, within the shaded area, sending out signals to target genes. The shaded area indicates only the *logical* clustering of the transcriptional regulators. There is no

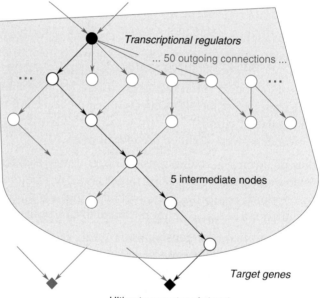

Fig. 7.15 Simplified sketch illustrating some features of an 'average' segment of the pathways in the yeast interaction network. Transcriptional regulators appear as circles. Target genes appear as squares. A transcriptional regulator typically has direct influence over about 50 genes, indicated by multiple connections from the filled black circle to the circles on the line below it. Roughly 1 in 10 of the neighbours of any node is connected to another neighbour, indicated by the horizontal arrow on the second row. The ultimate receptor of the signal lies at the end of a pathway typically containing about five intermediate nodes (shown in black). This ultimate target gene receives on average about two inputs. This diagram shows only a small fragment of a network that is in fact quite dense.

suggestion about physical localization; indeed, transcriptional regulators interact with DNA, and almost never interact physically with the proteins, the expression of which they control.

Each transcriptional regulator directly influences approximately 50 genes on average, although, as with other 'small-world' networks following power-law distributions of connectivities, the distribution is very skewed—some 'hubs' have very many output connections, but most nodes have very few. A few of the inter-regulatory connections between transcription factors are shown in blue. In about 10% of the cases, two neighbours of the same transcription factor interact with each other. A pathway from one regulator (filled black circle) to one ultimate receptor (filled black square), through five intermediate nodes, is shown in black. The intermediate nodes are other transcriptional regulators, connected both within the path drawn in black, and off this path. Even the transcription factor used as the origin of the path receives input connections. Although it is possible to identify target genes from the absence of outgoing connections, it is more difficult to identify ultimate initators of signal cascades.

The ultimate receptor is a target gene that receives regulatory input but itself has no output links. This target is expected to receive (on average) a second control input. The black target node receives input via a black arrow, along the selected path, and via a blue arrow suggesting the second input. Of course the second input may arrive via a path that shares common nodes with the black path, including other routes from the filled black circle.

The dense forest of additional pathways, from which this fragment is extracted, is not shown. Some 'back-of-the-envelope' calculations: there are ~3500 nodes, each receiving on average two input connections. There are ~140 transcription factors, making an average of 50 output connections. The number of input connections must equal the number of output connections, and indeed $3500 \times 2 = 140 \times 50 = 7000$.

Given the complexity, it is difficult to illustrate larger segments of the network in more detail than the simplified version appearing in Fig. 7.15. However, dissections of yeast and other regulatory networks have defined certain recurrent motifs that serve as building blocks. These might be considered the 'secondary structures' of network architectures. (See Box: *Common motifs in biological control networks*).

The high ratio of interactions to transcription regulators implies that we cannot expect to associate individual regulatory molecules with single, dedicated, activities (as we can, for the most part, with metabolic enzymes). Instead, the activity of the network involves the coordinated activities of many individual regulatory molecules.

Common motifs in biological control networks

Within the high complexity of typical regulatory networks, certain common patterns appear frequently. In the architecture of networks, these form building blocks which contribute to higher levels of organization. Shen-Orr et al.* have described examples including: the *fork*, the *scatter* and the *'one–two punch'* (a phrase from the boxing ring):

$\longrightarrow$

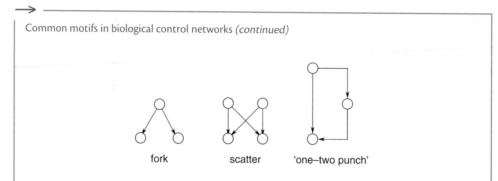

Common motifs in biological control networks *(continued)*

fork   scatter   'one–two punch'

The **fork**, also called the single-input motif, transmits a single incoming signal to two outputs. Successive forks, or forks with higher branching degrees, are an effective way to activate large sets of genes from a single impulse. Generalizations of the binary fork include more downstream genes under common control (more tines to the fork), and autoregulation of the control node. Forks can achieve general mobilization. Moreover, if the regulatory genes have different thresholds for activation, the dynamics of building up the signal can produce a temporal pattern of successive initiation of the expression of different genes.

The **scatter** configuration, also called the multiple input motif, can function as a logical 'or' operation: both downstream targets become active if *either* of the input impulses is active. Generalizations of the square scatter pattern shown may contain different numbers of nodes on both layers. Note that scatter patterns are superpositions of forks.

The **'one–two punch'**, also called the 'feed-forward loop', affects the output both directly through the vertical link, and indirectly and subsequently through the intermediate link. This motif can show interesting temporal behaviour if activation of the target requires simultaneous input from both direct and indirect paths (logical 'and'). Because build up of the intermediate requires time, the direct signal will arrive before the indirect one. Therefore, a short pulsed input to the complex will not activate the output—by the time the indirect signal builds up, the direct signal is no longer active. The system can thereby filter out transient stimuli in noisy inputs. Conversely, the active state of the system can shut down quickly upon withdrawal of the external trigger.

* Shen-Orr, S.S., Milo, R., Mangan, S. and Alon, U. (2002). Network motifs in the transcriptional regulation network of *Escherichia coli. Nature Genetics*, **31**, 64–68.

Adaptability of the yeast regulatory network

The yeast regulatory network achieves versatility and responsiveness by reconfiguring its activities. This is seen by comparing the changes in the activities of networks controlling yeast gene expression patterns in different physiological regimes: cell cycle, sporulation, diauxic shift (the change from anaerobic fermentative metabolism to aerobic respiration as O_2 levels increase), DNA damage and stress response. Cell cycling and sporulation involve the unfolding of endogenous gene expression programmes; the others are responses to environmental changes.

Different states are characterized both by similarities and differences in gene expression patterns, and by the components of the regulatory network that are active. There is considerable shift in expression of target genes. About a quarter of the target genes are specialized to individual physiological states. That is, of the total of 3420 target genes, the expression of almost half (1514) do not show major changes in the different states. Of the 1906 that show altered expression levels in different states, almost half of them (803) are specialized to a single physiological state.

In contrast, different states show much more overlap in the usage of transcriptional regulators. For instance, for cell cycle control, 280 target genes (8%) are differentially regulated by 70 (49%) of the transcription regulators. Clearly there is a much greater degree of specialization in the target genes. In general, half the transcription factors are active in at least three out of the five physiological regimes. However, in contrast to the high overlap of usage of the transcriptional regulators (the nodes), the overlap of the activities within the network (the connections) is relatively low. Different components of the interaction network organize the different gene expression patterns in different states.

Whereas different physiological states are characterized by substitutions of different sets of synthesized proteins, the regulatory network uses much of the same structure but reconfigures the pattern of activity. Think of the transcription factors as 'hardware' and the connections as reprogrammable 'software'. The molecules do not change but the interactions do: in different states, many transcription regulators change most, or a substantial part, of their interactions. In particular, the set of transcription regulators that form the hubs of the network—those with many outgoing nodes that form foci of control—are not a constant feature of the system. Some hubs are common to all states, but others step forward to take control in different physiological regimes. The result of the reconfiguration of activity is that over half of the regulatory interactions are *unique* to the different states.

The effect of the changes in the active interaction patterns is to alter the topological characteristics of the network in different states. For instance, under panic conditions—DNA damage and stress—the average number of genes under control of individual transcriptional regulators increases, the average minimal path length between regulator and target decreases, and the clustering becomes less dense (i.e. there is less inter-regulation among transcription factors). This can be understood in terms of a need for fast and general mobilization—the equivalent of broadcasting 'Go! Go! Go!' over the radio. Normal circumstances—cell cycle control for instance—allow for a more dignified and precise regulatory state, which permits finer control over the temporal course of expression patterns. In cell cycle control and sporulation, there is a much denser inter-regulation among transcription factors, and longer minimal path lengths between transcriptional regulators and target genes.

Different physiological states also differ in their usage of the common motifs—fork, scatter and 'one–two punch'. (See Box: *Common motifs in biological control networks*.) Scatter motifs are more used in conditions of stress, diauxic shift and DNA damage. They are appropriate to the need for quick action. Requirements for build up of intermediates would delay the response. Conversely, the 'one–two punch' motif is more common in cell cycle control. This is consistent with the need for a signal from one stage to be stabilized before the cell enters the next stage.

Much of evolution proceeds towards greater specialization. The human eye is a classic example. It is an intricate and fine-tuned structure, features that were once adduced as evidence *against* Darwin's theory. Many evolutionary pathways show a trade-off between specialized adaptation and generalized adaptability.

Regulatory networks are an exception. Evolution has produced structures that are both specialized *and* versatile. The reconfigurability of regulatory networks allows them to respond robustly to changes in conditions, by creating many different structures, specialized to the conditions that elicit them.

Recommended reading

Genomics and proteomics

Tyers, M. and Mann, M. (2003). From genomics to proteomics. *Nature*, **422**, 193–197.

Albert, R. and Barabási, A.-L. (2002). Statistical mechanics of complex networks. *Reviews of Modern Physics*, **74**, 47–97.

Barabási, A.-L. (2003). *Linked: How Everything is Connected to Everything Else and What it Means*. Plume Books, New York.

Ideker, T. (2004). A systems approach to discovering signaling and regulatory pathways—or, how to digest large interaction networks into relevant pieces. *Advances in Experimental Medicine and Biology*, **547**, 21–30.

Babu, M.M., Luscombe, N.M., Aravind, L., Gerstein, M. and Teichmann, S.A. (2004). Structure and evolution of transcriptional regulatory networks. *Current Opinion in Structural Biology*, **14**, 283–291.

Bechhoefer, J. (2005). Feedback for physicists: a tutorial essay on control. *Reviews of Modern Physics*, **77**, 783–836.

Phage λ

Ptashne, M. (2004). *A Genetic Switch: Phage Lambda Revisited*. Cold Spring Harbor Laboratory Press, Cold Spring Harbor, NY.

Court, D.L., Oppenheim, A.B. and Adhya, S.L. (2007). A new look at bacteriophage λ genetic networks. *Journal of Bacteriology*, **198**, 298–304.

Ptashne, M. (2005). Regulation of transcription: from lambda to eukaryotes. *Trends in Biochemical Sciences*, **30**, 275–279.

Dodd, I.B., Shearwin, K.E and Egan, J.B. (2005). Revisted gene regulation in bacteriophage λ. *Current Opinion in Genetics and Development*, **15**, 145–152.

Exercises, Problems and Weblems

Exercises

Exercise 7.1 Hen egg white lysozyme has a relative molecular mass of about 14 300. If mass spectroscopy can measure mass to within 0.01%, could the following be confidently distinguished from the unmodified protein: (a) N-terminal acetylation? (b) phosphorylation of a single serine residue? (c) a single Lys→Gln substitution?

Exercise 7.2 On photocopies of Fig. 7.5, indicate the positions of the peaks if the sequence were: (a) MNLVQVR. (b) GNLQVVR. (c) MNLQVVG.

Exercise 7.3 (a) What is the sequence of the fragment y_6 in Fig. 7.5b? (b) To which peak in Fig. 7.5b does the fragment NH_3^+LQVVR correspond?

Exercise 7.4 Oligonucleotide samples may vary by the binding of an Na^+ or K^+ ion to a phosphate, instead of a proton.

(a) What is the difference in mass between an oligonucleotide binding a proton or an Na^+ ion at a single site?

(b) What base change has the closest mass difference to the H^+–Na^+ mass difference?

(c) Would measuring mass to within 1 D be sufficient accuracy to distinguish this base change from the binding of a Na^+ ion instead of a proton, at a single site?

(d) In a mass spectrum of an oligonucleotide, what is the difference in mass between an oligonucleotide with a proton or a Mg^{2+} ion at a single site?

(e) What base change has the closest mass difference to the H^+–Mg^{2+} mass difference?

(f) Would measuring mass to within 1 D be sufficient accuracy to distinguish this base change from the binding of an Mg^{2+} ion instead of a proton, at a single site?

Exercise 7.5 Assuming a typical single-nucleotide polymorphism (SNP) density of 1 SNP/5 kb in a human genome, and only two possible bases observed at the position of any SNP, how many sequences could you expect to find throughout a population, within a 100 kb region, if recombination were common at every position in the region? If only three of the possible combinations of SNPs, i.e. three haplotypes, are observed, what fraction of possible sequences does this represent?

Exercise 7.6 For which of the methods for determining interacting proteins (page 416) (a) must one of the proteins be purified? (b) Must both of the proteins be purified?

Exercise 7.7 In the undirected, unlabelled graph in the Box on page 420:

(a) Name two vertices such that if you add an edge between them at least one vertex has exactly four neighbours. (Note that two edges may cross without making a new vertex at their point of intersection.)

(b) Name two vertices such that if you add an edge between them to the original graph, the graph remains a tree.

(c) Name two vertices (neither of them V_1) such that if you add an edge between them to the original graph, the graph does not remain a tree.

(d) Name two vertices such that if you add an edge between them to the original graph, there are alternative paths, of lengths 3 and 4, between V_1 and V_5, with no vertices repeated. (In determining the length of a path, you have to count the initial and final vertices. A path of length 3 between V_1 and V_5 contains one intermediate vertex.)

(e) Name two vertices such that if you add an edge between them to the original graph, there is exactly one path between V_1 and V_3, with no vertices repeated, and it has length 4.

Exercise 7.8 Of the examples of graphs in the box on page 421, (a) Which are directed graphs? (b) Which are labelled graphs? (c) In each example, what is the set of nodes? (d) In each example, what is the set of edges?

Exercise 7.9 In a typical protein–protein interface of area 1700 Å^2: (a) How many intermolecular hydrogen bonds would you expect to be formed? (b) How many fixed water molecules would you expect to find in the interface? (c) If the entire buried area were hydrophobic, what contribution to the free energy of stabilization would you estimate it to make?

Exercise 7.10 In the London Underground:

(a) What is the shortest path between Moorgate and Embankment stations? Note that, considered as a graph, the shortest path between two nodes is the path with the fewest intervening nodes, not the path that would take the minimal time or fewest interchanges.

(b) What is the shortest cycle containing Kings Cross, Holborn and Oxford Circus stations?

(c) The clustering coefficient of a node in a graph is defined as follows: suppose the node has k neighbours. Then the total possible connections between the neighbours is $k(k-1)/2$. The clustering coefficient is the observed number of neighbours divided by this maximum potential number of neighbours. If the neighbours of a station are the other stations that can be reached without passing through any intervening stations, what is the clustering coefficient of the Oxford Circus station? (If necessary, see http://www.bbc.co.uk/london/travel/downloads/tube_map.html)

Exercise 7.11 In the London Underground: (a) What is the maximum path length between any two stations? That is, for which two stations does the shortest trip between them involve the maximum number of intervening stops? (b) If the District Line were not active, what stations if any would be inaccessible by underground? (c) If the Jubilee line were not active, what stations if any would be inaccessible by underground?

Exercise 7.12 From the fragment of the B. subtilis protein interaction network shown in Fig. 7.6, what is the clustering coefficient of DnaC? (See Exercise 7.10 for the definition of the clustering coefficient.)

Exercise 7.13 On a photocopy of the three common combinations of network control motifs (Box, page 454) (a) indicate which nodes are controlled by only *one* upstream node; (b) indicate which node exerts control over only *one* downstream node.

Exercise 7.14 On a photocopy of the simplified fragment of the yeast regulatory network (Fig. 7.15) indicate examples of the network control motifs (a) fork, (b) 'one-two punch.' (c) Add one arrow to create a scatter motif.

Exercise 7.15 In the dimer between syntrophin and neuronal nitric oxide synthase (Fig. 7.16), (a) is the dimer structure open or closed? (b) What secondary structure element is shared between the two domains?

Exercise 7.16 In the overall yeast transcriptional regulatory network the number of incoming connections to target genes follows an exponential distribution. That is, the probability that a gene is controlled by k transcriptional regulators is proportional to $e^{-\alpha k}$, with $\alpha = 0.8, k = 1, 2, \ldots$. What is the ratio of the number of target genes receiving four input connections to the number receiving two input connections?

Problems

Problem 7.1 (a) How many positions in all are there in the microarray in Plate IX? (b) How many are complementary to RNAs from liver? (c) How many are complementary to RNAs from brain? (d) How many are complementary to RNAs from liver and brain? (e) How many are complementary to neither?

Problem 7.2 For dissociation of a complex involving a simple equilibrium: $AB \rightleftharpoons A + B$, the equilibrium constant, $K_D = \frac{[A][B]}{[AB]}$, is equal to the ratio of forward and reverse rate constants: $K_D = k_{off}/k_{on}$.
For avidin–biotin, $K_D = 10^{-15}$. Suppose k_{on} were as fast as the diffusion limit, $\sim 10^{-9} M \cdot s^{-1}$. (a) What is the value of k_{off}? (b) What would be the half-life of the avidin-biotin complex? (c) Suppose k_{on} for avidin-biotin were $10^{-7} M^{-1} \cdot s^{-1}$. What would be the half-life of the complex?

Problem 7.3 The anti-tuberculosis drug isoniazid requires activation by the *M. tuberculosis* enzyme KatG (a catalase-peroxidase), but the related drug ethionamide does not require activation. Suppose expression profiles were measured for the following:

(a) A strain with active KatG, not exposed to either drug.

(b) A strain with active KatG, exposed to isoniazid.

(c) A strain without active KatG, exposed to isoniazid.

(d) A strain with active KatG, exposed to ethionamide.

(e) A strain without active KatG, exposed to ethionamide.

The genes for which expression was enhanced, relative to (a), would be the same for which two conditions? Why would you expect the enhancement pattern to be similar in: (b), (d) and (e) but not (c)?

Problem 7.4 J. Foote and G. Winter compared the dissociation constants of a natural mouse antilysozyme antibody (D1.3), an engineered 'humanized' antibody in which the antigen-binding site was grafted onto a human framework (Human-original) and several mutants of the 'humanized form', including Human-mutated. The antigen was hen egg white lysozyme.

Antibody	Number of sequence differences from D1.3	k_{on} $M^{-1}s^{-1}$	K_D
D1.3	0	1.4×10^{-6}	3.7×10^{-9}
Human-original	48	0.7×10^{-6}	260×10^{-9}
Human-mutated	44	1.3×10^{-6}	14×10^{-9}

(a) Calculate the 'off-rate' k_{off} for each antibody. (b) Which has the major effect on the dissociation constant: differences in 'on-rate' or differences in 'off-rate'?

Problem 7.5 In the map of the London Underground, what is the distribution of numbers of neighbours of vertices?

Problem 7.6 Analyse the map of the London Underground by counting the number of connections made from each station in zone 1 (the central portion). Count connections to stations inside and outside zone 1 as long as they originate within zone 1. Count only one connection if two stations are connected by more than one line; in other words, for each station, the question is: how many other stations can be reached without passing through any intermediate stops?

(a) What is the maximum number of connections of any station?

(b) For each integer k from 1 to this maximum number, how many stations have k connections?

(c) Plot these data on a log–log plot. Does the relationship appear reasonably linear?

(d) If so, fit a straight line to the log–log plot and determine the exponent. Results of network analysis of this sort are more significant if the data cover several orders of magnitude, but this is not possible for this example.

Problem 7.7 In the overall yeast transcriptional regulatory network the number of incoming connections to nodes follows an exponential distribution. That is, the probability P_k that a gene is controlled by k transcription regulators is given by $P_k = Ce^{-\alpha k}$, $k = 1, 2, \ldots$, with $\alpha = 0.8$.

(a) Determine the constant of proportionality C in terms of α, by summing the series $\sum_{k=1}^{\infty} Ce^{-\alpha k} = 1$.

(b) If $\alpha = 0.8$, what is the maximum value of k for which at least 1% of the nodes would be expected to have at least k incoming connections?

(c) If $\alpha = 0.8$, plot the expected histogram for $1 \le k \le 7$.

(d) Determine the mean value of k in terms of α. [Hint: in the solution of (a) you expressed $\sum_{k=1}^{\infty} e^{-\alpha k}$ as a function $f(\alpha)$. Differentiate this relationship with respect to α to produce the equation: $\sum_{k=1}^{\infty} -ke^{-\alpha k} = f'(\alpha)$. Then the mean value of k is given by $-f'(\alpha)/f(\alpha)$.]

(e) What is the mean value $< k >$ corresponding to $\alpha = 0.8$?

(f) What is the median value of k? This is the value k of such that half the nodes have $\le k$ incoming connections, and half the nodes have $\ge k$ incoming connections. Find

k in terms of α. [Hint: if $\sum_{k=1}^{\infty} Ce^{-\alpha k} = 1$, then $\sum_{k=\kappa+1}^{\infty} Ce^{-\alpha k} = \frac{1}{2}$. But $\sum_{k=\kappa+1}^{\infty} Ce^{-\alpha k} = e^{-\alpha \kappa} \sum_{k=\kappa+1}^{\infty} Ce^{-\alpha k}$. In general, this approach will provide a non-integral estimate of k, just round this result to the nearest integer.]

(g) If $\alpha = 0.8$, what is the median value k? How does it compare with the average value $< k >$? Are the two values approximately equal?

Problem 7.8 Indicate how to connect a selection of the three common network control motifs so that a single input node can influence three output nodes.

Problem 7.9 What is the minimum number of 'yes-or-no' questions required to identify a specific letter of the English uppercase alphabet: ABC...Z? Assume a random text with equal distribution of all letters.

Problem 7.10 On a photocopy of the diagram on page 450, add the interactions involving viral proteins cII and cIII, and bacterial protease HflB.

Problem 7.11 On a photocopy of the diagram on page 450, indicate which if any of the regulatory interactions would be altered (and how they would be affected) by mutations in OR1, OR2 or OR3 that destroyed their affinities for (a) cro and (b) cI.

Problem 7.12 Generalize the dotplot program of Chapter 4 to metabolic pathways. Represent glycolysis and gluconeogenesis pathways by an interspersed sequence of metabolites and enzymes. (Represent the enzymes by EC numbers.) Do dotplots of (a) metabolites only, and (b) enzymes. Do not filter the results.

Problem 7.13 The network of metabolic pathways must obey constraints of thermodynamics and physical–organic chemistry. E. Meléndez-Hevia and colleagues suggested the principle that metabolic pathways are optimized, subject to the constraints, for the minimum number of steps.
The non-oxidative phase of the pentose phosphate pathway converts six 5-carbon sugars to five 6-carbon sugars:

$$6 \text{ ribulose-5-phosphate} \rightarrow 5 \text{ glucose-6-phosphate}$$

A simplified model of a pathway for this conversion is a series of steps, each of which is either:

1. Transfer of a 2-carbon unit from one sugar to another (a transketolase reaction)

2. Transfer of a 3-carbon unit from one sugar to another (a transaldolase or aldolase reaction).

Represent each sugar only by a number of carbon atoms. Starting with five 5-carbon sugars, one possible initial step would be a transketolase step converting two 5-carbon sugars to a 3-carbon sugar and a 7-carbon sugar. Assume that all intermediates must have *at least* three carbon atoms.
Create a tableau with the following initial and final states (an initial transketolase (TK) step is also shown):

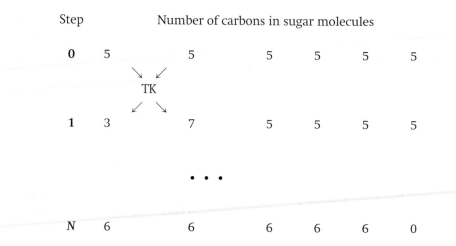

Step		Number of carbons in sugar molecules				
0	5	5	5	5	5	5
1	3	7	5	5	5	5
		⋯				
N	6	6	6	6	6	0

Copy and fill in the tableau to find the shortest route from top (step 0, six 5-carbon sugars) to bottom (five 6-carbon sugars). Identify the intermediates created. Compare with the observed metabolic pathway.

Weblems

Weblem 7.1   Define the following terms: (a) interactome, (b) metabolome, (c) signalome. (d) More difficult: can you think of, and define, a reasonable 'ome' that has not yet been proposed?

Weblem 7.2   After separation of a mixture of proteins by 2-D gel electrophoresis, the mass spectrum of an eluted spot showed y-ions of the following masses:

y_1	147.11
y_2	260.16
y_3	374.20
y_4	487.28
y_5	544.30
y_6	681.35
y_7	768.30
y_8	881.46
y_9	1010.48
y_{10}	1123.56

(a) What are the possible sequences of these fragment? (b) Identify the protein that contained this fragment.

Weblem 7.3   The catalase-peroxidase KatG of *M. tuberculosis* activates the drug isoniazid. The most common mutation in KatG associated with resistant strains is S315T. From a model of *M. tuberculosis* KatG based on the structure of the homologue from *Burkholderia pseudomallei* (PDB entry [1MWV]), suggest a mechanism for the reduced activity of the mutant to activate isoniazid while retaining activity against smaller substrates.

Weblem 7.4   Find a fragment of the genealogy of the royal families of Europe containing a family tree that is not a graph.

Weblem 7.5 One model for the growth of a scale-free network suggests that if new nodes and edges are added according to the rule that the probability of adding a new edge is proportional to the number of edges that a node already has, the network will remain scale free and retain the same exponent. Using historical maps of the London Underground, check whether earlier networks are scale-free. Test whether addition of edges to the network has followed the rule that edges have been added preferentially to more highly connected nodes.

Weblem 7.6 From the *B.subtilis* protein interaction database SPiD (see Fig. 7.7), (a) what type of experimental evidence links DnaC to DnaG? (b) What type of experimental evidence links DnaC to DnaI?

Weblem 7.7 From the *B.subtilis* protein interaction database SPiD (see Fig. 7.7), print a graph showing not only the immediate neighbours of DnaC, but the second neighbours (the neighbours of the neighbours).

Conclusions

How can we extrapolate from the current state of play to the bioinformatics of the future? Clearly, data collection will proceed and continue to accelerate. New high-throughput techniques will provide additional types of data, including information about the integration and control of life processes. Computing facilities of increasing power will be applied to the storage, distribution and analysis of the results. New databases will appear on the web, and links between databases will become more effective. Improved algorithms will be devised to analyse and interpret the information given to us and to transmute it from data to knowledge to wisdom.

Sequencing power will continue to increase, and the amount of sequence data will attain immense proportions. It is not too soon to plan for the time when a large fraction of people will have their genomes sequenced completely. Metagenomics will provide another prolific source of data.

One threshold will be reached when our knowledge of sequences and structures becomes more nearly complete, in the sense that a fairly dense subset of the available data from contemporary living forms has been collected. (Of course there is no question of being able to know everything.) This will be recognized operationally when a random dip into the pot of a genome, or the isolation of a new protein structure, is far more likely to turn up something already known, rather than to uncover something new. Nature is, after all, a system of unlimited possibilities but finite choices.

Applications will become more feasible, and mature ever more quickly from 'blue-sky' research to standard industrial and clinical practice. Some of the higher levels of biological information transfer—such as the programs of genetic development during the lifetime of individuals, and the activities of the human mind—will come to be included in the processes we can describe quantitatively and analyse at the level of molecules and their interactions.

In Michaelangelo's frescos on the ceiling of the Sistine Chapel, the serpent offering Eve the fruit of the tree of knowledge is represented with its legs coiled around the tree in the form of a double helix. We can hope that our new temptation to knowledge embodied in another double helix will have more fortunate consequences.

Index

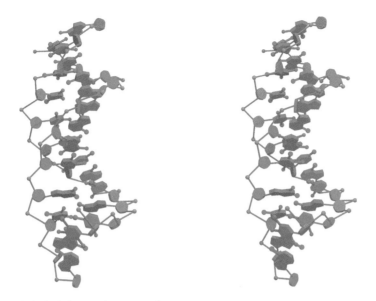

Plate I Double helix of DNA (see page 6).

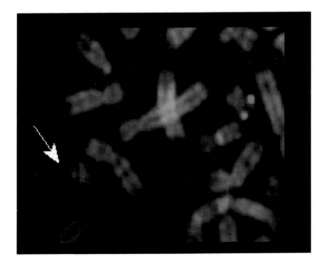

Plate II FISH (fluorescent *in situ* hybridization) can detect the presence of locus-specific probes and visualize their chromosomal positions. Shown in red is a probe for the centromeric region of chromosome 20, to identify the two homologous copies of the chromosome that appear at metaphase. Shown in green is a probe for D20S108, within the region 20q11.2–13.1, which is present in one copy of chromosome 20 and deleted from the other (see arrow). This cell came from a patient suffering from polycythaemia rubra vera (an abnormal increase in blood cells, primarily erythrocytes, arising from abnormality in bone marrow). The region deleted in the long arm of chromosome 20 is believed to contain tumour suppressor gene(s), the loss of which contributes to the development of leukaemias (see page 77). (Courtesy of Dr E Nacheva, Department of Academic Haematology, Royal Free & University College London School of Medicine, London.)

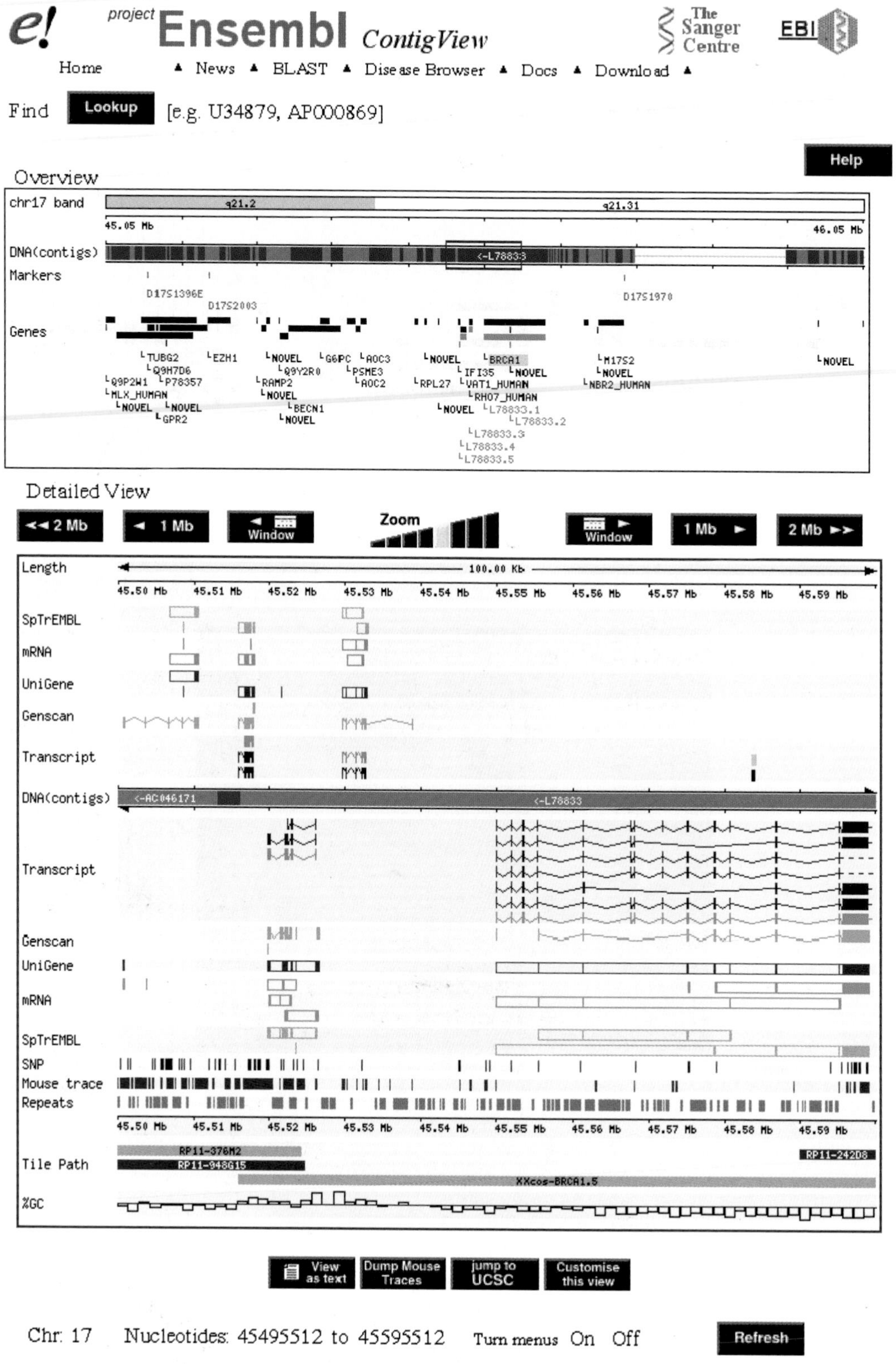

Plate III Ensembl genome browser showing the region surrounding the human *BRCA1* locus (see page 193).

Mammalian elastases

```
                     10        20        30        40        50        60        70        80        90       100
CFAD_RAT    MHSSVYLVALVVLEAAVCVAQPRG-----RILGGQEAMAHARPYMASVQVNG----THVCGGTLVDEQWVLSAAHCMDGVTKDEVVQVLLGAHSLSSP-
CLCR_RAT    MLGITVLAAILACASCCGNPAFPPN-LSTRVVGGEDAVPNSWPWQVSLQYLKDDTWRHTCGGSLITTSHVLTAAHCIN---KDFTYRVGLGKYNLTVED
EL1_BOVIN   MLRLLVFTSLVLYGHSTQDFPETN----ARVVGGTAVSKNSWPSQISLQYKSGSSWYHTCGGTLIKQKWVMTAAHCVD---SQMTFRVVLGDHNLSQN-
EL1_PIG     MLRLLVVASLVLYGHSTQDFPETN----ARVVGGTEAQRNSWPSQISLQYRSGSSWAHTCGGTLIRQNWVMTAAHCVD---RELTFRVVVGEHNLNQN-
EL1_RAT     MLRFLVFASLVLYGHSTQDFPETN----ARVVGGAEARRNSWPSQISLQYLSGGSWYHTCGGTLIRRNWVMTAAHCVS---SQMTFRVVVGDHNLSQN-
EL2A_HUMAN  MIRTLLLSTLVAGALSCGDPTYPP--YVTRVVGGEEARPNSWPWQVSLQYSSNGKWYHTCGGSLIANSWVLTAAHCIS---SSRTYRVGLGRHNLYVA-
EL2B_HUMAN  MIRTLLLSTLVAGALSCGVSTYAP--DMSRMLGGEEARPNSWPWQVSLQYSSNGQWYHTCGGSLIANSWVLTAAHCIS---SSRIYRVMLGQHNLYVA-
EL2_BOVIN   MIRALLLSTLVAGALSCGVPTYPP--QLSRVVGGEDARPNSWPWQVSLQYSSSGQWRHTCGGSLIEQNWVLTAAHCIS---SSRTYRVVVGRQSLSTV-
EL2_MOUSE   MIRTLLLSALVAGALSCGYPTYEVEDDVSRVVGGQEATPNTWPWQVSLQVLSSGRWRHNCGGSLVANNWVLTAAHCLS---NYQTYRVLLGAHSLSNP-
EL2_PIG     MIRALLLSTLVAGALSCGLPANLP--QLPRVVGGEDARPNSWPWQVSLQYDSSGQWRHTCGGTLVDQSWVLTAAHCIS---SSRTYRVVLGRHSLSTN-
EL2_RAT     MIRTLLLSALVAGALSCGYPTYEVQHDVSRVVGGQEASPNSWPWQVSLQYLSSGKWHHTCGGSLVANNWVLTAAHCIS---NSRTYRVLLGRHSLSTS-
EL3A_HUMAN MMLRLLSSLLLVAVASGYGPPSSHS---SSRVVHGEDAVPYSWPWQVSLQYEKSGSFYHTCGGSLIAPDWVVTAGHCIS---RDLTYQVVLGEYNLAVKE
EL3B_HUMAN MMLRLLSSLLLVAVASGYGPPSSRP---SSRVVNGEDAVPYSWPWQVSLQYEKSGSFYHTCGGSLIAPDWVVTAGHCIS---SSRTYQVVLGEYDRAVKE
ELNE_HUMAN MTLGRRLACLFLACVLPALLLGGTA--LASEIVGGRRARPHAWPFMVSLQLRGG----HFCGATLIAPNFVMSAAHCVAN-VNVRAVRVVLGAHNLSRR-
                                           G         P   S Q      H CG  L     V  A HC         V  G

                   110       120       130       140       150       160       170       180       190       200
CFAD_RAT   -EPYKHLYDVQSVVLHPGSRPDSVED--DLMLFKLSHNASLGPHVRPLPLQREDREVKPGTLCDVAGWGVVTHAGRRPDVLQQLTVSIMDRNTCNLRTYH
CLCR_RAT   -EEGSVYAEVDTIYVHEKWN--RLFLWNDIAIIKLAEPVELSNTIQVACIPEEGSLLPQDYPCYVTGWGRLWTNGPIAEVLQQGLQPIVSHATCSRLDWW
EL1_BOVIN  -DGTEQYISVQKIVVHPSWNSNNVAAGYDIAVLRLAQSATLNSYVQLGVLPQSGTILANNTPCYITGWGRTKTNGQLAQTLQQAYLPSVDYATCSSSSYW
EL1_PIG    -DGTEQYVGVQKIVVHPYWNTDDVAAGYDIALLRLAQSVTLNSYVQLGVLPRAGTILANNSPCYITGWGLTRTNGQLAQTLQQAYLPTVDYAICSSSSYW
EL1_RAT    -DGTEQYVSVQKIMVHPTWNSNNVAAGYDIALLRLAQSVTLNNYVQLAVLPQEGTILANNNPCYITGWGRTRTNGQLSQTLQQAYLPSVDYSICSSSSYW
EL2A_HUMAN -ESGSLAVSVSKIVVHKDWNSNQISKGNDIALLKLANPVSLTDKIQLACLPPAGTILPNNYPCYVTGWGRLQTNGAVPDVLQQGRLLVVDYATCSSSAWW
EL2B_HUMAN -ESGSLAVSVSKIVVHKDWNSNQVSKGNDIALLKLANPVSLTDKIQLACLPPAGTILPNNYPCYVTGWGRLQTNGALPDDLKQGRLLVVDYATCSSSGWW
EL2_BOVIN  -ESGSLTIAVSKSVIHEKWNSNQLAQGNDIALLKLASSVPLTDKIQLGCLPAAGTILPNNYVCYVTGWGRLQSNGALPDILQQGKLLVVDYATCSNPSWW
EL2_MOUSE  -GAGSAAVQVSKLVVHQRWNSQNVGNGYDIALIKLASPVTLSKNIQTACLPPAGTILPRNYVCYVTGWGLLQTNGNSPDTLRQGRLLVVDYATCSSASWW
EL2_PIG    -EPGSLAVKVSKLVVHQDWNSNQLSNGNDIALLKLASPVSLTDKIQLGCLPAAGTILPNNYVCYVTGWGRLQTNGASPDILQQGQLLVVDYATCSKPGWW
EL2_RAT    -ESGSLAVQVSKLVVHEKWNAQKLSNGNDIALVKLASPVALTSKIQTACLPPAGTILPNNYPCYVTGWGRLQTNGATPDVLQQGRLLVVDYATCSSASWW
EL3A_HUMAN GPEQVIPINSEELFVHPLWNRSCVACGNDIALIKLSRSAQLGDAVQLASLPPAGDILPNKTPCYITGWGRLYTNGPLPDKLQQARLPVVDYKHCSRWNWW
EL3B_HUMAN GPEQVIPINSGDLFVHPLWNRSCVACGNDIALIKLSRSAQLGDAVQLASLPPAGDILPNETPCYITGWGRLYTNGPLPDKLQEALLPVVDYEHCSRWNWW
ELNE_HUMAN -EPTRQVFAVQRIFEN---GYDPVNLLNDIVILQLNGSATINANVQVAQLPAQGRRLGNGVQCLAMGWGLLGRNRGIASVLQELNVTVVTSLCR------
                                   D     L                              C   GWG          L

                   210       220       230       240       250       260       270       280       290
CFAD_RAT   DGAITKNMMCAESN-RRDTCRGDSGGPLVCGDA-----VEAVVTWGSR-VCGNRRKPGVFTRVATYVPWIENVLSGNVSVNVTA-----------
CLCR_RAT   FIKVRKTMVCAGGDGVISACNGDSGGPLNCQAEDGSWQVHGIVSFGSSSGCNVHKKPVVFTRVSAYNDWINEKIQL-------------------
EL1_BOVIN  GSTVKTTMVCAGGDGVRAGCQGDSGGPLHCLVN-GQYAVHGVTSFVSSLGCNVSKKPTVFTRVSAYISWINNAIASN------------------
EL1_PIG    GSTVKNSMVCAGGDGVRSGCQGDSGGPLHCLVN-GQYAVHGVTSFVSRLGCNVTRKPTVFTRVSAYISWINNVIASN------------------
EL1_RAT    GSTVKTTMVCAGGDGVRSGCQGDSGGPLHCLVN-GQYSVHGVTSFVSSMGCNVSKKPTVFTRVSAYISWMNNVIAYT------------------
EL2A_HUMAN GSSVKTSMICAGGDGVISSCNGDSGGPLNCQASDGRWQVHGIVSFGSRLGCNYYHKPSVFTRVSNYIDWINSVIANN------------------
EL2B_HUMAN GSTVKTNMICAGGDGVICTCNGDSGGPLNCQASDGRWEVHGIGSLTSVLGCNYYYKPSIFTRVSNYNDWINSVIANN------------------
EL2_BOVIN  GSTVKTNMICAGGDGVTSSCNGDSGGPLNCQAANRQWQVHGIVSFGSSLGCNYYRKPSVFTRVSNYNDWISSVIENN------------------
EL2_MOUSE  GSSVKSSMVCAGGDGVTSSCNGDSGGPLNCRASNGQWQVHGIVSFGSSLGCNYPRKPSVFTRVSNYIDWINSVMARN------------------
EL2_PIG    GSTVKTNMICAGGDGIISSCNGDSGGPLNCQGANGQWQVHGIVSFGSSLGCNYYHKPSVFTRVSNYIDWINSVIANN------------------
EL2_RAT    GSSVKTNMVCAGGDGVTSSCNGDSGGPLNCQASNGQWQVHGIVSFGSTLGCNYPRKPSVFTRVSNYIDWINSVIAKN------------------
EL3A_HUMAN GSTVKKTMVCAGGY-IRSGCNGDSGGPLNCPTEDGGWQVHGVTSFVSGFGCNFIWKPTVFTRVSAFIDWIEETIASH------------------
EL3B_HUMAN GSSVKKTMVCAGGD-IRSGCNGDSGGPLNCPTEDGGWQVHGVTSFVSAFGCNTRRKPTVFTRVSAFIDWIEETIASH------------------
ELNE_HUMAN ----RSNVCTLVRGRQAGVCFGDSGSPLVCNGL-----IHGIASFVRG-GCASGLYPDAFAPVAQFVNWIDSIIQRSEDNPCPHPRDPDPASRTH
                                C GDSG PL C          C                P    F    V     W
```

Plate IV Alignment of amino acid sequences of mammalian elastases (see page 234).

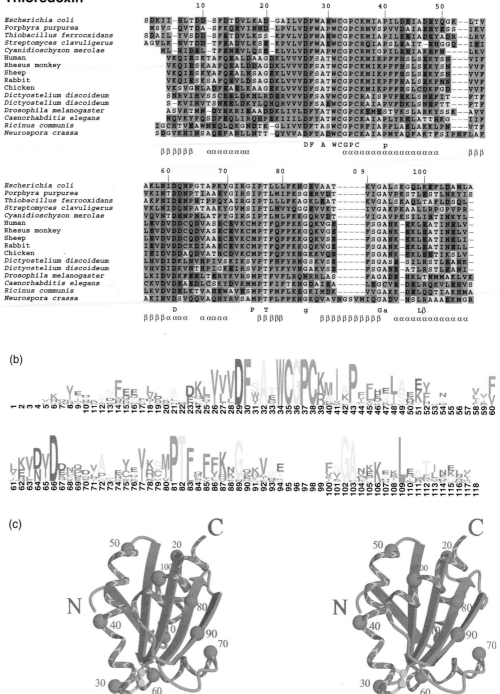

Plate V (a) Alignment of amino acid sequences of *E. coli* thioredoxin and homologues. Some of the sequences have been trimmed at their termini. Residue numbers in this table corresponds to positions in the *E. coli* sequence (top line). Helix (α) and strand (β) assignments for *E. coli* thioredoxin are from PDB entry 2TRX. (b) Sequence logo derived from this multiple sequence alignment. (c) The structure of *E. coli* thioredoxin [2TRX] contains a central five-stranded β-sheet flanked on either side by α-helices. Residue numbers correspond to those in the multiple sequence alignment table. The N- and C-termini are also marked. Spheres indicate positions of the Cα atoms of every tenth residue. The reactive disulphide bridge between Cys32 and Cys35 appears in yellow (see page 272).

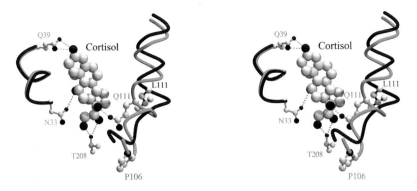

Plate VI Effect of mutations S106P and L111Q in evolution of differential specificity to aldosterone and cortisol in vertebrate steroid receptors. Green: experimental structure corresponding to predicted sequence ancestral to all jawed vertebrates. Yellow: experimental structure corresponding to predicted sequence ancestral to all bony vertebrates. A region from the binding site is shown. The mutation S106P reconforms the loop between helices 6 and 7 (H6 and H7). The mutation L111Q creates a hydrogen bond to the C17 hydroxyl which is present in cortisol but not in aldosterone nor in 11-deoxycorticosterone (see page 293). From: Ortlund, E.A., Bridgham, J.T., Redinbo, M.R. and Thornton, J.W. (2007). Crystal structure of an ancient protein: evolution by conformational epistasis. *Science*, **317**, 1544–1548. Copyright AAAS. Reproduced by permission.

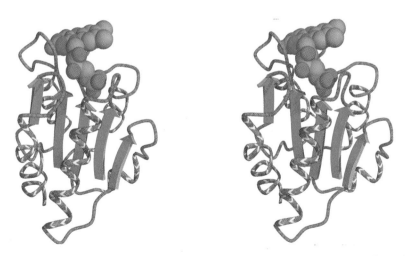

Plate VII Flavodoxin from *Clostridium beijerinckii*, binding cofactor FMN [5NLL]. Large arrows represent strands of sheet. Placement of this structure in a hierarchical classification of protein structures according to the SCOP database is described on page 332.

(a)

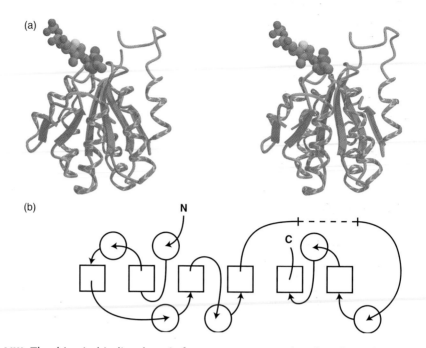

(b)

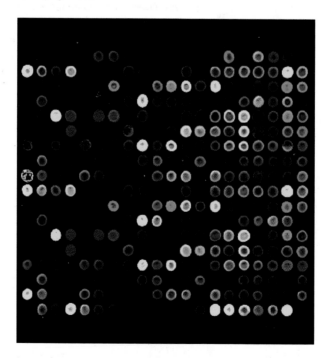

Plate VIII The thiamin-binding domain from yeast pyruvate decarboxylase. Thiamin-binding domains, identified by M. Gerstein as one of the five most common folding patterns, have been found in archaea, bacteria and eukarya. (a) Three-dimensional structure. (b) schematic topology diagram (see page 359).

Plate IX Comparison of gene expression patterns in liver (red) and brain (green). The liver RNA is tagged with a red fluorophore, the brain RNA with a green one, then both are exposed to the array. Red spots correspond to genes active in the liver but not in the brain. Green spots correspond to genes active in the brain but not in the liver. Yellow spots correspond to genes active in both brain and liver (see pages 391–392). (Courtesy Dr P. A. Lyons.)

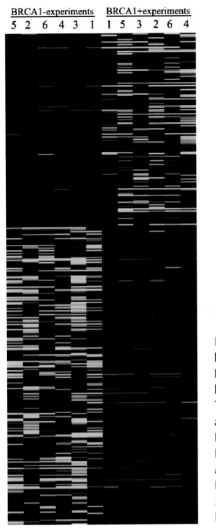

BRCA1-experiments BRCA1+experiments
5 2 6 4 3 1   1 5 3 2 6 4

Plate X Clustering of gene expression data in cells expressing high and low levels of BRCA. BRCA1⁻ experiments, with low expression levels of BRCA1, appear in the left-hand six columns. BRCA1⁺ experiments, with high expression levels of BRCA1, appear in the right-hand six columns. The intensity of the colour reflects the ratio of the expression to that of a control. Red reflects genes with higher expression levels in response to BRCA1. Green reflects genes with lower expression levels in response to BRCA1. From: Welcsh, P.L., Lee, M.K., Gonzales-Hernandez, R.M., Black, D.J., Mahadevappa, M., Swisher, E.M., Warrington, J.A. and King, M.-C. (2002). BRCA1 transcriptionally regulates genes involved in breast tumourigenesis. *Proceedings of the National Academy of Sciences, USA,* **99**, 7560–7566. Reproduced by permission (see page 397).

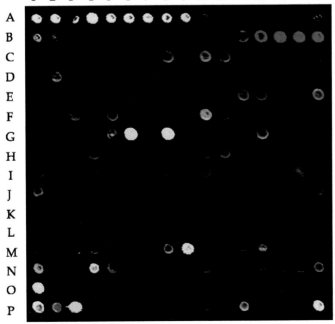

Plate XI The effect of 4-hour treatment by isoniazid on the mRNA expression profiles of 203 ORFs from *M. tuberculosis.* Red = expressed in cells treated with isoniazid, green = expressed in untreated cells, yellow = expressed in both treated and untreated cells. The row of red spots at the upper right corresponds to genes of the FAS-II gene cluster. (From: Wilson, M., DeRisi, J., Kristensen, H.H., Imboden P., Rane, S., Brown, P.O. and Schoolnik, G.K. (1999). Exploring drug-induced alterations in gene expression in *Mycobacterium tuberculosis* by microarray hybridization. *Proceeding of the National Academy of Sciences, USA,* **96**, 12833–2838.) (See page 407.)

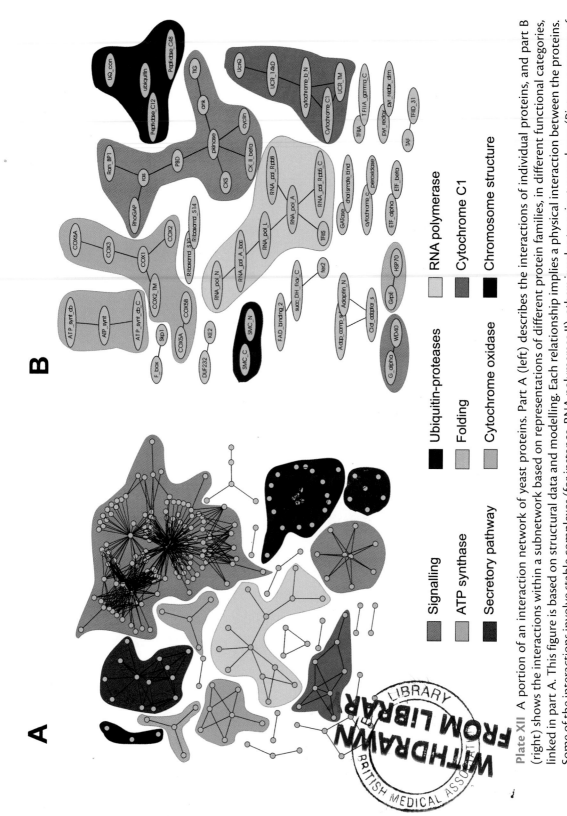

Plate XII A portion of an interaction network of yeast proteins. Part A (left) describes the interactions of individual proteins, and part B (right) shows the interactions within a subnetwork based on representations of different protein families, in different functional categories, linked in part A. This figure is based on structural data and modelling. Each relationship implies a physical interaction between the proteins. Some of the interactions involve stable complexes (for instance, RNA polymerase II); others involve transient complexes. (Picture courtesy of P. Aloy and R.B. Russell.) (See page 418.).

Legend (part B):

- Signalling
- ATP synthase
- Secretory pathway
- Ubiquitin-proteases
- Folding
- Cytochrome oxidase
- RNA polymerase
- Cytochrome C1
- Chromosome structure